HANDBOOK OF ENERGY FOR WORLD AGRICULTURE

in collaboration with

L. W. Faidley
Food and Agriculture Organization of the United Nations, Via delle Terme di Caracalla, Rome, Italy

C. A. Myers
University of California Cooperative Extension Service, Small Farms Center, Davis, California, USA

W. T. Rose
Hueber Hares Glavin, 726 West Onondaga Street, Syracuse, New York, USA

HANDBOOK OF ENERGY FOR WORLD AGRICULTURE

B. A. STOUT

Agricultural Engineering Department,
Texas A & M University, College Station, Texas, USA

ELSEVIER APPLIED SCIENCE
LONDON and NEW YORK

ELSEVIER SCIENCE PUBLISHERS LTD
Crown House, Linton Road, Barking, Essex IG11 8JU, England

Sole Distributor in the USA and Canada
ELSEVIER SCIENCE PUBLISHING CO., INC.
655 Avenue of the Americas, New York, NY 10010, USA

WITH 171 TABLES AND 263 ILLUSTRATIONS

British Library Cataloguing in Publication Data

Stout, B. A.
Handbook of energy for world agriculture.
1. Energy. Consumption by agricultural industries
I. Title
338.1′6

ISBN 1-85166-349-5

Library of Congress Cataloging-in-Publication Data

Stout, B. A.
Handbook of energy for world agriculture / B. A. Stout
p. cm.
Bibliography: p.
Includes index.
ISBN 1-85166-349-5 (U.S.)
1. Agriculture and energy—Handbooks, manuals, etc. I. Title.
S494.5.E5S77 1989 89-7775
338.1′4—dc60 CIP

Printed in Great Britain by Galliard (Printers) Ltd, Great Yarmouth

Foreword

Adequate food supplies and a reasonable quality of life require energy—both noncommercial and commercial forms. This book is intended to be a reference for individuals who want a comprehensive overview of energy for agriculture, many of whom are located in remote areas with limited library resources. The purpose is to put energy for agriculture in perspective by presenting numerous national and regional examples of energy usage.

Since the early 1970s, world petroleum prices have fluctuated from $3/barrel to more than $40/barrel in 1981 and then back to less than a half of the peak price today. Commercial energy is expensive at any price for rural villagers in developing countries, and energy prices in these countries do not necessarily correspond with world prices. For example, diesel fuel prices in some developing countries have increased, while world oil prices have decreased.

The rural sector of developing countries depends heavily on noncommercial energy sources. Availability of noncommercial energy is highly site-specific. This book deals extensively with noncommercial energy—its sources, the technologies for converting energy to more useful gaseous and liquid forms, and its ultimate end uses.

Chapter I is an introduction and overview. It emphasizes the essential nature of energy to supplement the limited work output of human beings.

Chapter II describes energy use in world agriculture with particular emphasis on fertilizer, machinery, and irrigation. It tabulates cereal grain yields as related to commercial energy input for various regions.

Chapter III presents numerous national and regional illustrations of energy use.

Chapter IV addresses energy management in agriculture and is organized by crop production operation. It focuses on efficient use of energy, that is, deriving the maximum benefit from each unit of energy invested in agriculture and the food system.

Renewable energy alternatives include all forms of solar energy. Chapters V, VI, VII and VIII address some of the major alternative energy forms: biomass for fuel, solar heating and other solar applications, small-scale windpower, and hydropower.

Chapter IX concludes with a discussion of various strategies to meet agricultural energy needs in developing countries.

An extensive reference list is included for readers who want more detailed information. Complete agency names and addresses for further information are provided in the Appendix, along with conversion units.

Acknowledgments

Many individuals and agencies have contributed to this handbook. It is not possible to acknowledge individually the hundreds of authors of references cited, although their work, collectively, made this book possible. The sources of tables and figures and text citations are indicated throughout the book.

The first draft of the manuscript was prepared with the support of the Food and Agriculture Organization of the United Nations. FAO also provided many excellent photographs and references. Dr LeVern W. Faidley, Senior Officer, Research and Technology Department Division, FAO, prepared the first draft of Chapter II.

Claudia Myers, Associate Director, Small Farm Center, University of California Cooperative Extension Service, wrote the first draft of Chapter VI and William T. Rose, PE, Rose Engineering, authored the first draft of Chapter VII.

The entire manuscript was smoothed and polished by Ms Terri Huff, a very capable editor at Texas A&M University who helped weed out many errors and made the text much more readable.

A project of this type would have been impossible without many dedicated and competent research assistants, typists, reviewers and other helpers. The authors extend a sincere thanks to everyone who contributed to this book.

B. A. Stout, PE
Texas A&M University

Contents

Chapter III **Energy Flow**

Chapter IV **Energy Management**

Chapter V **Energy from Biomass**

Chapter VI **Solar Energy**

CHAPTER I

Overview

INTRODUCTION

Life could not exist without energy, most of which comes directly or indirectly from the sun. Solar radiation warms the earth and maintains a temperature range that living plants and animals (including humans) can endure. Plants convert solar energy to stored carbohydrates through photosynthesis and, in the process, release oxygen, an essential ingredient for the existence of human and animal life.

Plants and animals are also the basis for fossil fuels which supplement direct solar radiation and provide an energy subsidy to increase yields and thereby provide food for all people. Figure I.1 illustrates the various energy sources that have fueled human development throughout the centuries.

Energy is defined as the capacity to perform work, and work is the result of a force moving a body some distance. Energy is available in many forms; however, some are more useful than others. The total amount of energy in the universe is constant, but energy is frequently converted from one form to another through natural and man-made processes.

Although numerous energy studies have been conducted in recent times, the amount of energy that is actually consumed in developing countries is difficult to determine. What is this energy used for? In what form? At what time of year? How much is renewable? Nonrenewable? Commercial/noncommercial? And most importantly, what is the potential for substituting one form for another, and what are the consequences of energy shortfalls?

Dozens of authors have tried to answer these questions. Only a few are mentioned here. Makhijani & Poole (1975) described life in energy-poor developing countries and provided a variety of examples to illustrate that people in developing countries were not receiving their share of energy. The Food and Agriculture Organization of the United Nations (FAO) devoted a major section of one of its annual reports (1976) to an analysis of energy and agriculture.

Stout *et al.* (1979) wrote *Energy for World Agriculture*, a predecessor to this book. The purpose was to put energy and the food system in perspective, and to provide a manual for individuals interested in maintaining or increasing world food supplies. Numerous studies, conferences, papers, and books followed, each contributing additional information and understanding about the energy problem.

For example, *Energy and the Developing Countries* (Auer, 1981), a book based on a conference at Stanford University in 1980, contains 31 chapters on various development issues related to energy. Another publication, also released in 1980, is

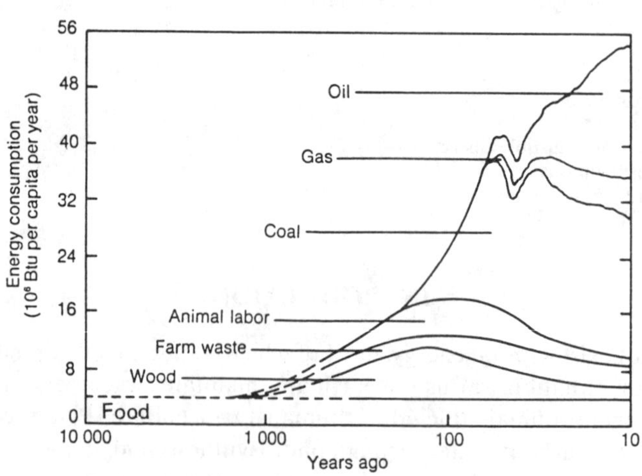

Fig. I.1. Consumption of the world's energy resources through the ages. Humankind has already experienced several energy epochs; and the transitions altered our lifestyles immeasurably (source: Loftness, 1984).

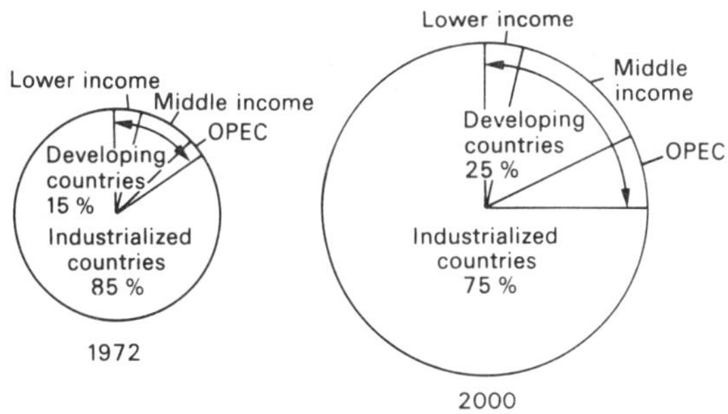

Fig. I.2. Energy shares for developing and industrialized countries (source: Smil & Knowland, 1980).

Parikh's *Energy Systems and Development*. This book presents current energy consumption statistics, projects energy requirements until the year 2000, and examines various supply options. Still another book published in 1980, *Energy in the Developing World* (Smil & Knowland, 1980), focuses on the largest and most populous developing countries where populations total more than 2·2 billion,† or nearly 75% of the Third World. The book includes an analysis of current energy use and a discussion of alternative strategies. More recently, Goswami (1986) edited a series of two volumes on alternative energy in agriculture which dealt primarily with the situation in the United States.

Readers who want to collect information directly from the source will find the United Nations Educational, Scientific and Cultural Organization (UNESCO)/Solar Energy Research Institute (SERI) directory of information sources and research centers useful (UNESCO, 1986). It lists hundreds of information sources from around the world.

These books and references serve as examples of the large collection of literature that has emerged in the 1970s and 1980s, but they only scratch the surface. More than 600 references are cited in this publication in an attempt to grasp the essence of energy problems facing rural areas in developing countries. Studies are often difficult to compare because of variations in data quality and quantity, time periods, and analysis techniques; nevertheless, they combine to improve our level of understanding.

The developing countries where most of the world's population lives—Asia, Africa, Latin America, the Middle East, and elsewhere—consume only about 15% of the world's energy (Fig. I.2 and Tables I.1–I.4). As these countries develop, they will require more energy. One scenario portrays the energy share of the developing countries growing to 25% of the world's total by the year 2000. This increased energy demand is necessary for development, but is certain to further deplete the poorer countries' monetary resources.

The International Institute for Applied Systems Analysis (IIASA) report, *Energy in a Finite World* (IIASA, 1981), states, 'During this period (the next 50 years) the

† See Appendix.

TABLE I.1
World commercial primary energy consumption, 1970–95 (source: World Bank, 1983)

	Million toe[a]			*Growth rate (percentage per year)*	
	1970	*1980*	*1995*	*1970–80*	*1980–95*
Oil	2 311	3 067	3 355	2·9	0·6
Coal	1 475	1 825	2 821	2·2	2·9
Natural gas	889	1 241	1 930	3·4	3·0
Primary electricity[b]	328	611	1 423	6·4	5·8
Total	5 003	6 744	9 529	3·0	2·3
	Percentage of total			*Percentage of increase*	
	1970	*1980*	*1995*	*1970–80*	*1980–95*
Oil	46·2	45·5	35·2	43·4	10·3
Coal	29·5	27·1	29·6	20·1	35·8
Natural gas	17·8	18·4	20·3	20·2	24·7
Primary electricity[b]	6·5	9·0	14·9	16·3	29·2
Total	100·0	100·0	100·0	100·0	100·0

Note: Energy consumption includes bunkers.
[a] Toe, Tons of oil equivalent.
[b] Primary electricity comprises electricity generated from hydropower, nuclear energy, or geothermal resources. Primary electricity is converted into tons of oil equivalent (toe) at thermal replacement value, assuming an average conversion efficiency of about 34%.

worldwide population will reach 8 billion, and even with only modest economic growth and extensive conservation, global energy demand is likely to expand to three to four times today's level' (Fig. I.3).

Will the technology and resources be available to satisfy this dramatically increased demand? While expert opinions vary, IIASA says yes, but it emphasizes that to meet the growing worldwide demand, unlimited use of all available resources—oil and gas, solar energy, renewables, and nuclear energy—will be required. Later chapters in this book discuss the technical aspects of increasing the renewable energy supply.

The developing countries face an energy crisis of staggering proportions. In most developing countries, annual commercial energy consumption is barely a quarter of a tonne of coal equivalent per capita. In the industrialized countries of Europe, as well as in Australia, New Zealand, and Japan, the annual energy consumption exceeds 5 tonnes (t) of coal equivalent; in North America, it is more than 10 t per capita.

Energy is a prime mover of economic growth and development. Although the linkages between energy and development are complex and still imperfectly

TABLE I.2

Commercial primary energy production in developing countries, 1970–95 (source: World Bank, 1983)

	Million toe			*Growth rates (percentage/year)*	
	1970	*1980*	*1995*	*1970–80*	*1980–95*
Oil[a]	774	919	1 375	1·7	2·7
Coal	294	502	886	5·5	3·9
Natural gas	52	116	424	8·4	9·0
Primary electricity[b]	56	130	396	8·8	7·7
Total[c]	1 176	1 667	3 081	3·6	4·2

[a] Includes natural gas liquids and oil production from secondary recovery techniques.
[b] Includes hydropower, nuclear, and geothermal electricity.
[c] Excludes alcohol, oil shale, tar sands, and other nonconventional primary energy sources which may add a small amount (up to 10 million toe, or less than 0·5%) to developing country energy production by 1995, but of which the prospects are too uncertain to quantify.

TABLE I.3

Commercial primary energy production and consumption in oil-importing developing countries, 1970–95 (source: World Bank, 1983)

	Million toe			*Growth rates (percentage/year)*	
	1970	*1980*	*1995*	*1970–80*	*1980–95*
Production					
Oil	63	65	145	0·3	5·5
Coal	118	192	384	5·0	4·7
Natural gas	14	27	115	6·8	10·1
Primary electricity	41	98	306	9·1	7·9
Total	236	382	950	4·9	6·3
Consumption					
Oil	223	360	531	4·9	2·6
Coal	121	186	442	4·4	5·9
Natural gas	12	26	120	8·0	10·7
Primary electricity	41	98	306	9·1	7·9
Total	397	670	1 399	5·4	5·0
Oil Imports	160	295	386	6·3	1·8

understood, energy undoubtedly fuels economic development. Each developing country needs to harness as much energy as it can afford and effectively utilize. The bulk of available energy in many developing countries includes firewood, organic residues, and dung—the so-called noncommercial energy forms. Energy problems in developing countries are just as severe for noncommercial fuels as they are for oil and other commercial fuels.

TABLE I.4
Commercial energy typology of developing economies (source: World Bank, 1983)

Energy resources or options (relative to country size)	*Oil exporters*		*Oil importers (Net oil imports as a percentage of primary commercial energy consumption in 1980)*			
	Large	*Small or medium*	*0–25*	*26–50*	*51–75*	*76–100*
Limited			Middle income Lesotho Namibia		Low income Burundi Kampuchea Lao PDR **Nepal** Rwanda	Low income Bhutan **Ethiopia** Guinea Bissau Haiti Niger Somalia **Sri Lanka** Togo **Upper Volta** Middle income Barbados Cuba **Dominican Republic** Hong Kong Israel **Jamaica** **Jordan** Lebanon Liberia Mauritania Singapore **Uruguay** **Yemen AR** Yemen, PDR
Moderate		Middle income *Syria, AR*	Low income Zaire Middle income Botswana [b] Korea, PDR [b] Vietnam Zambia [b] Zimbabwe	Low income Ghana [a] **Pakistan** Middle income [a,b] *Brazil* [a] Chile Guatemala Ivory Coast Mongolia	Low income [a] **Bangladesh** Central Afr. Rep. Chad Equatorial Guinea Malawai Mozambique **Uganda** Middle income Costa Rica El Salvador Honduras [b] **Korea, Rep. of** **Paraguay** **Portugal** [b] **Turkey**	Low income **Benin** Guinea Madagascar Mali Sierra Leone **Sudan** **Tanzania** Middle income [b] **Greece** **Kenya** **Morocco** **Nicaragua** **Panama** Papua New Guinea **Philippines** **Senegal** **Thailand**

TABLE I.4—*contd.*

Energy resources or options (relative to country size)	*Oil exporters*		*Oil importers (Net oil imports as a percentage of primary commercial energy consumption 1980)*			
	Large	*Small or medium*	*0–25*	*26–50*	*51–75*	*76–100*
Substantial	Low income [a,b] *China* Middle income [a] *Indonesia* [a] *Iran* [a] *Iraq* [a,b] *Mexico* [a] *Nigeria* [a] *Venezuela*	Middle income [a] *Algeria* *Angola* Congo, PR *Ecuador* [a] *Egypt* *Gabon* [a] *Malaysia* [a] *Peru* [a] *Trinidad & Tobago* *Tunisia*	Low income Burma [a,b] *India* Middle income [a] *Argentina* Cameroon [a,b] *Colombia*	Low income [a] Afghanistan Middle income [a,b] Yugoslavia	Middle income [a] Bolivia	

Not shown are economies with less than 1 million population and without production (or prospects of future production) of oil, gas, or coal. The economies included in this table are classified according to their energy resource potential (oil, gas, coal, and primary electricity) that could be economically developable during the next decade. Oil exporters are countries in which official earnings from net oil exports exceed 10% of their total export earnings in 1980–81. *Large oil exporters* refers to countries that produced more than 70 million toe during 1980.
[a] Produced 1 or more million toe of gas in 1980.
[b] Produced 2 or more million toe of coal in 1980.
Economies shown in *italics* produced more than 5 million toe in 1980. Economies shown in **bold** print had net energy imports amounting to 30% or more of their merchandise exports in 1980. (Information is not available for all countries.)

> Ample energy supply is not an automatic guarantee of smooth economic advancement, social progress, and stability; it is, indisputably, their essential precondition. The future of our increasingly interdependent world will thus be very much influenced by the success or failure of the developing countries to assure sufficient and sustainable flows of energies. (Smil & Knowland, 1980)

Leach (1979) listed several main features of energy use in the Third World. Limited commercial energy (coal, oil, gas, and electricity) is used; particularly in rural villages or on farms; and, unfortunately, a significant amount of traditional or noncommercial energy (wood, charcoal, animals, and crop wastes) is used with dreadfully low efficiencies and high ecological and human costs.

In addition, modern energy is rarely used in rural areas because it is usually extremely expensive, not just relative to low incomes, but in global terms as well. Because of high transportation or transmission costs, fuel and electricity prices in most developing country rural areas are several times higher than prices in the United States or Europe. The cost of purchasing and maintaining equipment which operates on these fuels also prevents most people from using them.

The global inequity in the use of commercial fuels is familiar. About 1·5 billion people live in countries where per capita consumption is less than 7 gigajoules (GJ)/year; another 1·1 billion is situated between this and the 20 GJ level (see the Appendix for metric units used throughout this book). Per capita consumption is more than 300 GJ/year in the United States and more than 50 GJ/year in Western Europe. In developing countries, as in industrialized countries, the bulk of commercial fuel is used in urban industrial centers and by the urban, developed class. Common users include cars and public transportation, heated or air-conditioned buildings, lighting, and industrial plants.

Electricity, in particular, is primarily confined to urban and industrial centers. Since millions of rural inhabitants use virtually no commercial fuel, this imbalance within countries is at least as great as the international inequities in commercial fuel consumption (Leach, 1976). How then is it possible for rural residents to survive on so little energy? The answer is noncommercial energy, i.e. firewood, animal dung, agricultural residues, and other indigenous materials.

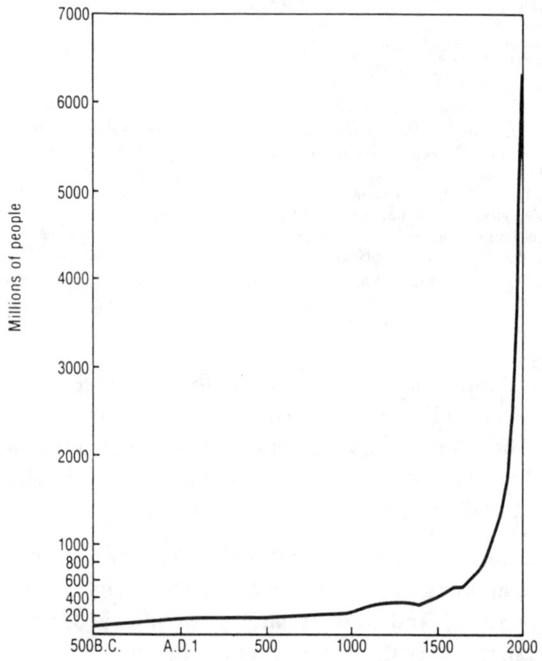

Fig. I.3. World population growth, 500 BC–AD 2000 (source: Loftness, 1984).

Competition for commercial fuels is likely to become even more intense in the future. During the late 1970s and early 1980s, energy analysts were forecasting reduced energy consumption, a natural reaction to high energy prices. Recently, however, world energy prices have been decreasing and consumption has been predicted to reach 485 exajoules (EJ) by the year 2000, compared with 300 EJ in 1982 (Table I.5). Another 50 EJ will be used in noncommercial energy. Chandler (1985) wrote, 'The physical magnitude of this scenario numbs the mind. If it comes to pass, the oil output of two new Saudi Arabias will be needed. In addition, the coal production of this world will almost double, and three times as many rivers must be impounded to produce hydropower.'

TABLE I.5
Energy consumption in selected countries, 1982 (source: Chandler, 1985; © 1985 Worldwatch Institute)

Country	*Population (millions)*	*Commercial energy consumption*[a] *(EJ)*[c]	*Per capita energy consumption (GJ)*[c]	*Energy imports as share of exports*[b] (%)
Argentina	28·4	1·7	61	11
Brazil	126·8	4·0	32	52
Canada	24·6	9·7	395	11
China	1 008·2	17·9	17	n.a.
East Germany	16·7	3·8	231	n.a.
France	54·4	8·5	156	33
India	717·0	4·9	7	81
Italy	56·3	6·2	110	41
Japan	118·4	15·8	134	48
Mexico	73·1	4·2	58	−76
Poland	36·2	5·0	138	20
Soviet Union	270·0	55·0	204	−77
United Kingdom	55·8	7·7	152	14
United States	231·5	75·1	324	36
West Germany	61·6	11·5	187	23
Total	2 880·0	232·0	80	—
World Total	4 585·0	300·0	65	

[a] Commercial energy consumption figures are Worldwatch estimates from 1981 data, assuming little change.
[b] A negative figure indicates the percent of exports earned from energy sales.
[c] The units are 0·95 million and 0·95 quadrillion (quad: see Appendix) Btu, respectively.

COMMERCIAL AND NONCOMMERCIAL ENERGY†

Commercial energy, a term almost synonymous with conventional energy excludes wood, dung, agricultural residues, and other traditional fuels, although these are widely traded. The balance of energy supplies, or noncommercial energy, is composed primarily of renewable energy—the supply of which is partly or entirely regenerated in the course of the annual solar cycle. Renewables include biomass which may be converted to liquid, gaseous, or solid fuel with higher energy densities, and other properties more suitable for agricultural use and engine operation.

Distinguishing between commercial and noncommercial or non-traded energy sources is sometimes difficult. In areas of intensive land use or scarcity of new arable land where competition for energy supplies is high, 'free' gathering of energy

† Adapted from World Bank (1985).

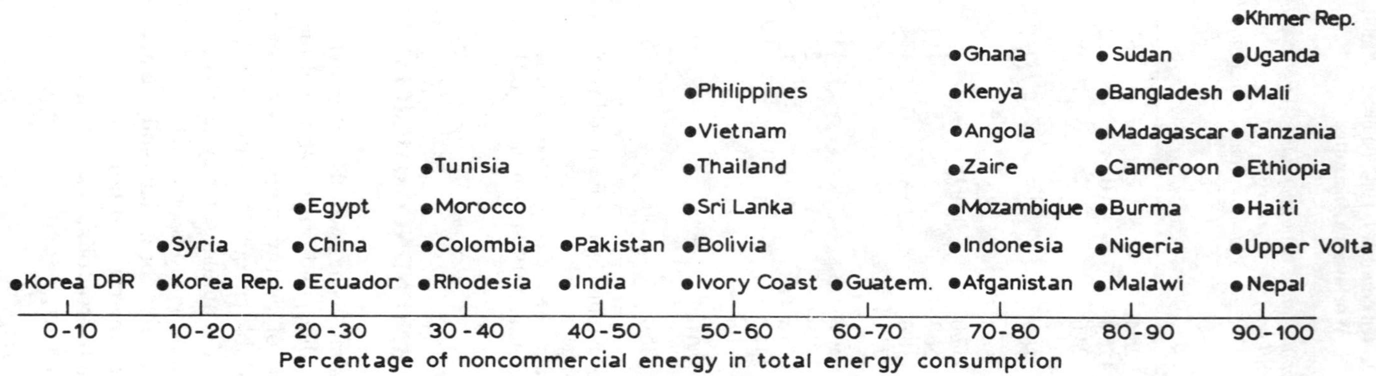

Fig. I.4. Ordered histogram showing the percentage of noncommercial energy consumption in the total energy consumption of developing countries (source: Parikh, 1980).

TABLE I.6
Energy consumption from commercial and noncommercial sources in kgce[a] 1973 (source: Parikh, 1980)

	Per capita				*Percentage*		
	Commercial	*Fuelwood*	*Agricultural waste*	*Total*	*Commercial*	*Fuelwood*	*Agricultural waste*
Cameroon	93·2	462·8	48·4	604·4	15·4	76·6	8·0
Ghana	161·5	360·9	30·0	552·4	29·2	65·3	5·4
Ivory Coast	372·2	491·5	46·6	910·3	40·9	54·0	5·1
Kenya	173·3	384·1	67·9	625·2	27·7	61·4	10·9
Madagascar	66·4	237·7	108·6	412·7	16·1	57·6	26·3
Malawi	55·2	380·4	107·1	542·8	10·2	70·1	19·7
Mali	24·7	205·5	62·3	292·6	8·5	70·2	21·3
Morocco	240·1	63·9	83·2	387·2	62·0	16·5	21·5
Mozambique	170·2	434·8	41·8	646·8	26·3	67·2	6·5
Rhodesia	754·1	419·3	76·5	1 249·9	60·3	33·5	6·1
Sudan	119·8	494·8	49·3	663·8	18·0	74·5	7·4
Tanzania	79·2	957·7	39·0	1 075·9	7·4	89·0	3·6
Uganda	64·8	635·8	59·9	760·6	8·5	83·6	7·9
Zaire	115·8	297·6	13·8	427·2	27·1	69·7	3·2
Cuba	1 162·7	71·7	11·8	1 246·2	93·3	5·8	0·9
Guatemala	273·8	481·4	68·9	824·0	33·2	58·4	8·4
Mexico	1 132·2	66·5	56·0	1 254·8	99·2	5·3	4·5
Brazil	626·4	600·9	50·8	1 278·1	49·0	47·0	4·0
Chile	1 327·9	124·2	26·5	1 478·7	89·8	8·4	1·8
Peru	643·2	163·4	20·1	826·8	77·8	19·8	2·4
Afghanistan	70·2	130·7	110·1	311·0	22·6	42·0	35·4
Bangladesh	28·1(?)	50·1	94·7	172·9	16·3	29·0	54·7
Burma	61·3	271·4	125·1	457·8	13·4	59·3	27·3
India	137·1	134·1	87·9	359·8	38·1	37·5	24·4
Khmer Rep.	29·6	211·8	54·9	296·4	10·0	71·5	18·5
Korea D.P.R.	2 284·7	127·5	94·3	2 506·5	91·2	5·1	3·8
Korea Rep.	868·4	94·5	103·1	1 066·1	81·5	8·9	9·7
Nepal	12·9	315·2	135·7	463·8	2·8	68·0	29·3
Pakistan	180·6	50·5	78·8	309·9	58·3	16·3	25·4
Philippines	289·5	217·4	80·2	587·1	49·3	37·0	13·7
Sri Lanka	144·7	132·0	44·4	321·2	45·0	41·1	13·8
Thailand	317·5	164·1	180·0	661·7	48·0	24·8	27·2
Vietnam	273·7	172·9	113·9	560·5	48·8	30·8	20·3
Turkey	598·8	116·0	87·2	801·9	74·7	14·5	10·9

[a] kgce, kilograms of coal equivalent.

supplies may not be allowed. In these circumstances, a hidden market may be operating in which permission to gather agricultural residues or fuelwood is granted in exchange for a service or gift which does not necessarily have to be paid at the moment of gathering.

Noncommercial energy constitutes significantly more than half the total energy used in many developing countries (Fig. I.4). Estimates indicate that in the 1970s,

TABLE I.7
Summary of selected national energy consumption surveys (source: Goldemberg *et al.*, 1987)

Country	*Commercial energy (kW per capita)*	*Noncommercial energy (kW per capita)*	*Total energy (kW per capita)*	*Percentage of energy from noncommercial sources*
Bangladesh	0·038	0·095	0·133	71
Niger	0·035	0·254	0·289	88
Gambia	0·098	0·222	0·320	69
Morocco	0·267	0·073	0·340	21
India	0·165	0·190	0·355	54
Ethiopia	0·019	0·371	0·390	95
Nepal	0·009	0·429	0·438	98
Somalia	0·092	0·476	0·568	84
Bolivia	0·340	0·263	0·603	44
Sudan	0·159	0·635	0·794	80
Thailand	0·305	0·524	0·829	63
Tanzania	0·060	0·810	0·870	93
China	0·778	0·317	1·10	29
Brazil	0·737	0·371	1·11	34
Mexico	1·29	0·127	1·43	9
Libya	1·76	0·095	1·86	5
All developing countries (average)	0·550	0·416	0·966	43

more than 60, 50, and 20% of all energy consumed in Africa, the Far East, and Latin America, respectively, came from noncommercial energy sources.

The heavy reliance of developing countries on noncommerical energy is shown further in Tables I.6 and I.7. Note that many African countries use less than 10% commercial energy (Mali uses 8·5%; Tanzania, 7·4%; and Uganda, 8·5%). Fuelwood is by far the most important form of noncommercial energy, accounting for 89% of the total energy used in Tanzania, 84% in Uganda, and 77% in Cameroon. Agricultural wastes, including animal dung, are also important noncommercial energy sources.

ENERGY FOR SUBSISTENCE AND DEVELOPMENT

The 'subsistence level' is that which barely sustains life, i.e. food production and preparation, protection from the elements, transportation of water and food, etc. The energy necessary to maintain subsistence varies with geography and climate. Development implies movement above the subsistence level, signifying more food, better balanced diets, and improved distribution of food to consumers, as well as elevated housing, education, sanitation, and health care. Development requires more energy—both commercial and noncommercial.

Figure I.5 shows the Gross National Product (GNP) as a function of commercial

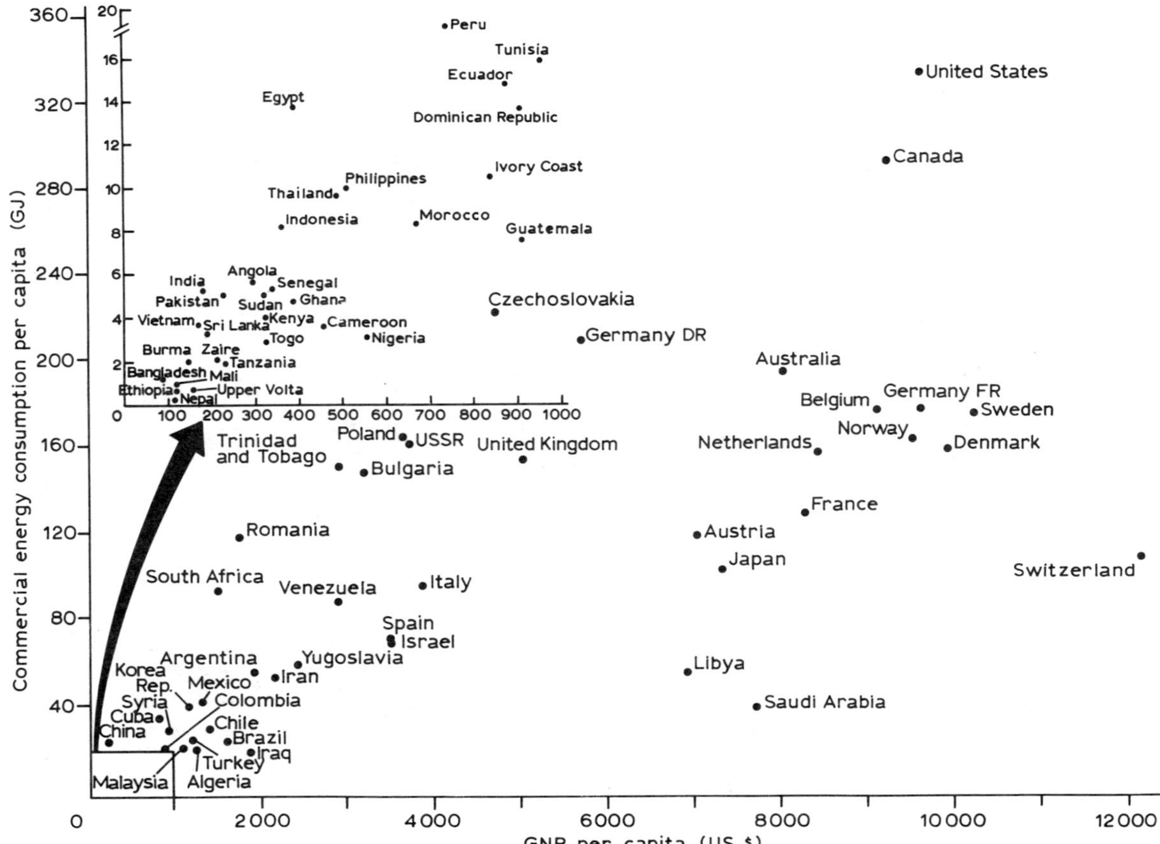

Fig. I.5. Relationship between commercial energy consumption and GNP for selected countries during 1978 (source: Mudahar & Hignett, 1982; World Bank, 1984*c*).

energy use for various countries. While the relationship is not as firm as once believed, the developing countries (smaller box in Fig. I.5) definitely have a lower per capita GNP coupled with lower energy input. Conversely, energy is required in large quantities to generate high per capita GNP and the associated high standard of living.

In 1980, the World Bank initiated an Energy Sector Assessment Program to diagnose the most serious energy problems faced by individual developing countries (World Bank, 1984). More than 50 country assessments have been completed. Such assessments include:

—analyzing the potential for changes in pricing, institutional arrangements, and other policies to encourage economically profitable production from indigenous sources, fuel substitution, and more efficient energy use;
—assessing the investment priorities in the energy sector;
—providing a framework for official and private assistance to this sector.

Emphasis on energy assessments in the rural and household sector is growing. On the demand side, the reports evaluate existing programs for dissemination of more efficient cooking stoves and charcoal kilns, and make recommendations based on cross-country experiences. On the supply side, more detailed analyses of fuelwood production options and the prospects for alternative household fuels are underway. Rural energy was a priority area for action cited by the United Nations 1981 Nairobi Conference on new and renewable sources of energy.

TABLE I.8
Human energy expenditures and returns for gardening by the Tsembagas (source: Rappaport, 1971)

Human energy input	*MJ/ha*
Clearing underbrush	293
Clearing trees	117
Fencing garden	177
Weeding and burning	98
Placing soil retainers, etc.	75
Planting and weeding until end of harvest	933
Other maintenance	238
Sweet potato harvest	232
Taro harvest	29
Cassava harvest	11
Yam harvest	81
Cartage	619
Total	2 903
Human food output	31 808
Pig feed output	18 760
Total	50 568

$$E = \frac{50\,568}{2\,903} = 17{\cdot}4$$

Despite the heavy reliance on noncommercial energy by rural areas in developing countries, the tenfold increase in petroleum prices from 1973 to 1983 caused significant alarm. Lower world oil prices in the mid-1980s were slow to reach rural areas of developing countries. Because the margin between energy surplus and shortfall is small, temporary gluts and periods of scarcity are likely to continue.

Finite fossil fuel resources are spread unevenly around the world. Some regions are blessed with sizable petroleum or natural gas reserves, whereas other regions have more coal. In time, industries and individual consumers can shift, within limits, from one energy form to another. The ultimate survival of humankind will depend on our ability to learn to utilize the inexhaustible energy supply emanating from the sun in a wide variety of forms. Subsequent chapters focus specifically on solar heating, wind power, hydropower, and biomass fuels.

As an example of the energy requirements of subsistence agriculture, consider the primitive New Guinea farmers known as the Tsembagas. Rappaport (1971) measured the human energy inputs and agricultural outputs of their farming operations. Table I.8 presents typical values for performing various tasks entirely by

TABLE I.9
Energy efficiency ratios for various agricultural systems (source: Leach 1976)

Agricultural system	*Energy ratio*
Hunting–gathering	7·8
iKung bushmen	
Herders	
Dodo tribe, Uganda	5·0
Shifting cultivators	
Congo	65
Tsembaga	20·3
Subsistence and shifting	
Rice, Dayak	16·5–18·2
Rice, Iban	14·2
Rice, Tanzania	23·4
Maize, Africa	37·7
Millet, Africa	36·2
Sweet potato, Africa	31·3
Cassava, Africa	22·9
Yams, Africa	22·9
Groundnut, Africa	12·8
Subsistence	
India (from Odum)	14·8
China (1935–37)	41·1
Corn, Mexico (axe and hoe)	30·6
(oxen)	14·15
Guatemala (axe and hoe)	13·6
(oxen)	3·95
Nigeria (axe and hoe)	10·5
Philippines (Carabao)	5·07
Wheat, India (bullock)	1·69
Rice, Philippines (Carabao)	5·51

hand. The output/input energy ratio achieved in this example was 17·4, not atypical of primitive agriculture.

Leach (1976) compiled energy budgets from a number of sources for various agricultural systems and then calculated the output/input energy ratios (Table I.9). As expected, the range is wide, presumably due to inaccuracies in the original data, but many examples show energy returns that are 10, 20, and 30 times more than the input.

HUMAN ENERGY CAPACITY

Life as we know it today could not exist without energy to supplement the meager work capacity of humans. A healthy adult can produce about 75 W or 1/10 horsepower (hp) continuously. Assuming a steady output of 75 W for a 10-h day, the work produced is about 750 watt-hours (Wh), which is equal to 3/4 kilowatt-hours (kWh) or 645 kilocalories (kcal). Depending on the location, electrical power costs from \$0·06/kWh to \$0·36/kWh. Thus, a human's work for a day is worth from \$0·04 to \$0·25. Clearly, this level of work output will not generate a reasonable standard of living.

However, food production in most developing countries is labor-intensive. Tillage, planting, weeding, and other tasks are often performed by hand labor or animal

Fig. I.6. Malawi woman preparing the soil to plant groundnuts (source: FAO photo 5954/J/31).

power (Fig. I.6). Tractors and other forms of advanced mechanization are the exception—not the rule.

SUPPLEMENTAL ENERGY NEEDS†

Energy for Rural Development

Energy is a critical aspect of the rural development process. It is expended in agricultural operations; food processing and transportation; fertilizer production; pesticide and farm equipment usage; industrial operations that provide jobs; household cooking, heating, and lighting; and the construction and operation of the infrastructure needed for schools, health centers, and water supplies.

Rural development is affected by the severe problems faced by many developing countries that rely heavily on imported petroleum. As prices rise, governments face the necessity of using increasingly large sums of foreign exchange for imported oil, thus diverting resources that might be used for other development activities. In Kenya, for example, 85% of the commercial energy is derived from imported petroleum at a cost of 25% of the country's foreign exchange earnings.

In Jamaica, 90% of the total energy supply comes from petroleum imports. Although noncommercial fuels supply an estimated 73% of the total energy consumption in Bangladesh, the cost for petroleum imports from 1977 to 1978 amounted to 33% of the country's total export earnings. About 36% of the 1979 export revenues of the Philippines were allocated to oil imports, accounting for almost 92% of the total energy requirements.

Even with lower world energy prices in the 1980s, the energy situation was very troublesome in many developing countries, especially in relation to rural development. Many developing countries have important characteristics that make increased use of renewable and nonconventional energy sources attractive. They lack the massive investments in the petroleum-oriented energy infrastructure that are typical of industrialized countries. And many developing countries have more sunlight than industrialized countries, so solar energy is a more attractive alternative. Furthermore, their climatic conditions permit faster vegetative growth for biomass fuels, and the decentralized nature of renewable energy makes it an ideal energy source for small-scale rural development programs.

Nonconventional energy sources will not, in themselves, provide enough energy to power rural development in most countries. They can, however, provide a substitute for imported fuels and play an increasingly important role in national energy balances.

Rural application of nonconventional energy resources and technologies now receive considerable attention, and they will continue to benefit from ongoing research and development. For example, direct solar energy can be used for heating water and buildings, drying agricultural and animal products, and producing salt. Wind energy can be harnessed by small-scale wind machines for use in pumping

† Adapted from Hurley (1981).

TABLE I.10
Quality of life model. Weighted categories and subcategories of variables (source: Loftness, 1984)

Social (24·0)

Satisfaction of basic human needs: standard of living (1·2)
+ Index of food consumption per capita
+ Steel consumption per capita (kg)
+ Cotton yarn production per capita (ton)
+ Motor vehicles in use per 1 000 population
+ Rail traffic, passenger kilometers per capita
– Relative living-cost index

Informed citizenry with modern conveniences (0·78)
+ Television sets per 1 000 population
+ Radio sets per 1 000 population
+ Telephone sets per 1 000 population
+ Newspaper circulation, copies per 1 000 population
+ Pieces of domestic mail sent per 1 000 population

Welfare and independent status (1·02)
+ Ratings of social security and insurance system
+ Average annual rate of growth in private consumption, 1970–1976
+ Ratio of economically active to economically inactive population
– Adult illiteracy rate
– Total fertility rate from desired fertility rate, absolute values

Economic (20·0)

Individual economic well-being: flow and stock measures (1·27)
+ GNP per capita (1976 $US), adjusted by cost-of-living index
+ Motor vehicles per 1 000 population
+ Television sets per 1 000 population
– Average annual rate of inflation, 1970–1976

Economics, structure and productivity (0·73)
+ Cement production per 1 000 economically active population (ton)
+ Cotton yarn production per 1 000 economically active population (ton)
+ Acres of arable land per 1 000 agricultural workers
+ Change in agricultural productivity per worker
+ Percentage of labor force employed by manufacturing industries
– Ratio of armed persons to economically active population
+ Average annual growth rate of gross domestic investment, 1970–1976

Health and education (22·0)

Individual health status (1·32)
+ Life expectancy at birth, 1975
– Infant mortality rate per 1 000, 1975
– Crude death rate per 1 000 population, 1975
+ Per capita food consumption index, 1974

Community health conditions (0·86)
– Population (1 000) per physician, 1974
– Population (1 000) per nurse, 1974
+ Percentage of population with access to safe (or piped) water, 1975
– Population per hospital bed

Educational attainment (0·82)
– Adult illiteracy rate, 1974
+ Numbers enrolled as percentage of age group, 1975: in primary school; in secondary school
+ Numbers enrolled as percentage of age group 20–24 in higher education, 1975
+ Percentage of GNP spent on education, 1975

Environmental (15·0)

Natural environment and utilization (1·16)
+ Arable land, acres per capita
– Average annual rate of total population growth, 1970–1975
– Rate of urbanization and urban growth, 1970–1975

Manmade environmental problems (0·84)
– Population density (population per km^2)
– Number of motor vehicles in use per km^2
– Percentage of labor force employed by manufacturing industries
– Percentage of population in urban areas

National vitality and security (19·0)

National carrying capacity (1·16)
Human capital
+ Percentage of economically active population
+ Average annual rate of labor force growth, 1970–1975

National and capital resources
+ Acres of available land per capita
+ Average annual growth rate of gross domestic investment, 1970–1976

Technology
+ Percentage of exports in manufactured products
+ Steel consumption per capita

Market stability and price mechanism
– Index of relative living cost
– Average annual rate of inflation

International security and independence (1·0)
Economic and financial considerations
+ Export and import ratio
+ Percentage of labor force employed by manufacturing industries
+ Gross international resources in months of import coverage
– Energy independence, percentage of imports for fuel
– Percentage share of imports for food
– Index of resources imbalance (gross domestic savings less gross domestic investment)
– Dependence on foreign trade

Military power
+ Armed forces per 1 000 population
+ Military expenditures as percentage of GNP
+ Military expenditures per capita adjusted by cost-of-living index

water and irrigation; in grinding, threshing, and winnowing; and in generating electricity for various purposes.

Small-scale, low-head hydroelectric units can generate electricity or provide power for grain mills, pumps, wood- and metal-working machinery, and other types of machines. Fermentation of plant products or animal wastes can be used to produce alcohol fuels or methane gas.

None of these alternative energy sources and technologies offer a complete solution to the energy requirements of rural development. However, all countries must use the broadest possible range of available energy resources and devote considerable effort to identifying, developing, and disseminating the technology needed to take advantage of these resources.

Energy for Improved Quality of Life

Energy availability and effective use affects the quality of life (QOL). Positive and negative indications of QOL are tabulated in Table I.10. Many of these indicators are energy-related. Using this model, the QOL in 50 countries was ranked as shown in Fig. I.7.

GNP is another indicator of QOL (Fig. I.5). Regardless of its shortcomings as an index of human well-being, a high GNP is often associated with a high standard of living. Figure I.8 illustrates a more generalized relationship between standard of living and effective energy consumption per person. The less developed countries are striving to increase their standard of living through the development and more effective use of energy.

D. E. Morrison (1978) expressed quality of life as a function of energy use (Fig. I.9). At low levels of energy use (quadrant III), basic need satisfaction is linearly related to energy use. As the amount of energy increases (quadrant II), two paths may be hypothesized: Option 'A' projects a linear relationship between QOL and energy use; whereas, Option 'B' suggests an optimum QOL at a moderately high level of energy use, followed by a deterioration for excessively high energy-use rates.

Energy-intensive industrialized countries like the United States are, no doubt, operating in quadrant II, and excessive energy use has often resulted in environmental deterioration and, subsequently, a lower quality of life. The developing countries, however, are unquestionably operating in quadrant III, at least as far as their rural populations are concerned. Thus, a formula for agricultural and rural development is to supplement the meager work capacity of humans with as much energy as is efficient and cost-effective.

> For more than a third of the world's people, the real energy problem is a daily scramble to find the wood they need to cook dinner. Nine-tenths of the people in most poor countries depend on firewood as their chief source of fuel. And all too often, the growth of human population is outpacing the growth of new trees—not surprising when the average user burns as much as a tonne of firewood a year. Firewood prices have risen much faster than kerosene prices, a rise that reflects the growing difficulty with which wood is being procured. (Eckholm, 1976)

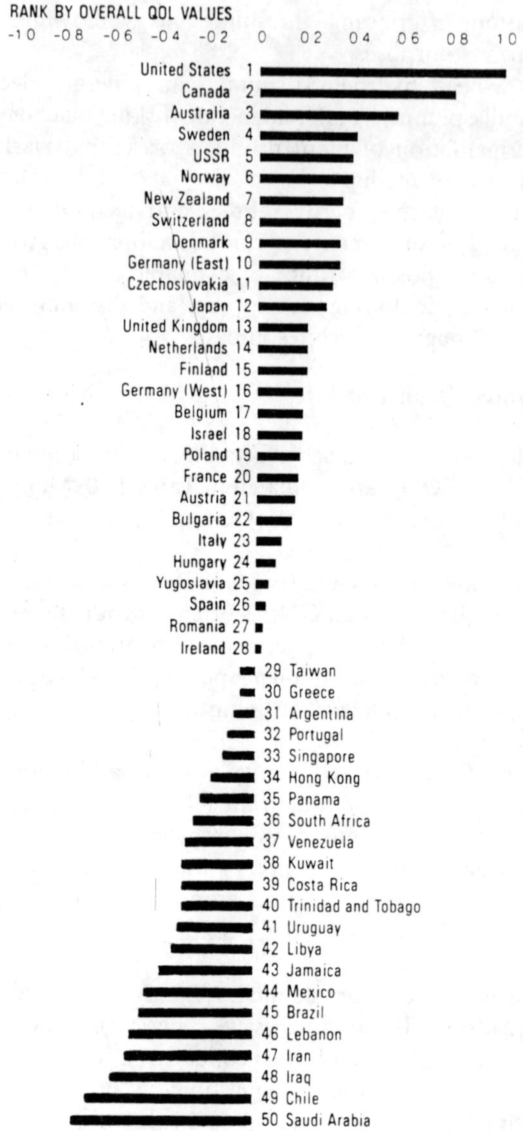

Fig. I.7. Energy and the quality of life (source: Loftness, 1984).

In Asia and many areas in Africa and Latin America, animal dung provides a major source of cooking fuel. However, its use is controversial because fertilizers and organic matter need to be returned to the soil. The National Commission on Agriculture in India declared that 'the use of cow dung as a source of noncommercial fuel is virtually a crime' (Smil & Knowland, 1980). Conversion of dung to biogas is a popular concept because it retains the fertilizer value while producing a useful fuel at the same time.

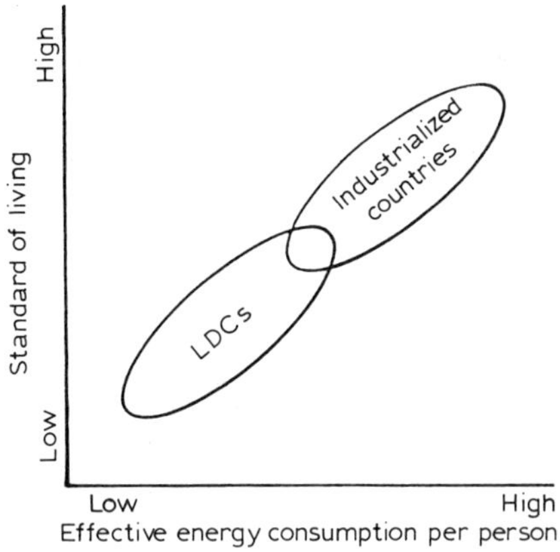

Fig. I.8. Relationship between standard of living and effective energy consumption per person (source: Stout, 1982).

Energy for Agriculture

Agriculture is essentially an energy conversion process—the conversion of solar energy through the photosynthetic process to food energy for humans and feed for animals. Primitive agriculture involved little more than scattering seeds on the land and accepting the scanty yields that resulted. Modern agriculture, on the other hand, is the application of science and technology to enhance yields. An energy subsidy such as fossil fuels for fertilizers, farm machinery, and water pumps, or electricity for certain tasks, greatly enhances yields and makes farm life more tolerable.

Before 1950, increases in food output came largely from expanding cultivated area, but with the scarcity of fertile new land and the advent of cheap chemical

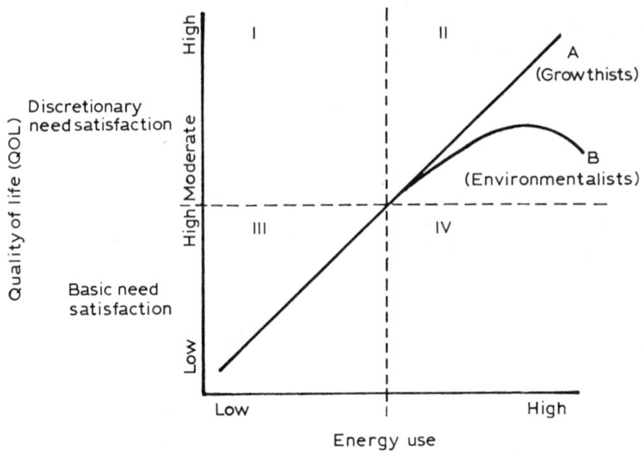

Fig. I.9. Conflicting growthist and environmentalist perspectives on the relationship of energy use and quality of life (source: Morrison, 1978).

> fertilizer, this changed. Between 1950 and 1983, world fertilizer use climbed from 15 million to 114 million tonnes, nearly an eightfold increase within a generation. In effect, as fertile land became harder to find, farmers learned to substitute energy in the form of chemical fertilizer for land. Fertilizer factories replaced new land as the principal source of growth in food production. (L. R. Brown, 1984)

The result, depicted in Fig. I.10, shows a dramatic increase in energy inputs into agriculture from 1940 to 1970.

This substitution of energy for land is graphically evident in Fig. I.11. In 1950, when world population totaled 2·51 billion, the harvested area of cereals per person

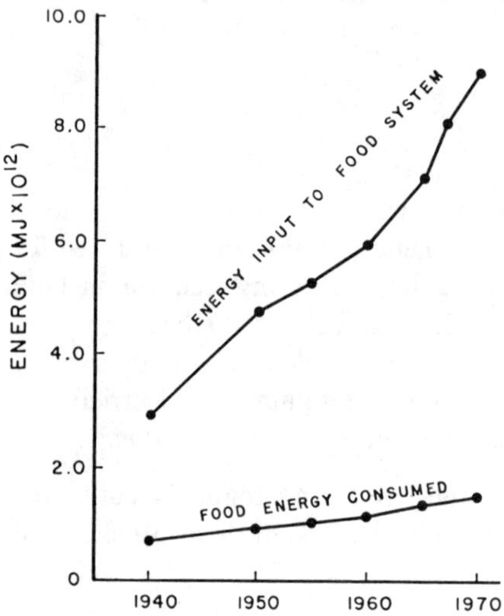

Fig. I.10. Energy use in the food system, 1940–70, compared to the caloric content of food consumed (source: Steinhart & Steinhart, 1974).

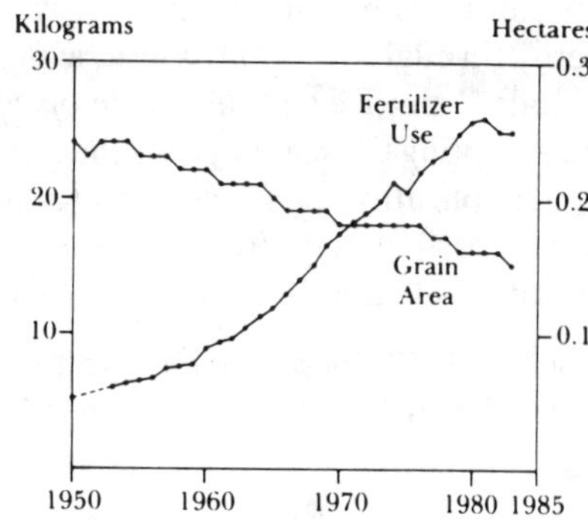

Fig. I.11. World fertilizer use and grain area per person, 1950–83 (source: L. R. Brown, 1984).

was 0·24 hectares (ha). As population growth outstripped cultivated area, the area per person dropped steadily to 0·15 ha by 1983. While the amount of cropland per person declined by one third, fertilizer consumption per person increased by a factor of five, climbing from just over 5 kilograms (kg) in 1950 to 25 kg in 1983.

Agricultural modernization requiring increased energy use is, and will continue to be, essential in providing enough food for growing populations, but it must also provide sufficient employment for a rapidly expanding rural labor force. New, non-farming jobs must be created to raise incomes and employment levels. One approach is to develop light industries in towns and small cities. A considerable increase in rural energy supplies is a prerequisite for such industrial development, however (Revelle, 1978).

Energy supplies must grow more rapidly than populations to elevate the quality and quantity of human diets, increase income and employment, and relieve human drudgery. Greater energy efficiency, as well as additional sustainable energy supplies, are needed for cooking, space heating, and other domestic activities of rural households.

Most available statistics on energy use refer to commercial energy. But, on the average, total noncommercial energy use for developing countries is twice that of commercial energy. Biomass fuels—wood, charcoal, agricultural residues, and cattle dung—comprise most of the noncommercial energy consumed. Fuel energy for cooking and other household needs is just as essential as food, and often, almost as expensive (Revelle, 1978). Cereal grains and grain legumes—wheat, rice, maize, sorghum, millets, beans, and lentils—are principal foods of the poor, and they usually cannot be eaten uncooked.

Four specific needs for additional energy in agriculture outlined by Revelle (1978) include:

—providing a larger and more stable water supply;
—allowing application of commercial fertilizers, especially nitrogen;
—allowing more rapid seedbed preparation;
—improving transportation.

Linkages between energy and agriculture are complex and not always completely defined. Figure I.12 illustrates some conceptual linkages between energy, fertilizers, and the agricultural system.

No single or short-term solutions exist for energy problems in poor countries; however, many partial solutions can gradually bring energy supplies into balance with energy needs (Cervinka & DeForest, 1980). For many countries without domestic resources, future agricultural development cannot be based on nonrenewable fossil fuel supplies. The diminishing availability of these supplies results in high prices and unreliable supplies.

Furthermore, dependence on fossil fuels increases the political and economic vulnerability of both industrial and developed nations. This situation, however, should be approached as a challenge rather than a crisis. All nations can benefit from the development of renewable energy resources—a realistic means for long-term economic development. Agricultural development based on renewable energy establishes a sound and sustainable foundation for food production. Development of renewable energy resources will lead to increased energy and food, self-reliance, and a re-creation of viable political, economic, social, and cultural trends (Cervinka & DeForest, 1980). Nevertheless, the technology for utilizing renewable energy resources is imperfect and still evolving.

Energy—A Vital Resource for Food Systems

The food system includes all aspects of food production, processing, storage, transportation, and final preparation. Production agriculture includes all farming

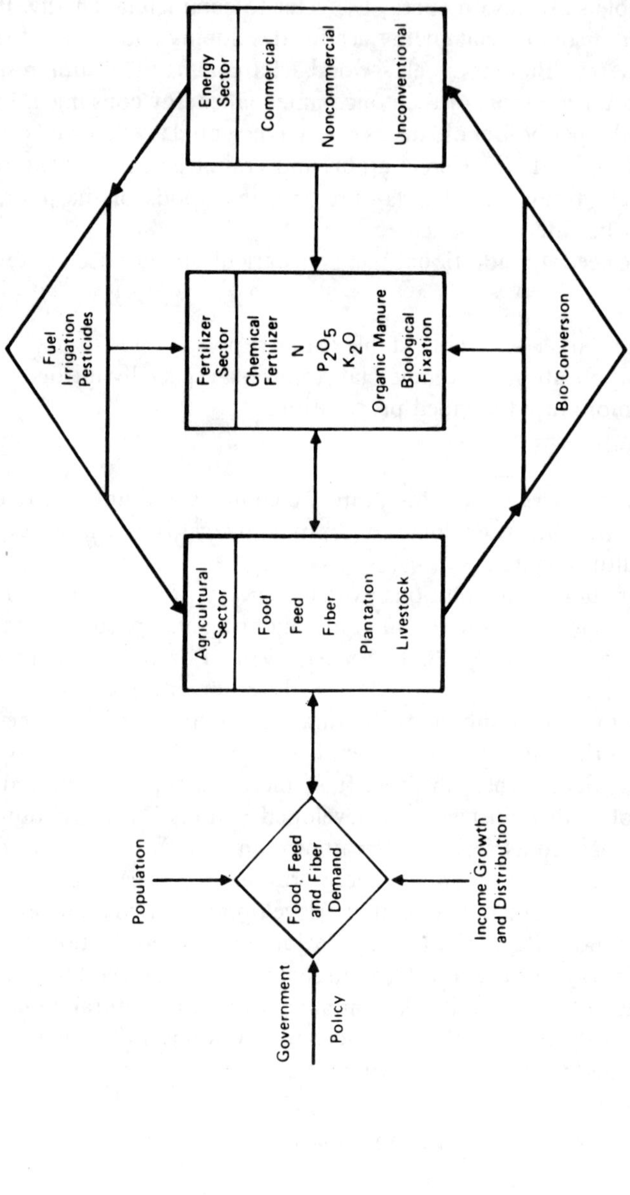

Fig. I.12. Conceptualization of linkages between energy, fertilizer, and agricultural sectors (source: Mudahar & Hignett, 1982).

operations that occur after the land is cleared and developed, such as tillage, planting, fertilizing, pest control, harvesting, and farm-level drying or processing—up until the product leaves the farm. The energy needed for production agriculture is about 3% of the national energy consumption in developed countries and perhaps 5 or 6% in developing countries. The entire food system may require 15%, 20%, or more of the developing nation's energy consumption.

The challenge ahead is to ensure that agricultural energy needs are met in a timely fashion to fulfill world food requirements. Most farm operations must be performed within a narrow time frame to maximize yields. For example, irrigating after the crop has withered and almost died of drought is practically useless, just as is applying fertilizer too late. Thus, energy in its various forms must be available to agriculture when needed, and more effort must be undertaken to increase the efficiency of energy utilization.

This book primarily focuses on two questions: (1) How much energy is needed for agriculture in the developing countries? and (2) How will these energy needs be met? Chapter II deals with the first question—energy flow in agriculture. Primary attention is focused on production agriculture, such as farming in developing countries. General energy flows are tabulated; individual farm or country needs are site-specific and require individual study. Three options for meeting agriculture's energy needs are apparent: conservation, substitution of alternative energy forms, and priority allocation. Subsequent chapters address methods for achieving these options.

Conservation and More Efficient Energy Use

Much information has been published about the potential of energy conservation for saving money and stretching scarce energy supplies. In some cases, particularly in the developed countries where energy use is relatively high, conservation measures have been stimulated by higher prices.

Figure I.13 indicates that world oil production has dropped somewhat since it peaked in 1979. Energy consumption decreased in the United States, Western

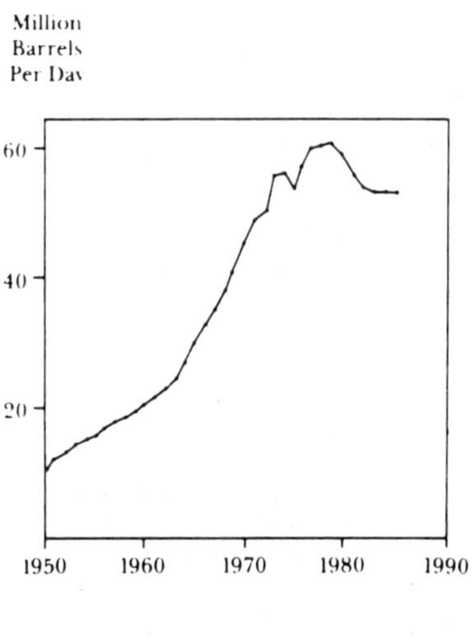

Fig. I.13. World oil production, 1950–85 (source: L. R. Brown *et al.*, 1986).

Europe, and Japan in the decade following the OPEC oil embargo in 1973 (Fig. I.14). The rest of the world increased energy consumption only slightly during the same period, so total world consumption of oil and gas changed very little during this time.

In developing countries, however, the energy input to agriculture is small, so the potential for conservation is limited. This is not to suggest that energy wastefulness should be condoned, however. Certainly, waste should be eliminated and every effort should be made to optimize energy use. But energy conservation on farms in developing countries does not offer much opportunity to provide the energy needed to produce high yields.

Alternative Energy Forms
In areas where fossil energy is available, it will continue to be used as long as it is the least expensive form of energy. If fossil energy is not available in the needed form—liquid or gaseous fuels or fertilizer—or if cheaper alternatives are available, then these alternatives will be used.

Priority Allocation
While this book is not policy-oriented, the author is compelled to argue that if all other mechanisms fail (i.e. conservation, substitution of alternatives, etc.), the public welfare and its need for an adequate food supply demands that the energy needs of agriculture be met. Energy inputs to agriculture in terms of increased yields are so

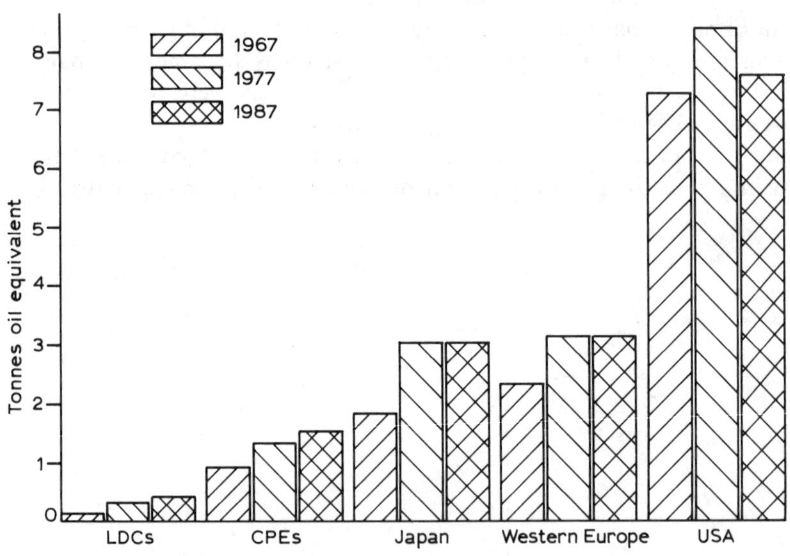

Fig. I.14. Energy consumption per capita, 1967–87. The LDCs use only a small fraction of the energy consumed in Japan, Western Europe, or the USA. Although energy consumption per capita in the United States has fallen since 1977, it is still more than twice that in Western Europe and Japan (source: British Petroleum Company, 1988).

beneficial that a priority allocation system should be devised, if necessary, to meet the energy needs of agriculture.

Energy and Labor Intensity

Relationships among labor, fossil fuel use, and food production are of great importance in developing countries. Table I.11 presents energy budgets for 16 countries, including three developing countries—Pakistan, China, and Egypt. The data show a wide range of fossil fuel-based inputs (0·18 GJ/ha per year for Australia to 57·6 GJ/ha per year for the Netherlands); labor input (0·000 02 GJ/ha per year for Australia to 2·82 GJ/ha per year for Japan); and food output (0·55 GJ/ha per year for Australia to 49·4 GJ/ha per year for Japan). Fossil fuel input varied from 11 GJ/man per year in Pakistan and Egypt to 671 GJ/man per year in California in the United States. Food output ranged from a low of 26 GJ/man per year for Japan and Egypt to a high of 663 GJ/man per year in the United States (Stanhill, 1984). Another example is provided in Table I.12 which tabulates commercial energy requirements for rice and maize by modern, transitional, and traditional agricultural methods (Loftness, 1984).

Figure I.15 indicates a positive correlation between fossil fuel and labor use. Two discrete linear regressions are shown, one for the industrialized nations of Europe, North America, Israel, Australia, and New Zealand, and a second for the more labor-intensive agricultural systems in Asia (Stanhill, 1984).

Clearly, agriculture in developing nations will need more energy in the future. Table I.13 projects total commercial energy input in 90 developing countries,

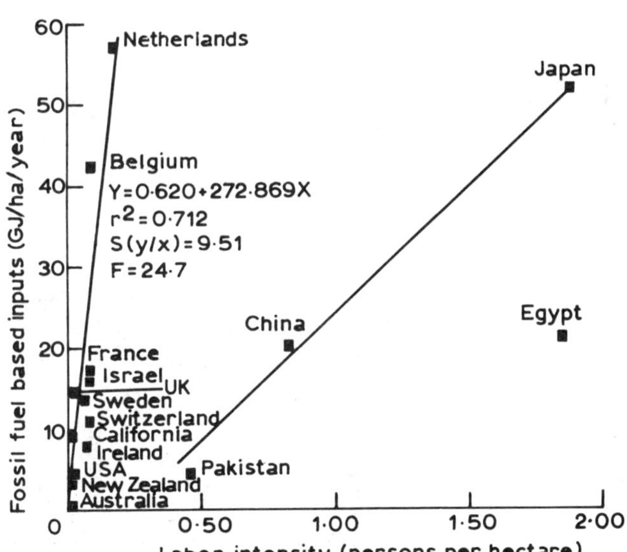

Fig. I.15. Relationship between intensities of fossil fuel and labor use in 16 national agricultural systems—third quarter of the 20th century. (For data sources, see Table I.11.) (Source: Stanhill, 1984; © 1984 Springer-Verlag.)

TABLE I.11
Energy balance of 16 national agricultural systems (source: Stanhill, 1984; © 1984 Springer-Verlag)

Country	*Year*	*Output in food (GJ/ha per year)*	*Inputs (GJ/ha per year)*			*Labor intensity (GJ/man per year)*		
			Fossil fuel base	*Human labor*	*Total*	*Fossil fuel*	*Total input*	*Food output*
Japan	1970	49·4	52·3	2·82	55·1	28	28	26
Egypt	1972–1974	47·5	21·3	1·38	25·9	11	14	26
The Netherlands	1964–1965	37·1	57·6	0·359	79·0	322	441	207
Belgium	1975	32·9	43·1	0·146	94·9	474	1 043	362
China	1978	29·0	20·1	0·533	24·0	24	29	35
Sweden	1972	17·0	13·7	0·54	18·4	268	361	333
Pakistan	1977	13·9	4·2	0·53	4·8	11	9	31
France	1970	11·7	16·8	0·147	17·5	199	209	140
Switzerland	1976	9·3	10·6	0·131	28·2	113	301	99
California	1974	8·5	9·4	0·015	9·4	671	671	607
Ireland	1974	7·5	8·1	0·113	10·1	140	174	129
Israel	1969–1970	7·4	16·2	0·082	26·7	208	343	78
United Kingdom	1968	6·7	14·6	0·041	19·5	604	839	287
United States	1974	5·3	4·9	0·005	5·0	613	625	663
New Zealand	1978–1979	4·5	3·11	0·012	3·21	346	357	504
Australia	1965–1969	0·55	0·18	0·000 02	0·18	194	194	591

TABLE I.12

Commercial energy required for rice and corn (maize) production, by modern, transitional, and traditional methods (source: Chancellor, 1978)

Input	*Rice*						*Maize*			
	Modern (United States)		*Transitional (Philippines)*		*Traditional (Philippines)*		*Modern (United States)*		*Traditional (Mexico)*	
	Quantity per ha	*Energy per ha (10^6 J)*	*Quantity per ha*	*Energy per ha (10^6 J)*	*Quantity per ha*	*Energy per ha (10^6 J)*	*Quantity per ha*	*Energy per ha (10^6 J)*	*Quantity per ha*	*Energy per ha (10^6 J)*
Machinery and implements[a]	$4{\cdot}2 \times 10^9$ J	4 200	335×10^9 J	335	173×10^9 J	173	$4{\cdot}2 \times 10^9$ J	4 200	173×10^6 J	173
Fuel[b]	224·7 liters	8 988	40 liters	1 600	—	—	206 liters	8 240	—	—
Nitrogen fertilizer[c]	134·4 kg	10 752	31·5 kg	2 520	—	—	125 kg	10 000	—	—
Phosphate fertilizer[d]	—	—	—	—	—	—	34·7 kg	586	—	—
Potassium fertilizer[e]	67·2 kg	605	—	—	—	—	67·2 kg	605	—	—
Seeds[f]	112·0 kg	3 360	110 kg	1 650	107·5 kg	—	20·7 kg	621	10·4 kg	—
Irrigation[b]	683·4 liters	27 336	—	—	—	—	351×10^6 J	351	—	—
Insecticides[g]	5·6 kg	560	1·5 kg	150	—	—	1·1 kg	110	—	—
Herbicides[g]	5·6 kg	560	1·0 kg	100	—	—	1·1 kg	110	—	—
Drying	$4{\cdot}6 \times 10^9$ J	4 600	—	—	—	—	$1\,239 \times 10^6$ J	1 239	—	—
Electricity	$3{\cdot}2 \times 10^9$ J	3 200	—	—	—	—	$3\,248 \times 10^6$ J	3 248	—	—
Transport	724×10^6 J	724	31×10^6 J	31	—	—	724×10^6 J	724	—	—
Total		64 885		6 386		173		30 034		173
Yield (kg/ha)		5 800		2 700		1 250		5 083		950
Energy input yield per unit (10^6 J/kg)		11·19		2·37		0·14		5·91		0·18

[a] Energy input to produce 1 kg of equipment assumed to be $68{\cdot}7 \times 10^6$ joules.
[b] One liter of fuel assumed to contain 40×10^6 joules.
[c] Production of 1 kg of nitrogen fertilizer assumed to require 80×10^6 joules.
[d] Production of 1 kg of phosphate assumed to require 14×10^6 joules.
[e] Production of 1 kg of potassium fertilizer assumed to require 9×10^6 joules.
[f] Production of 1 kg of high-quality seed assumed to require 30×10^6 joules in the United States and 15×10^6 joules in the Philippines and Mexico.
[g] Production of 1 kg of pesticide assumed to require 100×10^6.

TABLE I.13
Agricultural energy use in developing countries, 1980–2000 (source: *Agriculture: Toward 2000,* FAO, Rome, 1979)

Aggregate grouping and year	*Machinery*		*Irrigation*		*Fertilizer*		*Pesticides*		*Total*	
	Amount (10^3 mtoe)	*%*	*Amount (10^3 mtoe)*	*%*	*Amount (10^3 mtoe)*	*%*	*Amount (10^3 mtoe)*	*%*	*Amount (10^3 mtoe)*	*Amount/ha (kgoe)*
90 developing countries										
1980	11 119	31	4 259	12	18 781	53	1 142	3	35 301	47
2000	54 947	35	7 200	5	92 988	59	2 795	2	157 930	169
African developing countries										
1980	1 011	41	122	5	1 128	46	188	8	2 449	12
2000	5 111	44	232	2	5 811	50	488	4	11 642	46
Latin American developing countries										
1980	5 107	45	451	4	5 317	47	410	4	11 285	59
2000	27 018	57	761	2	18 745	40	929	2	47 452	163
Near East developing countries										
1980	2 803	43	1 096	17	2 535	39	145	2	6 580	78
2000	7 041	36	1 734	9	10 351	53	333	2	19 458	210
Far East developing countries										
1980	2 199	14	3 011	20	9 802	64	397	3	15 409	58
2000	15 777	20	5 135	6	58 081	73	1 046	1	80 039	269

mtoe, million tons of oil equivalent.
kgoe, kilograms of oil equivalent.

increasing from 35 million tonnes in 1980 to 157 million or almost 158 million tonnes in the year 2000—more than a fourfold increase. An even greater increase is projected for countries starting from a very low energy-use base.

ENERGY BALANCE

A myriad of information has been published about energy balance. So much so, that some danger exists of getting lost in the numbers and failing to comprehend the trends and implications. Unifying fundamental principles should be recognized and structured to provide a framework for evaluating agricultural practices and renewable energy forms.

Laws of Thermodynamics

The First Law of Thermodynamics tells us that energy is neither created nor destroyed. If all forms of energy entering a process and all forms of energy coming out are accounted for, the input and output will be equal.

The Second Law of Thermodynamics states that the entropy increases as an energy conversion process occurs. This means that the amount of usable or available energy coming out of a conversion process will be less than the input; therefore, the efficiency of the conversion of useful energy is always less than 100%.

Agriculture—An Energy Conversion Process

As stated previously, agriculture is an energy conversion process. Its primary purpose is to convert inedible forms of energy into forms of biomass—carbohydrates, proteins, fats, etc., that are digestible as food energy for humans. If all energy inputs are considered, the energy ratio of food production is one and energy ratio has no meaning.

Boundaries of Agricultural Energy Analysis

Most agricultural energy analysts consider the solar energy input that drives the hydrologic cycle to be energetically free. The hydrologic cycle produces rainfall, which allows plants to produce biomass through photosynthesis. If we neglect the solar input and consider only the cultural energy input, plants produce energy gains as illustrated by Heichel (1973) in Fig. I.16. Some ecologists, however, such as Odum (1984) and Gilliland (1975), do not accept the concept that solar energy, or other energy forms that cannot be monetized, are energetically free.

Purpose of Agricultural Energy Analysis

Undoubtedly, agricultural energy analysis serves many purposes. Energy is the only commodity for which a substitute cannot be found, and energy cannot be recycled

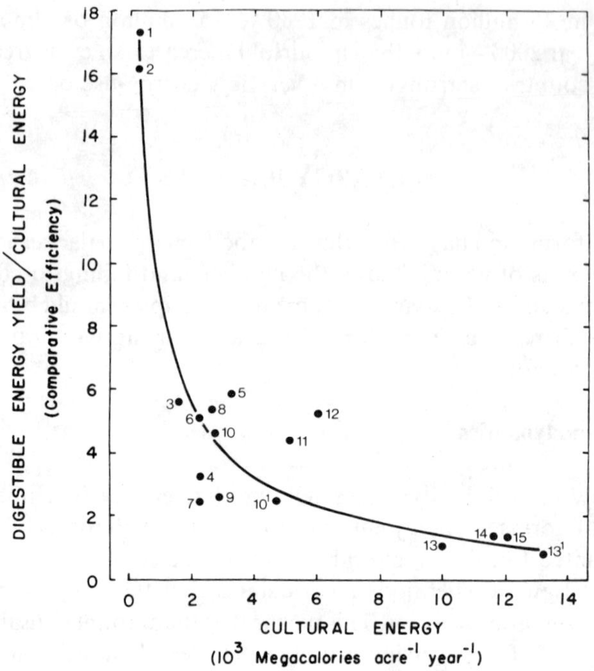

Fig. I.16. Efficiency of energy use in 15 agricultural systems. Cultural energy is the energy input to produce crops, except solar energy and including human and animal work; fossil fuel used in production, processing, and transport; and energy required to grow seeds, construct buildings, and produce machinery, chemical, and fertilizer. 1. Paddy rice, Philippines, 1970. 2. Vegetable garden, New Guinea, 1962. 3. Corn for grain, Iowa, *c.* 1915. 4. Corn for grain, Pennsylvania, *c.* 1915. 5. Corn silage, Iowa, *c.* 1915. 6. Alfalfa-brome hay, Missouri, 1970. 7. Oats, Minnesota, 1970. 8. Sorghum for grain, Kansas, 1970. 9. Soybean, Missouri, 1970. 10. Sugar cane, Hawaii, 1970, cultural energy excludes processing. 10^1. Sugarcane, Hawaii, 1970, cultural energy includes processing. 11. Corn for grain, Illinois, 1970. 12. Corn silage, Iowa, 1970. 13. Sugar beets, California, 1970, cultural energy excludes processing. 13^1. Sugarbeets, California, 1970, cultural energy includes processing. 14. Peanuts, North Carolina, 1970. 15. Irrigated rice, Louisiana, 1970 (source: Heichel, 1973).

without violating the Second Law of Thermodynamics. Thus, energy analysis can be a tool for policy formulation. For example, it might lead to public policies discouraging the production and consumption of meat because of the low energy ratio associated with its productivity.

Another benefit of agricultural energy audits and subsequent analyses is to provide a basis for energy conservation and guidelines for farm management decisions. In this case, economic considerations must be coupled with energy analysis. Farming is a business which must be profitable to continue. Thus, farmers will not voluntarily adopt practices that are economically unattractive, regardless of their energy implications. Many examples of energy analysis in various agricultural systems are provided in Chapter III.

WORLD ENERGY OUTLOOK

Several excellent world energy surveys have been compiled in the last decade (Exxon Corporation, 1980; Sivard, 1981; *UN Yearbook of World Energy Statistics*, 1981; British Petroleum Company, 1984). They indicate that fossil fuels will probably be available for many years because no one really knows how much of this resource is still undiscovered. World petroleum production is approaching its peak, however, and supplies will be insufficient to meet all demands at some point in the future. The complete production cycle for a finite resource is illustrated by the characteristic bell-shaped curve shown in Fig. I.17.

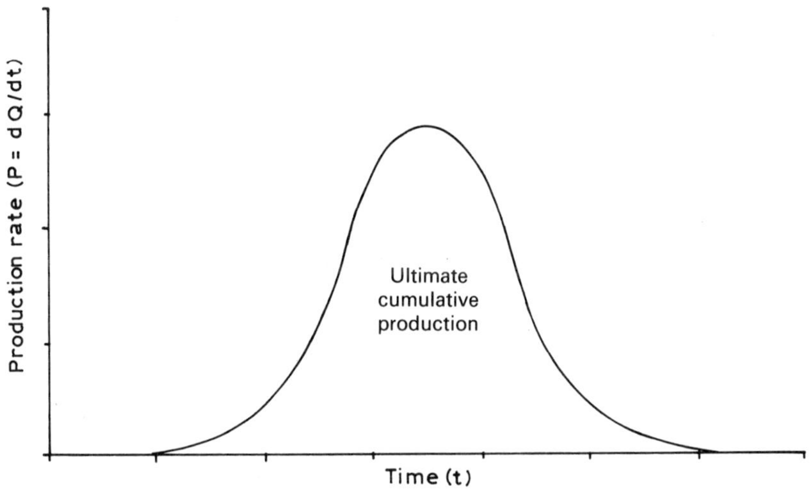

Fig. I.17. Complete production cycle of an exhaustible resource (source: Hubbert, 1956).

Figure I.18 illustrates the world production of primary energy. Prior to 1960, coal was the largest source of commercial energy, but since the 1960s, petroleum has become the number one energy source, since 1980 petroleum production has levelled off, whereas production of other energy forms continues to increase.

The appetite of developing countries for oil is growing. Energy needs for development are well-documented. Eventually, the demand for oil will presumably overtake the supply, and the upward price trend that resulted in a tenfold price increase in the 1970s, will likely resume.

Numerous examples of energy needs in developing countries could be cited (Tables I.5 and I.14, and Fig. I.19). Many argue that developing countries need more energy to feed their growing populations, to continue their development, and to raise the overall standard of living. Efficient energy use makes good economic sense, but conservation and efficient use alone will not fulfill the developing countries' need for increased energy supplies. Before considering the agricultural energy situation in detail, let's look broadly at the world energy picture.

Energy Density

A primary focus of this book is energy both for and from agriculture in the developing countries. We must distinguish between rural and urban energy needs when seeking energy solutions for rural areas. Goldemberg (1980) emphasized that urban areas could have population densities of 500–1000 or more people per square kilometer (km^2), while significantly fewer people reside in rural areas. Energy requirements for rural areas, therefore, are generally lower, i.e. 0·04 W/m^2 in rural India compared to 12 W/m^2 in Indian cities.

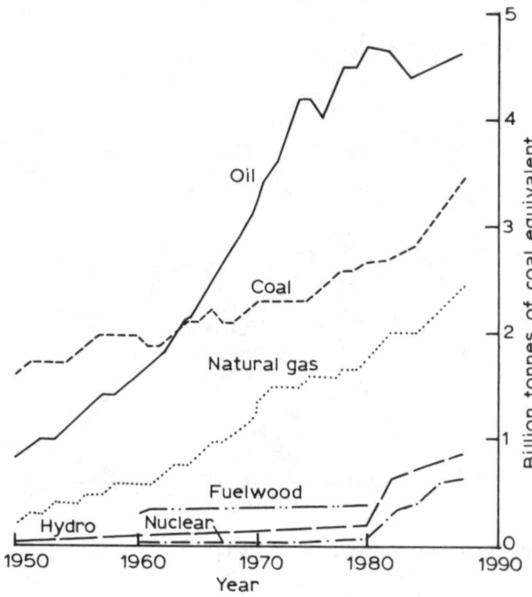

Fig. I.18. World production of primary energy (sources: Sivard, 1981 (to 1979) and USDOE, 1988 (after 1979)).

In rural areas of developing countries, the energy requirements (W/m^2) are often 100 times less than in the cities. Thus, the implementation of renewable, small-scale, and decentralized energy systems makes more sense in rural areas. Direct solar energy, for example, provides approximately 100 W/m^2; whereas, wind, biomass, and hydropower have lower energy densities (0·6, 0·2, and 0·02 W/m^2, respectively).

Nonrenewable Energy Production and Consumption

Table I.15 summarizes world primary energy consumption in 1982. Total consumption was $6{\cdot}9 \times 10^9$ t of oil equivalent (approximately 42·2 GJ/t), with oil constituting 40% of the total followed by coal with 30%, natural gas with 19%, water power with 7%, and nuclear power with 3%.

TABLE I.14

Estimated fossil fuel, hydropower, and forest energy resources in developing countries (source: Revelle, 1981; © 1981 Pergamon Press plc)

	Coal and lignite[a]	*Petroleum & natural gas liquids*	*Natural gas*	*Hydro*[b]	*Forests*[c]	*Total*	*Total per capita (tonnes/person)*	
							Total	*Less forests*
Bangladesh	1 187		290	10	270	1 753	20	17
Burma	NA	12	NA	NA	6 920	>6 932	>217	?
China	719 000[d]	17 300	26 500	18 000	19 850	800 650	890	868
India	85 800	3 000	9 800	730	11 680	110 010	172	154
Indonesia	2 520	2 300	NA	410	21 620	26 850	196	38
Japan	8 760	11	150	7 950	4 350	21 221	182	145
Korea	1 515	—	NA	615	1 170	3 300	89	58
Malaysia	NA	630	NA	4 915	4 170	9 715	747	426
Nepal	—	—	—	400	800	1 200	92	31
Pakistan	650	42	690	470	460	2 312	30	24
Philippines	920[e]	400[e]	NA	240	2 470	4 030	88	34
Sri Lanka	—	—	—	45	230	275	20	3
Taiwan	450	2·5	—	1 280	NA	>1 733	>102	102
Thailand	235	0·4	190	825	2 840	4 090	91	28
Vietnam	>12	NA	NA	>335	2 400	>2 747	>55	>7
Total	821 049	23 687	37 620	36 225	79 230	996 818	449	413
% of estimated world total	16·2	5·3	10·9	30·7	12·4	16·5		
% world total, less China	2·0	1·4	3·2	15·4	9·3	3·2		

[a] Total known resources of coal and lignite; economically recoverable resources may be less.

[b] Hydropower resources taken as 100 times potential annual production of electric energy, converted to coal equivalents, assuming one equivalent ton of coal $= 3 \times 10^{10}$ J.

[c] Forest resources taken as 100 times potential annual yield (=22% of estimated net annual primary production of 8·07 metric tons of coal equivalent/ha).

[d] 50% of total estimated resources, based on assumed recoverability.

[e] Solid fuel and petroleum resources from data in Republic of the Philippines, Ministry of Energy, 'Ten-Year Energy Program, 1979–1988.' The Philippines also have significant geothermal resources. The planned electric generating capacity for geothermal sources may be nearly 1900 megawatts by 1988, equivalent to annual combustion of 6 million tonnes of coal equivalent.

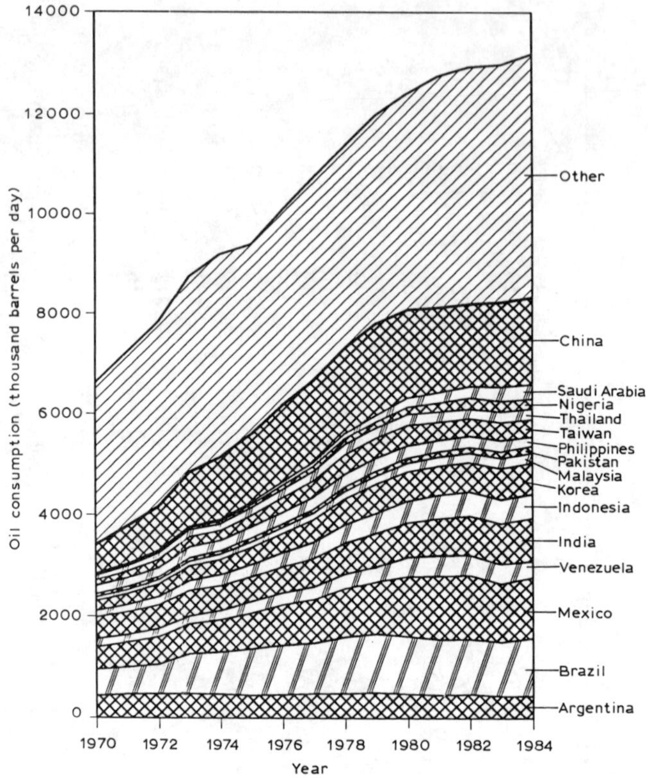

Fig. I.19. LDC oil consumption (source: Sathaye *et al.*, 1987).

Petroleum

For about the past 30 years, petroleum has been a major fossil fuel. World oil demand since 1955 is illustrated in Fig. I.20, while the geographic distribution of proved oil and gas reserves is outlined in Figure I.21. Although world oil production peaked about 1979 and has been declining since, oil is expected to remain the predominant form of fossil fuel for the next 20 or 30 years, especially for mobile vehicles.

The main oil shipping routes are shown in Fig. I.22. Since most oil originates in the Middle East, instability in that region may cause unpredictable disruptions anywhere in the world.

Coal

Production of this fossil fuel is increasing. Coal is widely used for industrial process heat, space heating, and electric power generation. Because reserves are extensive (Fig. I.23), coal is expected to play a more prominent role in meeting *world* energy needs in the future, although many *developing nations* have only limited reserves.

TABLE I.15

Primary energy/consumption (1982) by fuel million tonnes oil equivalent (source: British Petroleum Company, 1984)

	Oil	*Natural gas*	*Coal*	*Hydro-electric*	*Nuclear energy*	*Total*
North America						
United States	700·3	432·3	400·3	95·4	80·0	1 708·3
Canada	66·5	46·7	28·3	55·7	11·5	208·7
Total North America	766·8	479·0	428·6	151·1	91·5	1 917·0
Latin America	217·9	60·7	17·2	52·6	0·6	349·0
Western Europe						
Austria	10·0	3·9	2·8	7·1	—	23·8
Belgium & Luxembourg	21·2	8·2	9·1	0·1	5·3	43·9
Denmark	10·4	—	5·4	—	—	15·8
Finland	10·4	0·6	2·2	3·3	4·2	20·7
France	89·3	24·8	24·3	15·2	27·4	181·0
Greece	11·5	—	4·8	1·1	—	17·4
Iceland	0·5	—	—	1·0	—	1·5
Republic of Ireland	4·2	1·9	2·2	0·2	—	8·5
Italy	88·7	22·6	13·0	11·1	2·1	137·5
The Netherlands	29·1	32·3	5·4	—	0·9	67·7
Norway	7·6	—	0·6	25·3	—	33·5
Portugal	9·4	—	0·4	2·3	—	12·1
Spain	47·6	2·3	18·9	6·8	2·4	78·0
Sweden	20·0	—	1·6	10·9	7·0	39·5
Switzerland	12·3	0·9	0·3	9·3	3·8	26·6
Turkey	16·7	—	8·6	3·4	—	28·7
United Kingdom	72·4	43·3	65·6	1·4	10·7	193·4
West Germany	109·8	38·8	80·5	6·2	14·8	250·1
Yugoslavia	13·0	3·9	16·7	6·0	0·9	40·5
Cyprus/Gibraltar/Malta	1·6	—		—	—	1·6
Total Western Europe	585·7	183·5	262·4	110·7	79·5	1 221·8
Middle East	90·4	38·5		1·6	—	130·5
Africa	80·3	17·6	74·4	17·2	—	189·5
Japan	205·8	25·2	63·0	19·5	27·5	341·0
South East Asia	114·5	7·8	24·1	4·1	7·2	157·7
South Asia	46·4	13·0	97·0	16·0	0·7	173·1
Australasia	33·2	13·4	32·3	9·2	—	88·1
Total NCW	2 141·0	838·7	999·0	382·0	207·0	4 567·7
Centrally-planned economies (CPEs)						
China	84·7	10·7	436·0	21·9	—	553·3
USSR	450·5	405·0	358·0	51·0	22·0	1 286·5
Others	117·8	74·5	304·1	14·3	7·3	518·0
Total CPEs	653·0	490·2	1 098·1	87·2	29·3	2 357·8
Total world	2 794·0	1 328·9	2 097·1	469·2	236·3	6 925·5

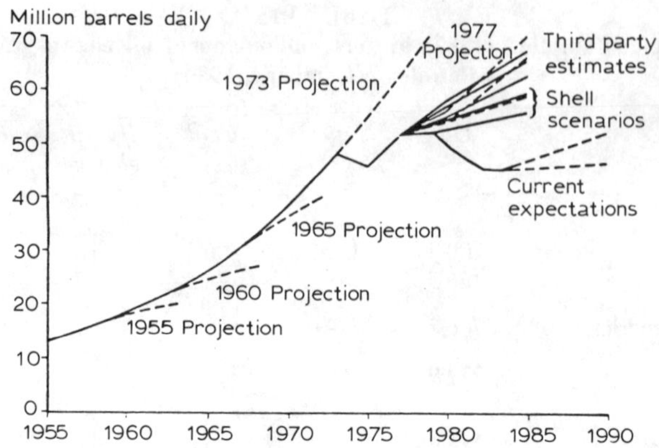

Fig. I.20. Estimates of non-communist world oil demand. Note the pattern of underestimation followed by overestimation just as demand growth slowed down (source: Schwartz, 1987).

Natural Gas

In 1981, natural gas production was about one-third that of oil on an energy basis, increasing from 1009 million tonnes of coal equivalent (mtce) in 1971 to 19 378 mtce in 1981 (British Petroleum Company, 1981).

Electricity Production

Electricity is an essential ingredient for development. It is expensive compared to other energy forms, but is convenient and easily controllable. Electricity is most appropriate for low-energy end uses such as lighting and communications, or situations in which automatic control is critical. Heavy energy users, such as resistance heating, may not be a good end use for electricity.

Electricity can be produced from various energy sources, including hydro, geothermal, nuclear, and thermal energy (oil, coal, solar, etc.). Table I.16 shows the relative cost of various methods of generating electricity. Figure I.24 illustrates ways to generate electricity from solar energy.

TABLE I.16

Average system costs per kilowatt of generating capacity in 97 developing countries (source: J. J. Fritz, 1984)

System type	*$US, 1979*
Hydro	1 296
Geothermal	1 564
Nuclear	1 436
Thermal	836

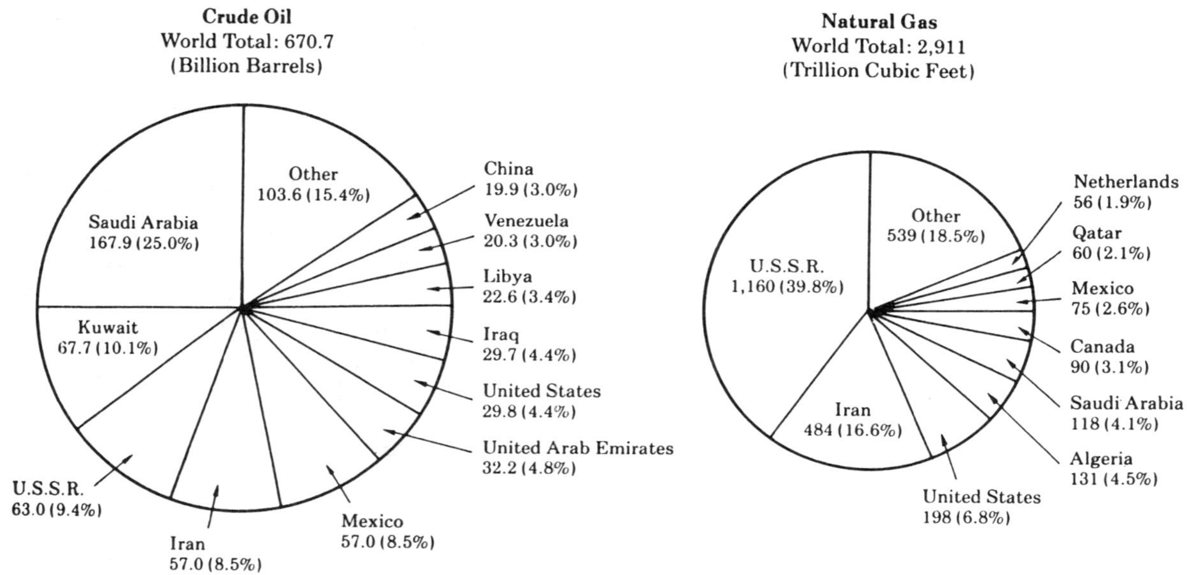

Fig. I.21. Estimated international crude oil and natural gas proved reserves, 31 December 1981 (source: Loftness, 1984). Note: Quantities are scaled in proportion to area according to the Btu content of the reserves. One billion barrels of crude oil equals approximately 5·7 trillion cubic feet of natural gas (1980).

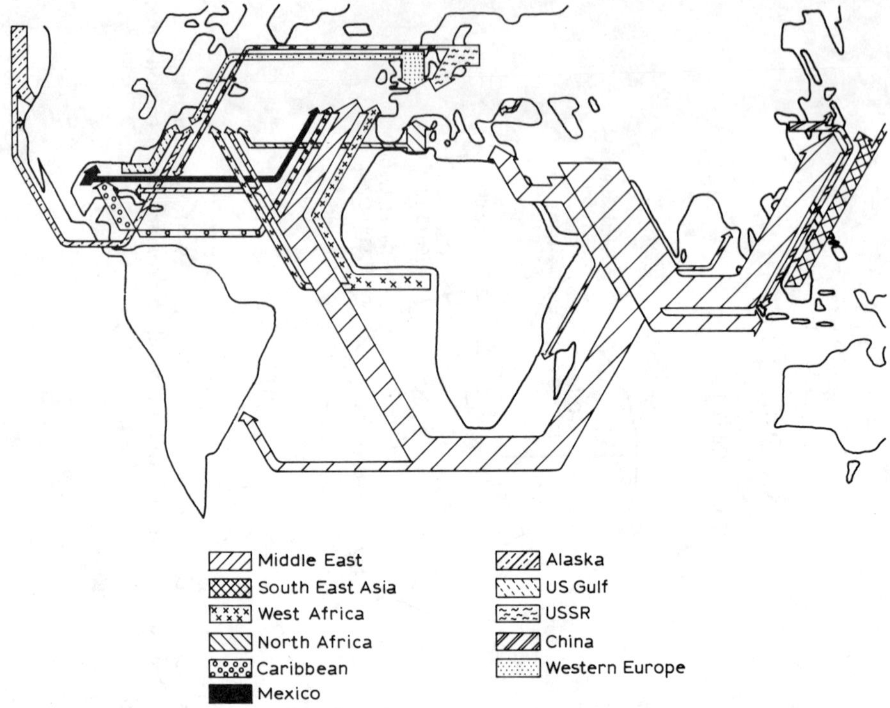

Fig. I.22. Main oil movements by sea, 1987. The Middle East dominates oil exports by sea. The width of the arrows is proportional to the quantity shipped (source: British Petroleum Company, 1988).

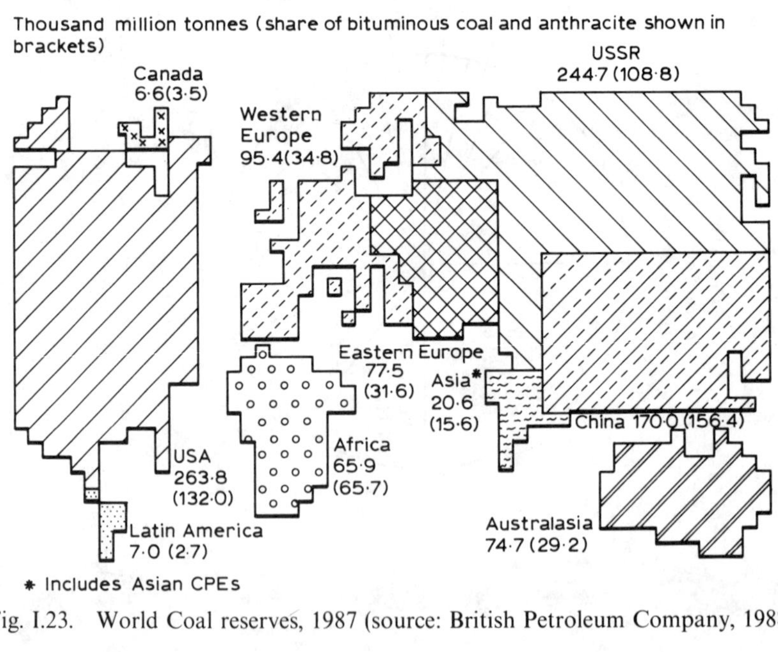

Fig. I.23. World Coal reserves, 1987 (source: British Petroleum Company, 1988).

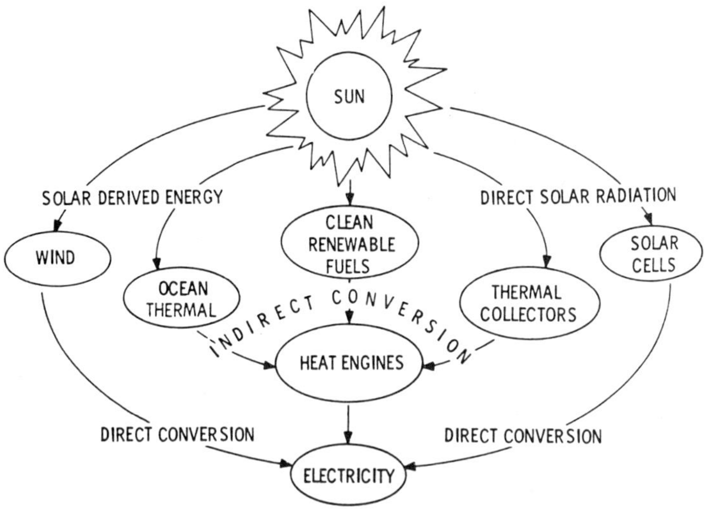

Fig. I.24. Electricity from the sun (source: Loftness, 1984).

Nuclear Energy

By 1984, 40 nations around the globe had a total of 555 commercial nuclear-electric generating plants in operation (Kiely, 1981), under construction (World Bank, 1985), actively being planned (World Bank, 1984*d*), or on order (USAID, 1981). The United States operates 75 plants and France has 61. Brazil became the twenty-fourth nation to have nuclear plants actively on-line and The People's Republic of China has ordered its first reactor (*Houston Chronicle*, 26 January 1984). The total world nuclear energy consumption in 1983 was 236 million tonnes of oil equivalent (mtoe) (Table I.15).

TABLE I.17
Renewable energy resource base (source: Loftness, 1984)

Continuous supply (10^{12} *W*)	*World*	
	Maximum	*Possible by 2000*
Solar radiation	28 000	
Fuelwood	3	1·3
Farm waste	2	0·6
Photosynthesis fuel	8	0·01
Hydropower	3	1·0
Wind power	0·1	0·01
Direct conversion	?	0·01
Space heating	0·6	0·006
Nonsolar		
Tidal	1·0	0·06
Geothermal	0·06	0·006
Total	18+	3
Annual demand year 2000 (10^{12} W)		~15

Renewable or Inexhaustible Energy

Ultimately, the future of mankind rests with the development of renewable or inexhaustible sources of energy. The sun is the source of virtually all forms of renewable energy. Solar energy comes in many forms—the major sources of which include direct radiation, wind, hydropower, and biomass (Table I.17). Chapters V, VI, VII, and IX deal with these renewable energy forms in detail.

In many developing countries, human and animal energy provide much of the power for agriculture. Thus, a section is devoted to human and animal power in Chapter II.

ECONOMIC CONSIDERATIONS

The price of a commodity is normally thought of as a mechanism to allocate supply; however, energy prices are often quite unrelated to energy costs because of taxes and regulated or arbitrary pricing policies. Gasoline is probably the most extreme example of this situation because tax policy varies so radically from one country to another (Table I.18). Kerosene prices are somewhat more uniform as indicated in Table I.19.

World oil prices have increased substantially throughout the past decade as shown in Fig. I.25. Two major international shocks occurred during this period—the 1973 oil embargo and the 1978–9 Iranian hostage crisis. Although oil prices have declined since 1981, the decline has not reached the rural areas of many developing countries.

TABLE I.18
Prices of petroleum products in selected countries, 1987 (source: L. R. Brown, 1988)

Country	*Gasoline* (*dollars per gallon*[a])	*Diesel* (*dollars per gallon*[a])	*Kerosene* (*dollars per gallon*[a])
Italy	3·71	1·55	1·56
Denmark	3·58	2·29	1·82
France	2·95	1·91	n.a.
Brazil	2·94	1·92	1·07
Japan	2·89	1·80	1·08
Ethiopia	2·40	1·44	1·20
United Kingdom	2·24	1·77	n.a.
India	2·17	1·01	0·65
West Germany	2·09	1·67	n.a.
Thailand	1·30	0·92	0·89
United States	0·82	0·87	0·83
Mexico	0·71	0·61	0·30

[a] Figures are for average consumer prices as of 1 January 1987. They include taxes, which account for the largest differences in price.

TABLE I.19
Kerosene prices in selected Third World cities, 1981 (source: L. R. Brown, 1984)

City	*Price per gallon (dollars)*
Mexico City	0·24
New Delhi	0·73
Buenos Aires	0·94
Islamabad	1·05
Bangkok	1·10
Accra	1·20
Addis Ababa	1·20
Nairobi	1·20
Singapore	1·21
Santiago	1·29
Manila	1·49
San Salvador	1·58

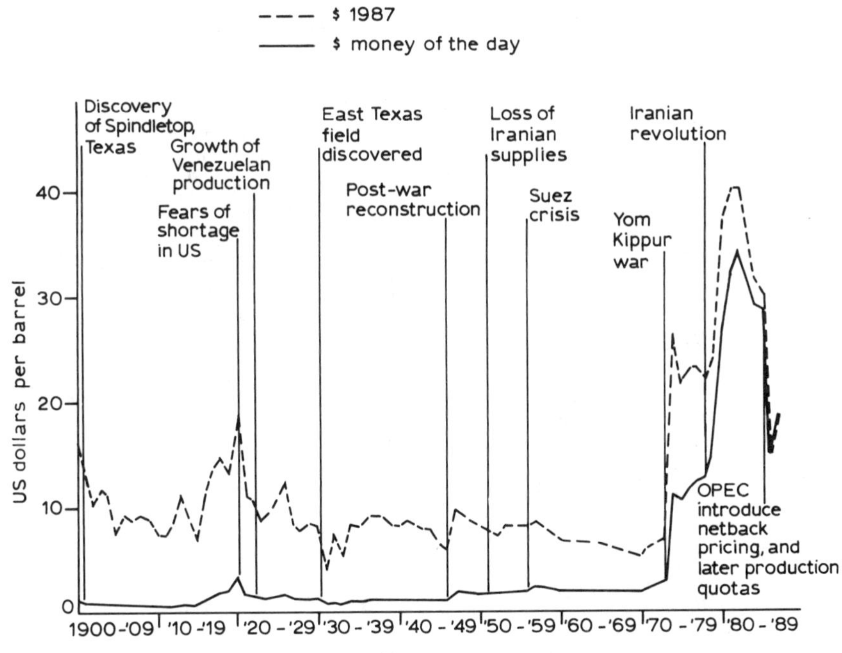

Fig. I.25. Crude oil prices since 1900 (source: British Petroleum Company, 1988).

The substantial increase in the price of oil during the 1970s:

...contributed to a massive redistribution of wealth from oil-importing countries to those that export petroleum. The dimensions of this shift are perhaps illustrated by the change in the relative prices of grain and oil, both widely traded commodities. Between 1950 and 1973, a bushel [27 kg] of wheat could be traded for a barrel of oil [Table I.20]. Then, as the price of oil jumped, the relationship began to shift. By 1977, 4·4 bushels [120 kg] of wheat were required to purchase a barrel of oil; in 1982, nearly 8 [218 kg] were required. Clearly, oil has become very expensive in relation to wheat. The same is true of other agricultural commodities such as tea, coffee, and sugar. During the early 1970s, when a barrel of oil was still priced at $2, it could be purchased with 4 pounds [1·8 kg] of tea; in 1983, more than 30 pounds [13·6 kg] were required. For coffee, the terms of trade went from 4 pounds [1·8 kg] in the early 1970s to 22 pounds [10 kg] in 1983. With sugar, the most widely exported farm commodity in the Third World, the 'sugar price' of a barrel of oil increased from just above 30 pounds [13·6 kg] to nearly 200 pounds [90·7 kg] during the same period.

These price increases for oil relative to agricultural commodities raise questions of the sustainability of energy-intensive agricultural systems and place added emphasis on the need for reasonably priced renewable energy alternatives. (L. R. Brown, 1984)

TABLE I.20
The international terms of trade between wheat and oil, 1950–83 (source: L. R. Brown, 1984)

Year	*World price (dollars)*		*Amount of wheat to buy a barrel of oil (bushels)*
	Bushel of wheat	*Barrel of oil*	
1950	1·91	1·71	0·9
1955	1·77	1·93	1·1
1960	1·58	1·50	1·0
1965	1·62	1·33	0·8
1970	1·50	1·30	0·9
1971	1·68	1·65	1·0
1972	1·90	1·90	1·0
1973	3·81	2·70	0·7
1974	4·90	9·76	2·0
1975	4·06	10·72	2·6
1976	3·62	11·51	3·2
1977	2·81	12·40	4·4
1978	3·48	12·70	3·6
1979	4·36	16·97	3·9
1980	4·70	28·67	6·1
1981	4·76	32·50	6·8
1982	4·36	33·47	7·7
1983	4·35	28·50	6·6

The impact of the energy price increases of the 1970s on less developed countries has been well documented. In 1981, the World Bank prepared a 201-page review of the literature related to energy pricing in developing countries (World Bank, 1981). More than 1000 articles were reviewed and 450 were cited in the text.

What will energy cost in the future? No one really knows how many factors other than predictable physical forces, such as politics and war, are at work. Many diverse forecasts have been made. One is depicted in Fig. I.26. The price of oil has already fallen below the lowest scenario. Each country must analyze its own situation and develop energy policies designed to optimize its economic position.

The free market should, in principle, be allowed to allocate finite resources such as energy. Market mechanisms through higher prices stimulate the substitution of expensive forms of energy with more economical forms. Higher prices also encourage conservation and more efficient use of resources as well as research and development designed to control costs.

The free market, however, does not normally project very far into the future. The energy needs of future generations are not factored into today's energy prices. Consequently, renewable energy resources are often more expensive than nonrenewable energy forms such as petroleum products. In some areas, nonrenewable energy forms are not available in quantities necessary to meet demands, and, therefore, need to be supplemented with renewable energy.

A sustainable future depends on ample energy supplies at affordable prices. Many research and development projects are underway to help reduce the cost of renewable energy, thereby making it available for developing countries to use in agriculture and other areas.

Developing countries face difficult energy choices: capital is scarce; the number of trained specialists is inadequate; the infrastructure is deficient; nonrenewable fuel reserves are low; population pressures are great; and food reserves are limited.

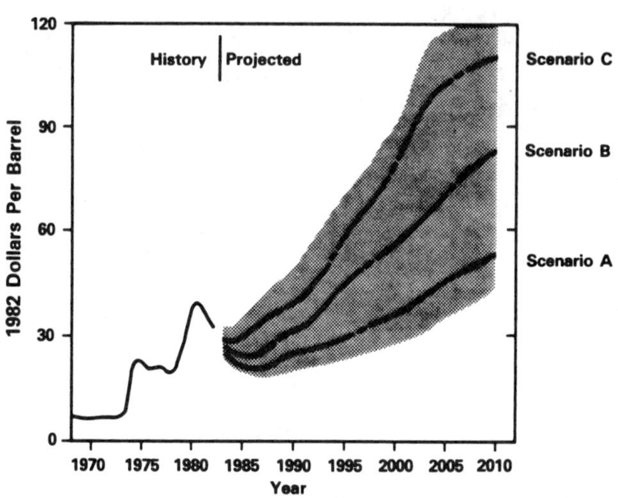

Fig. I.26. World oil prices (source: USDOE, 1983).

Commitment of resources to one energy option deny them to another. If food supplies are to meet the needs of growing populations, the food system must have energy available in the right form at the right time.

Third World energy problems are complex and diverse in nature, and show major differences, not only *across* nations, but *within* individual nations. The plight of the poorest less developed countries is particularly distressing, because to advance economically, they need to expand and modernize industry, transportation, agriculture, and commerce. In other words, they need to consume more commercial energy. But increases in oil prices in the 1970s severely limited the ability of many of these countries to continue high levels of oil imports and forced them to look for indigenous sources of energy (Pachauri, 1982).

ENVIRONMENTAL CONSIDERATIONS

Energy, other natural resources, and the environment are interwoven in a complex fashion. When social and economic considerations are superimposed, the global interdependence becomes overwhelming. Disturbances introduced anywhere in the system may have immediate or delayed repercussions elsewhere. As an example, Table I.21 lists several energy forms and energy policy issues along with associated environmental and socioeconomic concerns.

Environmental quality is not easy to quantify, but several indicators may be used to show trends. In 1950, worldwide fossil fuel combustion emitted 1·6 billion tonnes of carbon. By 1979, carbon emissions had climbed to more than 5 billion tonnes (Fig. I.27). Since 1979, carbon emissions have fallen off somewhat, perhaps due to reduced oil consumption and improved emission controls (L. R. Brown *et al.*, 1985).

Future environmental quality will be influenced significantly by the level of world energy consumption. Table I.22 presents three scenarios of world energy consumption for the years 2000 and 2025 along with associated levels of carbon and sulfur emissions. Acid-rain-forming sulfur emissions are projected to increase from 100 million tonnes in 1984 to 170 million tonnes in the year 2000, according to the high-energy use scenario. Reduced energy consumption and implementation of improved pollution control technology would limit sulfur emissions to a 20%

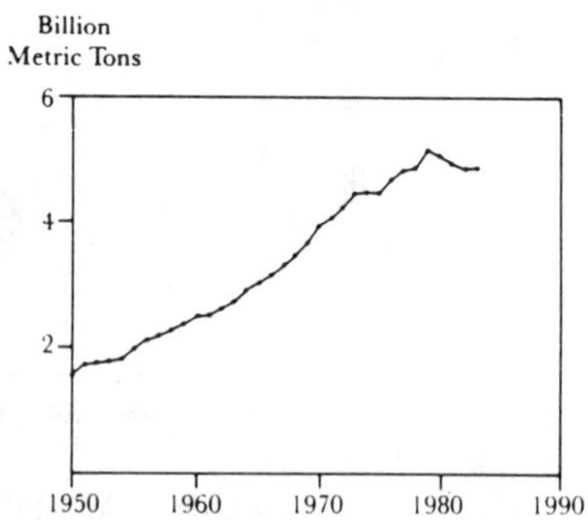

Fig. I.27. World carbon emissions from fossil fuel combustion, 1950–83 (source: L. R. Brown *et al.*, 1985).

TABLE I.21

Environmental concerns related to energy (source: W. H. Matthews & Siddiqi, 1981; © 1981 John Wiley & Sons, Inc.)

Energy form	*Energy policy issue*	*Possible action*	*Environmental concern*	*Socio-economic concern*
Fuelwood	Should fuelwood be used as a major source of energy?	1. Continue present practices.	Deforestation leading to loss of topsoil and soil productivity, silting of dams, increased runoff, etc.	What to do about loss of food production and income? Excessive.
		2. Introduce rapid growing tree plantations.	Changes in other parts of ecosystem.	Will food production be affected? Pest control? Availability of water?
		3. Substitute with energy from animal manure and agricultural wastes.	Loss of fertilizer from ecosystem. Reduced organic matter in soil.	Where will farmers get fertilizer? How to pay for it?
Crude oil	Reduce dependence on oil imports.	1. Increase exploration, including offshore and wilderness areas.	Possible damage to coastlines, wildlife areas, fishing, etc.	Will oil spills ruin beaches and send tourists elsewhere? Will oil spills ruin fishing grounds?
		2. Reduce consumption by increasing prices.	General positive environmental impact, but secondary effects uncertain.	How to pay for increased costs? Hardship caused by lack of energy for those who cannot afford higher prices.
		3. Reduce consumption by rationing.		How to enforce? Possible black market. What is the basis for allocation?
		4. Switch to other fuels, i.e., electricity, coal, etc.	Dependent on new energy source. New sources of pollution; i.e., mine runoff, air pollution from particulates, SO_x, radioactive wastes, etc. CO_2 build-up.	Relate cost-benefits from various energy sources. Would CO_2 buildup cause climate change?
		5. Shift to hydropower.	Changes in land use patterns and local ecosystem.	Where to relocate. Displaced people? Productive land may be flooded. How will irrigation be affected?

TABLE I.22
World energy consumption and carbon dioxide and sulfur emissions in 1984, with alternative projections for 2000 and 2025 (source: L. R. Brown *et al.*, 1985)

Scenario	*Assumed annual energy efficiency improvement (%)*	*Annual energy use (EJ)*	*Annual carbon dioxide emissions (billion tons)*	*Annual sulfur emissions (million tons)*
1984 (estimated)	2·3	300	5·0	100
Year 2000				
US DOE medium	0·8	460	7·2	170
Worldwatch available technology	1·8	360	5·8	120
Worldwatch new technology	1·8	360	5·8	120
Year 2025				
US DOE medium	0·8	675	10·3	265
Worldwatch available technology	1·2	500	7·9	170
Worldwatch new technology	1·8	450	7·0	135

increase by the year 2000 and provide comparable control of carbon dioxide emissions. Although the subject is complex and controversial, many scientists have strong convictions that dramatic changes in the global climate will occur if carbon dioxide emissions reach 600 parts per million (ppm) (from 343 ppm in 1980).

The United Nations Environment Program (UNEP) was created in 1972 to address all aspects of the environment. UNEP has considered the interactions of energy and the environment and prepared a report discussing the environmental aspects of renewable energy (UNEP, 1980). It includes chapters on geothermal energy, solar energy, wind energy, energy from the sea, hydropower, and energy from biomass. It acknowledges that renewable sources of energy vary widely in their impact on the environment and presents a detailed analysis of each energy form.

An adequate energy supply is necessary for human welfare and economic development. A clean and healthy environment is also necessary. Trade-offs may be required to meet different needs. Matthews and Siddiqi (1981) reported the views of several prominent authorities and groups.

> A prime goal of the energy sector should be to ensure a balance between the adverse environmental effects of energy development and their benefits. (New Zealand Minister of Energy, 1978).

An even stronger statement of the environmental concerns in energy policy formulation was voiced by the Australian delegation at an energy meeting in Bangkok in 1978.

> Energy goals, of course, are subservient to national objectives for increased economic and social welfare and for protection of the environment and security

of the country. Progress toward energy conservation or energy self-reliance is often achieved only at the expense of living standards or environmental quality, and an energy policy which fails to recognize and measure these interactions could be damaging or ineffectual.

The more affluent nations are not the only ones expressing such views on reconciling energy needs with environmental considerations. The policy objectives of the Philippines' 10-Year Energy Development Program, for example, are 'to support national and international economic and social development by providing adequate and secure energy supplies in forms that the markets require, promoting judicious and efficient consumption of energy, and ensuring that supply levels and utilization activities are environmentally acceptable' (Republic of Philippines, Ministry of Energy, 1978).

One of the basic objectives of energy planning in India is 'to ensure that energy production and utilization do not lead to adverse impact on ecology and to pollution of the environment' (India, Ministry of Energy, 1978).

Similarly, the conclusion of the Indonesian National Committee to the World Energy Conference states that 'in the implementation of using all forms of energy, attention needs to be paid and sufficient steps need to be taken to protect and conserve the soil and environment' (Indonesian National Committee, 1978).

These policy statements are typical of many similar views expressed by most countries of the world. Matthews and Siddiqi (1981) conclude that 'it is clear that the concern to assure adequate energy supplies at reasonable cost is a global one, and it is increasingly evident that people in developing, as well as developed, countries feel that goals of environmental improvement and protection should be pursued in parallel.'

SOLUTIONS TO DIFFICULT PROBLEMS

There are no easy remedies for the energy dilemmas of developing countries, either on the side of supply or in opportunities for conservation. Solutions will be composites of a wide variety of supply and demand measures. Reduction of oil imports is one central objective. On the supply side, clearly fossil fuels and conventional hydropower will be the main source of commercial energy for several decades. But tropical conditions may also afford special opportunities for exploiting direct and indirect forms of solar energy. On the conservation side, the generally low levels of per capita energy consumption and the limited use of energy for space heating allow developing countries to achieve significant energy savings, especially in the rural areas (Auer, 1981).

By necessity, each country is planning for its own energy needs in terms of its national interests, but all nations must cooperate on the analysis of overall energy issues and the strategies to cope with energy problems and make recommendations for the future. People everywhere must be informed about the possible effects of governmental decisions on energy development and use, as well as the need for a co-operative energy effort between nations.

CHAPTER II

Energy Use

INTRODUCTION

Rapid increases in agricultural production, particularly in developing countries, are required if the demand for food and other agricultural products is to be met. Although many developing countries still have some potentially productive land reserves, most of them will be able to meet the food demands of their growing populations only by increasing yields on both existing and new land.

The FAO study, *Agriculture: Towards 2000* (1981), estimates that up to the year 2000 the expansion of arable land in developing countries can be expected to account for only about 28% of the required increase in agricultural production; the remaining 72% must come from cropping intensity and yield increases on land already under cultivation. More intensive use of commercial energy inputs, particularly mineral fertilizer, farm machinery, pump irrigation, and chemical pesticides, as well as the appropriate application of renewable sources of rural energy, will be required to achieve these increases.

AGRICULTURE'S SHARE OF COMMERCIAL ENERGY

As shown in Table II.1, agricultural production is responsible for only a small portion of total commercial energy: 4·2% in 1972 and 5·0% in 1982. Although total consumption is small, its crucial importance is illustrated by the increasing amount of energy devoted to agricultural production during the period of rapidly increasing energy prices which occurred between 1972 and 1982. The growth rate in total commercial energy was 5·5% in developing countries and 1·0% in developed countries between 1972 and 1982. During this same period, commercial energy for agricultural production grew at a rate of 8·6% per year in developing countries and 2·2% in developed countries.

In developing countries, commercial energy use in agriculture increased from 33·5 mtoe in 1972 to 76·5 mtoe in 1982 with increases occurring in every region. Agriculture's share of total energy in developing countries increased from 4·9% in 1972 to 6·5% in 1982 (Table II.1). During this period, the share of agriculture in total energy consumption decreased only in the Near East region, mainly because of very rapid urban and industrial development financed from petroleum income. It remained constant in Latin America and increased marginally in Africa. However, major increases in agriculture's share of total commercial energy occurred in both the Asian centrally planned economies and in the Far East.

In developed countries, agriculture's share of total energy increased from 4·2% in 1972 to 4·9% in 1982. Western Europe had the largest increase, while agriculture's share of total commercial energy use remained fairly constant in North America, Eastern Europe, and the USSR, and declined in Oceania. North America was the only region where the commercial energy use in agricultural production declined between 1972 and 1982.

Table II.1 also shows the commercial energy consumption per caput of the total population and the commercial energy used per agricultural worker in agricultural production. Between 1972 and 1982, commercial energy per agricultural worker increased in every region of the world. In developed countries, the increase was from about 2000 kilograms of oil equivalent (kgoe) per agricultural worker in 1972 to nearly 3300 kgoe in 1982. In developing countries, the average increased from 47 kgoe per worker in 1972 to about 100 kgoe per worker in 1982. While significant increases in commercial energy per agricultural worker were made in developing countries during this period, the agricultural energy consumption per worker was still more than 30 times higher in the developed countries in 1982. For the population as a whole, the 1982 per caput energy consumption of 3866 kgoe in the developed countries was about 11 times the per caput energy consumption of 343 kgoe in developing countries.

In 1972, North America and Oceania were the only regions where the energy used per agricultural worker was larger than the per caput consumption of the population as a whole, while in 1982, Western Europe was also in this group. In developing countries, total commercial energy use per caput in 1972 was more than five times the commercial energy use per agricultural worker. However, between 1972 and 1982 commercial energy use per agricultural worker grew at a faster rate

TABLE II.1

Total commercial energy and the commercial energy used in agriculture (sources: UN, 1984; FAO estimates)

Region	Total commercial energy (×1000 mtoe)		Commercial energy used in agriculture				Per caput energy consumption (kgoe)		Energy per agricultural worker (kgoe)	
	1972	1982	1972 (×1000 mtoe)	1982 (×1000 mtoe)	1972 (%)	1982 (%)	1972	1982	1972	1982
North America	1 763 212	1 668 188	67 973	66 161	3·9	4·0	7 609	6 492	18 929	25 744
Western Europe	954 024	1 004 146	51 654	67 912	5·4	6·8	2 654	2 682	2 453	4 387
Oceania	53 456	82 312	3 479	3 582	6·5	4·4	3 366	4 503	6 361	7 786
Other developed countries	295 573	354 249	6 104	16 204	2·1	4·6	2 201	2 321	491	1 789
Developed market economies	3 066 265	3 108 895	129 211	153 859	4·2	4·9	4 136	3 875	3 433	5 581
Eastern Europe, USSR	1 080 113	1 474 460	39 574	56 510	3·7	3·8	3 052	3 847	851	1 557
Total developed countries	4 146 378	4 583 355	168 786	210 369	4·1	4·6	3 786	3 866	2 006	3 294
Africa	35 342	51 212	1 783	2 774	5·0	5·4	117	125	20	26
Latin America	187 326	297 730	7 158	11 203	3·8	3·8	628	785	194	286
Far East	135 069	146 930	8 789	20 738	6·5	14·1	131	113	33	72
Near East	53 931	231 675	3 989	9 946	7·4	4·3	319	1 046	123	285
Other developing countries	1 000	2 843	38	57	3·8	2·0	231	523	27	33
Developing market economies	412 668	730 390	21 757	44 718	5·3	6·1	229	315	51	95
Asian centrally planned economies	276 278	444 425	11 743	31 844	4·3	7·2	295	400	40	106
Total developing countries	688 946	1 174 815	33 500	76 561	4·9	6·5	251	343	47	99
Total	4 835 324	5 758 170	202 285	286 931	4·2	5·0	1 260	1 248	252	344

than the per caput energy consumption of the population as a whole in all developing regions but the Near East. As a result, per caput energy consumption of the general population in developing countries was about 3·5 times the commercial energy use per agricultural worker in 1982.

Agriculture's share of commercial energy is small, but vital. An energy shortfall at a critical time in the production cycle can seriously depress yields or lead to unduly high losses. Developing countries use significantly less commercial energy in agricultural production than developed countries. If agricultural production in developing countries is to increase to meet growing food needs, more commercial energy will be needed.

COMMERCIAL ENERGY USE FOR AGRICULTURAL INPUTS

Commercial energy in agricultural production is mainly used for the manufacture and operation of farm machinery and equipment for pump irrigation and for the production and application of mineral fertilizers and chemical pesticides. Table II.2 shows the estimated consumption of commercial energy for these inputs in agricultural production in various regions in 1972 and 1982. The share of agricultural energy devoted to each of these inputs in the various regions of the world is shown in Table II.3.

The commercial energy consumption for agricultural production was estimated at 202 mtoe in 1972, and increased to 287 mtoe in 1982, with more than 50% of this increase occurring in the developing countries. As a result, the developing countries' share of commercial energy used for agricultural production increased from 16·6% in 1972 to 26·7% in 1982. Although increases occurred in each of the developing regions, the primary increases were in the Asian centrally planned economies, where the annual growth rate in agricultural energy was 10·5% between 1972 and 1982, followed by the Near East with an annual growth rate of 9·6%, and the Far East with an annual growth rate of 9·0%. In the developing regions of Africa and Latin America, the growth rate in commercial energy for agricultural production was about 4·5% per year.

While the use of agricultural energy has increased substantially in the developing countries, it is still low considering that, in 1982, developing countries had 54% of the world's arable land and land under permanent crops, and produced 48% of the world's cereals and 62% of the world's roots and tubers.

Farm Machinery

As shown in Table II.3, the manufacture and operation of farm machinery accounted for more than 50% of total commercial energy used in agricultural production in both 1972 and 1982. However, the growth rate of energy for farm machinery during this period was 2·5% per year, in comparison to 5·0% per year for mineral fertilizers. As a result, the proportion of agricultural energy used for fertilizers increased from 38·5% in 1972 to 44·3% in 1982, while the proportion used

TABLE II.2

Use of commercial energy for inputs to agricultural production (sources: UN, 1984; FAO estimates)

Region	Farm machinery (× 1000 mtoe)		Pump irrigation (× 1000 mtoe)		Mineral fertilizer (× 1000 mtoe)		Chemical pesticides (× 1000 mtoe)		Commercial energy in agriculture (× 1000 mtoe)		Share of each region (%)	
	1972	1982	1972	1982	1972	1982	1972	1982	1972	1982	1972	1982
North America	48 853	44 212	747	943	17 102	19 630	1 271	1 377	67 973	66 161	33·6	23·1
Western Europe	33 937	45 123	335	404	16 535	21 485	847	900	51 654	67 912	25·5	23·7
Oceania	2 567	2 585	31	36	866	945	16	16	3 479	3 582	1·7	1·2
Other developed countries	3 630	13 336	84	86	2 372	2 764	18	18	6 104	16 204	3·0	5·6
Developed market economies	88 988	105 256	1 197	1 469	36 875	44 824	2 152	2 311	129 211	153 859	63·9	53·6
Eastern Europe, USSR	18 854	25 396	302	461	19 571	29 753	847	900	39 574	56 510	19·6	19·7
Total developed countries	107 842	130 653	1 499	1 929	56 446	74 576	2 999	3 211	168 786	210 369	83·4	73·3
Africa	827	1 138	88	117	842	1 366	27	152	1 783	2 774	0·9	1·0
Latin America	3 306	4 515	119	157	3 611	6 300	123	231	7 158	11 203	3·5	3·9
Far East	1 533	4 552	627	1 236	6 596	14 809	33	142	8 789	20 738	4·3	7·2
Near East	1 382	4 269	527	676	2 049	4 888	32	113	3 989	9 946	2·0	3·5
Other developing countries	19	24	0	0	19	29	1	5	38	57	0·0	0·0
Developing market economies	7 066	14 497	1 361	2 187	13 115	27 391	215	643	21 757	44 718	10·8	15·6
Asian centrally planned economies	2 052	5 004	839	958	8 323	25 246	530	635	11 743	31 844	5·8	11·1
Total developing countries	9 118	19 501	2 200	3 145	21 438	52 637	744	1 278	33 500	76 561	16·6	26·7
Total	116 960	150 154	3 699	5 075	77 884	127 213	3 743	4 489	202 285	286 931	100·0	100·0

TABLE II.3

Commercial energy used in agricultural production and share of input in each region (sources: UN, 1984; FAO estimates)

Region	*Commercial energy in agriculture (×1000 t)*		*Farm machinery (%)*		*Pump irrigation (%)*		*Mineral fertilizer (%)*		*Chemical pesticides (%)*	
	1972	*1982*	*1972*	*1982*	*1972*	*1982*	*1972*	*1982*	*1972*	*1982*
North America	67 973	66 161	71·9	66·8	1·1	1·4	25·2	29·7	1·9	2·1
Western Europe	51 654	67 912	65·7	66·4	0·6	0·6	32·0	31·6	1·6	1·3
Oceania	3 479	3 582	73·8	72·2	0·9	1·0	24·9	26·4	0·5	0·4
Other developed countries	6 104	16 204	59·5	82·3	1·4	0·5	38·9	17·1	0·3	0·1
Developed market economies	129 211	153 859	68·9	68·4	0·9	1·0	28·5	29·1	1·7	1·5
Eastern Europe, USSR	39 574	56 510	47·6	44·9	0·8	0·8	49·5	52·7	2·1	1·6
Total developed countries	168 786	210 369	63·9	62·1	0·9	0·9	33·4	35·5	1·8	1·5
Africa	1 783	2 774	46·4	41·0	4·9	4·2	47·2	49·2	1·5	5·5
Latin America	7 158	11 203	46·2	40·3	1·7	1·4	50·4	56·2	1·7	2·1
Far East	8 789	20 738	17·4	21·9	7·1	6·9	75·0	71·4	0·4	0·7
Near East	3 989	9 946	34·6	42·9	13·2	6·8	51·4	49·1	0·8	1·1
Other developing countries	38	57	49·6	41·6	0·1	0·1	48·9	50·0	1·4	8·3
Developing market economies	21 757	44 718	32·5	32·4	6·3	4·9	60·3	61·3	1·0	1·4
Asian centrally planned economies	11 743	31 844	17·5	15·7	7·1	3·0	70·9	79·3	4·5	2·0
Total developing countries	33 500	76 561	27·2	25·5	6·6	4·1	64·0	68·8	2·2	1·7
Total	202 285	286 931	57·8	52·3	1·8	1·8	38·5	44·3	1·9	1·6

for farm machinery decreased from 57·8% in 1972 to 52·3% in 1982. The growth rate for pump irrigation was about 3·2% per year and it maintained its share of agricultural energy at 1·8%. For pesticides, the growth rate was 1·8% per year, and its share decreased from 1·9% in 1972 to 1·6% in 1982.

These statistics emphasize different commercial energy use patterns. In developed countries, farm machinery is an important user of commercial energy, whereas developing countries use relatively more commercial energy in the form of fertilizers.

Fertilizers, Pesticides, and Irrigation

In developing countries, mineral fertilizer is by far the most important user of agricultural commercial energy, and, because of a 9·4% average annual growth, the proportion of agricultural energy used for fertilizer increased from 64·0% in 1972 to 68·8% in 1982. During this same period, the annual growth for the other inputs was 7·9% for farm machinery, 5·6% for pesticides, and 3·6% for pump irrigation. The low growth rate in pump irrigation, in comparison to the other inputs, caused its share of agricultural energy to decrease in developing countries from an estimated 6·6% in 1972 to 4·1% in 1982. The most significant change occurred in the Near East, where the proportion of agricultural energy devoted to irrigation decreased from 13·2% in 1972 to 6·8% in 1982, and in the Asian centrally planned economies, where it decreased from 7·1 to 3·0%.

COMMERCIAL ENERGY USE AND CEREAL OUTPUT

Commercial energy use in agriculture is compared with cereal output per hectare and cereal output per agricultural worker in Table II.4. There is a close relationship between commercial energy input and cereal output per agricultural worker in nearly all regions. For both 1972 and 1982, the largest input was in North America, where more than 18 000 kgoe per worker corresponded with the largest cereal output of more than 200 t per worker. Similarly, the rank of the other developed regions in the use of commercial energy per worker corresponded with their rank in output per worker. In developing regions, Latin America had the largest input of commercial energy per agricultural worker with 194 kgoe in 1972 and 286 kgoe in 1982, and the largest output per worker with 6·1 t per worker in 1972 and 9·0 t per worker in 1982. With the exception of the Asian centrally planned economies, the rank of other developing regions in commercial energy per worker also corresponded with their rank in output per worker.

The area of arable land per farm worker is also provided in Table II.4 and shows a very different trend between developed and developing countries. In all of the developed regions, the area per farm laborer increased between 1972 and 1982 as workers left agriculture for other occupations. This was particularly evident in North America and Oceania, where in excess of 90 ha per farm worker existed in 1982, an increase of more than 25 ha per worker from the 1972 level. In developing countries, the area per worker decreased in all regions except Latin America because

TABLE II.4

Commercial energy use and cereal output per hectare and per agricultural worker (sources: FAO, 1983; FAO estimates)

Region	*Arable area per agricultural worker (ha)*		*Cereal yield (t/ha)*		*Cereal production per agricultural worker (t)*		*Energy per hectare of arable land (kgoe)*		*Energy per tonne of cereal (kgoe)*		*Energy per agricultural worker (kgoe)*	
	1972	*1982*	*1972*	*1982*	*1972*	*1982*	*1972*	*1982*	*1972*	*1982*	*1972*	*1982*
North America	64·5	92·0	3·4	3·7	219·3	340·3	293	280	86·3	75·7	18 929	25 744
Western Europe	4·6	6·1	3·2	3·8	14·7	23·3	535	716	167·2	188·3	2 453	4 387
Oceania	75·4	99·7	1·2	1·3	90·5	129·6	84	78	70·3	60·1	6 361	7 786
Other developed countries	1·5	2·1	2·5	2·6	3·8	5·4	321	858	128·5	330·1	491	1 789
Developed market economies	10·3	14·4	3·1	3·4	32·0	48·8	333	389	107·3	114·3	3 433	5 581
Eastern Europe, USSR	6·0	7·7	1·8	1·8	10·8	13·8	142	203	78·8	112·9	851	1 557
Total developed countries	7·9	10·6	2·5	2·7	19·8	28·5	253	312	101·2	115·6	2 006	3 294
Africa	1·6	1·4	0·8	0·9	1·2	1·3	13	18	15·8	20·3	20	26
Latin America	4·1	4·5	1·5	2·0	6·1	9·0	48	64	31·7	31·8	194	286
Far East	1·0	0·9	1·4	1·8	1·5	1·7	34	77	24·3	42·7	33	72
Near East	2·6	2·4	1·2	1·5	3·2	3·6	46	120	38·7	80·3	123	285
Other developing countries	0·8	0·7	1·9	2·1	1·5	1·4	34	49	18·1	23·5	27	33
Developing market economies	1·5	1·4	1·3	1·6	1·9	2·3	34	66	26·3	41·0	51	95
Asian centrally planned economies	0·4	0·4	2·2	3·3	0·9	1·3	103	278	46·9	84·3	40	106
Total developing countries	1·0	1·0	1·5	2·0	1·6	2·1	45	96	29·7	48·1	47	99
Total	1·8	1·8	1·9	2·3	3·4	4·0	143	195	75·1	84·9	252	344

the rural population increased faster than new land entered production and non-farm employment opportunities were created.

Table II.4 also provides the energy input per hectare of arable land and the energy per tonne of cereal production. As illustrated, in general, regions with higher energy consumption also have higher yields. For example, the four regions with the highest yield per hectare in 1982 (Western Europe, North America, Asian centrally planned economies, and the other developed countries) also used the most energy per hectare. These regions used more than 275 kgoe/ha and achieved yields of 2·6–3·8 t of cereal/ha. Africa, on the other hand, had the lowest energy inputs with 18 kgoe/ha and 20 kgoe/t of cereal and the lowest yield with 0·9 t/ha. However, the relationship between energy input and output per hectare is also affected by varying ecological and environmental conditions in different regions. North America and Western Europe achieved about the same yield in both 1972 and 1982, although North America used only half as much energy per hectare and energy per tonne of cereal. This appears mainly to be the result of the high natural fertility of many North American soils. As another example, the Asian centrally planned economies achieved high yields per hectare through a combination of high labor inputs and high levels of energy input per hectare, particularly in the form of mineral fertilizer, which accounted for almost 80% of the commercial energy used for agricultural production in this region in 1982.

The preceding analysis indicates a number of trends in the transition of agricultural production in both developed and developing countries. In general, more commercial energy is being devoted to agriculture by nearly all countries except the countries in North America. This increase in agricultural commercial energy use is accompanied by an increase in energy per agricultural worker as well as yield per hectare and production per agricultural worker. One of the primary differences between developed and developing countries is the continued increase in the number of agricultural workers in developing countries and their continued decline in developed countries.

In developed countries, one of the most interesting trends has been increased efficiency in the use of agricultural energy in both North America and Oceania, where yields have increased, while energy per hectare and energy per tonne of cereal have decreased. In other developed countries, however, both energy per hectare and energy per tonne of cereal has continued to increase.

In developing countries, three main groups can be identified. The first group consists of the countries with limited land, limited capital resources, and abundant labor, such as many of those in the Far East and Asian centrally planned economies. Here, emphasis in agricultural development has been on increasing yields through extended use of mineral fertilizer in combination with continued use of noncommercial forms of energy and labor-intensive methods of production. Agricultural development has been emphasized within the overall development efforts in these regions, and both commercial energy use in agriculture and agriculture's share of total commercial energy has increased significantly. As a result, per caput food production has also been increasing in these regions.

A second group of countries are those with relatively abundant land and labor resources, but increasingly limited capital. Many countries in Latin America fall into this category. In these countries, farm machinery and irrigation, which have a high initial investment and require several years for cost recovery, have accounted for about half of the commercial energy used in agriculture. However, between 1972 and 1982, the emphasis changed toward greater use of mineral fertilizer, which allows quick cost recovery. Also, during this period, the rate of agricultural development in these countries grew at about the same rate as the economy as a whole.

The third group are the countries with relatively limited land resources but large capital resources. This describes a number of Near East countries between 1972 and 1982. Under these conditions, investments in agricultural development were significant, although less than in other types of development, with the main increases occurring in farm machinery and mineral fertilizer.

In Africa, the application of energy-intensive forms of agricultural production has been lower than in other regions and the result of this agricultural transition is not yet clear. However, the decline in per caput food and agricultural production during the past decade requires that individuals give particular attention to agricultural development in this region.

Because of the increasing importance of commercial energy inputs to agricultural production, we must look more closely at the energy associated with their use, as well as ways in which they can be used more effectively and supplemented with renewable sources of rural energy.

COMMERCIAL ENERGY FOR FERTILIZER PRODUCTION AND USE

Mineral fertilizer is the fastest growing user of commercial energy in agricultural production. As noted earlier, it is the predominant consumer in developing countries, accounting for nearly 70% of the commercial energy used in agriculture. This reflects its essential role in the technologies currently used to increase agricultural production by raising crop yields. The types of energy used in the fertilizer sector are illustrated in Fig. II.1.

Commercial energy use for fertilizer production in 1972 and 1982 is shown in Table II.5. Nitrogen fertilizer is by far the most important mineral fertilizer, both in the amount of plant nutrient used in world agriculture and in energy requirements. World consumption of nitrogen fertilizer increased from 36·0 million tonnes of nutrient in 1972 (30% in developing countries) to 61·2 million tonnes of nutrient in 1982 (43% in developing countries). Nitrogen fertilizer is very energy-intensive, with 1 kg of nutrient requiring about 1·85 kgoe for its manufacture, packaging, transportation, distribution, and application (Leach, 1976). On the other hand, phosphate (P) and potash (K) are mined and do not require feedstock energy.

Nitrogen (TRW, 1980; Lewis, 1982; Mudahar & Hignett, 1982)

Nitrogen fertilizer is manufactured from synthetic ammonia, the production of which requires pure gaseous nitrogen and hydrogen. The primary sources of energy

TABLE II.5
Commercial energy for fertilizer production and application (source: FAO estimates)

Region	Nitrogen[a] (×1000 mtoe)		Phosphate[b] (×1000 mtoe)		Potassium[c] (×1000 mtoe)		Total (×1000 mtoe)		Share of each region (%)	
	1972	1982	1972	1982	1972	1982	1972	1982	1972	1982
North America	14 576	17 222	1 616	1 426	910	982	17 102	19 630	22·0	15·4
Western Europe	13 484	18 655	1 962	1 710	1 090	1 119	16 535	21 485	21·2	16·9
Oceania	365	553	448	341	52	50	866	945	1·1	0·7
Other developed countries	1 876	2 209	345	399	151	156	2 372	2 764	3·0	2·2
Developed market economies	30 300	38 639	4 371	3 877	2 203	2 308	36 875	44 824	47·3	35·2
Eastern Europe, USSR	16 451	25 163	1 862	2 914	1 258	1 675	19 571	29 753	25·1	23·4
Total developed countries	46 752	63 803	6 233	6 791	3 461	3 983	56 446	74 576	72·5	58·6
Africa	707	1 165	96	146	38	54	842	1 366	1·1	1·1
Latin America	3 018	5 314	433	697	159	289	3 611	6 300	4·6	5·0
Far East	6 121	13 745	339	801	136	262	6 595	14 808	8·5	11·6
Near East	1 898	4 454	141	417	9	16	2 049	4 888	2·6	3·8
Other developing countries	17	27	1	1	0	1	19	29	0·0	0·0
Developing market economies	11 762	24 705	1 011	2 063	343	622	13 115	27 391	16·8	21·5
Asian centrally planned economies	7 814	24 046	482	1 084	27	116	8 323	25 246	10·7	19·8
Total developing countries	19 576	48 752	1 493	3 147	370	738	21 438	52 637	27·5	41·4
Total	66 327	112 554	7 726	9 938	3 831	4 721	77 884	127 213	100·0	100·0

[a] Production of 1 kg (nutrient content) of nitrogen fertilizer assumed to require 1·85 kgoe.
[b] Production of 1 kg (nutrient content) of phosphate fertilizer assumed to require 0·33 kgoe.
[c] Production of 1 kg (nutrient content) of potassium fertilizer assumed to require 0·21 kgoe.

Fertilizer type	Fertilizer activity			
	Production (Pr)	Packaging (P)	Transportation (T)	Application (A)
N	Feedstock (H) Fuel (E)	Feedstock (H) Fuel (E)	Fuel (H)	Fuel (H)
P_2O_5	Fuel (E)	Feedstock (H) Fuel (E)	Fuel (H)	Fuel (H)
K_2O	Fuel (E)	Feedstock (H) Fuel (E)	Fuel (H)	Fuel (H)
NPK	Feedstock (H) Fuel (E)	Feedstock (H) Fuel (E)	Fuel (H)	Fuel (H)

Where: H = All or major share from hydrocarbons (solid, liquid, and gas).
E = All or major share from electricity and steam.

Fig. II.1. Types of energy used in the fertilizer sector (source: Mudahar & Hignett, 1982).

for ammonia production are hydrocarbons such as natural gas, naphtha, or fuel oil, whereas the energy sources for phosphate and potash are steam and electricity, which can be generated from any convenient fuel.

Figure II.2 presents a typical fertilizer production flowsheet showing natural gas as the feedstock for ammonia synthesis. Figure II.3 provides a more complete description of the ammonia manufacturing process and identifies the various inputs and outputs. The energy requirements for manufacturing ammonia are given in Fig. II.4 and the requirements for converting ammonia to nitric acid are shown in Fig. II.5. Ammonia is also the basic form of nitrogen for other fertilizer formulations as illustrated in Figs II.6 and II.7. Significant progress has been made in improving the

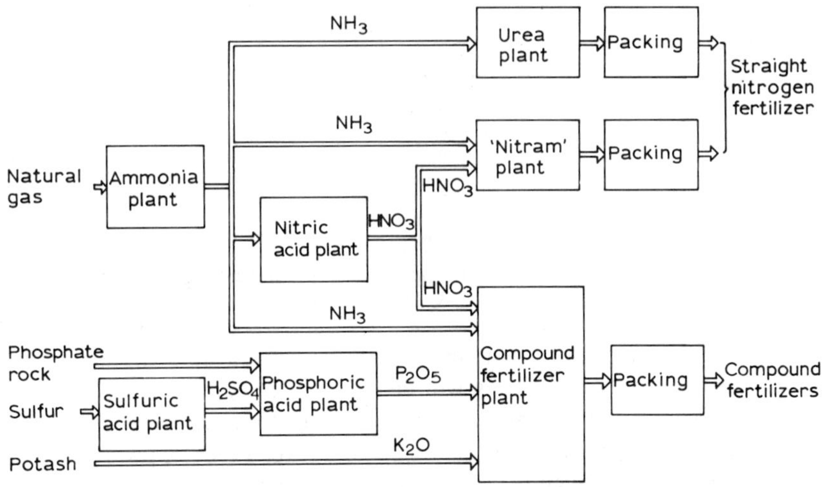

Fig. II.2. Fertilizer production flowsheet (source: Lewis, 1982).

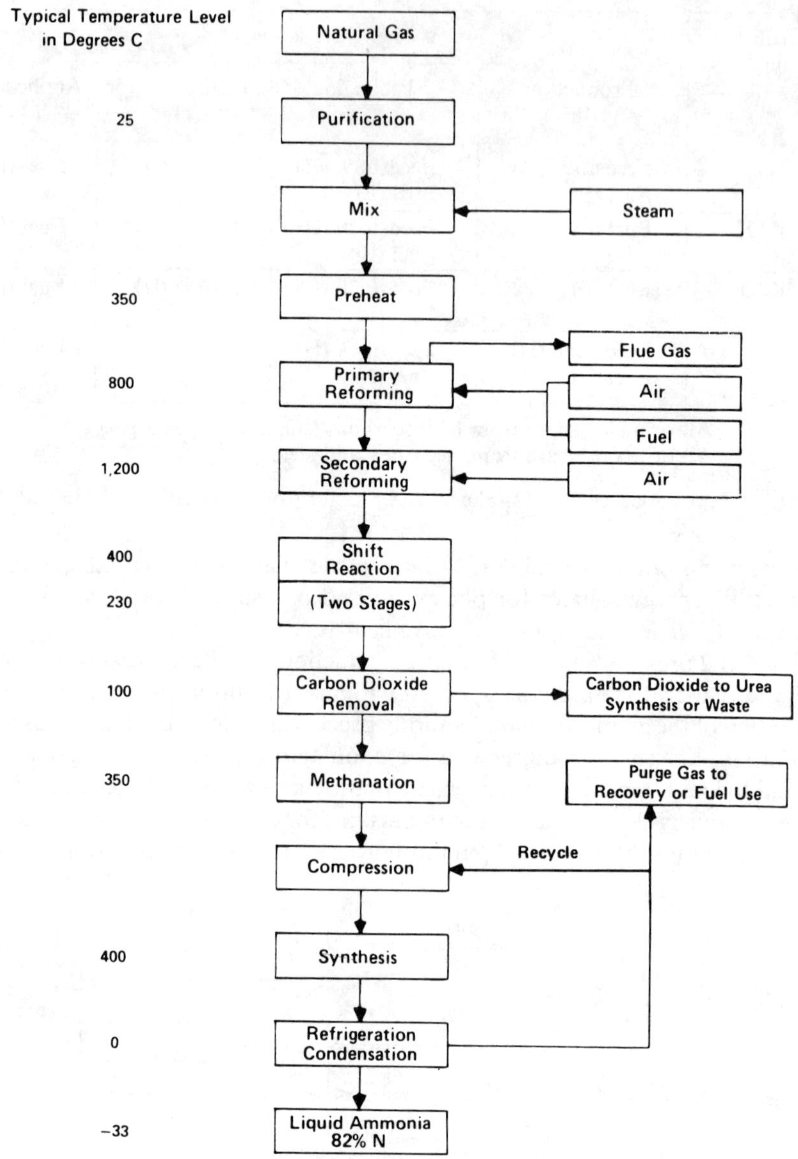

Fig. II.3. Typical ammonia production process using natural gas feedstock (source: Mudahar & Hignett, 1982).

energy efficiency of ammonia synthesis as illustrated in Table II.6 and Fig. II.8. Further improvements in efficiency will be difficult; however, older plants can be updated to incorporate more modern and efficient processes.

Pure gaseous nitrogen is fairly simple and inexpensive to produce, but the production of pure hydrogen is more difficult. The main sources of hydrogen for fertilizer production are natural gas and coal. Hydrogen can also be extracted from

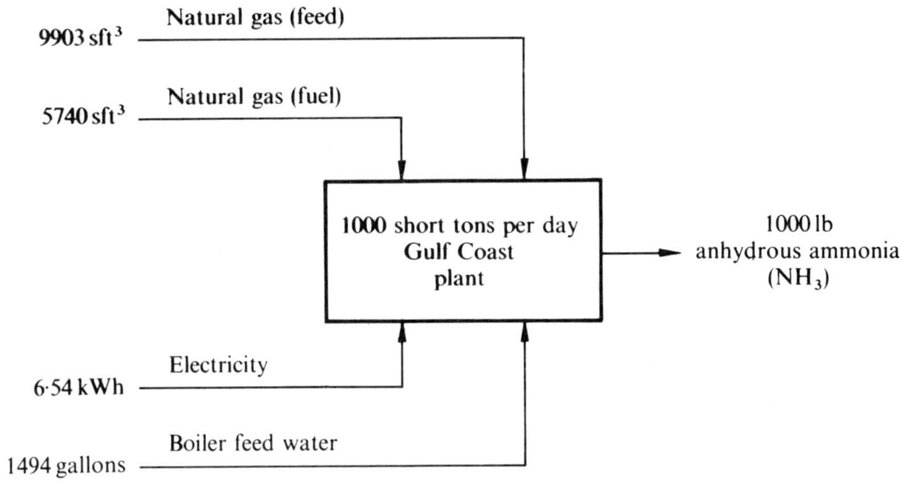

Fig. II.4. Energy investments in manufacturing of ammonia; sft^3 = standard cubic feet (source: TRW, 1980).

water, but this process requires large quantities of electrical energy. Because the simplest and cheapest source of hydrogen in many areas is natural gas, the price of nitrogen fertilizer is often closely related to the price of natural gas. Fertilizer production techniques are discussed in more detail in Chapter IV.

As shown in Table II.5, nitrogen accounted for more than 88% of the commercial energy devoted to mineral fertilizer in 1982, up from 85% in 1972. Nitrogen increased its share in all regions except the Near East, with the largest change occurring in Oceania, where it increased from 42·2% in 1972 to 58·2% in 1982. However, with the exception of Oceania, nitrogen already accounted for more than

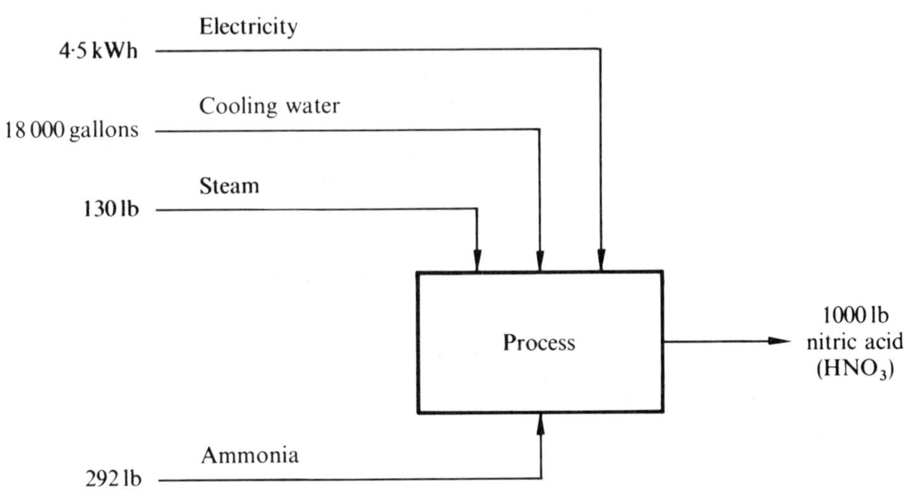

Fig. II.5. Energy investments in manufacturing of nitric acid (source: TRW, 1980).

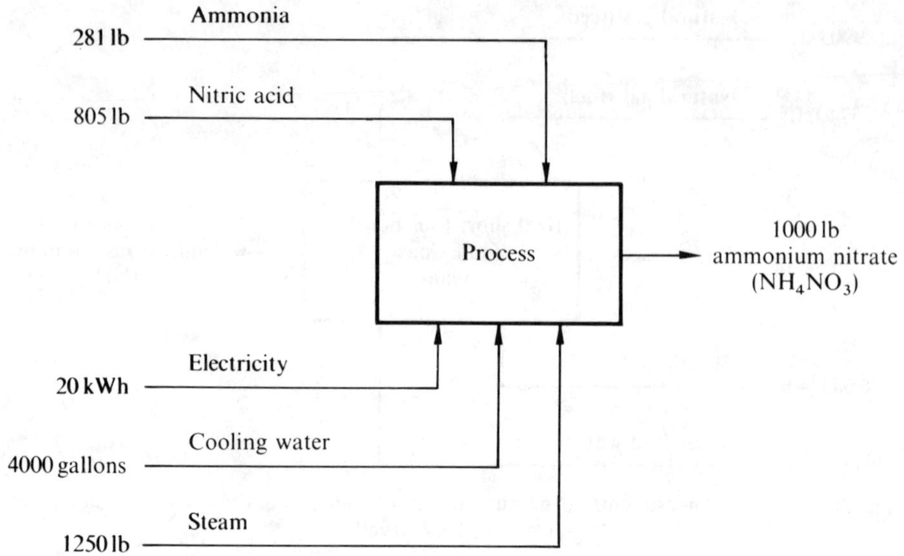

Fig. II.6. Energy investments in manufacturing of ammonium nitrate (source: TRW, 1980).

TABLE II.6

Reduction in energy use in ammonia synthesis through time (source: Mudahar & Hignett, 1982)

Time period	*Synthesis process*	*Average energy use ($GJ/t\ NH_3$)*
Early 20th century (1910)	Electric arc process	574[a]
Early 20th century (1910)	Cyanamide process	190
Early 20th century (1915)	Water electrolysis	117
Mid 20th century (1940)	Classical H–B process: coke	88
Mid 20th century (1960)	Modern H–B process: naphtha	55
1970s	Modern H–B process: natural gas	47[b]
Practical minimum	Modern ammonia plants	32
Theoretical minimum	Modern ammonia plants	23[c]

[a] The electric arc process produced nitric acid, not ammonia. The energy use is about 700 GJ/t of N, equivalent to 574 GJ/t of ammonia.

[b] Average of energy use for ammonia plants (both reciprocating and centrifugal) in operation in the United States.

[c] Some other calculations of the theoretical minimum energy requirement range from 19·2 to 21·2 GJ/t, of NH_3, depending on the assumptions made about state (gas or liquid) of reactants and reaction product, amount of oxygen introduced as air in secondary reforming, reference temperature, and natural gas composition.

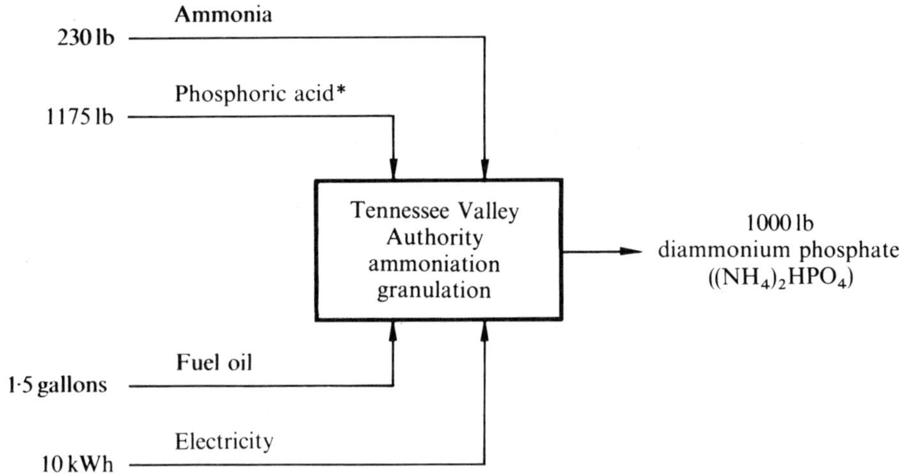

*40% P_2O_5

Fig. II.7. Energy investments in manufacturing of diammonium phosphate (source: TRW, 1980).

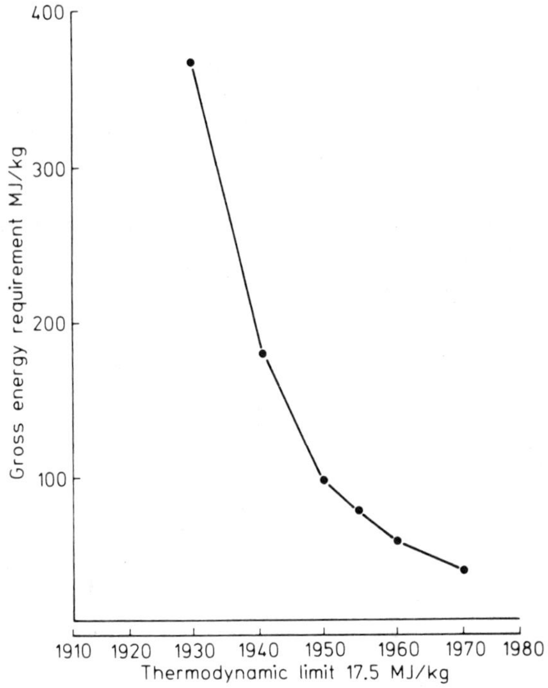

Fig. II.8. Improvement in gross energy requirement of ammonia production through time (source: Stanhill, 1984; © 1984 Springer-Verlag).

80% of the energy devoted to mineral fertilizer in all regions, with the highest proportion in the Far East, Near East, and Asian centrally planned economies at more than 92%.

The dramatic increases in nitrogen fertilizer use in developing countries increased their average annual application per hectare of arable land and permanent crops to 33·3 kg of nutrient in 1982, up from 14·2 kg in 1972. This compares with an average use in developed countries of 38·1 kg/ha in 1972 and 51·5 kg/ha in 1982. However, the average use in 1982 continued to vary considerably, from a low of 4·2 kg/ha in Africa and 6·8 kg/ha in Oceania to a high of 107 kg/ha in Western Europe and 116 kg/ha in the Asian centrally planned economies.

Nitrogenous fertilizers are energy-intensive, but their use is directly correlated with high yields. Thus, commercial energy must be available to meet each country's specific fertilizer needs.

Phosphate and Potash

Phosphate and potash are obtained through mining operations. Energy requirements for surface mines are provided in Fig. II.9. Converting phosphate rock to phosphoric acid requires additional energy as illustrated in Fig. II.10. Table II.7 provides information concerning the amount of energy invested in phosphate and potash fertilizers. Figures II.11 and II.12 show the phosphoric acid and potash production processes from mining to the finished fertilizer product.

Phosphate fertilizer consumption increased from 24·0 million tonnes of nutrient in 1972 (19% in developing countries) to 30·5 million tonnes in 1982 (32% in developing countries) (Table II.5). The major known deposits of phosphate ore are in Morocco and in the United States in the state of Florida. The ore, which contains 12–15% phosphate (P_2O_5), is concentrated to 35%, and most of it is then reacted with sulphuric acid to product superphosphate. Phosphate fertilizer is much less energy-intensive than nitrogen, and the energy requirement to mine, concentrate, process, package, transport, distribute, and apply 1 kg of nutrient is estimated at about 0·33 kgoe (Leach, 1976).

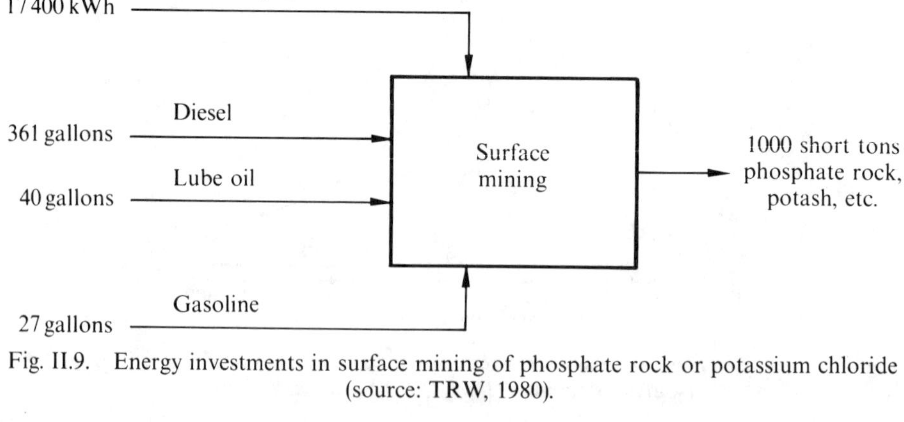

Fig. II.9. Energy investments in surface mining of phosphate rock or potassium chloride (source: TRW, 1980).

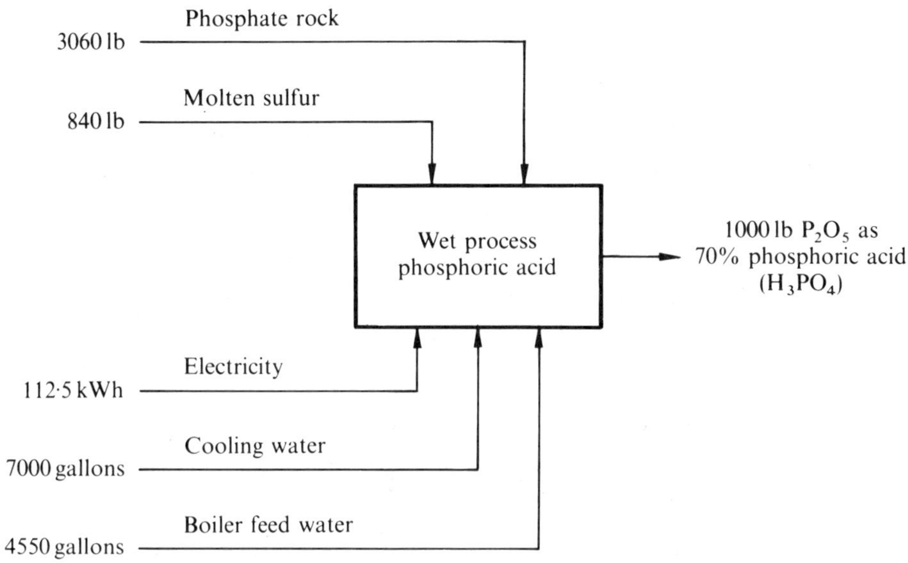

Fig. II.10. Energy investments in wet process for manufacturing of phosphoric acid (source: TRW, 1980).

Phosphate comprised about 10% of the commercial energy consumed by mineral fertilizers in 1972, and because of the rapid growth in nitrogen, its share decreased to 7·8% in 1982. Average use per hectare of arable land and permanent crops in developing countries increased from 6·2 kg in 1972 to 12·3 kg in 1982. However, this increase remained well below the average use in developed countries which was 29 kg/ha in 1972 and 31·3 kg/ha in 1982.

Potassium

World consumption of potassium fertilizer increased from 18·5 million tonnes of nutrient in 1972 (10% in developing countries) to 22·8 million tonnes in 1982 (16%

TABLE II.7

Energy inputs for phosphate and potash fertilizers (Mcal/kg of P_2O_5 or K_2O) (source: Pimentel, 1980; © 1980 CRC Press, Inc., Boca Raton. FL)

Type	*Production*	*Transportation of raw materials*	*Transportation and distribution of final product*
Phosphate rock	0·4	—	0·9
Normal superphosphate (0-20-0)	0·6[a]	0·2	1·5
Triple superphosphate (0-46-0)	2·2[a]	0·2	0·6
Muriate of potash (0-0-60)	1·1	—	0·5

[a] Assumes that byproduct sulfuric acid is used for normal superphosphate, and Frasch process sulfur is used for triple superphosphate.

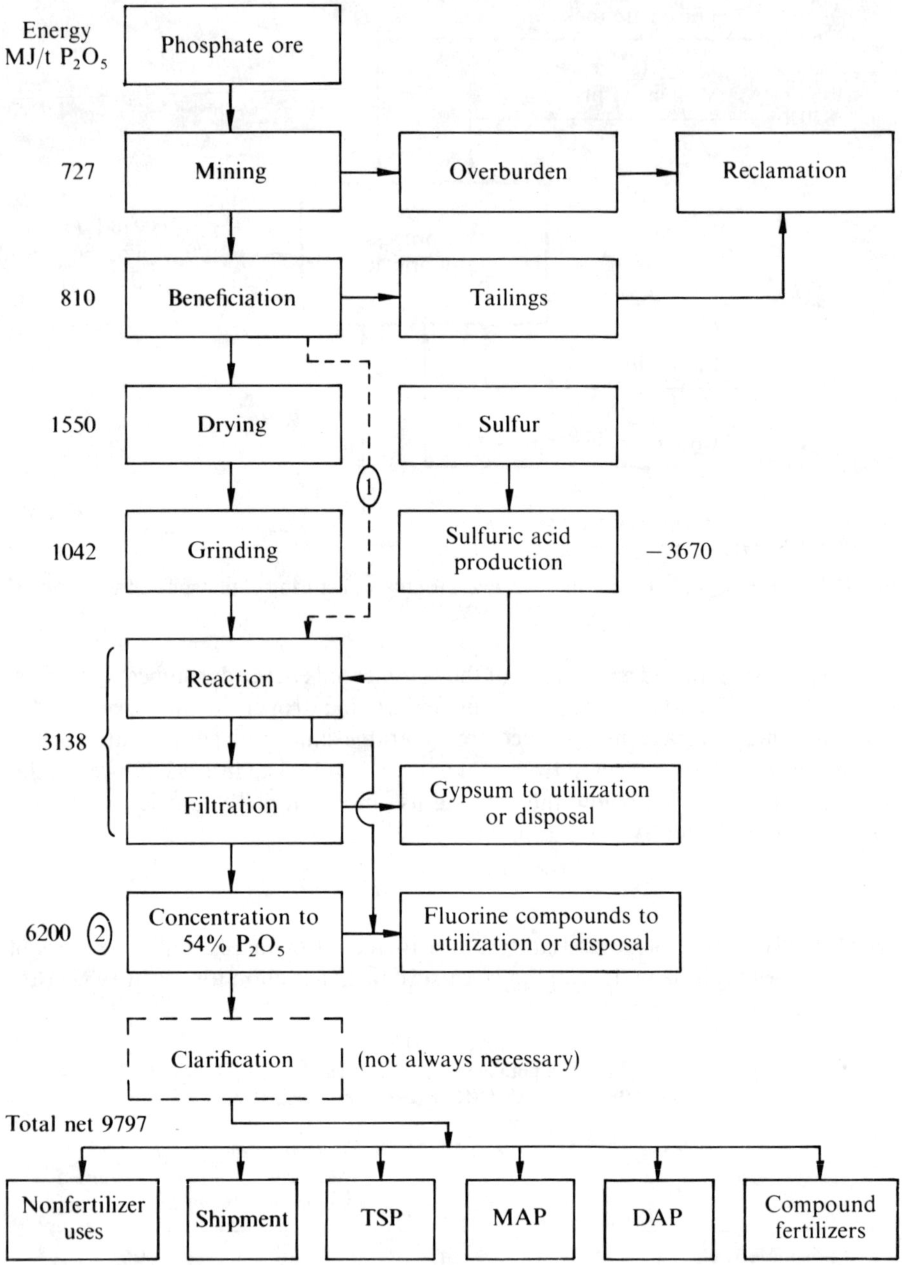

Fig. II.11. Typical phosphoric acid production process. Energy-saving options: (1) In some processes, drying or grinding or both can be omitted. (2) For some uses, 40% P_2O_5 concentration is sufficient, thus saving more than half of the energy for concentration. In hemihydrate processes, concentration can be omitted or significantly reduced. TSP, triple superphosphate; MAP, mono-ammonium phosphate; DAP, diammonium phosphate (source: Mudahar & Hignett, 1982).

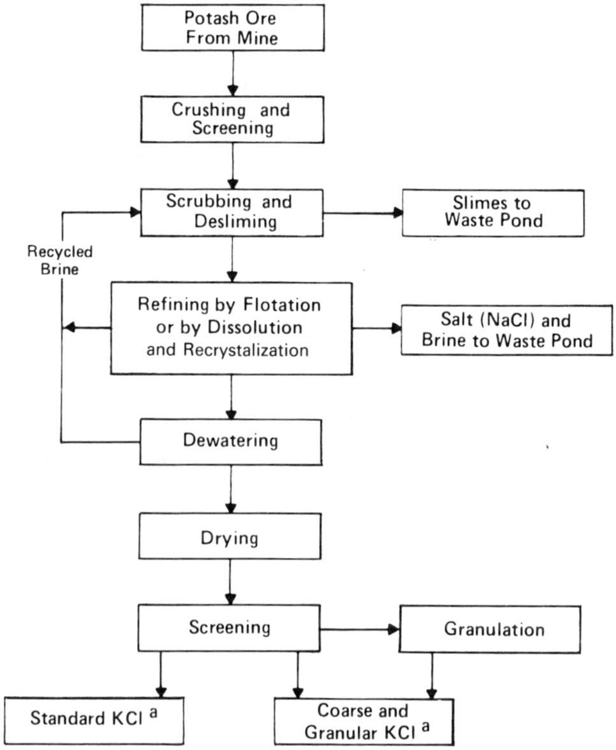

Fig. II.12. Typical potash refining process. [a]KCl = muriate of potash, 60–62% K_2O. Typical particle sizes: standard, 0·15–0·70 mm; coarse, 0·70–2·00 mm; granular, 1·00–3·30 mm (source: Mudahar & Hignett, 1982).

in developing countries) (Table II.5). Potassium or potash fertilizer is usually manufactured from salts (i.e. potassium chloride), which are available in an almost pure state in many parts of the world. Since the ores are generally soft and are often found near the surface, the energy requirements for mining are less than for phosphate, although there are usually subsequent energy requirements for enrichment. The total energy required to mine, concentrate, package, transport, distribute, and apply 1 kg of nutrient is estimated to be about 0·21 kgoe (Leach, 1976).

Potassium constituted about 5% of the commercial energy used for mineral fertilizers in 1972 and decreased to 3·7% in 1982. Average use of arable land and permanent crops increased slightly from 13·0 kg/ha in 1972 to 15·5 kg/ha in 1982. However, use in developing countries increased from 2·4 kg/ha in 1972 to 4·5 kg/ha in 1982, but still remained substantially below the average use in developed countries, which was 25·1 kg/ha in 1972 and 28·5 kg/ha in 1982.

In summary, nitrogen fertilizers are unquestionably the most energy-intensive fertilizers produced. Ammonia, the basis for most nitrogen fertilizers, requires 55–65 MJ/kg of nitrogen to manufacture, phosphorous requires 11–18 MJ/kg of

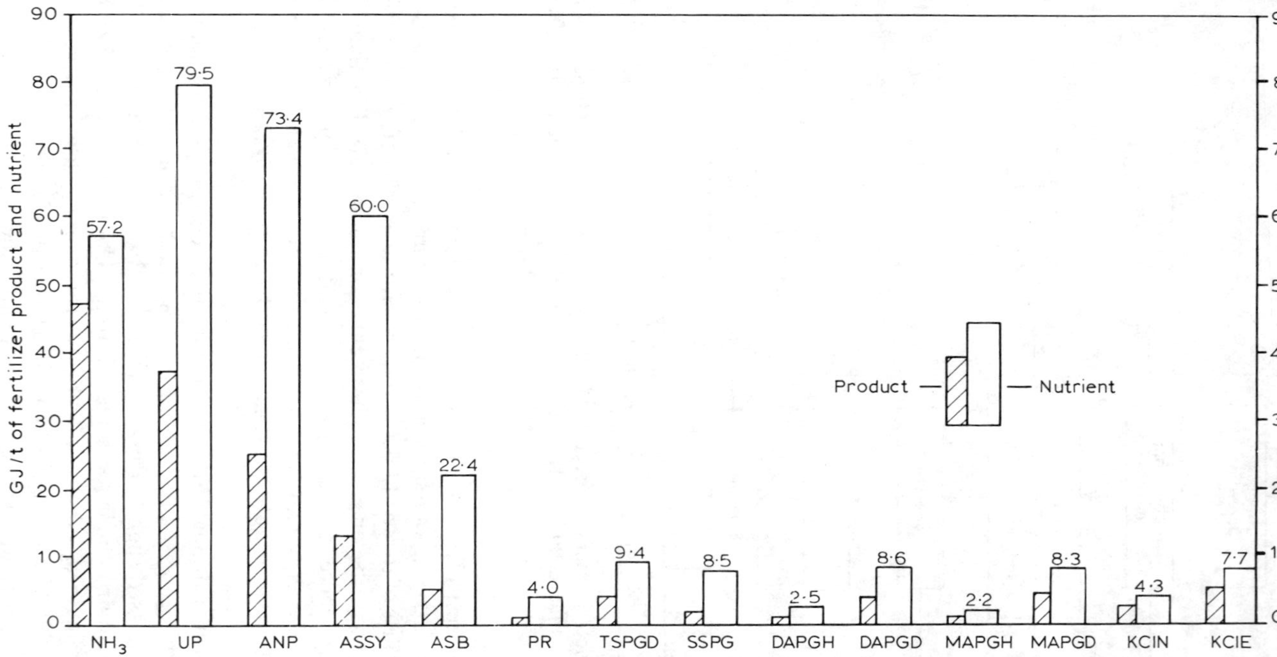

Fig. II.13. Average energy requirements for manufacturing selected nitrogen, phosphate, and potash fertilizers. NH_3—ammonia; UP—urea, prilled; ANP—ammonium nitrate, prilled; ASSY—ammonium sulfate, synthetic; ASB—ammonium sulfate, byproduct; PR—phosphate rock; TSPGD—triple superphosphate, granular dihydrate; SSPG—single superphosphate, granular; DAPGH—diammonium phosphate, granular hemihydrate; DAPGD—diammonium phosphate, granular dihydrate; MAPGH—mono-ammonium phosphate, granular hemihydrate; MAPGD—mono-ammonium phosphate, granular dihydrate; KClN—potassium chloride, North America; KClE—potassium chloride, Europe (source: Mudahar & Hignett, 1982).

TABLE II.8
Inputs of typical fertilizers (source: Lewis, 1982)

Ratio	*Ammonium nitrate 34·5:0:0*	*Urea 46:0:0*	*Compound fertilizers*			
			No. 1 15:15:21	*No. 2 22:11:11*	*No. 3 9:24:24*	*No. 5 17:17:17*
GJ/t bagged fertilizers[a]	22·5	36·1	14·9	17·5	13·1	16·0
MJ/kg N	65·1	77·8	70·1	66·6	73·4	68·4
MJ/kg P_2O_5			17·9	18·0	18·7	17·8
MJ/kg K_2O			7·9	7·9	8·2	8·1

[a] For comparison: the inherent energy content per tonne of coal is 25·6 GJ; the inherent energy content per liter of petrol is 34·9 MJ.

P_2O_5, and potash requires 7–9 MJ/kg of K_2O. Table II.8 and Fig. II.13 present energy values for various fertilizer formulations. The energy requirements for various fertilizers are substantially different, ranging from 2·2 MJ/kg of nutrients for granular monoammonium phosphate to 79·5 MJ/kg for prilled urea (Fig. II.13). These numbers do not include the energy for nitrogen input. The production of urea requires 8·5 times as much commercial energy as the production of triple superphosphate and 18·5 times as much as the production of potassium chloride.

Alternative Energy Sources for Ammonia (Mudahar & Hignett, 1982)

Additional alternatives are available other than natural gas for the primary energy input to ammonia manufacture. For example, coal can be used as a source of hydrogen; depending on the relative costs and availability of energy supplies, ammonia from coal may become economically viable in the future.

Technically, manufacturing ammonia from biogas, which contains 50–60% methane, is feasible. Ammonia may also be generated by the steam reforming of ethanol produced from biomass. In addition, biomass can be used directly for ammonia feedstock through partial oxidation or gasification methods.

In addition, electrical energy produces hydrogen for ammonia production. As early as the 1930s, a number of ammonia plants were designed based on electrolytic hydrogen. The electrolytic ammonia process is appropriate for small plants wherever electricity is available.

Another approach to nitrogen fertilizer production is the electric arc process. This process requires establishment of a high-temperature electric arc between electrodes inside an iron pipe. Nitrogen oxide gases are formed inside the pipe and are pumped with water through a limestone bed to form calcium nitrate fertilizer.

Fertilizer Use Efficiency

Because mineral fertilizers make up a large and increasing proportion of total commercial energy use in agricultural production, and are so important in raising crop yields with existing technology, we must examine how they can be used more efficiently. In general, the first 15–30 kg/ha of nitrogen fertilizer bring an increased yield of 10–15 kg/ha of grain per kg of nitrogen, after which the response slowly declines. Thus, in many developed countries, additional applications of fertilizer will bring much smaller marginal returns than in developing countries, where both yields and fertilizer use remain low. Mineral fertilizers can be used more efficiently in both developed and developing countries in a number of ways. For example, timely sowing and improved water management can significantly enhance the efficiency of fertilizer use. Improved methods of fertilizer application, including proper soil placement, the timing of applications to coincide with the nutrient demands of crops, and split doses to reduce leaching losses, as well as better choice of fertilizer materials, can also increase efficiency considerably. Such improved methods are labor-intensive, and thus correspond with the rural employment situation in most developing countries. The breeding of high-yielding varieties of cereals and other crops, which are very responsive to fertilizer use, has also contributed to more efficient use of fertilizers, and has been a major factor in increasing fertilizer demand.

Research currently underway to improve the efficiency of the conversion of solar energy by plants should make higher yields with less fertilizers possible. Recent research indicates that the biological fixation of atmospheric nitrogen through a symbiotic relationship with certain bacteria, which is now found only in legumes, could possibly be genetically transferred to cereal and other crops.

The use of mineral fertilizers is still a comparatively recent phenomenon in the long history of agriculture. Until Chilean nitrate and Peruvian guano were first introduced into European agriculture in the 1830s, and until the manufacture of superphosphate began in the 1840s, the limited use of 'artificial' fertilizers was confined to such substances as soot, bones, hoofs and horns, saltpetre, and lime. The main reliance for the maintenance and restoration of soil fertility was on such practices as shifting cultivation, fallowing, crop rotation, catch-cropping (especially with nitrogen-fixing legumes), and recycling crop and animal residues.

Mineral fertilizers are in abundant supply at relatively low prices. They possess the advantages of concentration, portability, and adaptability to different soil conditions and different crop requirements. They have also led to a steady decline in the use of crop and animal residues to return fertility to the soil. Major exceptions are China, Japan, and the Republic of Korea, where, although the use of mineral fertilizers has increased rapidly, substantial use of crop, animal, and human residues as fertilizer is still carried out.

Undoubtedly, food for the immense increases in population in the last century, and particularly in the last quarter-century, could never have been adequate (even at the present levels, which are inadequate in many countries) without the rapid increase in reliance on mineral fertilizers. In addition, continued growth in the use of mineral fertilizers is both desirable and necessary. However, to ensure maximum usage of locally available materials, we must examine the possibility of intensifying the usage of crop and animal residues as fertilizer.

The amount of plant, human, and animal wastes available in developing countries that could potentially be used for organic manuring was estimated to be about 3·5

TABLE II.9
Percentage of fertilizer nutrients in manure

Nutrient	Cattle (1)	Cattle (2)	Cattle (3)	Dairy cow (4)	Beef yearling (4)	Livestock and poultry (5)	Horse (6)	Dairy (6)	Beef (6)
Nitrogen, N	2·2	3·7	3·1	3·9	3·5	3·6	1·7	2·7	3·5
P_2O_5	1·3	1·1	0·6	0·7		0·9	0·3	0·5	1·0
K_2O	2·8	3·0	2·2	2·6		2·2	1·5	2·4	2·3

(1), El Amani & Bahri (1980). (2), Taiganides & Hazen (1966). (3), Stout *et al.* (1978). (4), ASAE (1981). (5), Safley & Westerman (1981). (6), Azevedo & Stout (1976).

times the consumption of chemical fertilizer in these countries in 1982. By far the largest potential source of organic fertilizer is animal manure, which accounts for about one-third of the plant nutrients estimated to be available from all sources.

Energy Value of Animal Manure as a Fertilizer

The energy content of manure used as a fertilizer is determined in the same way as the energy content of commercial fertilizers. It can be evaluated similar to a mixed fertilizer once the content of N, P, and K is known.

Table II.9 shows the percentage dry basis of N, P_2O_5, and K_2O found in various manures. Nitrogen is between 7 and 3·9% dry basis, P_2O_5 is between 0·3 and 1·3%, and K_2O is between 1·5 and 3%.

The chemical composition of manure is influenced by the quality of feed and the method of collection and conservation. On the average, about 75% of the nitrogen, 80% of the phosphorous (as P_2O_5), 85% of the potassium (as K_2O), and 40–50% of the organic matter of the feed is in the manures (Taiganides & Hazen, 1966). All of the values in Table II.9 are for United States livestock except those calculated by El Amani and Bahri (1980), who were working with Tunisian livestock. Table II.10 lists the calculations by El Amani and Bahri in more detail. Their figures are used as the

TABLE II.10
Tunisian manure characteristics (source: El Amani & Bahri, 1980)

Moisture as % gross product	78
pH	7·8
C/N	16
% dry material	
carbon	36·2
nitrogen	2·2
P_2O_5	1·3
K_2O	2·8
Ca	2·6
Mg	0·7
Na	0·3

basis for determining the energy content of manure because they are more typical of developing countries than the livestock figures from the United States.

Fresh, wet manure contains about 78% moisture, so 78 kg of water and 22 kg of solids or dry material (DM) are included in 100 kg of manure. In addition, dry material is 2·2% N. The amount of N in fresh, wet manure is as follows:

$$\frac{2{\cdot}2\ \text{kg N}}{100\ \text{kg DM}} = \frac{x\ \text{kg N}}{22\ \text{kg DM}} \qquad x = \frac{2{\cdot}2\ \text{kg N}}{100\ \text{kg DM}} \times 22\ \text{kg DM} = 0{\cdot}484\ \text{kg N}$$

Therefore, 0·484 kg N is contained in 100 kg of fresh manure. Similarly, the amount of P_2O_5 in 100 kg of fresh manure is 0·286 kg P_2O_5. For K_2O, 0·616 kg is included in 100 kg of fresh manure.

Using the energy value of fertilizer nutrients (55–65 MJ/kg N, 11–18 MJ/kg P_2O_5, and 7–9 MJ/kg K_2O), the energy content of fresh manure can be calculated as 0·34–0·42 MJ/kg of manure.

Many difficulties exist in the mobilization and use of organic fertilizers. The main constraints in developing countries that are limiting increased use of organic fertilizers include the heavy labor requirement, the lack of opportunities to profitably raise output, the need to develop skills, the need to change cultural attitudes, and the lack of mixed livestock and crop husbandry. Of special importance, therefore, is the further development of anaerobic fermentation, whereby methane (biogas) is produced as fuel, while the nutrient value of the organic material is retained for fertilizer. In addition, research is underway to supplement nitrogen fertilizer through the nitrogen fixation capacity of legumes and the blue-green algae in paddy fields.

COMMERCIAL ENERGY FOR FARM MACHINERY MANUFACTURE AND USE

Farm machinery manufacture and operation is currently the largest user of commercial energy in world agriculture. As shown in Table II.11, it consumed 117 mtoe in 1972 and 150 mtoe in 1982. However, during this period, its share of the total commercial energy used in agriculture decreased from 57·8% in 1972 to 52·3% in 1982. In addition, while farm machinery comprised more than 50% of the agricultural energy consumed in developed countries, it accounted for only 25% of energy consumed in developing countries.

Like chemical fertilizers, farm machinery in its most advanced form of mechanical power is a comparative newcomer to the traditional inputs used for agricultural production. In fact, human and animal energy still contribute a significant share of the total power used for traction in agriculture in developing countries.

The tremendous improvements that were made in farm machinery in Europe and North America in the 18th and 19th centuries, such as improved plows, seed-drills, and reapers, were initially designed for animal traction. Animal, wind, and water

TABLE II.11

Commercial energy for farm machinery manufacture and operation (source: FAO estimates)

Region	*Tractors in use* (× 1000)		*Energy for manufacture*[a] (× 1000 t)		*Energy for operation* (× 1000 t)		*Total farm machinery energy* (× 1000 t)		*Share of each region* (%)	
	1972	*1982*	*1972*	*1982*	*1972*	*1982*	*1972*	*1982*	*1972*	*1982*
North America[b]	5 810	5 258	19 989	18 090	28 864	26 122	48 853	44 212	41·8	29·4
Western Europe[c]	5 719	7 604	14 049	18 679	19 888	26 444	33 937	45 123	29·0	30·1
Oceania[d]	429	432	1 075	1 083	1 492	1 502	2 567	2 585	2·2	1·7
Other developed countries[c]	472	1 734	1 989	7 306	1 641	6 030	3 630	13 336	3·1	8·9
Developed market economies	12 430	15 028	37 102	45 158	51 886	60 098	88 988	105 256	76·1	70·1
Eastern Europe, USSR[e]	2 922	3 936	7 240	9 753	11 613	15 643	18 854	25 396	16·1	16·9
Total developed countries	15 352	18 964	44 343	54 911	63 499	75 741	107 842	130 653	92·2	87·0
Africa[f]	170	234	320	441	507	698	827	1 138	0·7	0·8
Latin America[f]	681	930	1 276	1 742	2 030	2 772	3 306	4 515	2·8	3·0
Far East[f]	262	778	752	2 233	781	2 319	1 533	4 552	1·3	3·0
Near East[f]	234	723	684	2 114	698	2 155	1 382	4 269	1·2	2·8
Other developing countries[f]	4	5	7	9	12	15	19	24	0·0	0·0
Developing market economies	1 351	2 670	3 039	6 538	4 027	7 959	7 066	14 497	6·0	9·7
Asian centrally planned economies[f]	367	895	958	2 336	1 094	2 668	2 052	5 004	1·8	3·3
Total developing countries	1 718	3 565	3 997	8 875	5 121	10 627	9 118	19 501	7·8	13·0
Total	17 070	22 529	48 340	63 786	68 620	86 368	116 960	150 154	100·0	100·0

[a] Average energy requirement for manufacture assumed to be 2 kgoe per kg of machine weight; replacement rate of 8% per year assumed.
[b] Average weight of tractor and associated equipment assumed to be 15 t and annual fuel consumption 5 t per tractor.
[c] Average weight of tractor and associated equipment assumed to be 8 t and annual fuel consumption 3·5 t per tractor.
[d] Average weight of tractor and associated equipment assumed to be 10 t and annual fuel consumption 3·5 t per tractor.
[e] Average weight of tractor and associated equipment assumed to be 8 t and annual fuel consumption 4 t per tractor.
[f] Average weight of tractor and associated equipment assumed to be 6 t and annual fuel consumption 3 t per tractor.

power were used to operate stationary machinery such as threshers and grinders. Steam power was introduced for stationary farm machinery early in the 19th century, and was applied to field operations beginning about the middle of the century. However, no effective replacement for animal traction emerged in field operations until the introduction of the internal combustion engine, in the form of the agricultural tractor, around the turn of this century.

Except at the most primitive level, in which only wooden tools are used, both hand tool and animal draft technology require some commercial energy input for the manufacture of implements. In addition, improvements could be made in the efficiency of hand tool application. Labor productivity could often be increased by more efficiently designing and using hand tools, and by planning farm operations to avoid unnecessary seasonal peaks in labor requirements. The effectiveness of these power sources may also be increased by combining their use in developing countries. Mechanized power is often the most effective method for tilling the soil, but animal power may then be used for planting and secondary cultivation, and human labor for inter-row cultivations and harvesting. Power mechanization in developing countries should be used to complement rather than replace human and animal power.

But human and animal power consume a negligible amount of energy in comparison with the requirements of mechanical power technology, which uses extensive quantities of commercial energy for both the manufacture and operation of farm machinery. Thus, the rapid mechanization of agriculture which has characterized developed countries in the last 60 years, combined with the more recent increase in the use of mechanical power technology in developing countries, has resulted in a substantial rise in the use of commercial energy in agriculture.

The number of tractors (excluding two-wheel hand tractors) in use in agriculture increased from 17·1 million, or a 10% usage in developing countries in 1972, to 22·5 million, or a 16% usage in 1982. The manufacture of tractors and other farm machinery is fairly energy-intensive, requiring about 2 kgoe to produce 1 kg of machinery (Leach, 1976).

The size of tractors and the total weight of other farm machinery associated with these tractors (including tillage, planting, cultivation, harvesting, and other types of equipment) varies from region to region. For example, in North America, the number of tractors began to decrease about 1968 as older tractors were replaced with fewer but more powerful new ones. Tractors and other machinery are used to perform nearly all crop production operations in North America. Thus, the total weight of a tractor and associated equipment in North America is estimated to be about 15 t. In developing countries, tractors are mainly used for tillage and transportation. Therefore, the average tractor size is smaller and the weight of the associated equipment is less, so the total weight is estimated to be only 6 t. Other regions fall within these extremes and are described in the footnotes to Table II.11.

Table II.11 also provides estimates of the commercial energy that is required as fuel for the operation of farm machinery. The annual fuel consumption of a tractor varies with its size and with the agricultural operations it performs. Thus, in North America, where large tractors are used for most crop production operations, fuel consumption is assumed to be 5 t per tractor, while in developing countries, consumption is estimated at 3 t per tractor. Within the energy used for farm machinery, about 60% is consumed for operation and 40% for manufacture.

The importance of mechanically powered farm machinery to world agriculture is threefold. In the first place, mechanically powered machinery must be used if certain operations, such as deep plowing and land clearance, are to be performed effectively. It is, thus, particularly important in the development of new land. Secondly, its greatest advantage is probably its ability to perform crucial operations, such as tillage and planting, quickly. If these and other operations are performed at the correct time, yields are increased and losses due to pests or weather are reduced. Timely tillage and planting are of key importance in semi-arid areas, where the total crop area depends on how much land can be prepared and planted in the brief period when the uncertain rains arrive. Similarly, in subtropical and tropical areas, timely tillage and planting are usually essential if the potential for multiple cropping is to be realized.

The third and most basic function of farm machinery is to replace human labor. This function is particularly important in developed countries, where agricultural labor has become scarce and costly. Except in limited areas, this function is unlikely to be as important in developing countries in the foreseeable future. Estimates indicate that the agricultural labor force as a whole will continue to grow in absolute numbers throughout the rest of this century. In the meantime, however, rapid mechanization of certain agricultural operations is urgently needed in developing countries, in some cases just in order to reduce the dreadful human drudgery involved in hand labor.

Because of the need to maximize employment opportunities in developing countries for a long time to come, it is generally agreed that these countries should pursue a selective and rational approach to using their scarce capital and foreign exchange resources for agricultural mechanization. However, farm machinery use should continue to be expanded in the many developing countries where a rapid rise in food and agricultural production is essential.

In areas where the expansion of farm machinery use is necessary, it is imperative to examine methods of improving efficiency. In developed countries, attempts have been made in recent years to promote minimum tillage practices, and energy requirements have also been reduced by combining operations such as planting and fertilization. For example, estimates show that fuel requirements for tillage can be reduced to 60% of the requirements necessary for conventional methods by using reduced tillage practices. When minimum and no-till methods are introduced, these requirements are reduced to 50 and 20%, respectively. Fuel conservation can also be accomplished by regularly maintaining and tuning engines, keeping implements in good condition, using diesel engines, and using the right equipment for each operation.

There are many situations in developing countries in which defective planning and management have resulted in as many as half of the country's tractors being out of operation, with the other half operating well below capacity. One factor, among many, is that spare parts are often subject to taxation and delays, from which the original imported machinery is exempt. More attention also needs to be directed to the selection of farm machinery that is best suited to the conditions in which it will be used in developing countries. In addition, more effective use of human and animal draft energy could reduce the need for mechanized power in many developing countries.

ENERGY FROM HUMAN LABOR

Energy Input

The first attempt at measuring human energy expenditure analyzed the diet of workers in different occupations in the late 1800s. Other methods were later developed to examine the energy consumed during specific activities. These included measuring heart rate and recording oxygen consumption. Table II.12 presents data for oxygen consumption and heart rate for several levels of work, ranging from light to extremely heavy.

Oxygen consumption (indirect calorimetry) was originally measured with a Douglas bag. Later, a lightweight, portable respirometer was developed that weighed less than 4 kg and could be worn like a backpack (Passmore & Durnin, 1955). Oxygen consumption corresponds closely to energy consumption. One liter per minute of oxygen consumption is equal to a chemical energy expenditure of 20·9 kJ (Astrand & Rodahl, 1977). This chemical energy expenditure, which includes both basal metabolism and work, corresponds to the total energy expended by the body while the work is being performed. The energy expended is essentially the food energy that is being utilized by the body. Oxygen uptake may be measured with an accuracy of ±0·04 liters/min. Methods based on heart rates have an error rate of as much as 15%. But, the Douglas bag method is 'unsurpassable in accuracy' (Astrand & Rodahl, 1977).

Basal energy requirements of people depend on body size, age, climate, and other environmental factors (Cullumbine, 1950; Quenouille *et al.*, 1951; Suggs & Splinter,

TABLE II.12
Oxygen consumption and heart rate for several work levels (source: Astrand & Rodahl, 1977)

In terms of oxygen uptake (liters/min)	
Light work	Up to 0·5
Moderate work	0·5–1·0
Heavy work	1·0–1·5
Very heavy work	1·5–2·0
Extremely heavy work	More than 2·0
In terms of heart rate responses (beats/min)	
Light work	Up to 90
Moderate work	90–100
Heavy work	100–130
Very heavy work	130–150
Extremely heavy work	150–170

1961; FAO & WHO, 1974). Body size is the most significant of these factors; in fact, a general formula for basal metabolism has been developed:

$$C = aW^b$$

when C = basal metabolism in kcal/day;
a = coefficient, constant for given age and sex, around 70;
W = body weight in kg; and
b = an exponent generally between 0·7 and 0·75.

Using $a = 70$, $b = 0{\cdot}73$, and $W = 55$ kg for a woman and 65 kg for a man, this formula provides a basal metabolism of 0·9 kcal/min (3·8 kJ/min) for the woman and 1·02 kcal/min (4·27 kJ/min) for the man. Other formulas have been developed using various combinations of height, weight, age, surface area, ambient temperature, and relative humidity (Cullumbine, 1950; FAO, 1950; Quenouille *et al.*, 1951), but the preceding formula is probably more widely used because of its relative simplicity.

Gross energy or calorie requirements for individuals based on activity and gender have been developed by the FAO and the World Health Organization (WHO) (FAO, 1950; FAO & WHO, 1974). Tables II.13 and II.14 provide the energy expenditures of a 'reference' woman and a 'reference' man, both 25 years of age, weighing 55 and 65 kg, respectively. While sleeping, the energy expenditure is approximately equal to the basal metabolic rate; about 4·18 kJ/min. This rate doubles for standing or moving slowly and quadruples for walking briskly. For moderate activity, defined as workers in light industry (some farm workers, homemakers, and students), the energy requirements are 9210 kJ/day for the 'reference' woman and 12 560 kJ/day for the 'reference' man (FAO & WHO, 1974).

The most exhaustive analysis of chemical energy expenditures by occupation and specific activity was undertaken by Passmore and Durnin (1955). They collected data from numerous studies worldwide that had utilized indirect calorimetry (oxygen consumption) to measure human energy expenditure. Data for agricultural tasks were collected in Hungary, the USSR, Italy, Gambia, Nigeria, and Germany. All of the information was collected between 1945 and 1954. In these studies, agricultural work was performed with little or no mechanical power. Midday temperatures reached 35°C in the Hungarian study. Table II.15 lists the chemical energy expenditures of agricultural tasks as provided by Passmore and Durnin (1955) and Duff (1978). These expenditures are grouped by major category, such as soil preparation, weeding, and harvest.

Energy Output

Only 20–30% of the chemical energy in food can be converted into mechanical work. For an entire workday, human power output has been estimated at about 0·1 hp (75 W or 1·07 kcal/min) (Astrand & Rodahl, 1977). Many agricultural tasks require power outputs of up to 0·5 hp (375 W or 5 kcal/min). Humans can perform these tasks satisfactorily for short periods of time, but require frequent rest periods when performing heavy work.

TABLE II.13
Energy expenditures of a 55-kg reference woman distributed during 24 h showing effect of occupation (source: FAO & WHO, 1974)

Distribution of activity	*Light activity*		*Moderately active*		*Very active*		*Exceptionally active*	
	(kcal)	*(MJ)*	*(kcal)*	*(MJ)*	*(kcal)*	*(MJ)*	*(kcal)*	*(MJ)*
At rest (8 h)	420	1·8	420	1·8	420	1·8	420	1·8
At work (8 h)	800	3·3	1 000	4·2	1 400	5·9	1 800	7·5
Non-occupational activities (8 h)	580–980	2·4–4·1	580–980	2·4–4·1	580–980	2·4–4·1	580–980	2·4–4·1
Range of energy expenditure (24 h)	1 800–2 200	7·5–9·2	2 000–2 400	8·4–10·1	2 400–2 700	10·1–11·8	2 800–3 200	11·7–13·4
Mean (24 h)	2 000	8·4	2 200	9·2	2 600	10·9	3 000	12·5
Mean (per kg of body weight)	36	0·15	40	0·17	47	0·20	55	0·23

TABLE II.14
Energy expenditures of a 65-kg reference man distributed during 24 h showing effect of occupation (source: FAO & WHO, 1974)

Distribution of activity	*Light activity*		*Moderately active*		*Very active*		*Exceptionally active*	
	(kcal)	*(MJ)*	*(kcal)*	*(MJ)*	*(kcal)*	*(MJ)*	*(kcal)*	*(MJ)*
At rest (8 h)	500	2·1	500	2·1	500	2·1	500	2·1
At work (8 h)	1 100	4·6	1 400	5·8	1 900	8·0	2 400	10·0
Non-occupational activities (8 h)	700–1 500	3·0–6·3	700–1 500	3·0–6·3	700–1 500	3·0–6·3	700–1 500	3·0–6·3
Range of energy expenditure (24 h)	2 300–3 100	9·7–13·0	2 600–3 400	10·9–14·2	3 100–3 900	13·0–16·3	3 600–4 400	15·1–18·4
Mean (24 h)	2 700	11·3	3 000	12·5	3 500	14·6	4 000	16·7
Mean (per kg of body weight)	42	0·17	46	0·19	54	0·23	62	0·26

TABLE II.15
Energy expenditure of agricultural tasks (source: Passmore & Durnin, 1955; Astrand & Rodahl, 1977; Duff, 1978)

Activity	*Country*	*kcal/min*
Soil preparation		
Plowing	Russia	6·9
Plowing	Russia	5·4
Clearing shrubs	Gambia	7·1
Ridging (deep-digging)	Gambia	9·5
Hoeing	Gambia	5·8
Bush-clearing	Nigeria	6·1
Hoeing	Nigeria	4·4
Average		6·5
Seeding or planting		
Planting groundnuts	Gambia	3·7
Transplanting	Philippines	3·2
Average		3·5
Weeding		
Weeding rake	Russia	3·3
Weeding	Gambia	5·3
Hoeing	Gambia	5·8
Hoeing	Nigeria	4·4
Weeding	Philippines	6·1
Average		5·0
Harvest		
Mowing wheat	Hungary	7·7
Mowing barley	Hungary	7·0
Setting up stocks	Hungary	6·6
Preparing stocks	Italy	5·5
Preparing stocks	Italy	4·8
Mowing with a scythe	Italy	6·8
Loading stocks onto carts	Italy	5·6
Grass cutting	Nigeria	4·3
Cutting and stacking	Philippines	4·9
Average		5·9
Post-harvest		
Binding wheat	Hungary	7·3
Thrashing rye	Russia	5·0
Thrashing rye	Russia	4·5
Binding oats	Russia	3·3
Binding oats	Russia	4·1
Binding rye	Russia	4·2
Binding rye	Russia	4·7
Threshing	Italy	6·0
Threshing	Italy	5·1
Threshing	Italy	5·8
Threshing	Italy	3·8
Threshing	Italy	3·5
Threshing	Philippines	4·8
Average		4·8
Pesticide application		
Herbicide application	Philippines	6·9
Tractor driving		
Plowing	Germany	4·2
Plowing	Germany	4·2
Average		4·2
Animal driving		
Horse plowing	Germany	5·9
Horse plowing	Germany	5·1
Average		5·5
Other		
Standing		2·0
Sitting		1·8

ENERGY FROM DRAFT ANIMALS

Developing countries have large numbers of cattle, buffalo, donkeys, camels, and other livestock that may provide draft power. Approximately 55% of all the world's cattle are found in the tropics, as are 60% of the buffalo. They supply meat, milk, and byproducts, and also provide a major portion of the draft power for the developing world. In India and Pakistan, for example, cattle and buffalo supply as much as 85% of the total draft power (Goe & McDowell, 1980; Ward *et al.*, 1980; FAO, 1982; FAO, 1984*a*; Van Vaerenbergh, 1985).

A major advantage of bullocks and other draft animals is their ability to utilize grasses, forage, and agricultural wastes as fuel. Much of the world's land is too steep, wet, dry, or cold to produce food crops for humans, but it can sometimes produce food for ruminant animals, which, in turn, serve as a source of draft power. Maintaining animals during the off-season is a disadvantage, however, as is their competition for land that might be used for other purposes. In engineering terms, the efficiency of the conversion of biomass energy to perform work is generally estimated to peak at 20–30% for short work periods, but averages significantly less during an entire day of work. Because draft animals use low-quality feed, however, the low conversion efficiency may not be a concern.

In many areas of the world, the use of animal power has not reached full potential. Some countries, such as the tsetse belt in Africa, do not have the animal resources necessary, but many countries do. With the increasingly high cost of fuel, which must be purchased in cash and often is difficult to obtain, using animals for draft work and transportation becomes even more important (A. J. Smith, 1981).

Nevertheless, animal draft power is restricted by ineffective yokes and harnesses, hitches, and implements, as well as operator skill (Watson, 1981). The relative merits of animals versus tractors for draft power has provoked significant controversy. In reality, both are needed. At millions of small farms in developing countries, tractors are not feasible, either in technical or economic terms. Fields may be too small, steep, or rocky. In addition, the essential prerequisites for the successful use of engine power, such as dealer services, capital availability, roads—or the overall infrastructure—may not be available (Stout & Downing, 1976). On the other hand, every country has substantial numbers of larger farms where conditions are indeed suitable for engine-powered mechanization. Development can be enhanced by an optimal mix of hand labor, animal draft power, engine power, and electrical power. Ensuring that sufficient energy is available in all forms at the right time to meet agriculture's needs is more important than the type of energy used.

The energy requirements of draft animals can be computed in several ways. One approach is to compute the actual work performed by the animal. This figure can be calculated by multiplying the power developed using the animal by the time the animal is working. Another approach to the energy requirement is to compute the feed eaten by the animal. This feed energy value is higher than the actual work performed because most of the feed energy is used to maintain body temperature and perform normal bodily functions such as digestion, growth, fetal development, and milk production. Energy losses also occur in the feces and urine. In this sense, the draft animal can be considered an energy conversion device. The fuel input is the feed and the work performed is the energy output.

Draft Animal Energy Output Based on Drawbar Pull

Drawbar pull and speed can be used to measure the power an animal can produce. The power available for pulling or developing tractive effort is measured at the location where the implement is attached. If the draft and speed of the work performed is known, the drawbar power can be calculated as follows:

$$\text{Drawbar power (kW)} = \frac{\text{draft (N)} \times \text{speed (km/h)}}{3600}$$

Draft (pull or tractive effort) is the total force parallel to the direction of travel required to pull an object (e.g. an implement) a given distance. It varies depending on the weight of the object and the soil's resistant forces. For pull-type, non-tillage implements, the draft is in the form of rolling resistance. An implement which penetrates the soil has additional soil forces affecting it. These include resistance of the soil to compression, shear, adhesion (attractive forces between the soil and other materials), and frictional resistance. These resistances are a result of soil particle-to-particle friction and the bonding of individual soil particles by water.

Draft can also be expressed as 'specific draft', which is the draft per unit area of tilled cross-section. Table II.16 shows the draft requirements of various agricultural implements. The draft of pulled carts is shown in Table II.17.

Speed is merely the speed of travel (distance over time). Table II.18 provides typical draft requirements and speeds of travel for several agricultural implements used in Africa.

The draft a given animal can sustain during a workday is frequently expressed as a % of body weight. It typically varies from 10 to 20%, depending on the type of animal. A 500-kg horse, for example, can exert a draft of 10–12% of its body weight for several hours. This amounts to a draft of 50–63 kg (Goe & McDowell, 1980). Table II.19 lists the draft and power capabilities of five different draft animals as indicated in four different sources of reference. Table II.19 also indicates typical work speeds and powers developed. The speed of work is generally between 2 and 4 km/h.

Draft Animal Energy Output Based on Feed Consumption

Another way to analyze the capacity of draft animals is through feed eaten. Many different energy values are assigned to feed, ranging from gross energy, to digestible energy, to net energy. Figure II.14 depicts the different classifications of energy in feed. Gross energy is the heat of combustion of the feed as determined with a bomb calorimeter. Digestible energy is the gross energy minus the energy in the feces. Metabolizable energy is a further breakdown of the gross energy excluding the energy of the feces, urine, and fermentation gases. Metabolizable energy is used by the animal for work, growth, fetal development, and milk and meat production. Finally, net energy is the metabolizable energy minus the work of digestion (Cullison, 1979).

- Gross energy
  - Energy of feces
  - Digestible energy
    - Energy of the urine
    - Energy of methane
    - Metabolizable energy
      - Heat increment
        - Heat of fermentation
        - Heat of nutrient metabolism
      - Net energy

Fig. II.14. Energy value of feed as used by an animal (source: Cullison, 1979).

TABLE II.16
Draft requirements of agricultural implements (source: Kepner *et al.*, 1972)

Machine	*Typical range of requirements*
Tillage	
Moldboard or disk plow	3–6, 5–9, 8–14 lb/in^2 [a]
Lister (in firm soil)	400–800 lb/bottom
Vertical-disk plow (one-way disk)	180–400 lb/ft
Disk harrow	
Single-acting	50–100 lb/ft
Tandem (light-duty)	100–200 lb/ft
Offset or heavy tandem	250–400 lb/ft
Subsoiler	70–110, 110–160 lb/in depth[b]
Chisel plow or chisel-type field cultivator	40–120 lb/ft per in depth
Field cultivator with sweeps, 3–5 in depth	100–300 lb/ft
Powered rotary tiller, 3–4 in bite	15–25, 25–35, 30–40 lb/in^2 [c]
Spring-tooth harrow	75–200 lb/ft
Spike-tooth harrow	20–60 lb/ft
Rod weeder	60–120 lb/ft
Roller or packer	20–150 lb/ft
Rotary hoe	30–100 lb/ft
Row-crop cultivator	
Shallow	40–80 lb/ft
Deep	20–40 lb/ft per in depth
Planting	
Row-crop planter, drilling seed only	100–180 lb/row
Grain drill	30–100 lb/ft
Harvesting	
Mower	0·4–0·8 PTO hp/ft
Hay baler	1–2·5 hp-h/ton
Field cuber, self-propelled	20–30 hp-h/ton
Flail-type field chopper	
Without recutter	1·3–2·5 hp-h/ton
With flywheel-type recutter	2–4 hp-h/ton[c]
Shear-bar-type field chopper	
Corn silage, $\frac{1}{2}$ in cut	0·8–2 hp-h/ton
Green grass/legume silage	1–2·5 hp-h/ton
Low-moisture grass/legume silage	2–5 hp-h/ton
Self-propelled combine in grain	2–3 hp-h/1 000 lb of non-grain material
Corn picker, two-row	12–20 hp

[a] Specific draft, lb/in^2 of furrow slice, for light, medium, and heavy soil, respectively.
[b] Ranges are for sandy loam and medium or clay loam soils, respectively.
[c] All rotary power, but expressed as equivalent specific draft, which is numerically equal to specific energy in ft-lb/12 in^3 of soil. Ranges are for light, medium, and heavy soils. Increasing the bite length to 6 in decreases energy requirements by 20–25%.

TABLE II.17

Draft for carts equipped with various wheel types (source: Goe & McDowell, 1980)

Type of cart	*Ground surface*	*Weight (kg)*			*Draft (kg)*
		Cart	*Load*	*Total*	
Wooden cart with pneumatic tires	Plowed	364	933	1 297	51
Iron cart with pneumatic tires	Plowed	495	1 866	2 361	140
Farm cart with iron-rimmed tires	Plowed	373	933	1 306	178
Country cart with wooden wheels	Plowed	252	597	849	127
Wooden cart with pneumatic tires	Dirt road	364	933	1 297	22
Iron cart with pneumatic tires	Dirt road	495	1 866	2 361	45
Farm cart with iron-rimmed wheels	Dirt road	373	933	1 306	27
Country cart with wooden wheels	Dirt road	252	597	849	36

Gross energy is a function of the feed only, and, as such, is the best indication of the inherent energy of the feed. Metabolizable energy, on the other hand, is dependent on both the feed and the animal type. Different types of animals are able to metabolize varying amounts of energy in a feed, depending on their gastro-intestinal tract. Thus, metabolizable energy represents a more accurate measure of a feed's energy value to an animal.

Feed given to draft animals in North Africa is approximately 2 kg of hay equivalent/100 kg of body weight per day (Kelsoe, B., 1985, pers. comm.). A proximate analysis of meadow hay is provided in Table II.20. Crude protein represents 7·8% of meadow hay as feed. The nitrogen-free extract essentially

TABLE II.18

Draft requirements and operation speeds for agricultural implements in equatorial Africa (sources: Kline *et al.*, 1969; Goe & McDowell, 1980)

Implement	*Draft*	*Speed (km/h)*
Plow		
Indigenous	0·14–0·70 kg/cm^2	1·61–2·42
Moldboard	0·21–1·12 kg/cm^2	2·42–4·84
Disk	0·31–1·00 kg/cm^2	2·42–5·65
Disc harrow		
Single-action	0·45–1·50 kg/cm	1·61–4·03
Double-action	1·20–2·70 kg/cm	1·61–4·03
Rotary tiller	0·70–3·50 kg/cm	0·8–2·42
Harrow		
Spike, or peg	1·80–2·70 kg/peg	1·61–4·84
Spring tine	10–25 kg/tine	1·61–4·84
Roller or puddler	0·15–0·90 kg/cm	0·8–4·03
Leveler, float	0·30–0·70 kg/cm	1·61–4·84
Row-crop planter	30–70 kg/row	1·61–4·84
Grain drill	6–22 kg/row	1·61–4·84
Transplanter	10–20 kg/row	0·8–2·42

TABLE II.19

Power capabilities of draft animals (sources: 1, Goe & McDowell, 1980; 2, CEEMAT, 1972; 3, Inns, 1980; 4, Hopfen, 1969)

	1	*2*	*3*	*4*
Horse				
Animal weight (kg)	500			400–700
Draft as percentage of body weight	10–12			11–15
Draft (kg)	50–63		51	60–80
Speed of work (km/h)	2·4–4·0		3·6	3·6
Power developed (hp)	0·6–0·7		0·67	1·0
Daily work hours			6	
Daily work (hp-h)			4·02	
Mule				
Animal weight (kg)	200			350–500
Draft as percentage of body weight	10–16			12–14
Draft (kg)	20–32			50–60
Speed of work (km/h)	2·4–4·0			3·2–3·6
Power developed (hp)	0·2–0·3			0·7
Daily work hours				
Daily work (hp-h)				
Donkey (ass)				
Animal weight (kg)	300			200–300
Draft as percentage of body weight	10–16	17–20		13–15
Draft (kg)	30–48		40·8	30–40
Speed of work (km/h)	2·4–4·0	2·5–2·8	3·6	2·5
Power developed (hp)	0·4		0·54	0·35
Daily work hours		3–3·5	4	
Daily work (hp-h)			2·16	
Dromedary				
Animal weight (kg)	600			
Draft as percentage of body weight	10–14			
Draft (kg)	60–84			
Speed of work (km/h)	3·5–4·0			
Power developed (hp)	0·9–1·1			
Daily work hours				
Daily work (hp-h)				
Bovine				
Animal weight (kg)	450			500–900
Draft as percentage of body weight	10–14	10–13		9–12
Draft (kg)	45–64		51	60–80
Speed of work (km/h)	2·4–4·0		3·6	2·2–3·1
Power developed (hp)	0·6–0·7		0·67	0·75
Daily work hours		5–6	5	
Daily work (hp-h)			3·35	

TABLE II.20
Proximate analysis of meadow hay (sources: Cullison, 1979; NRC, 1972)

	As fed (%)	*Dry (%)*
Dry matter	93·5	100·0
Ash	8·0	8·6
Crude fiber (carbohydrates)	30·5	32·7
Ether extract (fats)	2·3	2·5
Nitrogen-free extract (carbohydrates)	44·8	47·9
Crude protein (proteins)	7·8	8·3

TABLE II.21
Energy inputs for typical draft animals in North Africa (source: Kelsoe, B., 1982, pers. comm.)

	Dromedary	*Horse*	*Mule*	*Donkey*	*Bovine*
Weight (kg)	636	455	364	273	409
Daily feed requirement (kg hay equivalent)[a]	12·7	9·1	7·3	5·5	8·2
Gross energy of daily feed (kcal)[b]	47 500	34 000	37 300	21 000	30 700

[a] Based on 2 kg of hay equivalent/100 kg of body weight.
[b] Based on gross energy of 3738 kcal/kg of hay equivalent.

represents digestible carbohydrates, such as sugars and starches. Crude fiber signifies carbohydrates of low digestibility; therefore, carbohydrates total 75·3% of meadow hay. The heat content of proteins, carbohydrates, and fats, as determined by a bomb calorimeter, are as follows (Astrand & Rodahl, 1977; Church, 1977; Cullison, 1979):

Proteins	23·4 kJ/g
Carbohydrates	17·2 kJ/g
Fats	38·9 kJ/g

Using these values, the gross energy content of meadow hay is 15 650 kJ/kg.

Feed for an entire day can be assigned to work performance if the draft animal is assumed to have little or no other value (Table II.21). A draft animal may have value for the production of milk, meat, and offspring. However, milk production value only applies to cows and tends to be quite low for local breeds; meat production value only occurs at an animal's death; and reproduction value is difficult to determine. Therefore, assigning the entire feed value to the work is logical.

In many areas, the use of animal traction can be developed considerably, and animal draft equipment can be improved.

COMMERCIAL ENERGY FOR PUMP IRRIGATION EQUIPMENT MANUFACTURE AND OPERATION

Controlled irrigation consists of two types: large-scale, gravity-flow irrigation, by which dams or water diversion structures and channels are built to transfer water to

TABLE II.22

Commercial energy for pump irrigation equipment manufacture and operation (source: FAO estimates)

Region	*Area irrigated by pumps (× 1000 ha)*		*Energy for manufacture (× 1000 t)*[a]		*Energy for operation (× 1000 t)*		*Total irrigation energy (× 1000 t)*		*Share of each region (%)*	
	1972	*1982*	*1972*	*1982*	*1972*	*1982*	*1972*	*1982*	*1972*	*1982*
North America[b]	3 862	4 875	133	168	614	775	747	943	20·2	18·6
Western Europe[b]	1 741	2 103	58	70	277	334	335	404	9·0	8·0
Oceania[b]	161	187	5	6	26	30	31	36	0·8	0·7
Other developed countries[b]	449	456	13	13	71	73	84	86	2·3	1·7
Developed market economies	6 213	7 622	209	257	988	1 212	1 197	1 469	32·4	28·9
Eastern Europe, USSR[b]	1 520	2 316	61	93	242	368	302	461	8·2	9·1
Total developed countries	7 733	9 938	270	350	1 229	1 580	1 499	1 929	40·5	38·0
Africa[c]	389	523	10	14	77	104	88	117	2·4	2·3
Latin America[d]	553	732	20	26	99	131	119	157	3·2	3·1
Far East[c]	2 815	5 547	68	134	559	1 102	627	1 236	17·0	24·4
Near East[d]	2 562	3 288	69	88	458	588	527	676	14·2	13·3
Other developing countries[d]	0	0	0	0	0	0	0	0	0·0	0·0
Developing market economies	6 319	10 090	167	262	1 194	1 925	1 361	2 187	36·8	43·1
Asian centrally planned economies[e]	4 164	4 756	94	108	745	851	839	958	22·7	18·9
Total developing countries	10 484	14 846	261	370	1 939	2 776	2 200	3 145	59·5	62·0
Total	18 217	24 784	531	719	3 168	4 356	3 699	5 075	100·0	100·0

[a] Energy requirements for manufacture assumed to be 2 kgoe/kg of equipment with 10 years average life of pump and equipment.
[b] Weight of equipment assumed to be 140 kg/ha irrigated, and fuel requirement 160 kg/ha.
[c] Weight of equipment assumed to be 100 kg/ha irrigated, and fuel requirement 200 kg/ha.
[d] Weight of equipment assumed to be 140 kg/ha irrigated, and fuel requirement 180 kg/ha.
[e] Weight of equipment assumed to be 100 kg/ha irrigated, and fuel requirement 180 kg/ha.

the fields; and pump irrigation, by which water is pumped either from groundwater or surface sources. Agricultural production is often only a secondary use of water from large dams; the primary use is electrical power generation. Therefore, the energy used to construct dams is excluded from the estimates made within this text. Also, since the development and maintenance of irrigation canals are usually performed with agricultural machines, the energy for this purpose has been included in the farm machinery discussion. The irrigation equipment discussed in this section is limited to pumps, engines, pipes and, if applicable, other material, such as equipment for sprinkler and drip irrigation.

In 1972, the irrigation equipment in use in the world amounted to about 2·3 million tonnes, with 1·2 million tonnes (52%) located in developing countries. In 1982, the total irrigation equipment was estimated to have increased to 3 million tonnes, with 1·6 million tonnes (53%) located in developing countries. The annual production of irrigation equipment supplied to agriculture was about 265 000 t in 1972, and increased to about 360 000 t in 1982. The average energy required to produce this equipment is assumed to be 2 kgoe/kg of equipment.

In addition to the energy needed for the manufacture of irrigation equipment, energy, usually in the form of petroleum fuel, is required for its operation. The fuel required per hectare of irrigated land varies with the depth of the water being pumped, the type of irrigation system, and the water requirements of the crops. Fuel requirements are estimated to vary from 160 kg/ha in developed countries to 200 kg/ha in the developing countries of Africa and the Near East. Estimates of the commercial energy for pump irrigation based on these assumptions are shown in Table II.22.

In 1982, pump irrigation consumed 5·1 mtoe with developing countries accounting for 62%. However, while developing countries increased their share of all inputs between 1972 and 1982, the increase for irrigation was only 2·5% (from 59·5 to 62%), in comparison to a 13% increase (from 28 to 41%) for fertilizer, an 8% increase (from 20 to 28%) for pesticides, and a 5% increase (from 8 to 13%) for farm machinery.

The primary increase in pumped irrigation was in the Far East, where pumps were estimated to have been used on 2·7 million hectares (21%) of the 13 million hectares of the additional land brought under irrigation between 1972 and 1982. This area accounted for more than 62% of the estimated 4·4 million hectares of additional land brought under pump irrigation in the developing countries during this period. As a result, the Far East used 24·4% of the total irrigation energy in 1982, in comparison to 17% in 1972.

Irrigated areas need to be expanded in the developing countries. However, the increasing investment and operating costs of the present pump irrigation systems require that they be used as effectively as possible and that alternative sources of energy be examined. More efficient use of irrigation water is feasible in many areas and could result in savings in energy consumption. The interdependence of land development methods, irrigation practices, and systems of cultivation and crop production is seldom fully appreciated. Both increased efficiency in irrigation water use and higher cropping intensities can be achieved by the following methods: (1)

improving water distribution channels and providing appropriate drainage; (2) improving field layouts; (3) grading and leveling land before irrigation; and (4) using better implements and water application methods.

Improved maintenance and operation of mechanical equipment can also improve the efficiency of energy used in irrigation. However, fragmented land holdings and the difficulty of organizing group action required to implement new projects often impede irrigation improvement. The lack of an adequate infrastructure to support and maintain mechanical equipment may also limit its efficient use.

In many developing countries, the high capital cost and long lead time required by large irrigation schemes have slowed their development and increased the emphasis on small-scale systems. Many farms in developing countries are less than 2 ha in size, and the water table in some areas is within a few meters of the ground surface. While the quantity of water needed to irrigate a given area depends on soil type, land topography, climate, method of water distribution, and crop water requirements, the peak pumping capacity needed for 1 ha is typically in the range of 2–6 liters/s. For water within a few meters of the ground surface, this may require as little as 100–300 W of energy inputs/ha. Alternative energy sources seem to have the greatest potential for these small areas that require low power inputs.

People have irrigated small areas near water sources for centuries, and continue to do so in many developing countries. However, the work efficiency of people is relatively low, particularly in developing countries with high temperature and humidity. Therefore, while improved hand pumps and water lifting devices could reduce the drudgery and increase the irrigated area to a limited extent, other forms of power are needed to substantially expand irrigated areas.

The use of draft animals for irrigation is widespread in both Asia and the Near East, where a pair of bullocks operating a Persian wheel or other water-lifting device can irrigate 1–1·5 ha/h. Efficiency could improve with upgraded water-lifting devices and drive mechanisms, better harnesses, and improved animal nutrition and management. However, the additional area that could be irrigated through such improvements is limited because the need for draft animal power for other agricultural operations increases as the irrigated area increases.

Solar energy resources in developing countries are generally plentiful and the utilization of photovoltaic solar cells appears to be one of the most promising alternative energy technologies for water pumping. Several types of photovoltaic pumping systems are commercially manufactured and have been operating on a limited basis in developing countries for several years. However, while technical feasibility has been proven, photovoltaic systems are currently too expensive for pumping irrigation water in most developing countries. However, a reduction in the price of solar cells and improved system designs seem possible and could make photovoltaic systems economically viable for future small-scale irrigation.

Wind power is another alternative. Wind speeds are generally low in the tropics, where most of the developing countries are located. However, water-pumping windmills have been used to supply water for livestock and human consumption in some developing countries for many years, and most commercially available windmills are reliable, safe, and seldom require repair if properly maintained. Since topographical features can augment wind speed, and doubling the wind speed increases the power output eight times, windmills may be an appropriate source of power for irrigation water pumping, although they do not currently appear economically viable except at selected locations in developing countries.

The power of flowing water may also be used to pump water for irrigation when irrigated areas are near rivers and streams with an average water velocity of 0·75 m/s or more. In addition, opportunities exist in selected locations to substitute petroleum fuels in internal combustion engines with alternatives such as biogas, producer gas, alcohol, and vegetable oil, which can be derived from agricultural products and/or their residues.

In order to meet future food needs, irrigated areas must be expanded in developing countries. This expansion effort will require the appropriate application of both conventional and renewable sources of energy. The primary objective must be to ensure that the energy needs of agriculture are met by effectively using the energy sources available at a particular location.

COMMERCIAL ENERGY FOR PESTICIDE PRODUCTION AND APPLICATION

The use of agricultural pesticides increased from 1·6 million tonnes of active ingredient in 1972 (a value of $6·8 billion) to 1·9 million tonnes in 1982 (a value of $8·2 billion). Developing countries accounted for about 20% of the consumption of pesticides in 1972, and this increased to 28·5% in 1982, as shown in Table II.23.

The commercial energy required to produce a pesticide can be substantial. Raw materials for modern pesticides come primarily from the petrochemical industry, and increased energy inputs are required to manufacture them. A pesticide also contains a number of formulating agents and often a solvent, and these too require energy input. Smaller energy inputs are also required for packaging, transporting, distributing, and applying pesticides. The total energy required to provide 1 kg of pesticide has been estimated to be about 2·3 kgoe (Leach & Slesser, 1973).

Tables II.24 and II.25 list typical energy values for various pesticides. While the energy content of pesticides on a per unit weight basis is higher than fertilizers, typical quantities applied are considerably less for pesticides than for fertilizers. (The energy content for pesticides is per kilogram of active ingredient. When pesticides are sold, the active ingredient is mixed with an inert ingredient; therefore, 1 kg of product contains less than 1 kg of active ingredient. Usually, the percentage of active ingredient is indicated on the product label.)

Thus, pesticides are the most energy-intensive agricultural input. However, since they are mostly used in small quantities—seldom more than 1 kg active ingredient/ha—they represent only a small portion of the total commercial energy used in agricultural production (1·9% in 1972 and 1·6% in 1982) (Table II.23).

Nevertheless, continued growth in the use of pesticides appears unavoidable. Their expanded use is particularly needed in developing countries where crop losses (both pre- and post-harvest) are substantial because of inadequate pest control.

TABLE II.23

Commercial energy for pesticide production and application (source: FAO estimates)

Region	*Pesticide value (× $ million)*		*Pesticide active ingredient (× 1000 t)*		*Commercial energy (× 1000 t)*[a]		*Share of each region (%)*	
	1972	*1982*	*1972*	*1982*	*1972*	*1982*	*1972*	*1982*
North America	2 318	2 511	546	591	1 271	1 377	34·0	30·7
Western Europe	1 545	1 642	364	386	847	900	22·6	20·1
Oceania	29	29	7	7	16	16	0·4	0·4
Other developed countries	33	33	8	8	18	18	0·5	0·4
Developed market economies	3 925	4 215	924	992	2 152	2 311	57·5	51·5
Eastern Europe, USSR	1 545	1 642	364	386	847	900	22·6	20·1
Total developed countries	5 471	5 857	1 287	1 378	2 999	3 211	80·1	71·5
Africa	49	278	11	65	27	152	0·7	3·4
Latin America	224	421	53	99	123	231	3·3	5·1
Far East	60	258	14	61	33	142	0·9	3·2
Near East	58	207	14	49	32	113	0·8	2·5
Other developing countries	1	9	0	2	1	5	0·0	0·1
Developing market economies	391	1 172	92	276	215	643	5·7	14·3
Asian centrally planned economies	966	1 159	227	273	530	635	14·1	14·2
Total developing countries	1 357	2 331	319	549	744	1 278	19·9	28·5
Total	6 828	8 189	1 607	1 927	3 743	4 489	100·0	100·0

[a] Production of 1 kg of pesticide (active ingredient) assumed to require 2·33 kgoe.

TABLE II.24

Energy inputs for the basic production of various pesticides (source: Pimentel, 1980; © 1980 CRC Press, Inc., Boca Raton, FL)

Pesticides	*kcal for production (1 kg active ingredient)*	*Pesticides*	*kcal for production (1 kg active ingredient)*
Herbicides		Insecticides	
MCPA	30 952	DDT	24 200
Diuron	64 290	Toxaphene	38 100
Atrazine	45 240	Methyl parathion	13 810
Trifluralin	35 170	Carbofuran	108 100
Paraquat	109 520	Carbaryl	36 430
2,4-D	24 200	Fumigants	
2,4,5-T	56 700	Methyl bromide	15 950
Chloramben	71 400	Fungicides	
Dinoseb	19 080	Ferbam	15 250
Propanil	52 240	Maneb	23 570
Propachlor	69 050	Captan	27 380
Dicamba	70 240	Sulfur	26 620
Glyphosate	108 100		
Diquat	95 240		
		Pesticide average	49 020

The percentage of oil, natural gas, and coal involved in the production of pesticides has been calculated as 42% oil, 38% natural gas, and 20% coal.

TABLE II.25

Energy inputs for various pesticides (GJ/t) (source: M. B. Green, 1978)

Herbicides		*Fungicides*		*Insecticides*	
MCPA	130	Ferbam	61	Methyl parathion	160
Diuron	270	Maneb	99	Toxaphene	58
Atrazine	190	Captan	115	Carbofuran	454
Trifluralin	150			Carbaryl	153
Paraquat	460			Phorate	172
2,4-D	85				
2,4,5-T	135				
Chloramben	170				
Dinoseb	80				
Propanil	220				
Propachlor	290				
Dicamba	295				
Glyphosate	454				
Diquat	400				

However, public concern about possible detrimental environmental effects resulting from the use of chemical pesticides has intensified the search for ways to economize their use. Weed control, either through better tillage or by incorporating mechanical or hand weeding, is, in many cases, still the best alternative to herbicide use. These methods may be especially appropriate in developing countries where labor is usually abundant and inexpensive in relation to the cost of imported materials. The need for insecticides and fungicides can also be reduced by greater use of new pest control methods based on the principles of integrated pest management. This approach relies mainly on natural elements for the control of pests, including planting pest-resistant varieties of crops. These new techniques are gradually increasing in acceptance and application in a number of developing countries.

CHAPTER III

Energy Flow

INTRODUCTION

Crop yields and food supplies to consumers are directly linked to energy; that is, sufficient energy in the right form at the right time. Energy audits and surveys attempt to depict this flow of energy in an agricultural system. They provide a firm information base for analysis leading to decisions on energy and agricultural policy. Should more petroleum be imported? Should another refinery or electric power generating plant be built? Where? How should energy for rural use be taxed? Or subsidized? What prices should be charged to farmers and rural villagers? What can be done to reverse the deforestation problem? How can the energy distribution system be improved in rural areas? How much energy is needed by the rural sector? In what form and when? What are the energy use trends? These and many other questions must be answered. A solid data base is a prerequisite to wise and enlightened policy decisions.

Countries differ with respect to energy resource availability, demand patterns, and institutional capabilities to conduct, analyze, and utilize surveys. Thus, no single prescription or approach will fulfill all needs. A wide variety of approaches are illustrated in the examples in this chapter. Unfortunately, data on energy use in the rural sector of developing countries is fragmentary and incomplete, partially because the important component of noncommercial energy is often excluded.

In order to present a valid picture of the rural energy situation in developing countries, many factors must be considered, such as:

dependence on organic fuels;
seasonal variation in energy demand and supply;
proximity of fuel supplies;
high human energy input;
dwindling fuel supplies;
fuel scarcity perceived by local residents;
fuel in transition from 'free' goods to commercial;
increasing commercial fuel prices;
greater hardship on some people than on others;
restricted land access in certain regions.

Many publications have addressed the philosophy and techniques of energy analysis (Anon., 1980; Fluck & Baird, 1980; Gifford & Millington, 1975; Gilliland, 1978; Heichel, 1973; Odum, 1984; Pimentel *et al.*, 1973; Smil *et al.*, 1983; Thomas, 1977). Key elements of energy analysis are discussed in this chapter along with many regional and national examples. The focus of this chapter is on developing countries, although a few exceptions provide comparisons with developed countries.

ENERGY SURVEYS AND ANALYSES

Energy analysis (accounting) evolved as an alternative to monetary accounting for situations in which the energy value of inputs and outputs was the primary objective.

The monetary value does not necessarily accurately reflect energy value. For example, in 1983 in the United States, a dollar's worth of electricity contained less than half as much energy as a dollar's worth of gasoline. So if the major concern is the energy 'cost' of goods or a service, we must look at the energy value of the inputs required to produce the goods or service.

How is the energy value of an input determined? If the input is a fuel, such as gasoline or diesel fuel, the most obvious energy value to use is the heat of combustion, or bomb calorimeter value. This represents the energy content of the fuel if it is completely burned. But what about the energy required to drill, transport, and refine the petroleum? Should these also be included? If so, just what should be included? The fuel used to run the drilling equipment? The heat required in the refinery? What about the energy required to manufacture the drilling equipment or the gasoline used in the refinery employees' cars as they drive to work?

The situation becomes even more complicated when evaluating the energy content of non-fuel inputs such as animal draft. Is the energy input the feed eaten by the animal? If so, is it the bomb calorimeter value of the feed or the energy required to grow the feed? Should the energy required to raise the animal from birth be added also?

Clearly, boundaries of energy analyses must be established. Figures III.1, III.2, and III.3 provide three illustrations. The structure of the flow of agricultural inputs to food production is shown in Fig. III.1. The five biophysical inputs are land, seed, water, fertilizer, and agricultural chemicals, and the flow to and from these for the various processes of mining, transporting, spraying, cooking, etc., are indicated by solid vertical lines with arrows (Gifford & Millington, 1975). The four inputs that are universally required processes in an industrial society—fuel, machine hours, worker hours, and infrastructure—are shown in boxes at the top. The broken horizontal lines show that every process in the material flow requires the presence of all four universal inputs. Infrastructure includes such things as farm buildings, food wholesaling and retailing systems, agricultural research, credit agencies, etc. Another example is shown in Fig. III.2. Items outside the boundary line, such as mining and other aspects of energy procurement, were excluded (Smil *et al.*, 1983).

Figure III.3 graphically depicts one way to reduce energy values to a common denominator to provide an opportunity for analysis and comparison. The input categories are animal draft, human labor, power units (i.e. tractors, combines, vehicles, and liquid-fueled irrigation pumps), fertilizer, and pesticides. Animal draft, human labor, and power units are all direct energy inputs. Each requires some sort of fuel—either petroleum, gas, food, or feed—which is converted into useful work by the energy conversion device (i.e. the worker, animal, or engine). The energy output is the energy available for useful work. Fertilizer and pesticides are indirect energy inputs. They do not produce useful work but they should be included in order to determine how much energy is consumed by agriculture, because the energy required to manufacture fertilizers and pesticides is substantial. For some crops, the manufactured energy invested in the fertilizer and pesticides applied to the fields is more than all other energy inputs combined.

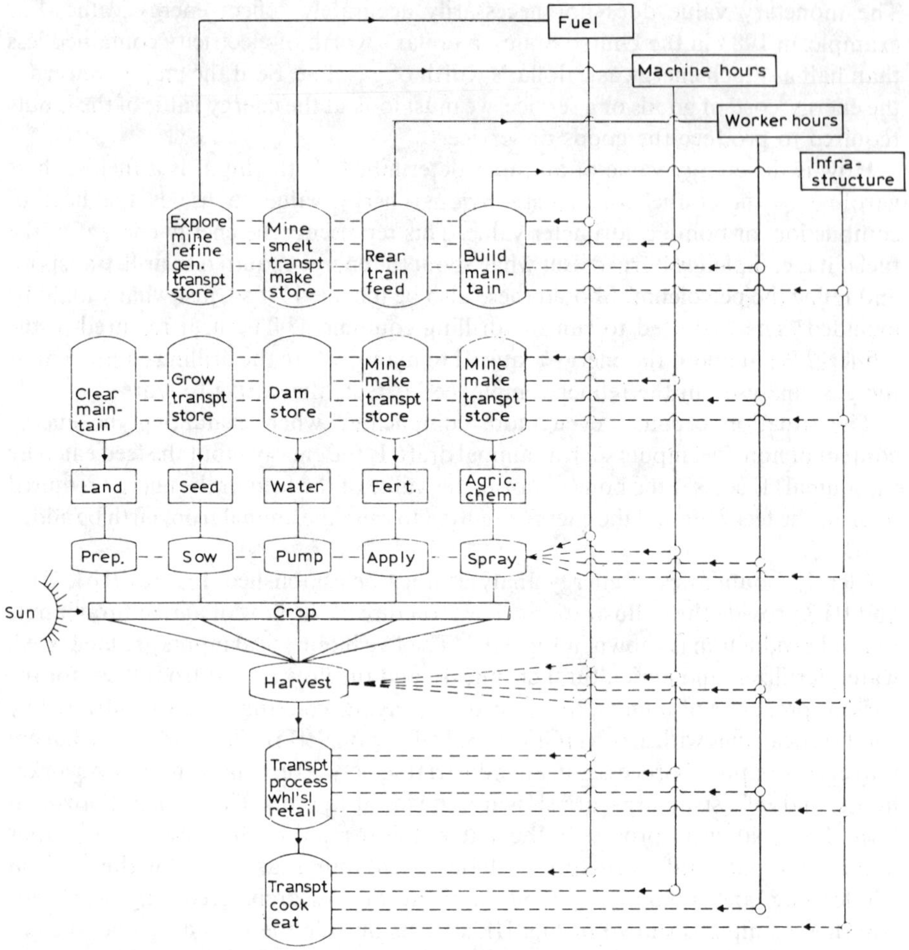

Fig. III.1. Structure of the flow of inputs to a food production/processing system in an industrial society (source: Gifford & Millington, 1975).

Only the energy to the right of the vertical line in Fig. III.3 was used to calculate the energy value of the input in that particular study. This is the fuel, food, or feed used directly by the input. These energy inputs were evaluated in terms of their heats of combustion or bomb calorimeter values. The energy required to mine, drill, and refine the fuel or grow the feed was not included. The further back one goes from the direct energy input, the more difficult it is to evaluate and the more subjective the decisions are on what to include. By assigning an energy value to an input, a basis is established for comparing them. To maintain the integrity of the common units, the inputs need to be evaluated in the same way and at the same place in the chain. But, at the same time, we must keep in mind that different energy forms have different

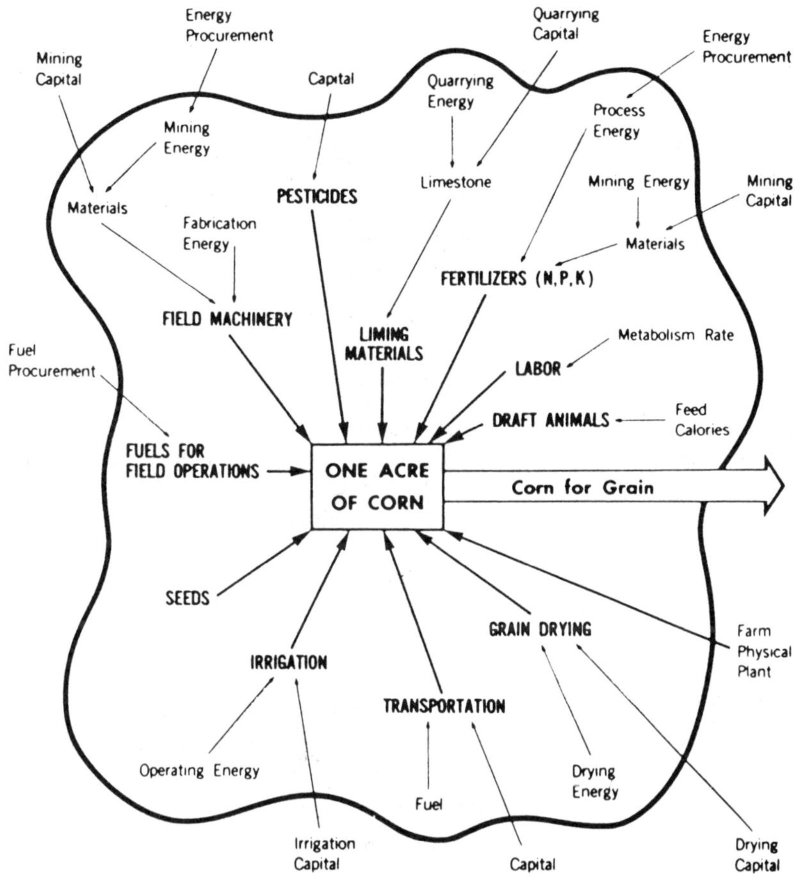

Fig. III.2. Boundaries of inputs included in an energy analysis of United States grain corn (source: Smil *et al.*, 1983)

values to mankind. A calorie of edible food is inherently different from a calorie of oil or fertilizer. These limitations of energy accounting need to be remembered.

Net energy analyses and output/input (O/I) ratios are interesting to examine and have their place in the overall analysis. A word of caution is warranted, however. Research has indicated that primitive agriculture produces a higher O/I ratio than modern agriculture (Heichel, 1973). For example, in one study, a vegetable garden in New Guinea with virtually no commercial energy input produced an O/I ratio of 16, whereas corn, soybeans, or irrigated rice with modern production techniques resulted in energy ratios in the range of 2–5. One might conclude that the primitive system is preferred because of the higher O/I ratio, but the goal of agriculture is not to produce a high energy ratio, but to produce food. Modern agricultural systems with substantial commercial energy input produce higher yields at lower energy ratios.

Direct energy inputs

	Energy input	*Conversion*	*Energy output*
Energy to grow the feed and raise the livestock	Feed	 Animal	Work done for field operations.
Energy to grow food, shelter the workers, etc.	Food	 Human	Work done for field operations.
Energy to drill and refine the oil	Fuel	 Mechanical	Work done for field operations, irrigation, and transportation.

Indirect energy inputs

Energy to drill, mine, and refine the fuels	Natural Gas Fuel Oil Electricity Coal	 Fertilizer	No direct energy output but potential for increased yield.
Energy to drill, mine, and refine the fuels	Natural Gas Fuel Oil Electricity	 Pesticides	No direct energy output but potential for increased yield.

Fig. III.3. Energy-bearing inputs to agriculture are animal draft, human labor, power units, fertilizer, and pesticides. The energy values of the inputs are evaluated at the level of the feed, food or fuel required to operate the direct inputs, or manufacture the indirect inputs (source: Myers, 1983).

REGIONAL AND NATIONAL EXAMPLES

Data in developing countries is fragmentary and incomplete and vastly different methodologies have been employed in various energy studies. Therefore, it is impossible to present a neat, concise, homogeneous global picture of rural energy use. As an alternative, a number of regional and national examples of rural energy analyses are presented in this section.

Latin America (Goldemberg, 1984)

The 'modern' sector of Latin America relies heavily on commercial energy sources such as oil, gas, coal, and hydropower. Rural areas are poor and largely dependent on noncommercial energy sources such as firewood, charcoal, and agricultural residues. Over-dependence on imported oil and inefficient use of locally available noncommerical energy sources are the basic energy problems of most Latin American countries.

The end use structure of energy in Latin America is shown in Table III.1. Agriculture uses only 3·8% of the total energy expended in Latin America; more than half is in the form of oil and gas, more than one-third is biomass, and about 7% is electricity. Bagasse is an important fuel in rural areas in many countries.

The oil producers (Venezuela, Mexico, Ecuador, Trinidad, and Tobago) face a somewhat different set of problems. Energy is only one dimension of the economic problems of Latin America, but is rapidly becoming a crucial one. Table III.2 shows commercial energy consumption in 1980. Figure III.4 shows the share of total exports represented by petroleum imports in Argentina (a moderate petroleum importer), Nicaragua and Brazil (heavy petroleum importers), and the average for all Latin American oil importers. In Brazil, oil imports reached 52% of the value of all exports in 1980. This pattern has led many countries to the brink of bankruptcy.

Nonconventional energy resources in Latin America include solar and wind energy, geothermal energy, and agricultural/urban residues. Because Latin America

TABLE III.1

Energy consumption structure by sector in Latin America in 1980. Values are percentages (source: Goldemberg, 1984)

Sector	*Energy consumption by sector*	*Coal*	*Biomass*	*Oil and gas*	*Other fuels*	*Electricity*
Residential, commercial, and public	28·6	0·21	57·24	28·15		14·40
Transportation	32·1	0·09		99·73		0·18
Agriculture	3·8		36·88	56·23		6·89
Industry	34·0	9·18	13·00	58·18	5·01	14·63
Other	1·5		4·79	90·07	0·35	4·79
Total	100·0					

TABLE III.2
Structure of primary commercial energy consumption in Latin America and the rest of the world in 1980. Noncommercial energy sources are not included (source: Goldemberg, 1984)

Energy type	*Latin America*		*Rest of the world*	
	× 10^6 *tep*	*%*	× 10^6 *tep*	*%*
Petroleum	222·8	65·3	2 778·7	42·4
Natural gas	53·0	15·6	1 253·1	19·1
Coal	16·6	4·9	1 989·9	30·4
Hydroelectricity	47·9	14·0	365·8	5·6
Nuclear	0·8	0·2	164·0	2·5
Total	341·1	100·0	6 551·5	100·0

is primarily in the tropics, it receives large amounts of solar radiation. Local conditions vary throughout Latin America, but the insolation varies from 8 to 16 kcal/cm^2 per month. Wind speeds are quite high in many coastal areas and have the potential to produce up to 10 W/m^2. Southern Chile and Argentina are especially windy with power potentials near 50–100 W/m^2; inland areas, on the other hand, experience winds of about 3 W/m^2. Geothermal energy is already in use in Mexico, Chile, and El Salvador and has good prospects for use throughout the Andean region.

Agriculture and urban residues have been estimated at 139 × 10^6 t of petroleum equivalent (tep) per year, which is about half of all energy consumed in Latin America in 1985.

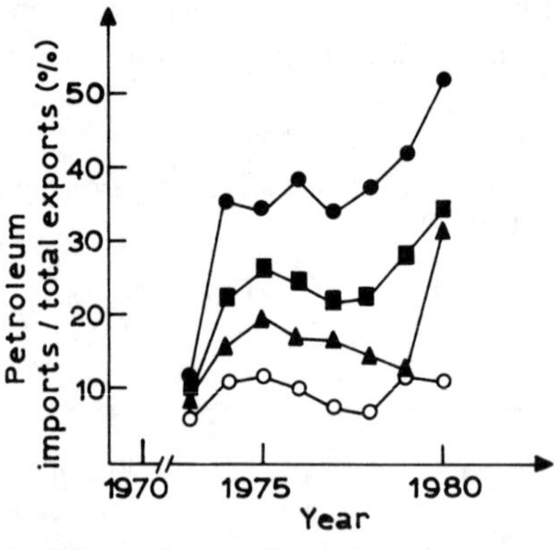

Fig. III.4. Share of total exports represented by petroleum imports, Latin America. ●, Brazil; ▲, Nicaragua; ○, Argentina; ■, average Latin American petroleum importers (source: Goldemberg, 1984).

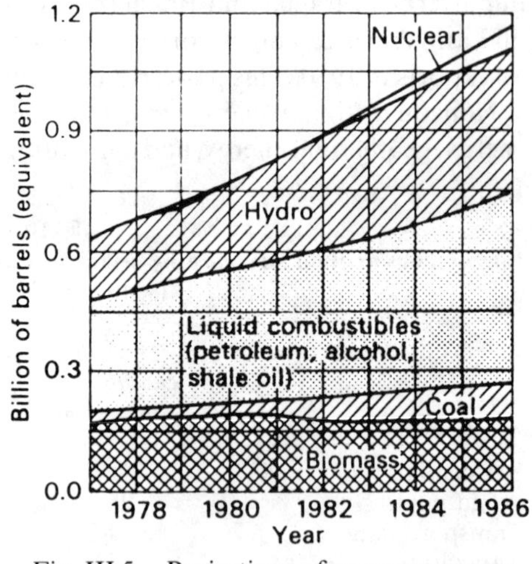

Fig. III.5. Projections of energy consumption of different sources in Brazil during 1977–86. All sources converted to oil equivalent (source: Goldemberg, 1980*a*).

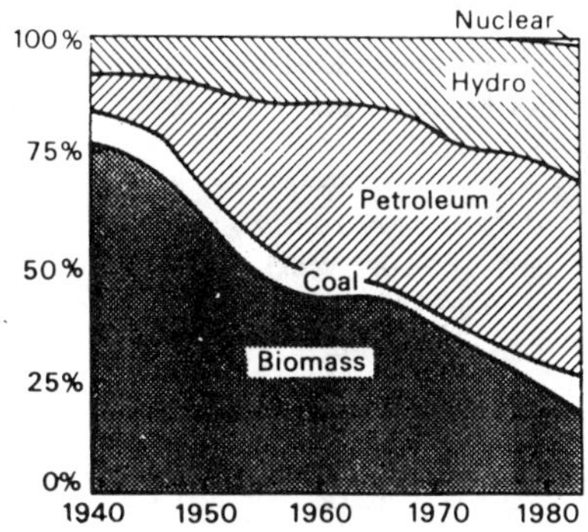

Fig. III.6. The energy consumption profile of Brazil indicating the rapid decline of biomass consumption (source: Goldemberg, 1980*a*).

Brazil (Goldemberg, 1980*a*, 1982)
Lack of abundant fossil fuels, a generous supply of land and forests, a highly developed urban sector, a skewed income distribution, and a mounting external debt are common characteristics of Brazil, as well as many other Latin American countries, and some African and Southeast Asian countries. However, emphasis on biomass as an energy source, mainly through the production of ethanol, is unique to Brazil.

In 1980, Brazil's total energy consumption was 5·1 quads† (compared to 75 quads for the United States). The extremes of energy use in Brazil are broad—the modern sector consumes commercial energy, whereas the traditional sector depends on noncommercial energy, primarily firewood. Projections of energy use in Brazil are provided in Fig. III.5.

The relative importance of biomass as a source of energy decreased dramatically from 1940 to 1975 (Fig. III.6), with a corresponding increase in petroleum consumption. In an effort to limit petroleum imports, while utilizing land resources to produce liquid fuel, the Brazilian alcohol program (PROALCOOL) was developed. By 1981, 4 billion liters of ethanol were produced, 1·9 billion liters of which were used straight as wet alcohol in more than 300 000 automobiles, with the remainder blended with gasoline.

Although the Brazil alcohol program has received the most publicity, other types of biomass fuels are also being developed. An oilseed program was established in 1980 with the original goal of substituting 16% of the diesel oil with vegetable oil by 1985 and 30% by 1990. Methanol produced from biomass is also being explored.

Africa

Ethiopia (Bernardini, 1983)
In the developing countries of East Africa, the relationship between energy and agriculture is intimate and complex. The land provides both energy and food, as well as building materials and other necessary goods.

The exhaustion of fuelwood has resulted in increasing use of agricultural residues

† See Appendix 2.

TABLE III.3

Estimates of consumption of traditional biomass fuel in Ethiopia in 1980 (source: Bernardini, 1983; excludes biomass intake by animals and humans)

	Total (Mt/year)	*Per capita (kg/year)*
Fuelwood	14·4	424
Charcoal (as wood)	1·7	50
Dung	12·7	373
Agricultural residue	7·9	232

and dung with consequent depletion of soil nutrients and organic matter. This, in turn, leads to a decrease in land productivity.

Table III.3 indicates a national average consumption of 424 kg of woodfuel per person per year and 373 kg of dung per person per year. The great diversity in local situations is illustrated in Table III.4. The town of Debre Markos is in an area of eucalyptus, but harvesting is strictly regulated so fuelwood is transported by donkey from some distance away. Chefe is located in one of the few remaining heavily forested areas of the country. Wood and charcoal are transported primarily by humans and horses. Moyale uses only fuelwood and in very low quantities.

A rural/urban household energy survey was conducted which included 88 settlements distributed throughout the entire country of Ethiopia and represented a

TABLE III.4

Inflow of traditional biomass fuel, Ethiopia (source: Bernardini, 1983)

Fuel and transport means	*Debre Markos*		*Chefe*		*Moyale*	
	Total (t/year)	*Per capita (kg/year)*	*Total (t/year)*	*Per capita (kg/year)*	*Total (t/year)*	*Per capita (kg/year)*
Fuelwood	5 660	352	5 138	1 618	792	417
Human load	392		2 974		792	
Mule load	6		485		—	
Donkey load	5 118		12		—	
Horse load	144		1 301		—	
Truckload	—		365		—	
Charcoal	281	18	64	20	—	—
Human load	217		—		—	
Mule load	—		—		—	
Donkey load	47		—		—	
Horse load	18		64		—	
Agriculture residue	1 402	87	9	3	—	—
Human load	1 402		9		—	
Dung	1 240	77	—	—	—	—
Human load	1 240		—		—	

wide diversity of agricultural and ecological situations. Figure III.7 gives an overview of the principal interrelationships.

The major dynamic impulse in the evaluation of local systems is population growth (Fig. III.8). In a closed system, population increases immediately result in an increase in cultivated land. If forest land is in ample supply, population increases do not significantly affect the fuel system, except to slowly increase the distance traveled to collect fuel. If forest land is in short supply, however, an increase in cultivation

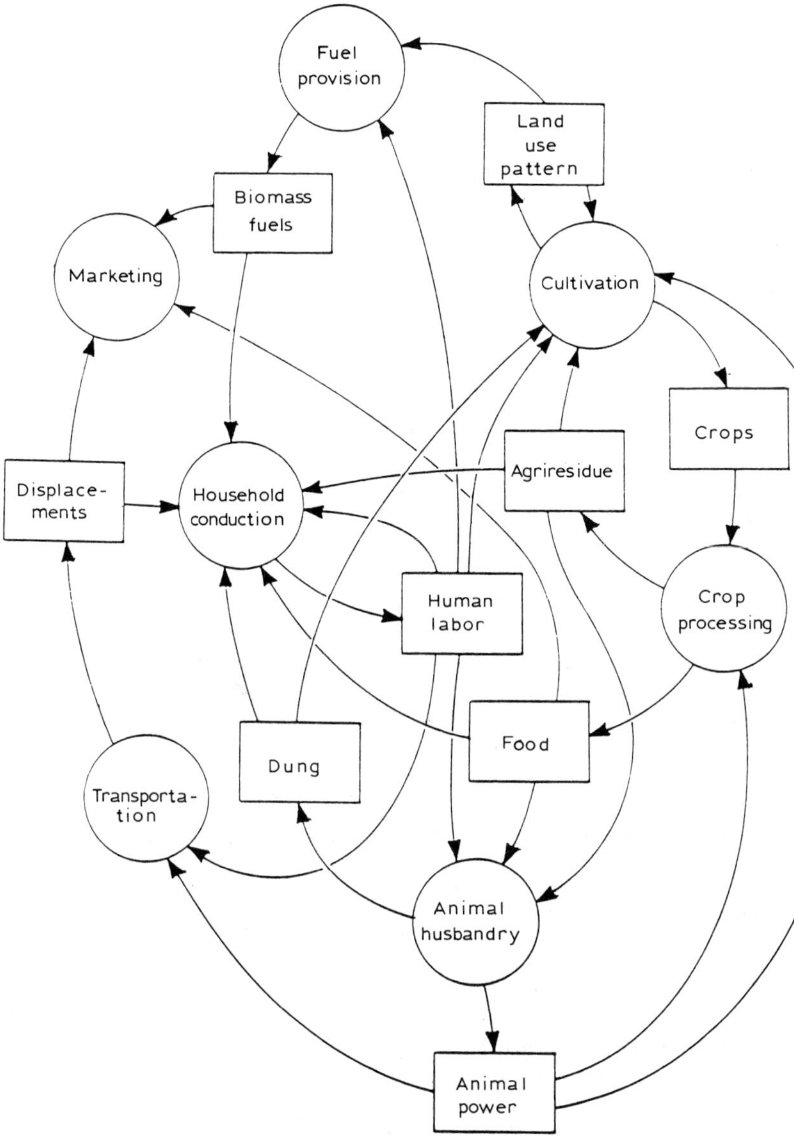

Fig. III.7. Principal interrelations of the food and fuel systems in Ethiopia (source: Bernardini, 1983).

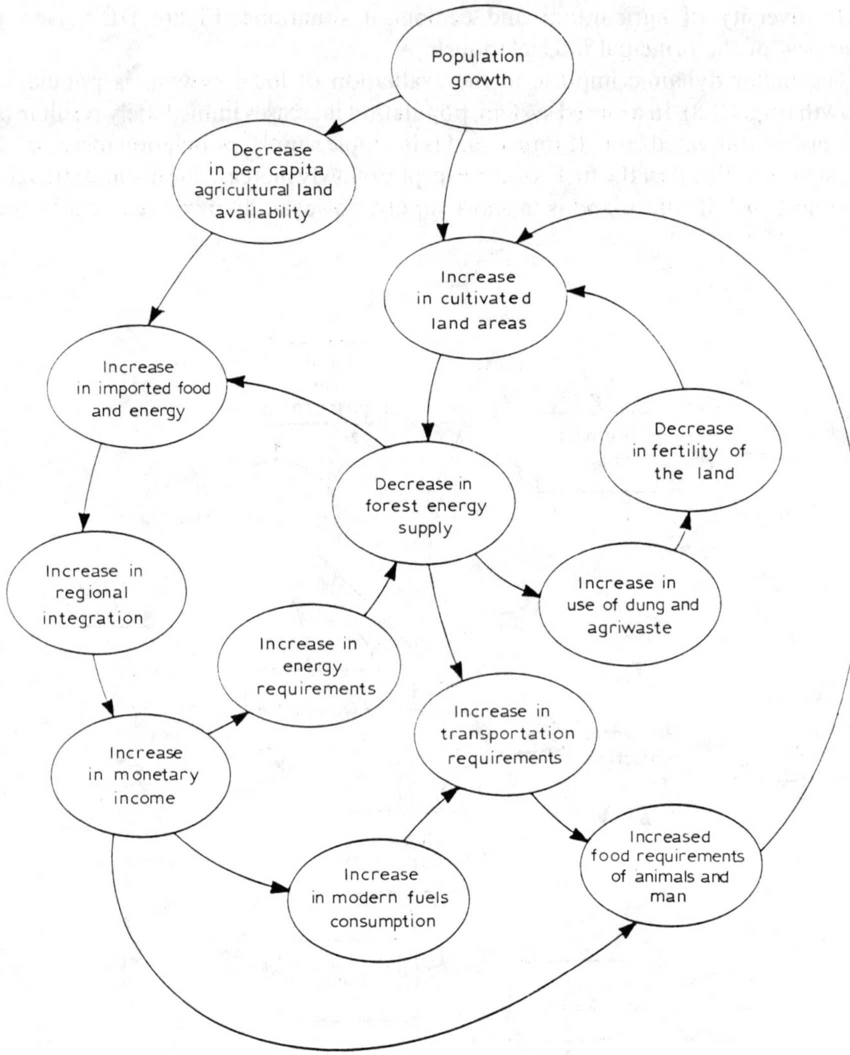

Fig. III.8. Some major dynamic system effects of population growth (source: Bernardini, 1983).

demands on the land can imply drastically increased impacts on both the supply and price of biomass fuel. As transportation requirements for collecting fuel from the receding forest encounter human and animal time constraints, the local system begins to increase the use of alternative biomass fuels from agriculture. Increasing the removal of dung and agricultural residue, however, leads to a decline in soil fertility and water retention capacity, which, in turn, results in the need for more cultivated land.

Table III.5 presents sample results for five Ethiopian rural settlements. The data reflect different responses to local energy needs and agricultural conditions.

TABLE III.5

Characteristics of energy and agriculture in selected rural settlements of Ethiopia (source: Bernardini, 1983)

	Lanch Sam	*Berie Dembie*	*Daka*	*Ruffo Chancho*	*Kerera*
Distance from nearest all-weather road (km)	4	36	2	2	56
Surface area of settlement (ha)	1 091	1 473	816	1 063	1 782
Number of households	177	313	422	619	489
Population density (inhabitants/km^2)	88·4	104	213	292	131
Cultivated area (%)	88	34	50	11	25
Animal density[a] (animals/km^2)	70·4	68·5	86·2	187	88·4
Work input into agriculture (10^3 kcal/ha per year)	525	785	1 074	1 910	930
Human labor	291	400	620	1 910	690
Animal draft power	234	385	454	0	240
Work input into local transport (10^3 kcal/ha per year)	158	213	408	548	295
Human labor	150	212	395	543	265
Animal draft power	8·3	0·8	13·0	5·2	29·8
Food energy[b] productivity of the soil (10^3 kcal/ha per year)	1 005	1 766	2 072	13 813	1 860
Production of agriwaste (10^3 kcal/ha per year)	8 275	13 227	20 471	1 906	12 574
Use of agriwaste (%)					
Fuel	0·0	36·5	52·0	21·1	28·7
Fodder	77·7	21·2	19·5	31·1	47·4
Fertilizer	22·2	33·3	14·1	47·8	12·8
Building	0·1	9·0	14·4		11·1
Production of dung[c] (10^3 kcal/ha per year)	1 334	1 299	1 636	3 069	1 677
Use of dung (%)					
Fuel	95·9	—	—	—	—
Fertilizer	—	76·2	76·4	100	76 1
Building	4·1	23·8	23·6	—	23·9
Local self-sufficiency of food staples	1·38	1·25	1·06	0·96	1·05
Woody biomass production (10^6 kcal/year)	752	7 623	3 660	4 279	12 091
Woody biomass offtake (10^6 kcal/year)	1 251	12 343	12 352	8 489	16 882
Local consumption	1 251	12 343	12 352	7 760	16 882
Exports				729	
Ratio of non-monetary to monetary income	1·46	3·21	1·46	0·85	2·27
Energy (output/input) ratio					
Food/agriculture input	1·91	2·25	1·93	7·23	2·00
Food + agriresidue)/agricultural input	14·2	5·82	5·65	7·54	8·41
(Food + agriresidue)/(agricultural + transportation) input	11·0	4·57	4·09	5·87	6·38

[a] Refers only to cows and oxen.

[b] Defined as nutritive value of human food staples and does not include the animal food value of agriwaste.

[c] Production of dung is referred to the whole settlement surface. In all other cases, the reference unit for production or input refers to cultivated land.

Sahel Village†

The Sahel is a vast expanse of arid land south of the Sahara Desert, stretching across Africa from Senegal to Sudan. The village described in this text is a composite of a typical Sahelian village. Although this description is not an energy survey in a quantitative sense, it conveys an understanding of the low level of energy use and the great extent to which human energy is used for survival in this and other regions of the world. This hypothetical village can be reached by large market trucks which bring in a few luxury items such as batteries, soap, razor blades, cigarettes, sugar, and tea. A produce market is open weekly, at which grain, beef, vegetables, animals, clothing, and handicrafts can be purchased. The market attracts a large number of nomads and farmers from the surrounding bush who may come as far as 25 km on foot, or on a donkey, camel, or horse.

The people in this village perform virtually all of their daily tasks by hand. Drawing water, constructing houses, processing and preparing foodstuffs, and farming are all performed manually. In many respects, life in the village has progressed unchanged for centuries. People travel around the village or out to their fields on foot. The village does not have electricity, although kerosene lamps and flashlights are widely used.

The prospects for expanding the area under cultivation is limited unless farmers are willing to walk considerable distances to new, unclaimed lands that are fragmented from their original holdings.

In general, there is no surface water in the Sahel; therefore, irrigation is limited. Irrigation is accomplished by mechanical pumps along the rivers, but is performed manually in other places, with the aid of watering cans, calabashes, gourds, or weighted, rotating scoops.

Villagers farm exclusively by hand, using rudimentary implements. They clear their fields before the rains by burning off brush and cutting down trees with small hatchets. Planting is performed with either long- or short-handled hoes. Animal traction for pulling plows is virtually nonexistent.

Farmers allow nomads to graze their animals on the stubble of their fields after harvest. Manure that is dropped by the animals is not wholly effective, for much of its value is lost as it lies exposed to the surface. A few farmers know how to use manure as compost or as fertilizer which is put underground before and during planting. Few farmers have access to chemical fertilizers or pesticides because of high cost and inadequate supply and distribution.

A small percentage of the farmers in the village can be classified as progressive. This 10–15% may use fertilizer, pesticides, anti-rodent treatment, or improved seed varieties. Progressive farmers might irrigate a small plot near or within their compound to serve as an additional source of income or as a dietary supplement. In rare instances, progressive farmers might experiment with animal traction or irrigation, assuming that they farm land near a standing body of water or in a region with shallow groundwater supplies.

† Abstracted from a section by John S. Holtzman, Peace Corps volunteer in the Niger, from J. Howe, *et al.* 1977.

Village women spend at least 2 h each day pounding millet grain into edible mash—a laborious physical task. A labor-saving device that has been introduced in the larger towns and a few of the villages is a two-horsepower, diesel-powered mill. Women in these towns with adequate financial resources are able to have their grain milled and processed into flour in a fraction of the time it takes to pound it. After the millet is prepared, it is cooked in a metal pot. This cooking process requires a large amount of firewood.

Drinking water is obtained either from rainwater caught and stored in drums or from subsurface water. Each family in the village spends about 1 h each day drawing and transporting water. In a few villages, hand pumps have been installed. In rare instances, a motor-driven well draws water from the depths of 75–800 m.

Firewood, the main source of fuel in the village, is used for cooking, heating, blacksmithing, and glazing pots. Deforestation is a severe problem in the Sahel. Some governments have enacted laws against gatherers of firewood who chop down trees or pull up living bushes. Loose wood can be collected as long as it is dead and lying on the ground. Enforcement of these regulations is, of course, difficult. Nevertheless, villagers are forced to search for firewood at greater and greater distances from the village because of the increasing demand and diminishing supply.

Because the primary method of transportation in and around the village is by foot, it is not unusual for someone to walk 20–25 km to a neighboring market in one day. When light to moderately heavy loads are transported any distance less than a few miles, villagers carry them on their heads. Animals are used for carrying goods greater distances. In areas that are not too sandy, the bicycle serves as a valuable source of transportation.

Electricity is a luxury in the Sahel. A few meandering rivers and temporary streams are the only sources of hydroelectric power. These flow rapidly during the rainy season, but become dry shortly after the onset of the dry season. Few have been harnessed to produce electricity.

Electricity is available in the larger towns using engine-powered generators. However, petrol-based rural electrification is neither feasible nor economic in most of the Sahel. Costly generators require skilled mechanics for maintenance, in addition to the expense of fuel which is in short supply.

At this point, little demand exists for electric lights. Very few villagers are able to read and those who do use cheap kerosene lamps or candles. Although kerosene lamps burn costly petroleum fuel, they are used universally in the bush.

Senegal (Finnell, 1984*a*)

The US Agency for International Development (USAID) has conducted a number of country assessments on energy, agriculture, and the food system including Morocco, Haiti, Sri Lanka, and Senegal.

Traditionally, the energy and agricultural sectors have been treated as separate and distinct; however, we are becoming increasingly aware that in order to have a more effective impact on agricultural productivity and food supply, these disciplines need to be examined more holistically to determine the linkages between them. In the past, energy assessments/surveys have not included an examination of energy used in production agriculture because production agriculture is a relatively minor consumer of commercial energy. Total energy use is generally understated in developing countries due to reliance on human and animal labor which is not included in the energy accounting system. When energy is examined in the entire food system, its usage increases considerably and can account for as much as one fifth of the total energy supply.

Conversely, energy is often overlooked in projects conducted by agricultural planners, i.e. planners sometimes simply assume that revenues from increased crop production will pay for recurring energy expenditures. Recurring energy costs are often glossed over and may serve as a bottleneck to successful implementation of a project. Often, project planners falsely assume that host governments and project beneficiaries can absorb these costs. Yet, if reliable fuel supplies are not available to operate tractors, or vehicles at research stations, etc., the entire project effort may be in vain. This situation seems unfortunate when fuel costs are relatively quite small in comparison to total project costs. Sometimes the *price* of energy is not the major factor inhibiting its use, but its *availability* soon enough to ensure that the timely operations of planting and harvesting can be performed can make the critical difference between crop failure or success. In some cases, it can also produce a double-cropping system.

Energy is, of course, only one of several critical inputs to agricultural production; thus, extreme care must be exercised in any analysis of the role of energy in agriculture. Neither farmers nor government planners make decisions about farm practices or associated activities on the basis of energy alone, especially given the relatively small (if not generally undervalued) contribution of energy to overall production costs.

However, Senegalese energy and agricultural planners clearly do need to consider a variety of ways to meet future energy requirements based on various senarios of future demand as they set goals for increased production, processing, and marketing of crops for both international and domestic markets. *More energy—not less*—is needed in the process of agricultural development, both for the production of food and the increase in food supply, if rural areas are to undergo a transformation.

The heart of Senegalese agriculture (almost 90% of the cultivated area) is dryland production—groundnuts for cash and millet for food—although a good potential exists for irrigation.

Biomass, mainly from forestry, provides the largest share of Senegal's energy, estimated at 55% in 1977. Fossil fuel resources are limited and no known coal or uranium deposits exist. Electricity is generated from residual fuel oils and hydropower resources are available to be developed. Senegal has abundant solar energy resources and a number of solar applications have been explored.

A Peace Corps Rural Energy Survey reported by Finnell (1984*a*), included 15 villages and a total of 137 families. The 15 villages studied rely primarily on human and animal labor. Only one village uses fuel in farming: it is used for a project which provides diesel tractors and irrigation pumps for tomato farming. Major crops grown include peanuts and millet.

The priority energy needs are to reduce manual labor and provide easier access to cooking fuels. The technologies most often mentioned are water pumps, grinding machines, and gas cook-stoves. The study coordinated needs with specific energy technologies and included animal power, wind, pedal/treadle mechanical devices, solar-powered pumps and grinders, and improved cook-stoves. The energy needs and technologies used in agriculture in these 15 villages are described below.

Seed bed preparation/planting/tillage. Clearing and preparing the land is usually performed by hand. Various designs of hoes and choppers are used in different regions of the country. Most have metal blades and wooden handles, and are used for hand-weeding as well as for clearing and breaking up soil.

Most of the farmers interviewed used horse-drawn machines for planting and some weeding. These machines, usually three-bladed, open a furrow in the ground, drop seeds into it, and close the furrow over the seeds. They can be used to plant both peanuts and millet, the major crops for most farmers in the study, through adjustment of the seed-spreading mechanism. They are also used throughout the growing period to cultivate between the rows of plants. Weeding between the plants is performed by hand, using hoes or other hand tools. The seeders themselves are also used by some farmers to plant beans, corn, sorghum, etc.; rice, cassava, yams, and other crops are planted by hand.

Irrigation. Most of the villages have no irrigation systems available and planting is coordinated with rainfall.

Fertilizers. Most of the farmers surveyed (77·8%) use fertilizer on their fields. Of these farmers, 10% rely on commercial fertilizers only, while slightly less than 25% use 'home-made' or natural fertilizers exclusively. The remaining two-thirds use a combination of the two. The commercial fertilizer is almost always bought through the national peanut marketing co-operative which advances fertilizer and seed peanuts to farmers on credit. The fact that commercial fertilizers are bought on credit means that they are used on cash crops, such as peanuts, in hopes of producing enough to repay the co-operative. Food crops are likely to benefit much less from these fertilizers. The most commonly used natural fertilizer is animal dung, but ashes are used in some areas as well, especially when fields are cleared by burning.

Pesticides. Among the farmers surveyed, almost all use a fungicide on their seed peanuts. Half of the farmers said that they occasionally use pesticides.

Harvesting. Routines are similar from village to village. The entire family usually participates, and other laborers may be hired as well. Peanut plants are harvested using another horse-drawn machine which has a single blade; the blade of the machine plows up a row of peanuts, turning the plants upside down to expose the tangle of roots and nuts to the air. In some cases, oxen are used for this task.

Harvesting of millet is not done by machine. The heads of the grain are cut from the stalks by hand, using a machete or other knife, only after the grain has been drying on the plant for approximately 15 days. Then the heads are left in the field to dry for another 2–4 weeks. Other crops are generally harvested by hand as well. Rice is cut with a small knife and dried in small bundles in the fields before threshing. Beans and corn are picked by hand, and cassava is dug only when needed for food because it stores well in the ground.

Drying and storage. Most of the farmers in the study report some crop loss during drying and storage. Most of this loss occurs while the crops are drying in the field. The farmers or their children try to keep birds and animals away by watching over the fields, erecting scarecrows, etc. In storage, the primary crop loss is from humidity, rats and mice, and insects.

Transportation. Some villages have taxis or small trucks to carry passengers, and some residents also own mopeds or motorcycles. Gasoline is used as a fuel for trucks, cars, and motorcycles. In one of the three communities where villagers fish from boats, a few men have gasoline-powered outboard motors.

However, most of the transportation available to villagers in the survey sites is non-motorized. The survey population relies on horse-drawn carts and oxen to carry goods and people to and from the village. Individuals either walk or ride horses and donkeys; bicycles are rare.

Processing. Diesel fuel was used to power millet grinders in seven villages, while an eighth had a gasoline-powered grinder. Even more than fetching water, the work of grinding/pounding grain is perceived by village women as a difficult task they would like to mechanize. Diesel or gasoline-powered grain grinding machines are available in most cities and towns, as well as in a growing number of villages in Senegal.

In every village without a machine, the women in surveyed households mentioned their desires and/or plans for obtaining one because the manual preparation of grain occupied 2–4 h of their time and caused numerous physical complaints, from blisters to aching backs and exhaustion. Nevertheless, even when a grinding machine is available, the grain must be pounded for some time before it can be processed by a machine.

Food preparation. Cooking fuels include fuelwood, agricultural residues, and dung. In selecting technologies which might prove to be workable as alternatives or as supplements to present energy use in these areas, planners must consider other major village concerns, as well as local resources for energy technologies.

Tunisia†

An energy use study of durum wheat production in Tunisia was conducted during one growing season. Data were collected from 23 small- and medium-sized farms in northern Tunisia (Myers, 1983).

† This case study was implemented jointly by the Tunisian Ministry of National Economy, Tunisian Ministry of Agriculture, and USAID. The author, Claudia Myers, was a consultant on agricultural energy use for two firms under subcontract to USAID; Ferguson-Bryan and Associates, Inc., Washington, DC; and Resource Management Associates, Inc., Madison, Wisconsin.

Tunisia is an Arabic-speaking country in North Africa. It is located on the Mediterranean Sea directly south of Italy. The total land area in Tunisia is 16 million hectares, of which agriculture comprises about 5 million hectares on 355 000 farms. The average farm size is just over 14 ha. Farms with less than 50 ha represent 95% of all farms and 54% of all cultivated land area.

The principal crops are tree crops (1 823 000 ha), cereals (1 198 000 ha), forages (285 000 ha), legumes (143 000 ha), and vegetables (117 000 ha).

The northern region contains the most fertile land in the country and is the only area where production patterns are fairly stable (except for olives in the Central Coast area). Average annual rainfall in the North is 400–600 mm.

Current energy use by yield classification for durum wheat. The case study farms were divided into three yield classifications. The lowest classification included farms with a yield of 600–1000 kg/ha, the middle group had 1010–1500 kg/ha, and the highest yields ranged from 1510 to 2500 kg/ha (Table III.6). The average number of ha of the farms in each of the classifications increased as yield increased, ranging from 10 ha for the lowest yield farms, to almost 15 ha for the middle group, to more than 19 ha for the highest yield category.

The energy inputs per hectare are summarized for all three classifications in Table III.7. Table III.8 summarizes energy use per tonne of production for the three yield categories. Total energy use per hectare increased and total energy use per tonne decreased as yield increased. In other words, the farms with the highest energy input per hectare have the highest production and the lowest input per tonne of production.

Overall energy use per hectare increased as yield increased. Human labor and animal draft both decreased as yield increased. Power unit use per hectare was lowest for the lowest yield farms but practically the same for the middle and highest yield farms. Note that the use of ammonium nitrate, phosphate fertilizer, and herbicides per hectare all increased as yield increased. Basically, then, as yield increased the use of modern inputs also increased. Figures III.9 and III.10 clearly depict these trends.

Current production and energy use at the national level. Table III.9 provides data on cereal production in 1981. For each type of cereal in each region, the table lists the

TABLE III.6
Yield classifications: Average size, production, and number of farms in Tunisian survey (source: Myers, 1983)

Classifications	*Average hectares*	*Average production (t/ha)*	*Number of farms*
Lowest yield: 0·60–1·00 t/ha	10·1	0·90	8
Middle yield: 1·01–1·50 t/ha	14·7	1·32	9
Highest yield: 1·51–2·50 t/ha	19·2	1·74	6

TABLE III.7

Summary of energy inputs per hectare for yield classifications, Tunisian farm energy survey (source: Myers, 1983)

	Lowest yield		*Middle yield*		*Highest yield*	
	Acutal units/ha	*10 000 kJ/ha*	*Actual units/ha*	*10 000 kJ/ha*	*Actual units/ha*	*10 000 kJ/ha*
Human labor (h)	102	14·2	81	10·6	53	7·1
Animal draft (days)	5	64·8	1	16·7	1	6·6
Power units (h)	5	139·9	11	319·5	11	312·7
Animal transport (km)[a]	9	3·3	13	5·0	6	2·4
Vehicle transport (km)[a]	15	12·4	12	11·4	7	7·4
Ammonium nitrate (kg)	104	203·4	109	212·7	139	271·3
Phosphate fertilizer (kg)	89	30·3	89	33·2	100	34·2
Manure (kg)	0	0·0	242	8·1	0	0·0
Herbicide (liters)	0·2	1·7	0·8	5·8	1·0	7·6
Total		470·1		622·9		649·4
95% Confidence interval		330–610		514–732		532–767

[a] Round trip, kilometers.

number of hectares planted, number of hectares harvested, yields, and total production. Durum wheat production in the North was almost 730 000 t in 1981. This is 91% of the country's total durum wheat production. It was harvested on 495 000 ha, so the average yield was 1·47 t/ha. (This study only considers durum wheat in the northern region.)

For cereal production in the North, about 56% of the land is farmed in a 'modern' fashion, i.e. use of tractors, with combines, fertilizers, and pesticides. This

TABLE III.8

Summary of energy inputs per tonne of production for yield classifications, Tunisian farm energy survey (source: Myers, 1983)

	Lowest yield		*Middle yield*		*Highest yield*	
	Actual units/t	*10 000 kJ/t*	*Actual units/t*	*10 000 kJ/t*	*Actual units/t*	*10 000 kJ/t*
Human labor (h)	113	15·7	61	8·0	30	4·0
Animal draft (days)	6	71·7	1	12·5	—	3·8
Power units (h)	5	154·9	9	241·5	7	179·3
Animal transport (km)[a]	10	3·6	10	3·7	4	1·4
Vehicle transport (km)[a]	17	13·7	9	8·6	4	4·3
Ammonium nitrate (kg)	116	225·3	82	160·8	78	155·5
Phosphate fertilizer (kg)	99	33·5	67	25·1	57	19·7
Manure (kg)	0	0·0	183	6·1	0	0·0
Herbicide (liters)	0·3	1·9	0·6	4·4	0·6	4·4
Total		520·5		470·7		372·3
95% Confidence interval		255–786		300–642		212–532

[a] Round trip, kilometers.

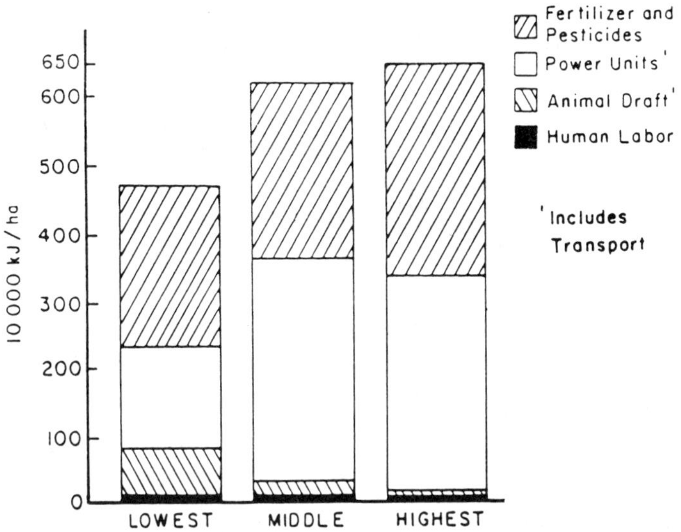

Fig. III.9. Energy use per hectare for three yield classifications (source: Myers, 1983).

would correspond with the highest yield category given in the previous section. The study also estimates that another 33% of the land is mechanized, but fertilizer and pesticide use is less than on the more 'modern' farms. This area would correspond to the middle yield group. The third category used little or no power units, but relied heavily on animal draft. These farms occupied about 11% of the cereal land area and correspond to the lowest yield group (El Amani & Bahri, 1980).

When applying these percentages of land area to the yield categories developed in this study, an average yield for the northern region is estimated to be 1·51 t/ha. In reality, the average yield for the region was 1·47 t/ha (Table III.9), a difference of 2·7%.

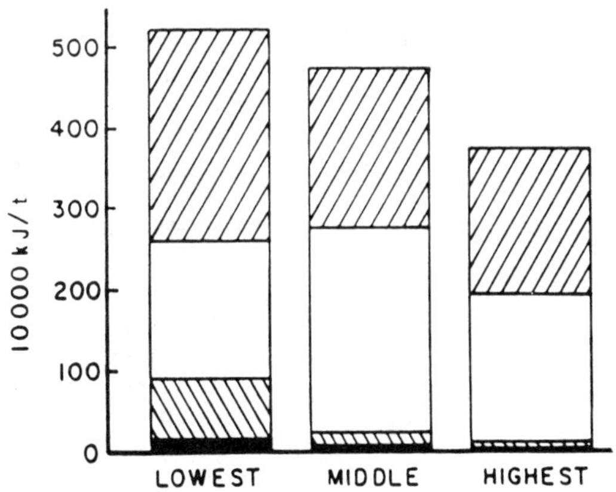

Fig. III.10. Energy use per tonne for three yield classifications, Tunisia; key as for Fig. III.9 (source: Myers, 1983).

TABLE III.9
Tunisian cereal production data for 1981 (source: Myers, 1983)

Region	*Cereal type*	*Land planted (1 000 ha)*	*Land harvested (1 000 ha)*	*Yield (t/ha)*	*Total production (1 000 t)*
North	Durum wheat	503	495	1·47	729·6
	Soft what	78	77	1·99	153·0
	Barley	174	170	1·10	186·2
Central and south	Durum wheat	318	200	0·37	74·8
	Soft wheat	16	11	0·55	6·0
	Barley	356	273	0·31	83·7
Total country	Durum wheat	821	695	1·16	804·4
	Soft wheat	94	88	1·81	159·0
	Barley	530	443	0·61	269·9

Using the total land planted with durum wheat in the North (503 000 ha), the number of hectares in each of the three categories can be calculated as follows: highest yield, 281 680 ha; middle yield, 165 990 ha; and lowest yield, 55 330 ha. Using the case study farm yields, the estimate of total production is 759 027 t.

Table III.10 details the results of extrapolating the survey farm data to include all durum wheat production in the northern region. Energy use totals about 75 000 tep. Almost 48% of it is used for fuel in tractors and combines; 39% is represented by ammonium nitrate.

Even though human labor is a small percentage of total energy use, it still represents a significant labor force—more than 4·25 million worker-days.

TABLE III.10
An estimate of energy use for durum wheat production in the northern region of Tunisia with current levels of mechanization and input use, 1981–2 (source: Myers, 1983)

	Actual units	*Total actual units*	*Energy (10 000 kJ)*	*Energy*
Human labor	8-h days	4 252 000	4 545 000	1·5
Animal draft	days	724 000	8 217 505	2·6
Power units	h	5 201 000 (35 577 tep)	148 856 000	47·7
Animal transport	km	4 346 000	1 689 000	0·5
Vehicle transport	km	4 794 000 (1,114 tep)	4 663 000	1·5
Ammonium nitrate	t	63 000	122 980 000	39·4
Phosphate fertilizer	t	48 000	16 821 000	5·3
Manure	t	40 000	1 345 000	0·4
Herbicide	liters	426 000	3 198 000	1·0
Total			312 312 000 (74 644 tep)	100

Projected production and energy use at the national level. Tunisia, like most developing countries, has a goal of increasing agricultural production and recognizes that additional energy will be required. This section estimates the amount of energy that will be required to meet production goals.

If each of the yield categories was moved up one level, i.e. the lowest yield farms were moved into the middle yield category, and the middle group was moved into the highest yield category, then the number of hectares in each group would be: highest yield, 447 670 ha; and middle yield, 55 330 ha. Thus, the extrapolated production is 851 981 t. This represents an average yield of 1·69 t/ha.

An estimate of total energy use for durum wheat production in the North is 77 700 tep as presented in Table III. 11. The estimate was formulated by using the number of hectares in each yield category as given in the preceding paragraph and the energy use per hectare for each category as noted in Table III.7.

Comparison between current and projected situations. Table III.12 shows the change in input use and mechanization between the current situation and the projected situation given the government's plans to increase production by modernizing small- and medium-sized farms. Production increases by 12% while total energy use increases by only 4%. This is accomplished with an increase in the use of power units, ammonium nitrate, phosphate fertilizer, and herbicides. The traditional, noncommercial inputs, such as human labor, animal draft and transport, and manure, decrease in usage.

Vehicle transport also decreases, but note that the percentage change in energy use is not as great as the percentage change in kilometers driven. This indicates that there was a switch from smaller, less fuel-intensive vehicles in the current situation to larger vehicles with higher fuel consumption in the projected situation. With the use

TABLE III.11

An estimate of energy use for durum wheat production in the northern region of Tunisia with projected levels of mechanization and input use (source: Myers, 1983)

	Actual units	*Total actual units*	*Energy (10 000 kJ)*	*Energy (%)*
Human labor	days	3 526 000	3 765 000	1·2
Animal Draft	days	503 000	3 879 000	1·2
Power units	h	5 533 000 (37 683 tep)	157 664 000	48·5
Animal transport	km	3 405 000	1 351 000	0·4
Vehicle transport	km	3 798 000 (943 tep)	3 944 000	1·2
Ammonium nitrate	t	68 000	133 222 000	41·0
Phosphate fertilizer	t	50 000	17 147 000	5·3
Manure	t	13 000	448 000	0·1
Herbicide	liters	492 000	3 723 000	1·1
Total			325 143 000 (77 711 tep)	100

TABLE III.12

Differences in energy use for durum wheat production in the northern region of Tunisia between projected and current levels of mechanization and input use (source: Myers, 1983)

	Actual units	*Projected versus current*			
		Change in actual units	*Change in actual units from current situation (%)*	*Change in energy use (10 000 kJ)*	*Change in energy use from current situation (%)*
Human labor	days	−726 000	−17·1	−780 000	−17·2
Animal draft	days	−221 000	−30·6	−4 338 000	−52·8
Power units	h	332 000 (2 105 tep)	6·4	8 809 000	5·9
Animal transport	km	−941 000	−21·6	−338 000	−20·0
Vehicle transport	km	−996 000 (−172 tep)	−20·8	−719 000	−15·4
Ammonium nitrate	t	5 000	8·3	10 242 000	8·3
Phosphate fertilizer	t	2 000	3·8	326 000	1·9
Manure	t	—27 000	−66·7	−896 000	−66·7
Herbicide	liters	66 000	15·6	526 000	16·4
Total				12 831 000 (3,067 tep)	4·1
		Change in actual units	*Change (%)*		
Land area		0	0		
Production		93 000 t	12·2		

of larger trucks, more grain could be handled at one time, thus resulting in fewer trips. The lower yielding farms generally drove longer distances to town; that is, they were more isolated than the higher yielding farms. Therefore, the model has simply moved the lower yield farms into the middle yield category. But, in reality, the lower yield farms have not changed their location and would still have to go just as far to market or distribution centers. So the percentage decrease in kilometers driven would probably not be as great in reality as it appears in Table III.12.

The decrease in human labor of 17% is significant. The government wants to provide increased employment opportunities in rural areas because, in many developing countries, rural people have flocked to the cities to seek employment and a higher standard of living. This migration has often created overpopulated cities where the infrastructure has not grown as fast as the population. The result is high unemployment, poor housing, and inadequate systems such as sewage and water. The Tunisian government is attempting to fight this trend in their own country, but the capital city of Tunis already has about one-sixth of the population of the entire country. It is possible, however, that increased farm profits resulting from higher production could keep people on the farms even though there are still fewer hours of field work to do.

The use of animal draft dropped by a substantial 30%. This decrease is not surprising since farmers tend to switch readily to tractors if they have the opportunity. Unfortunately, the use of manure decreased dramatically also. Commercial fertilizers are easier to use than manure and can more efficiently provide nutrients, but manure provides soil structure improvement characteristics that commercial fertilizers cannot. The largest percentage increase in input use was 16% for herbicides. This increase is reasonable considering the almost non-existent use of herbicides by the lowest yielding farms. The use of ammonium nitrate increased by about 8% while the use of power units increased about 6%.

Tunisia is an oil-producing country but does not have vast reserves. Officials are concerned that by the end of the 1980s, Tunisia will be a net oil importer. Hence, they wish to promote conservation in all sectors of the economy. Therefore, on the one hand, Tunisian residents have a desire to increase wheat production, at least partially through the use of imported oil, and, on the other hand, they want to conserve petroleum use in agriculture. What is the trade-off between the cost of imported oil versus the savings on imported grain?

The projected increase in production, if each farm was able to move up to the next highest yield classification, was almost 93 000 t (Table III.12). In 1980, 280 960 t of durum wheat were imported by Tunisia. This increased production is 33% of Tunisia's total durum wheat imports, a very substantial quantity. The value of the increased production, using 110 dinars/t [1982 price of imported wheat, Tunisian dinar = US $2·48 (1980 exchange rate)], is more than 10 000 000 dinars (Table III.13).

As was shown in Table III.12, this increased production was accomplished with an increase in the use of diesel fuel for power units, ammonium nitrate, phosphate fertilizer, and herbicides. All but the phosphate fertilizers are imported. The total

TABLE III.13

Value of increased production versus cost of increased input use for durum wheat production between projected and current levels of mechanization and input use, Tunisian farm energy survey (source: Myers, 1983)

	Quantity	*Unit price*	*Total value (dinars)*
Increased value of production	92 954 t	110 D/t	10 224 940
Increased cost of production			
Diesel fuel	2 105 tep	181 D/t (0·155 D/liter)	381 653
Ammonium nitrate	5 256 t	72 D/t	378 432
Phosphate fertilizer			
Super 45	347 t	53 D/t	18 391
Super 16	1 479 t	32 D/t	47 328
Herbicide	66 396 liters	1 D/t	66 396
Total increased cost			892 200
Increased cost as % of increased value			8·7%

value of these added inputs totals 892 000 dinars, which is not quite 9% of the value of the increased production.

In other words, this model shows that, theoretically, an increase in production worth over 10 million dinars can be attained by adding 892 000 dinars worth of inputs into durum wheat production.

Conclusions. The following conclusions can be reached from the Tunisian survey:

—Greater energy use per hectare corresponded to lower energy use per tonne of production for durum wheat because of increased yields.
—At the national level, a 12% increase in wheat production would result from a 4% increase in energy use if farms in each yield classification were moved up one level.
—Production increases could be accomplished with an increase in the use of power units, ammonium nitrate, phosphate fertilizer, and herbicides. The use of the traditional, noncommercial inputs, i.e. human labor, animal draft and transport, and manure would decrease.
—The total value of added production would be worth more than 10 million dinars while the total cost of the added inputs would be only 892 000 dinars. The increased cost is less than 10% of the increased value. Obviously, the cost of imported oil and other inputs is more than offset by the savings on imported grain.
—Policy-makers should be aware of the changes in farm labor caused by modernization and watch to see that farm income increases and rural employment opportunities rise.

Asia

Bangladesh (Hughart, 1979; Manibog, 1982)
The first nationwide estimate of traditional energy consumption in Bangladesh was made in 1973 (Table III.14). The per capita use of traditional energy was 0·07 t of oil equivalent (toe), a very low level of consumption (Burma and Nepal, 0·18 toe; and Indian Plains, 0·23 toe). More recent studies (Table III.15) show an increase in traditional energy use to 0·11 toe per person. Table III.16 puts agricultural energy use into perspective by including commercial energy as well as traditional energy use.

A shortage of animals for draft power is a major constraint to land preparation. Efforts to increase the number of draft animals are limited by a scarcity of good quality feed and fodder for buffalos and cattle (Manibog, 1982). Another energy problem is the low efficiency of the traditional chula (cooking stove). Improved cooking stoves have the potential to stretch fuelwood supplies from 25 to 50% (Manibog, 1982).

An example of the intricacy of the fuel supply and usage patterns in a nearly closed agricultural system is illustrated in Fig. III.11. Food and fuel balances for the village of 422 people are shown in the lower left corner. The village economy is approximately balanced at a subsistence level.

TABLE III.14
Estimated consumption of traditional energy sources in Bangladesh, 1973 and 1974 (source: Manibog, 1982)

Fuel type	*Amount used (10^6 Mt)*	*Conversion factor (kcal/kg[a])*	*Total energy (10^{12} kcal)*	*Percentage of total traditional energy*
Cow dung[b]	6·1	2 791	17·0	25
Rice husks	4·1	3 015	12·2	24
Rice straw	3·0	3 015	9·1	18
Small prunings, twigs and leaves	1·5	3 089	4·7	9
Jute sticks	1·0	3 015	3·0	6
Bagasse	1·5	1 787	2·8	5
Bulk fuelwood	0·5	3 089	1·6	4
Other wastes	1·5	3 015	4·6	9
Totals	19·2		55·0	100

[a] Factors account for energy expended in driving off excess moisture in the combustion process, and are lower than those used by BES.
[b] Dry weight.

Dry season rice production using pump groundwater in northwestern Bangladesh. In northwestern Bangladesh, conditions are favorable for agriculture and there is adequate drainage. Water tables range from 1 to 7 m below the surface. Rainfall in the area is quite seasonal. A number of pumping projects have been instituted. Some use deep bored wells equipped with electric or diesel driven pumps. Pumps are also used on surface water channels. This farming area grows dry season rice using high yielding varieties and substantial fertilizer inputs. Details of diesel fuel and labor requirements are provided in Table III.17.

In the cases examined, the average fuel and electrical energy input per hectare was the equivalent of 262 liters of diesel fuel (Table III.18). The new production activities

TABLE III.15
Comparative estimates of traditional energy consumption in Bangladesh, per capita/year (source: Manibog, 1982)

Year	*Fuelwood portion (m^3)*	*Total traditional energy use (toe)*	*Fuelwood as a percentage of total traditional energy use*
1975	0·44	0·07	100
1976	0·06	0·07	14
1978	—	0·13	—
1979	0·27	0·12	29
1979	—	0·17	—
1980	0·10	0·11	18

TABLE III.16

Rural energy sources and end-uses in Bangladesh, 1974 and 1975, in million toe (source: Manibog, 1982). (In brackets: % share of each source under each end-use category.)

	Agriculture	*Food processing*	*Cooking*	*Lighting*	*Pottery bricks, etc.*	*Transport*	*Total energy use*	*Percentage contribution of each source to total energy use*
Traditional source								
Human labor	0·93 (21)	0·12 (10)	0·49 (6)	—	0·01 (8)	0·14 (25)	1·69	12
Animal labor	2·13 (48)	—	—	—	—	0·41 (75)	2·54	17
Homestead wood and forest scraps	—	0·01 (1)	0·32 (4)	—	—	—	0·33	2
Cow dung	0·95 (21)	—	2·23 (28)	—	—	—	3·18	22
Crop residues	—	1·06 (86)	4·52 (58)	—	—	—	5·58	38
Subtotal	4·01 (90)	1·19 (97)	7·56 (96)	—	0·01 (8)	0·55	13·32	91
Commercial sources								
Cut fuelwood	—	0·01 (1)	0·24 (3)	—	0·05 (38)	—	0·30	2
Kerosene or diesel	0·03 (1)	0·02 (2)	0·05 (1)	0·49	0·07 (54)	—	0·66	4
Chemical fertilizer & pesticides	0·41 (9)	—	—	—	—	—	0·41	3
Subtotal	0·44 (10)	0·03 (3)	0·29 (4)	0·49	0·12 (92)	—	1·37	9
Total energy use	4·45	1·22	7·85	0·49	0·13	0·55	14·69	
(with human and animal labor excluded)	1·39	1·10	7·36	0·49	0·12	—	10·46	
Percentage claim of each end use to total energy consumption	30	8	53	3	1	4		100
(with human and animal labor excluded)	13	11	70	5	1			100

Note: at 70·7 million total population assumed for 1974 and 1975; apart from human and animal labor; in toe; per capita traditional energy consumption = 0·13; per capita commercial energy consumption = 0·02; total rural energy consumption per capita = 0·15.

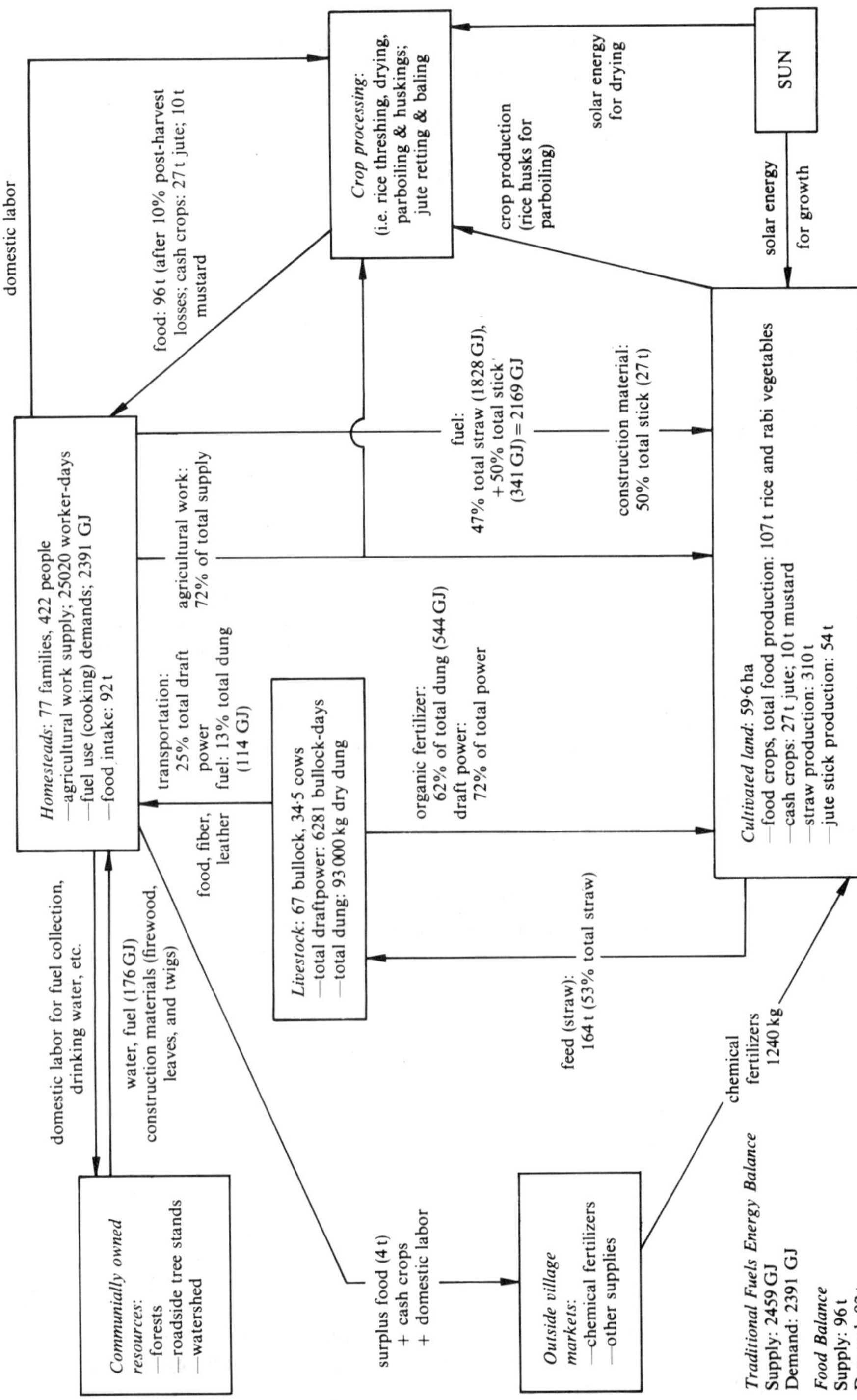

Fig. III.11. Aggregate annual village energy-related resource flows: Dhanishwar, Bangladesh (source: Hughart, 1979)

TABLE III.17

Quantitative parameters for dry season rice production in Bangladesh (source: Chancellor, 1978)

Operation	*Energy use (equivalent liters of diesel fuel/ha)*[a]	*Labor use (worker-days/ha)*
Pumping	103	7·4
Tillage		44·5
Seeding/transplanting		37·0
Weeding		30·9
Harvesting and threshing		31·6
Fuel hauling		0·5
Fertilizer	129	
Transporting paddy[b]		26·4
Parboiling paddy[b]		25·2
Milling	35	3·5
Totals	267	207·0
Household kerosene use[c]	36	

[a] The food energy equivalent of the crop produced (based on 4 401 kg/ha—a milling yield of 63·5%, and a food energy content of white rice of 3 621 kcal/kg is 1 086 liters/ha. The value of the crop (at 41·0 Taka per 82-lb maund and 15·25 Taka/$ US) is $317·42/ha.

[b] Computed on the basis (as reported by rice millers) that 25% of the rice received is parboiled at home and carried to the mill manually, while 75% is carried by bullock cart and parboiled at the mill in large batches. Rice husk is the fuel usually used for parboiling.

[c] Based on half the annual household use (since two crops are produced) and an average of 1·15 ha operated per household.

TABLE III.18

Data summary (source: Chancellor, 1978)

Site	*Energy use (equivalent liters of diesel fuel/ha)*	*Labor use (worker-days/ha)*	*Crop value ($/ha)*	*Energy ratio*[a] *(crop energy) to (input primary energy equivalent)*
Thailand	183	145	516·08	5·50
Philippines	310	94	420·22	1·65
Sri Lanka	286	109	512·13	2·72
Bangladesh	267	207	317·42	2·69
Average	262	139	441·46	3·14

[a] Computed by dividing energy content of the crop by [(energy use per unit land area in diesel fuel units) × 1·208 + energy equivalent of depreciation and repair + energy content of food for the workers]. The term 1·208 is the ratio between the primary energy that must flow from resource sites in order to produce one unit of refined petroleum, and the energy available from that one unit of refined petroleum.

requiring this modest energy input generated additional employment beyond traditional levels in the amount of 139 worker-days/ha and an average crop product value of $441/ha. The ratio of the crop caloric value to the primary energy required for fuel and electrical energy ranged from less than 2 to more than 5 (Table III.18). The total commercial energy input was 259 liters of diesel fuel equivalent per hectare in the form of fertilizer, fuel, and electrical input. It is interesting to note that this amount is about one-tenth the commercial energy input per crop area used in California (Table III. 19).

The agricultural production process is one in which various resources are combined to increase the photosynthetic output. The primary resource is the land area onto which a relatively constant amount of solar energy falls each day. High levels of productivity depend on using the land a large proportion of the time and on growing crops that are particularly efficient in solar energy use. Supporting resources such as water and nutrients from nature, as well as human-controlled operations for crop culture, harvest, and conservation, do not automatically flow to the production site in the proper amounts and on the proper schedule to obtain high productivity.

By using special technologies that require fuel and electrical energy inputs, these supporting resources can be made more adaptable or mobile so they can flow on different schedules, resulting in higher yields. The proportions in which production inputs of land, energy, water, etc., are combined to obtain a given amount of food energy are presented in Table III.20 for the average of the Asian sites examined and for California agriculture.

China (Smil & Knowland, 1980; Taylor, 1982)
The world's most populous nation, China, is also the fourth largest producer of primary energy and its third largest consumer (Smil & Knowland, 1980). China's

TABLE III.19
Input classification for modernized production achieved (Chancellor, 1978)

Item	*Labor*	*Input value*	*Commercial energy*[a]			
	(Worker-hours/ha)	*($/ha)*	*Asia*[b]		*California*	
			liters/ha	*%*	*liters/ha*	*%*
Fertilizer	—	32	108	42	322	13
On-farm operations	114·8	126	54	21	1 159	47
Transport	11·4	32	33	12	183	7
Processing	13·3	52	64	25	804	33
Total	139·5	242	259	100	2 468	100

[a] Energy equivalent to that contained in the specified number of volume units of diesel fuel per unit crop area.
[b] Average of values from the crops produced using modern technology in the four Asian sites studied.

TABLE III.20
Mix of production inputs to produce one unit of food energy (source: Chancellor, 1978)

Input	*Average of Asian sites studied*	*California*
Energy[a]—gallons (liters)	0·37 (1·40)	1·20 (4·54)
Labor—worker-days	0·56	0·05
Land—acre-days (ha-days)	1·96 (0·79)	0·96 (0·39)
Water—acre-ft (m^3)	0·045 (55)	0·013 (16)
Other purchased inputs—dollars	0·40	1·18
Processed crop food energy produced[a]—gallons (liters)	1·0 (3·78)	1·0 (3·78)

[a] Equivalent energy to that in a given volume of diesel fuel.

energy system is an intricate mixture of traditional and modern, of advanced technologies and inefficient practices, and of large and small enterprises. The reliance on small-scale energy technologies in the rural areas is particularly notable. Nearly half of China's huge coal output now originates in small local mines, more than 87 000 small hydropower stations generate one-third of the country's hydroelectricity, and biogas production is spreading rapidly, especially in the warm areas of the South and Southwest.

TABLE III.21
General structure of energy consumption in China, 1978 (source: Jin *et al.*, 1983)

Energy source	*Amount*	*Energy equivalent* ($\times 10^{14}$ *kJ*)	*Proportion* (%)
Commercial			
Coal[a]	$5{\cdot}295 \times 10^8$ t	110·8	67·6
Crude oil[b]	$1{\cdot}04 \times 10^8$ t	40·1	24·4
Electricity[c]	$2{\cdot}57 \times 10^{11}$ kWh	9·25	5·6
Natural gas[d]	$1{\cdot}04 \times 10^{10}$ m^3	4·02	2·4
Total		164·17	100·0
Noncommercial			
Crop residues[e]	$3{\cdot}89 \times 10^8$ t	71·64	93·1
Firewood	$0{\cdot}28 \times 10^8$ t	5·27	6·9
Total		76·91	100·0

[a] Average energy value of coal in China is 20 928 kJ/kg. With $0{\cdot}885 \times 10^8$ t used for generating electricity excluded.
[b] Energy value of crude oil is calculated based on 38 507 kJ/kg.
[c] 3 600 kJ/kWh.
[d] The output of natural gas energy is estimated as 10% that of crude oil.
[e] The output of crop residues is estimated based on grain/residue ratio = 1/1·5, 15% moisutre and 18 416 kJ/kg dry matter.

TABLE III.22
Commercial energy consumption in China's agriculture, 1978 (source: Jin *et al.*, 1983)

	Item	Amount	Energy equivalent ($\times 10^{14}$ kJ)	Proportion (%)
(1)	Chemical fertilizer[a]			
	Nitrogen[b]	$7{\cdot}03 \times 10^6$ t	5·148	
	Phosphorous[c]	$1{\cdot}46 \times 10^6$ t	0·038	
	Subtotal	$8{\cdot}48 \times 10^6$ t	5·186	19·4
(2)	Other farm chemicals[d]	$0{\cdot}533 \times 10^6$ t	0·536	1·9
(3)	Farm machinery[e]			
	Steels	$2{\cdot}89 \times 10^6$ t	8·674	30·8
	Pig Irons	$1{\cdot}26 \times 10^6$ t		
(4)	Fuel oil			
	Diesel oil	$8{\cdot}73 \times 10^6$ t	3·721	
	Gasoline	$0{\cdot}66 \times 10^6$ t	0·289	
	Lubri. oil	$0{\cdot}42 \times 10^6$ t	0·176	
	Kerosene	$0{\cdot}866 \times 10^6$ t	0·385	
	Subtotal	$10{\cdot}676 \times 10^6$ t	4·571	16·2
(5)	Electricity	$2{\cdot}53 \times 10^{10}$ kWh	0·912	3·2
(6)	Coal	$39{\cdot}60 \times 10^6$ t	8·287	29·5
Total			28·166	100·0

[a] Only the energy used in producing chemical fertilizers included. A small amount of potassium fertilizer was not included.
[b] An estimate of $7{\cdot}324 \times 10^7$ kJ was spent on the average for synthesizing each tonne of ammonia nitrogen in China.
[c] 31% was compound (or mixture of compounds) containing phosphorus, calcium, and magnesium, of which $8{\cdot}371 \times 10^6$ kJ was estimated to be spent in producing each tonne. The rest was calcium phosphate (primary), which needs much less energy to produce and was not taken into account in the energy expenditure.
[d] The energy expenditure for producing each tonne of farm chemicals was estimated as $10{\cdot}05 \times 10^7$ kJ according to FAO (1976).
[e] The average energy expenditure for producing, including shipment, each tonne of farm machinery was estimated to be about $20{\cdot}9 \times 10^7$ kJ.

Table III.21 shows the pattern of energy consumption for the entire country. Note that noncommercial energy comprised about one-third the total energy use in 1978. When traditional fuel consumption is included, energy consumption in rural households and agricultural production represents almost one-half of China's total energy consumption. Total rural household energy consumption for cooking and heating was about 370 million tonnes of coal equivalent (mtce) in 1978 or 14·2 GJ per rural citizen. Traditional fuel accounted for 85% of the energy consumed by rural households, while coal and some biogas made up the remaining 15%. Table III.22 provides a breakdown of commercial energy used in China's agriculture. When compared with Table III. 21, it shows that about 17% of the commercial energy is used in agriculture, including small industries in rural areas. Most of the farmer's daily energy use, however, comes from noncommercial sources, mainly crop residues.

The average commercial energy input in Chinese agriculture is about $1{\cdot}2 \times 10^7$ KJ/ha. As a primary target to modernize China's agriculture, an additional commercial energy input of $8{\cdot}7 \times 10^{14}$ KJ/ha is needed.

Two specific examples, the Tang-Jia-Ling brigade of Tong-Bei-Wang commune and the Do-Dian production brigade of Do-Dian commune are described in the following text. Both of the communes are located on the outskirts of Beijing. All cropland in these units is irrigated and the levels of mechanization and fertilizer input are relatively high compared with other parts of China. A double cropping system is used. Tong-Bei-Wang commune has 1700 ha of farmland and Tang-Jia-Ling brigade has 60 ha. The Do-Dian brigade has 349 ha. The distribution of commercial energy of these two units is shown in Tables II.23, III.24, and III.25.

Of the commercial energy directly used in crop production, chemical fertilizers comprise more than half, with machinery and fuel ranking second (Tables III.26 and III.27). All harvested grains are sun-dried. Except for plowing and threshing, all operations rely on human and animal power.

In recent years, Chinese efforts to develop biogas, small-scale hydroelectric plants, and other decentralized renewable energy systems have attracted significant interest. During the 1980s, China continues aggressive promotion of decentralized renewable energy technologies (Taylor, 1982).

Biogas. China is by far the leading country in the world in the popularization of biogas technology. Biogas has been introduced in more than one-half of China's provinces, municipalities, and autonomous regions, and in more than 1000 of the nation's 2100 counties. However, about 70% of China's biogas digestors are located in Sichuan Province, primarily in the western part of the Sichuan Plain. Climate is an

TABLE III.23

Total commercial energy consumption in Tong-Bei-Wang commune, 1978 (source: Jin *et al.*, 1983)

Item	*Amount*	*Energy equivalent* ($\times 10^9$ *kJ*)	*Proportion* (%)
Tractors[a]	10·2 t	5·36	
Farm implements[a]	15·3 t		4·0
Trucks[a]	14·0 t	2·93	
Fuel oil			
Diesel	326·5 t	13·81	13·4
Gasoline	315·7 t	13·81	
Chemical fertilizer	249·0 t	18·25	8·8
Other farm chemicals	19·9 t	2·01	1·0
Electricity[b]	$4{\cdot}845 \times 10^6$ kWh	17·45	8·4
Coal[c]	6 355·8 t	133·10	64·4
Total		206·72	100·0

[a] Ten years' depreciation is considered.
[b] Including $0{\cdot}21 \times 10^6$ kWh spent in farmers' daily use.
[c] Including those spent in farmers daily use.

TABLE III.24

Total commercial energy consumption in Tang-Jia-Ling brigade of Tong-Bei-Wang commune, 1978 (source: Jin *et al.*, 1983)

Item	*Amount*	*Energy equivalent* ($\times 10^9$ *kJ*)	*Proportion* (%)
Chemical fertilizer	11·90 t	0·871	19·8
Other farm chemicals	0·70 t	0·071	1·6
Electricity	38 600 kWh	0·139	3·2
Farm machinery[a]	0·90 t	0·188	4·3
Fuel oil	8·01 t	0·341	7·7
Coal[b]	133·20 t	2·788	63·4
Total		4·398	100·0

[a] Shared from the total amount of the commune based on the cultivated land area, with 10 years' depreciation taken into account.
[b] All the coal in this brigade was spent on farmers' daily use.

important factor that influences biogas development because of the temperature requirements for producing the gas. If above-ground temperatures drop below freezing, producing significant amounts of biogas is difficult using simple Chinese methods. Thus, as one moves north to colder temperatures, it is more difficult to produce biogas, and biogas production tends to be more of a seasonal operation. Biogas has been promoted in areas as far north as Manchuria, but conditions in the South are more advantageous for year-round production of the gas.

Biogas development is usually emphasized in areas where the provision of adequate supplies of household fuels has been particularly difficult. The Sichuan Plain is a good example. The area is densely populated, reaching 600 persons/km^2 in some rural districts, and coal. charcoal, and firewood are generally not locally produced.

TABLE III.25

Total commercial energy consumption in Do-Dian brigade of Do-Dian commune, 1979 (source: Jin *et al.*, 1983)

Item	*Amount*	*Energy equivalent* ($\times 10^9$ *kJ*)	*Proportion* (%)
Chemical fertilizer	170·42 t	12·473	17·5
Other farm chemicals	8·15 t	0·820	1·2
Electricity[a]	5·581 × 10^5 kWh	2·009	2·8
Coal[b]	2410 t	50·645	71·2
Farm machinery	7·79 t	1·632	2·3
Fuel oil[a]	84·0 t	71·158	100·0

[a] Including 3·337 × 10^5 kWh spent on sideline industry and farmers' daily use.
[b] Mainly for sideline industry and farmers' daily use.
[c] About half of the fuel oil was used for off-farm transportation.

TABLE III.26
Commercial energy consumption per hectare in grain production in Tang-Jia-Ling brigade of Ton-Bei-Wang commune, 1978 (source: Jin *et al.*, 1983)

Item	*Amount*	*Energy equivalent ($\times 10^6$ kJ)*	*Proportion (%)*
Chemical fertilizer[a]	146·4 kg	14·440	54·3
Other farm chemicals	11·7 kg	1·174	4·4
Farm machinery	15·0 kg	3·139	11·8
Fuel oil	133·5 kg	5·682	21·4
Electricity	570 kWh	2·160	8·1
Total		26·595	100·0

[a] Excluding small portion of vegetable land which consumes a substantially high level of chemical fertilizer.

The greatest number of China's biogas digestors are small, simple digestors, designed for the needs of a single household. Family-sized digestors generally range between 6 and 12 m^3 in volume, depending on the size of the family and climate conditions. In Sichuan, experience shows that each cubic meter of digestor volume produces 0·15–0·2 m^3 of gas per day during summer, and 0·1–0·15 m^3 of gas/day during winter. Because 1 m^3 of gas/day is usually considered sufficient to meet the cooking and lighting requirements of a family of five, a 10 m^3 digestor should be adequate for a family of five, if properly managed.

While the primary emphasis in China's biogas program is on the construction of family-sized digestors, larger, community-operated digestors have also been built. These digestors are usually 50–200 m^3 in volume. If larger capacities are desired, several large digestors are usually built together and interconnected. Designs vary, but when the gas is used for operating machinery or generating electricity, many systems employ plastic balloons to store the gas outside of the digestors.

TABLE III.27
Commercial energy consumption per hectare in grain production in Do-Dian commune, 1979 (source: Jin *et al.*, 1983)

Item	*Amount*	*Energy equivalent ($\times 10^6$ kJ)*	*Proportion (%)*
Chemical fertilizer[a]	358·5 kg	41·436	75·0
Other farm chemicals	23·3 kg	1·318	2·4
Farm machinery	22·4 kg	4·677	8·5
Fuel oil	120·0 kg	5·110	9·2
Electricity	887 kWh	2·681	4·9
Total		55·222	100·0

[a] Excluding small portion of vegetable land which consumes a substantially high level of chemical fertilizer.

In rural areas, community digestors are collectively owned and operated. They are usually built in conjunction with collective piggeries, but sometimes they are constructed at institutions such as schools, factories, or hospitals. In urban areas, large biogas systems are usually built in conjunction with night soil treatment and disposal schemes.

Currently, all available evidence points to continued, aggressive promotion of biogas development in China. Given the existence of rural traditions in the collection and fermentation of organic wastes and the fact that biogas production presents an effective way to use those wastes for both fertilizer and fuel, the Chinese have singled out biogas development as an especially good method of alleviating rural energy problems (Taylor, 1982).

Small-scale hydroelectric power plants. China's verified hydropower resources total 680 gigawatts (GW), the largest of any country in the world. The survey concludes that about 370 GW can be utilized. In the exploitation of their large hydropower resources, the Chinese have emphasized the policy of walking on two legs—both large or medium-sized hydropower stations and small hydropower stations have been constructed.

Small hydroelectric power stations have been defined as stations with single generating units rated at less than 6000 kW or total station capacities (including several generating units) of less than 12 000 kW. However, the vast majority of China's small hydropower plants are much smaller than this definition implies.

By the end of 1980, more than 90 000 small hydropower plants had been built in rural China, with a total installed capacity of 7·1 GW, up from 6·33 GW in 1979. Although most plants are located in the southern and southwestern parts of the country, where hydropower resources are greatest, at least a few can be found in nearly three-fourths of China's counties.

China plans to continue small hydropower plant construction in the future as one important way to expand rural electrification, alleviate rural power shortages, and provide more electricity for rural industry, agriculture, and rural homes. Tremendous potential exists for further development (Taylor, 1982).

Solar energy. Most areas of China have sufficient climatic conditions for the direct utilization of solar energy, but the most suitable climates for solar development are in the western areas of the country, which are sparsely populated. Insolation rates are highest in Tibet and western Qinghai, where the amount of solar radiation is between 8·0 and 8·4 GJ/m^2 per year. In southwest Xinjiang, eastern Qinghai, northern Ningxia, northern Gansu, and the western part of Inner Mongolia, annual solar radiation ranges between 6·3 and 8·0 GJ/m^2. In the remaining parts of China, annual solar radiation is less than 6·3 GJ/m^2, reaching the lowest levels in Sichuan and the Guizhou Provinces, where annual solar radiation is between 3·8 and 4·2 GJ/m^2 in most areas.

Compared to the development of forestry, biogas, and small hydropower stations, the development of solar energy has received relatively little attention in China. A variety of different solar cookers have been developed, including both box and focusing types, and a few models have been put into mass production. Several thousand cookers are reportedly in use. Recent reports state that more than 70 000 m^2 of solar collectors are currently used for heating water in China.

A limited amount of work has been conducted on solar refrigeration, air conditioning, and ice production. So far, only ammonia absorption cooling systems have been tested. In addition, little attention has been given to the development of solar irrigation pumps in China (Taylor, 1982).

A considerable amount of research, however, has been conducted on solar equipment for drying grain, other agricultural products, and wood, but few systems appear to have been popularized. Because grain drying is a seasonal operation, the Chinese have emphasized systems in which the solar collector and the grain drying equipment are separate, so that the collectors can be used for other purposes when grain drying is not necessary. One multi-purpose installation constructed in Daxing County, near Beijing, provides greenhouse facilities, grain dryers, and hot water for a shower room.

Wind energy. Wind energy has been used in China since ancient times for tasks such as pumping water, threshing grain, and powering boats. Wind-powered electric generators are still primarily in the research stage. Nevertheless, more than 200 small systems, with capacities ranging between 100 and 250 W, have already been installed on the plains of Inner Mongolia, where wind speeds exceed 3 m/s on more than 200 days/year. These winds have been used for lighting, projecting movies, and electrifying corral fences.

A few larger wind-powered electric generators have also been constructed in China. An 18-kW installation has been completed on Sijiao Island, off China's eastern coast. The three blades on the machine were made from old helicopter blades. A 20-kW wind generator has also been built near Harbin, in Manchuria, and a 40-kW installation was constructed in Zhejiang Province (Taylor, 1982).

Pakistan (Jameel, 1982)

Many modeling studies have been conducted to predict the impact of energy supply and price on future food production. One such study, conducted by IIASA, explored the agricultural energy requirements in Pakistan in the year 2000 with different levels of mechanization and fertilizer application.

Traditional agriculture relies heavily on animal energy, i.e. bullocks or horses to till the land and to draw water, manual labor for seeding and harvesting, dung for fertilizer, etc. The productivity of this mode of agriculture, in terms of yield per hectare per year, is proving insufficient to support the growing world population. Fortunately, the introduction of high-yielding varieties has made it possible to enhance land productivity substantially, but at a cost—heavy applications of chemical fertilizers, pumps to draw underground water, and mechanization to permit multi-cropping. All these measures require, directly or indirectly, energy from oil, gas, or electricity.

Two schools of thought exist regarding mechanization in developing countries: one for and one against. The 'pro' school maintains that farm machines: (a) help in the rapid preparation of land for multi-cropping; (b) enable a higher quality of tillage operation to kill weeds and improve soil structure; (c) reduce the number of animals competing for fodder; (d) improve water availability; and (e) release farm labor to other, more productive employment.

The 'con' group, on the other hand, disputes the validity or positive aspects of most of these claims and emphasizes the negative consequences, such as fewer animals would also result in less organic matter in the soil. While the controversy is unlikely to be resolved in the near future, it is an accepted principle that what people do over a long period of time must be to their advantage and must essentially be profitable to them. Therefore, since the trend in Pakistan is toward tractorization, it must be advantageous to farmers and the process is likely to continue. Figure III.12 indicates an interesting correlation between productivity and the number of tractors per hectare, although mechanization is often accompanied by other modern inputs and cannot be given full credit for yield increases.

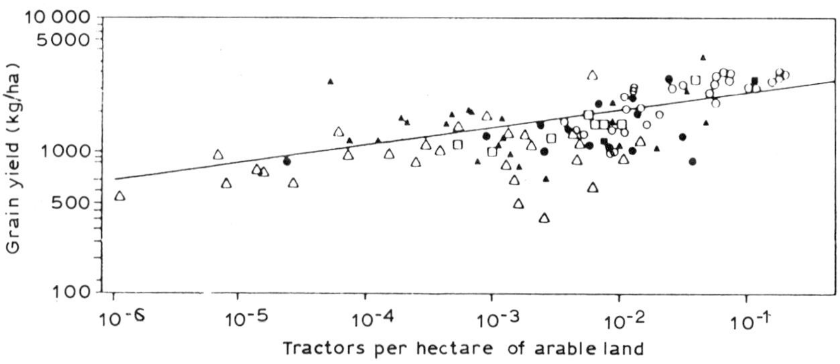

Fig. III.12. Yield of cereals per hectare versus tractor use per hectare of arable land (1970), using 5-year averages of 107 countries. △, Africa; ▲, Asia; ○, Europe; ●, North and Central America; □, South America; ■, Oceania (source: Jameel, 1982).

Table III.28 tabulates the average human and animal effort required for different farm operations. Table III.29 summarizes the energy requirements for farming in the year 2000 under three different mechanization scenarios. Future A (high mechanization) assumes approximately 75% of the area under major crops will be mechanized for both plowing and harvesting and 50% of the remaining crop area will be tilled by machine. Future B (low mechanization) assumes 25% of the area under major crops and 20% of the remainder will be mechanized. Future C (low mechanization and low tillage) assumes the same degree of mechanization as Future B, but with 33% of the nonmechanized area cultivated with low-tillage techniques to reduce soil erosion. Additional pesticides are assumed to accompany reduced tillage. As expected, Future B with a low level of mechanization will require the least energy and Future C does not appear feasible.

Two additional scenarios are related to fertilizer input. Future D assumes the entire fertilizer needs will be met by chemical fertilizers, whereas Future E assumes

TABLE III.28
Estimated average effort required per hectare for different farm operations (source: Jameel, 1982)

Operation	*Number of men*	*Pairs of bullocks*	*Hours worked*	*Total energy expenditure (10^3 kcal)*[a]
Plowing, planting, leveling	1	1	22	19·5
Sowing of wheat	1	1	5	4·4
Harvesting of wheat	4	—	12	1·2
Threshing of wheat	3	3	15	40
Winnowing of wheat	4	—	15	1·5
Sowing of rice	4	—	20	2·0
Harvesting of rice	5	—	20	2·0
Threshing of rice	4	—	12	1·2
Hoeing of sugar cane	4	—	20	2·0
Sowing of sugar cane	9	1	15	16
Cutting and stripping of sugar cane	25	—	32	20
Sowing of cotton	1	1	5	4·4
Interculture of cotton	4	—	20	2·0
Picking of cotton	5	—	10	1·0
Fertilizer spreading	1	—	3	0·15
Manure hauling and loading	2	1	5	4·6
Manure spreading	2	—	5	0·25

[a] Based on 25 kcal h^{-1} for humans and 430 kcal h^{-1} for bullocks.

TABLE III.29
Final energy demand for farm work in Pakistan, excluding irrigation (source: Jameel, 1982)

	Cropped area (10^6 ha)		*Final energy required for tilling/harvesting (10^{12} kcal)*			
	1977	*2000*	*1977*	*Future A*	*Future B*	*Future C*
Wheat	6·4	7·6		4·6	1·5	1·5
Rice	1·8	2·4		0·9	0·3	0·3
Sugar cane	0·8	0·95		0·3	0·1	0·1
Cotton	1·8	2·8		1·1	0·4	0·4
Others (tilling only)	6·5	8·2		1·6	0·7	0·7
Total	18·3	22	0·85 ($0·08 \times 10^6$ toe)	8·5 ($0·8 \times 10^6$ toe)	3·0 ($0·28 \times 10^6$ toe)	3·0
Additional pesticides[a]						1·4
Total (Future C)						4·4 ($0·41 \times 10^6$ toe)

[a] Calculated @ 10 kg of pesticide for each hectare under low-tillage techniques and 24 000 kcal/kg for pesticides.

most of the nutrients will be supplied by manure/crop residues with minimal use of chemical fertilizers. The resulting energy needs for fertilizers are:

Future D: $3{\cdot}4 \times 10^6$ toe/year
Future E: $2{\cdot}0 \times 10^6$ toe/year

Table III.30 presents several combinations that may characterize Pakistan's agriculture in the year 2000. Table III.31 presents the corresponding energy use. The largest amounts of commercial energy are expected to be used in fertilizer production, drainage and irrigation, tractors, seeds, and pesticides, respectively.

Philippines (Waddle, 1984)
In 1973, 95% of all energy consumed in the Philippines was imported. Thus, high oil prices in the 1970s were a serious blow to the Philippine economy. A national effort was undertaken to reduce oil imports. By 1983, imported oil was reduced to 64% of the nation's energy use (Fig. III.13), while various noncommercial sources comprised nearly 15%.

The effort to further reduce imported oil in the Philippines continues. Figure III.14 depicts about 35% imported oil in the energy mix by 1987, with corresponding increases in local and nonconventional sources.

Double-cropping of rice in the Philippines (Chancellor, 1978). Tractor hire services are used in Central Luzon for tillage. Large threshing machines are used for threshing the rice crop. Surface water is used for irrigating the dry season rice crop. Farmers used high yielding varieties, fertilizer, and tractors for tillage. A mix of

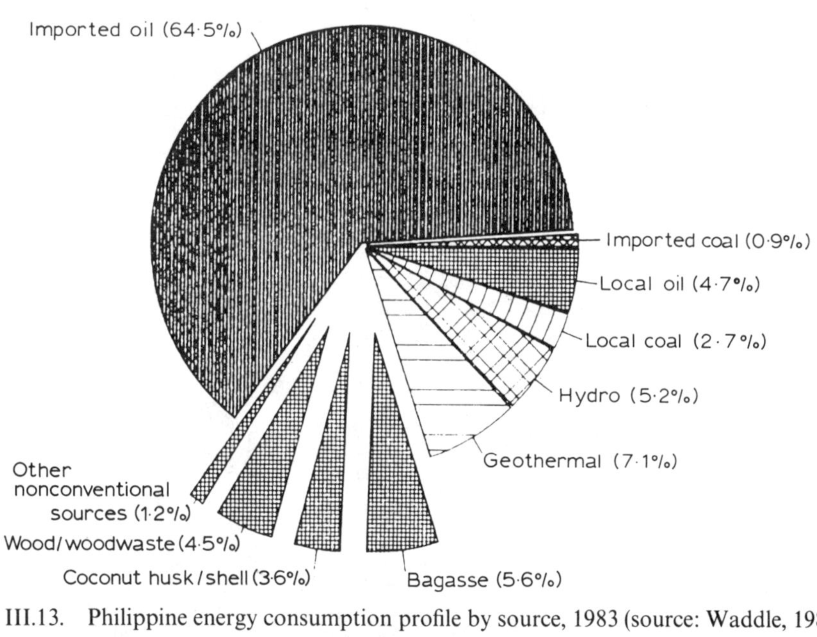

Fig. III.13. Philippine energy consumption profile by source, 1983 (source: Waddle, 1984).

TABLE III.30
Characteristics of possible evolution paths of Pakistan's agricultural system (source: Jameel, 1982)

	Parameter	*Base year, 1977*	*Year 2000*			
			Future AD	*Future AE*	*Future BD*	*Future BE*
(1)	Area cropped per year	$18{\cdot}3 \times 10^6$ ha	22×10^6 ha			
(2)	Irrigation (the per hectare figures pertain to irrigated area only)	$10{\cdot}8 \times 10^6$ ha-m per year of water available at farmgate ($\sim 0{\cdot}8$ ha-m/ha per year)	Farmgate availability of water increases to 18×10^6 ha-m per year ($\sim 1{\cdot}1$ ha-m/ha per year)			
(3)	Farm mechanization (the per hectare figures refer to entire cultivated area)	60 000 tractors ($3{\cdot}3 \times 10^{-3}$ tractors/ha)	3/4 of area under major crops mechanized; 1/2 of other area under tractors (20×10^{-3} tractors/ha)		1/4 of area under major crops mechanized; 1/5 of other area under tractors (7×10^{-3} tractors/ha)	
(4)	Supply of nitrogen from chemical fertilizers (per hectare figures pertain to total cropped area)	$0{\cdot}6 \times 10^6$ t nutrient (32·5 kg/ha per year)	$1{\cdot}9 \times 10^6$ t (86 kg/ha per year)	$1{\cdot}1 \times 10^6$ t (50 kg/ha per year)	$1{\cdot}9 \times 10^6$ t (86 kg/ha per year)	$1{\cdot}1 \times 10^6$ t (50 kg/ha per year)
(5)	P:N ratio	1:4	1:2	1:2	1:2	1:2

TABLE III.31
Evolution of commercial energy use in Pakistani agriculture (source: Jameel, 1982)

	Base year, 1977	*Year 2000*			
		Future AD (high mech, high urea)	*Future AE (high mech, low urea)*	*Future BD (low mech, high urea)*	*Future BE (low mech, low urea)*
(1) Drainage tube-wells					
electricity (MWyear)	118	250	250	250	250
(2) Irrigation tube-wells					
electricity (MWyear)	62	200	200	200	200
diesel (10^3 toe)	196	400	400	400	400
(3) Farm machines					
diesel (10^3 toe)	80	800	800	280	280
(4) Chemical fertilizers					
mainly gas (10^3 toe)	650	3 400	2 000	3 400	2 000
(5) Seeds and pesticides (10^3 toe)	300[a]	800	800	800	800
(6) Other requirements (10^3 toe)[b]	75	300	250	250	200
(7) Total non-animal energy					
electricity (MW year)	180	450	450	450	450
gas and oil (10^6 toe)	1·0	5·3	3·9	4·7	3·3
latent (10^6 toe)[c]	0·3	0·4	0·4	0·4	0·4
(8) Commercial energy use[d] (10^6 kcal/ha)	~1·0	3·2	2·5	3·0	2·3
(9) Persons supported per hectare	4·0	6·3	6·3	6·3	6·3

[a] Seeds only; pesticides estimate included in 'other requirements'.
[b] Other requirements include transportation, crop drying, fishing, etc.; estimated figures.
[c] Energy latent in seeds.
[d] Electricity converted to thermal equivalent at 35% efficiency.
MWyr = megawatt-years.

four-wheel tractors and Japanese-style power tillers was used. Transportation of the crop to buyers was accomplished by numerous modes such as water, buffalo cart, motorcycle-rickshaws, trailers pulled by tractors, jeeps, and large trucks.

Rains are common during the harvest operation. Consequently, a sizable portion of the crop is delivered to buyers with drying facilities in conjunction with storage structures and rice mills. Paddy and milled rice are handled manually in sacks.

The use of tractors and threshers was reponsible for a reduction in labor required for the specific tasks of tillage and threshing. High yielding varieties and favorable prices tended to justify increased manpower use in other phases of rice production, resulting in a net increase in labor requirement per crop area. Details of diesel fuel and labor requirements are provided in Table III.32.

Sri Lanka Rice Production with Collected Run-Off Water (Chancellor, 1978)
Surface run-off reservoirs have been constructed in an area in North Central Sri Lanka where rainfall is quite seasonal. Two crops of rice are grown each year. Powered pumps are used to supply water to the upland areas above the reservoir

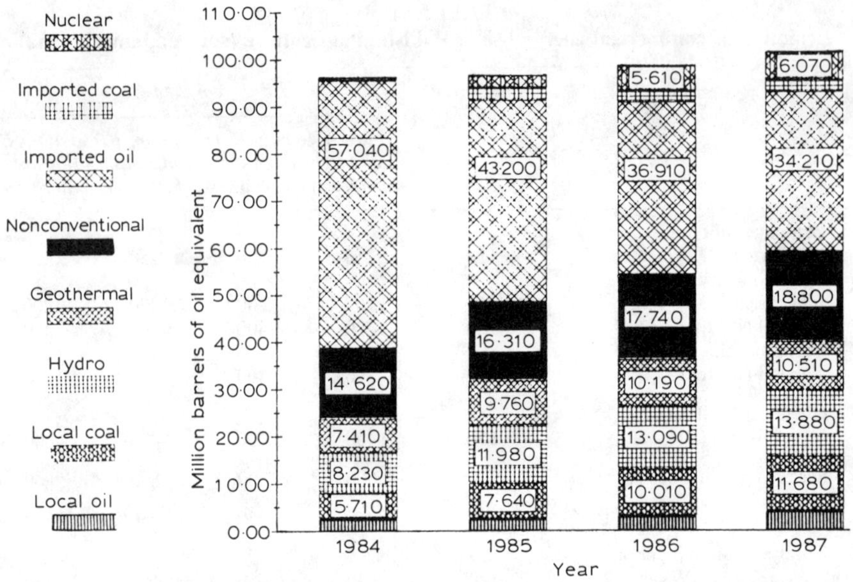

Fig. III.14. Energy consumption in the Philippines by source (source: Waddle, 1984).

where vegetables and upland crops are grown. The number of water buffalo in each area was inadequate for timely tillage of the cultivated land. Thus, 26 kW four-wheel tractors and two-wheel power tillers are now used. These machines are available on a hire basis. In addition to tillage, both two-wheel and four-wheel tractors are used for transportation, threshing, and powering winnowing fans.

Small-scale local rice mills are generally diesel engine driven, while some larger mills use electric power. Household use constituted the major on-farm application

TABLE III.32
Quantitative parameters for dry season rice production in the Philippines (source: Chancellor, 1978)

Operation	*Energy use (equivalent liters/ha of diesel fuel)*	*Labor use (worker-days per ha)*
Tillage	21·04	10·69
Fertilizer	162·02	
Cultural		51·40
Harvesting and threshing	6·17	19·49
Transport	3·27	4·00
Drying	67·31	2·00
Milling	30·85	4·99
Market transport	19·73	1·51
Total	310·39	94·08

TABLE III.33
Quantitative parameters for rice production in Sri Lanka (source: Chancellor, 1978)

Operation	*Energy use (equivalent liters/ha of diesel fuel)*	*Labor use (worker-days per ha)*
Tillage	36·55	3·21
Fertilizer	145·00	
Planting, weeding, and harvesting		80·99
Threshing	8·60	3·68
Winnowing	16·64	5·90
Paddy haulage (farm)	0·93	3·41
Paddy haulage (market)	3·46	4·45
Parboiling		5·80
Milling	41·60	4·07
Rice haulage	33·47	1·53
Total	286·25	113·04
Water lifting	516·0	3·63
Water application		61·75
Household fuel	198·66	

of fuel energy. Most households use kerosene, mainly for lighting. The diesel fuel and labor requirements are tabulated in Table III.33.

Thailand Cassava Production (Chancellor, 1978)
Energy and labor requirements were tabulated for the various operations in Cassava production. The general pattern of operation was as follows:

—Large obstructions in the brush-jungle land were cleared manually and 52 kW tractors with three-disc plows were hired to make one or two initial plowings of the remaining vegetation and rough soil.
—During subsequent years, tractors were also hired for plowing, usually a single pass, which was completed rapidly on the more regular field terrain.
—Family workers on small farms and hired workers on larger holdings planted, weeded, harvested, and loaded the roots onto trucks.
—Trucks carrying 3–7 t hauled the roots up to 10 km for chipping and drying.
—Roots were chipped by a locally manufactured machine driven by a 7-kW gasoline engine.
—Dry chips were collected, bagged manually, and loaded onto trucks for a 3-km haul to the pelleting mill.
—Bags of chips were unloaded manually and emptied into the conveying system. Most mills were electric, although some were powered by diesel engines.
—Pellets were bagged and manually loaded onto trucks for transportation to the port area where they were loaded onto ships.

The diesel fuel and manual labor requirements for this operation are tabulated in Table III.34.

TABLE III.34
Quantitative parameters for Cassava production in Thailand (source: Chancellor, 1978)

Operation	*Energy use (equivalent liters/ha of diesel fuel)*	*Labor use (worker-days per ha)*
Land clearing (including initial plowing)	6·17	2·47
Plowing	18·14	3·58
Planting		13·09
Hoeing		76·08
Harvest		37·42
Hauling roots	15·14	1·65
Chipping and drying	2·15	4·62
Hauling chips	0·93	0·74
Pelleting	86·38	3·38
Hauling pellets	54·22	1·60
Total	183·13	144·63

Australasia

New Zealand (McChesney *et al.*, 1981)

Unlike many nations, New Zealand relies on its agriculture to earn a large portion of the foreign exchange required to import the goods and services that have become part of a comparatively high standard of living. Because of a small population, a relatively large but often rugged land area, and soils with low natural fertility, New Zealand agriculture evolved as a predominantly livestock system based on extensive pastoralism. With the advent of refrigerated sea transport in 1882, trade in agricultural products was established, with the United Kingdom as the major market. Once the link with the distant British market was established, the animal industry flourished. Meat, cheese, and butter were the principal commodities exported.

Agricultural development in New Zealand has evolved in the last 150 years—an extremely short period when compared with European agriculture. Four phases of that development have been identified. First, a period of extensive pastoralism based on the large areas of indigenous grassland began about 1840. Inputs of energy and materials associated with this initial development were minimal, and by the 1870s, the natural limits to production were reached. The second period, beginning in the 1870s, was a time of rapid expansion. Forests were cleared in the North Island, while in the South Island, increases in cereal cropping for human and draft animal consumption quickly depleted the fertile lowland soils. Permanent grassland systems evolved slowly, and by the 1920s, the essential features of New Zealand's grassland economy had been established. The third period, from 1920 through 1970, was a time of intensification, during which the application of science and technology brought about major production increases. Particularly important were improvements to animal health, plant and animal breeding, and the use of corrective fertilizers. Meat and dairy production doubled and wool production tripled during this period. At the same time, the material and energy inputs necessary to sustain this expansion also significantly increased. Lastly, the decade of the 1970s was identified as a period of diversification from traditional pastoral production toward horticulture, and deer and goat farming.

Until the late 1960s, meat, wool, and dairy products consistently earned more than 80% of New Zealand's export income each year. Throughout the last decade, the situation has changed. The proportion of export income earned in 1980 from traditional animal products fell to 68%. During the same period, export earnings from manufactured goods more than doubled, now exceeding 25%. Nevertheless, agriculture still commands a central position in the country's economy. As shown in Fig. III.15, this heavy reliance on rural exports is unique among OECD countries.

Energy in New Zealand agriculture. Concern about the increasing reliance of food production systems on nonrenewable energy resources prompted the use of energy analysis techniques to highlight both the level of energy dependence and comparative energy efficiency of agricultural systems. Throughout the 1970s, a small rise in energy use was evident, although inputs fluctuated according to climatic conditions and farmers' relative financial situations. Gross energy inputs in 1978–9 were estimated at 44 petajoules (PJ). [This value is some 35% higher than earlier published estimates. The reasons for this increase are largely methodological. Past studies have been based on net energy values of energy inputs, which reflect delivered energy contents. In this publication, gross energy values have been used, which account for the 'energy requirement' of energy. The following gross energy values were used: diesel fuel, 45 MJ/liter; petrol, 41·3 MJ/liter; and electricity, 6·5 MJ/kWh (reflecting a predominantly hydroelectric base).] As shown in Table III.35, more than 80% was used for grain, fruit, and vegetable production. For the

TABLE III.35
Energy inputs to New Zealand farming, 1978–9 (source: McChesney *et al.*, 1981)

Input	%	*Commodity*	%
Direct		Meat[a]/wool	54
fuel	23	Milk	
electricity	9	processed	24
gas/coal	1	fresh market	3
Indirect		Pig meat	2
fertilizers	25	Poultry/eggs	2
farm services	12	Grains/seeds	7
animal feeds	4	Vegetables	4
chemicals	2	Fruit	3
repairs/maintenance	14	Other	1
capital goods	10		

[a] Includes mutton, lamb, and beef.

predominantly livestock farming systems in New Zealand, these energy requirements average 900 GJ per dairy farm.

Liquid fuels are particularly important inputs to New Zealand farms. Although direct fuel use accounts for only 23% of the energy input, many of the materials and services purchased by farmers have a high indirect liquid fuel requirement. This is especially true with fertilizers and transport services. The total direct and indirect input of liquid fuels is approximately 27 PJ, or more than 60% of the energy input to farming.

In 1978–9, agriculture accounted for about 18 and 9% of the national usage of diesel fuel and petrol, respectively. Three functional uses of fuel can be identified: transportation (transportation fuel includes that used by farmers and farm servicing organizations), which includes heavy goods, personnel movements off-farm, and all on-farm transport; field operations; and heating (Fig. III.16). More than 90% of the petrol and 45% of the diesel went for transportation. Most of the remaining diesel went for field operations, the most important of which were crop establishment and harvesting operations. A small but essential fuel input is the 12 million liters of aviation fuel required for fertilizer application to hill country. Only a comparatively small demand exists for heating fuels.

The heavy demand for transport fuels reflects two aspects of New Zealand and its

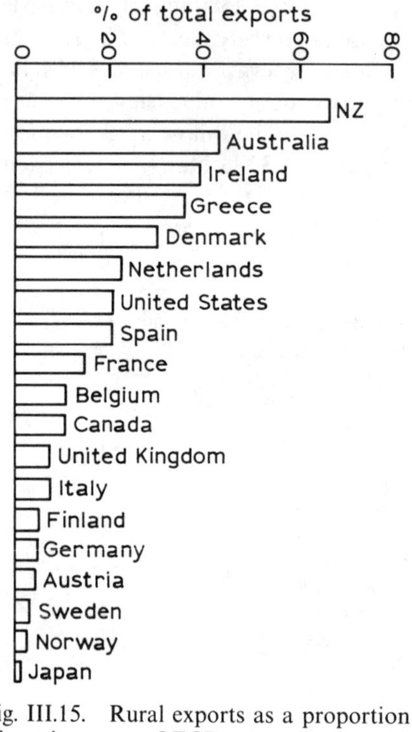

Fig. III.15. Rural exports as a proportion of total exports, OECD countries (source: Stoeckel & Patterson, 1980).

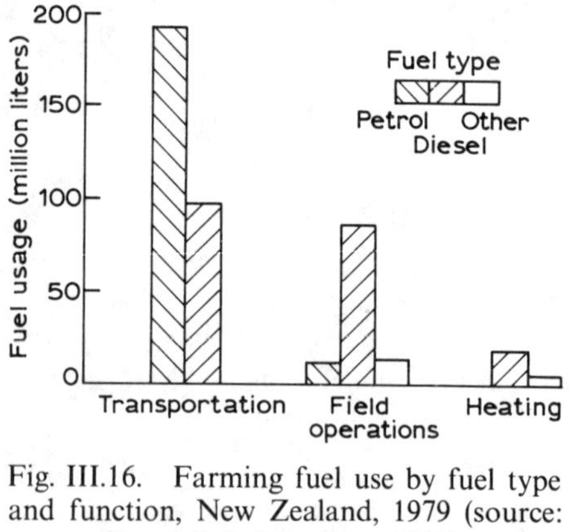

Fig. III.16. Farming fuel use by fuel type and function, New Zealand, 1979 (source: Stoeckel & Patterson, 1980).

farm. First, the country is elongated and population densities in rural areas are low. Processing facilities and servicing centers have become increasingly centralized, resulting in increased travel distances for farm servicing. Between 1975 and 1980, for example, a 27% decrease in the number of dairy factories resulted in a 10% increase in the fuel required per liter of milk collected. Second, farm properties are large when compared with European farms. This leads to a relatively high fuel demand associated with stock management, farm maintenance, and the feeding of supplements.

New Zealand agriculture relies significantly on phosphatic fertilizers. Most soils are phosphorus deficient and heavy applications of phosphorus (mostly applied as single superphosphate) have been necessary to achieve and maintain present production levels. While occupying less than 0·2% of total agricultural output, New Zealand used about 1·4% of world phosphate production. Current supplies are drawn from the high grade deposits of Nauru Island and Christmas Island. The gross energy requirement of phosphate rock landed in New Zealand has been estimated to be about 2·3 MJ/kg (14 MJ/kg P), a value similar to phosphate landed in the United Kingdom.

The energetics of farming. Figure III.17 shows average energy inputs and outputs per hectare for livestock and cereal production from New Zealand farms. Although substantial variations occur within farm types, the general relations illustrated provide a basis for comparing the energetics of agriculture in New Zealand with other temperate western countries. Energy use can also be expressed in relation to

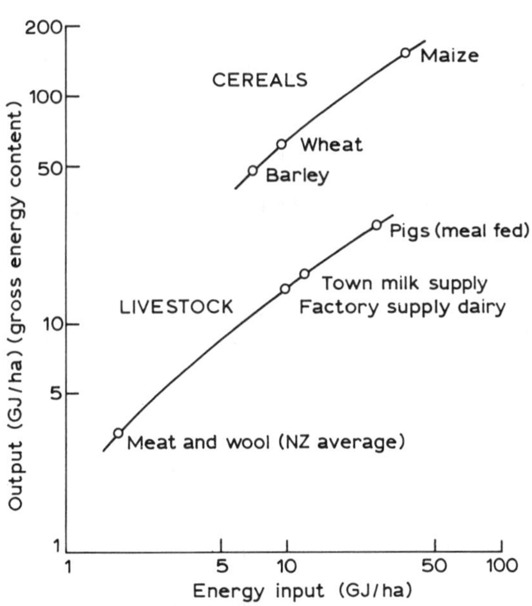

Fig. III.17. Average energy inputs and outputs for selected agricultural products in New Zealand (source: McChesney *et al.*, 1981).

protein output as shown in Table III.36, in which a comparison is made between United Kingdom and New Zealand products. For cereals, energy requirements in New Zealand are slightly over half of those in the United Kingdom and typically one-quarter to one-third of those in the United Kingdom for livestock. For the two vegetable crops shown, New Zealand energy requirements are about 50% higher. Several features of New Zealand farming systems contribute to these differences.

New Zealand has a temperate climate, without either hot or cold extremes and with adequate rainfall in most locations. As a result, some high energy modifications, which are features of agricultural production in European and North American countries, have not been widely adopted. Animal housing, for example, is confined to pigs, poultry, and some dairy herds, with only broiler chicken houses requiring extensive supplementary heating. Also, livestock directly harvests more than 99% of its feed intake from pasture and other crops, so that mechanical harvesting and feeding of fodder crops generally occurs only during seasonal troughs in pasture production.

The New Zealand pastoral system has been developed through a soil improvement program which incorporated heavy applications of phosphate fertilizer and lime, the selective addition of trace elements, and reliance on the biological fixation of nitrogen. The latter feature is a major factor in maintaining low energy livestock production relative to many other European countries. It has been estimated that the 10 million hectares of developed pastures in New Zealand fix about 200 kg N/ha per year, a total of about 2 million tonnes/year. In contrast, fertilizer nitrogen applied to pasture land was estimated to be only 12 500 t in 1978, less than 1% of total nitrogen input to New Zealand pastures.

The high rates of nitrogen fixation by pasture also enables the fertilizer nitrogen input to most cereal crops to remain relatively low. Cereal production is incorporated into a crop rotation, which usually involves a number of years in

TABLE III.36
Comparative energy requirements in relation to protein output (source: McChesney *et al.*, 1981)

Commodity	*NZ (MJ/kg protein)*	*UK (MJ/kg protein)*
Wheat	25	45
Barley	37	58
Potatoes	151	96
Fresh peas	69[a]	48
Milk	68	208
Mutton	116[b]	465
Beef	95	348

[a] Includes mechanical harvesting.
[b] Represents 53% of the energy input to sheep production. The remaining 47% is attributed to wool production. Inputs were partitioned on a dollar valuation. The energy requirement of wool was calculated to be 28 MJ/kg.

pasture. Hence, the cereals draw on a 'bank' of clover-fixed soil nitrogen. Nitrogen fertilizer inputs to wheat in 1977 averaged only 2 kg/t, compared with typical requirements of 20–30 kg/t in the United Kingdom.

General relationships showing the increasing energy inputs associated with higher land productivities have been described by many authors. If land availability is a constraint, food production can be increased by intensifying energy inputs. Under New Zealand conditions, diminishing returns to energy inputs are illustrated in Fig. III.17.

The land productivity of New Zealand farms is low compared with many other countries, a consequence of low population densities and extensive farming methods. Through these circumstances, New Zealand has been able to maintain a relatively low-energy type of agriculture.

Agriculture and energy use in the 1980s. New Zealand has an abundance of energy resources, including natural gas, coal, and hydropower, but it lacks any sizable oil fields. Like many other countries, New Zealand is dependent on imported oil for most of its transportation and automotive fuel requirements (Table III.37).

Policies aimed at achieving a greater degree of self-sufficiency in automotive fuels and a reduction in the cost of oil imports have been pursued. Two major industrial developments are involved: the construction of a plant to produce synthetic petrol from natural gas, and expansion of the country's single oil refinery to permit both a wider range of crude oil feedstocks to be refined, and a greater production of middle distillate fuels.

While energy use per unit of output in New Zealand increased until the late 1970s, energy requirements have since decreased as production shifts have occurred. These changes reinforce the lag between investment and production in the pastoral industry. However, energy costs in 1980 adversely affected farm profitability. In 1981, the farming sector purchased energy at an average estimated unit cost of just under \$8·00/GJ. Therefore, the total cost of energy was almost \$350 million,

TABLE III.37

New Zealand's energy supply, 1979 (source: McChesney *et al.*, 1981)

Type	*(PJ)*
Coal	47
Natural gas	47
Primary electricity[a]	70
Oil	
indigenous	18
imported	175
Subtotal direct energy	357
Indirect energy	125
Total direct and indirect energy	482

[a] Includes hydro and geothermal electricity generation.

accounting for approximately 20% of production costs. Prior to the 1973 oil embargo, energy costs contributed less than 19% to total production costs. Clearly, production methods which have lower energy requirements are of great interest when energy costs are high.

Lower energy requirements do not necessarily imply that production levels will be lower, however. A recent study on the effects of border dike irrigation development indicated that increases in livestock production of up to 90% could be achieved with a 35% reduction in the liquid fuel use per unit of output.

Changing land use toward higher value horticultural crops generally increases energy inputs. The main product of diversification has been kiwifruit (*Actinidia chinensis*). Energy inputs per hectare are typically 2–10 times higher than the land use displaced (typically dairying or maize), although the energy intensity of production at current prices is substantially lowered (Table III.38). Undoubtedly, this is one reason why kiwifruit production has become such a financially attractive land use alternative.

While New Zealand may enjoy a comparative energy advantage in production, it has a comparative disadvantage in marketing because of the long distances involved in transporting produce. Energy use in processing and transportation before delivery to the final consumer may be several times higher than the production energy input on the farm. For example, the energy required to process and transport lamb to the United Kingdom is about 150% higher than on-farm inputs (Table III.39).

The distance to markets for non-pastoral exports such as kiwifruit, mushrooms, and cut flowers, will alone make the role of energy a pivotal one. Although most horticultural products are exported by sea, air freight is an important transport mode for high quality, perishable items which often fit into off-season market gaps. For example, the delivery of kiwifruit to the early-season European market by air freight involves an energy input approximately 50 times that required to grow the fruit. (This is based on a fuel intensiveness of 15 Mj/t-km.) Energy costs of this magnitude will constrain market access in the future, and will limit the extent to which product diversification is possible.

Prevailing agricultural policies greatly influence energy use and energy use efficiency. The overriding policy objective in recent years has been to encourage increased output, and energy usage patterns have varied according to the methods

TABLE III.38
Energetics of production alternatives (source: McChesney *et al.*, 1981)

	Gross output (\$/ha)[a]	*Energy input (GJ/ha)*	*Energy intensity (MJ/\$)*
Kiwi fruit	40 000[b]	80–110	2·0–2·8
Factory supply dairy	650	10–11	15–17
Maize	1 150	34–38	30–33

[a] \$(NZ) 1980.
[b] Typical for developed orchard.

TABLE III.39
Energy requirements of New Zealand lamb landed in the United Kingdom (source: McChesney *et al.*, 1981)

Production/processing stage	*Energy requirements*	
	(MJ/kg[a] dressed weight)	*(MJ/kg protein)*
1. Farm		
on-farm	13·8	106
transport to slaughter	1·3	10
2. Processing	10·4	80
3. Transport to wharf	0·5	4
4. Shipping[b]	9·2	71
Total	35·2	271

[a] Assuming average dressed weight = 13·5 kg
[b] Average shipping energy requirement 0·48 MJ/t-km.

utilized to best meet this objective. Policies designed to increase production have tended to move away from subsidies on inputs in favor of output incentives. Specific examples include the reduction in the subsidy payable on fertilizers, the introduction of financial incentives to develop land and increase stock numbers, and product price stabilization measures.

Removing input subsidies on energy intensive items is a positive step toward redressing distortions in production costs. The output incentives, while they appear to have been initially successful in their prime objective, do not necessarily encourage more energy-efficient production systems. For example, incentives to increase stock numbers discriminate against production increases achieved by higher performance per stock unit.

Conclusions. As a small country with a relatively rugged topography which limits the area suitable for arable cropping, New Zealand's contribution to the world food supply is modest. A pastoral agricultural system has developed that is appropriate for its environment, requiring low energy inputs compared to many European countries. Biological fixation of nitrogen is a significant feature of this pastoral economy.

In 1980, agriculture earned 71% of New Zealand's export earnings. However, New Zealand has suffered from deteriorating terms of trade in recent years, primarily because of the high cost of oil imports. Nevertheless, lower oil prices have reduced the cost of oil imports, and plans have been made to produce synthetic petrol from natural gas, expanding the oil refining capacity. While cost savings may be achieved through eliminating imported refined products, New Zealand will still be heavily dependent on imported liquid fuel feedstocks in the mid-1990s. Servicing the capital invested in the liquid fuels program will place additional pressure on the agricultural sector to increase its foreign exchange earnings. While the potential exists for increases in agricultural output, the maintenance of profitable farming, including more energy-efficient production methods, will be necessary if this potential is to be achieved. Market expansion is also a prerequisite of increased farm output.

THE NEED FOR A SYSTEMS APPROACH

The preceding examples present data from many energy surveys and illustrate various types of energy analyses. In some cases, the data are fragmentary and, perhaps, not completely accurate. It is difficult to make any sweeping generalizations from these studies. Clearly, a more systematic approach is needed—one that follows a standardized methodology that permits a more scientific analysis of rural energy issues in developing countries.

The boundaries of energy studies must be clearly defined so that users of the data have a clear picture of the parameters that are included and not included. The methodology for including human and animal energy inputs must be clearly explained, and standard energy values must be developed for fuels, biomass, and other products. Some parts of these examples should be useful, and, hopefully, will lead to more comprehensive and useful rural energy studies that permit greater generalization and the development of empirical rules. Improved studies will be useful to planners, policy-makers, and others who face the challenge of increasing the rural energy supply and improving the efficiency of energy use in developing countries.

CHAPTER IV

Energy Management

INTRODUCTION

Increasing energy prices during the 1970s and 1980s stimulated worldwide interest in new technologies and strategies to make food systems more efficient and further utilize renewable energy alternatives.

Typical operations in agricultural production are outlined in Fig. IV.1. All agricultural operations require energy in one form or another. For example, fertilizer (especially nitrogen) and pesticides are heavily dependent on fossil fuel feedstocks. Mechanical work is necessary for all farming operations including tillage and water pumping. Numerous opportunities exist to conserve energy in agricultural production (Table IV.1). In developing countries, however, it is almost impossible to conserve commercial energy in production operations because so little is used now.

If the agricultural production sectors of developing countries receive the appropriate economic signals, and if the knowledge, skills, and credit access needed to take advantage of these signals are available, many of the adjustments necessary for more efficient agricultural energy use will not require government programs. For example, farmers will modify cropping techniques to reduce energy costs if modifications are clearly profitable for them, and if the required inputs are available on a regular basis. However, only limited improvements in many technical relationships—for example, the efficiency of a pump or improved plant growth resulting from various fertilizer applications—can be expected as a consequence of improvements in the functioning of economic markets. In many areas, additional research is needed to make significant progress in the long term, and this necessary research will not be implemented in the agricultural sector of most developing countries without direct government intervention.

Given the high cost and limited supply of energy in many developing countries,

AGRICULTURAL PRODUCTION

Preparing the land
- Tillage
- Contour and drainage grading and maintenance

Planting
- Seed genetics
- Seed production
- Germ plasm diversity

Irrigating
- Well drilling
- Water pumping and drainage
- Land leveling
- Canals, dams, and other structures

Applying nutrients
- Commercial N, P, K, Ca, Mg, S
- Nitrogen fixation
- Growth promoters
- Soil amendments
- Organic materials

Controlling pests
- Insects
  —Chemical
  —Non-chemical
- Weeds
  —Tillage and other non-chemical herbicides
- Predators
  —Rodents
  —Birds

Controlling disease
- Pesticides
- Non-chemical

Harvesting

Fig. IV.1. Operations in agricultural production requiring energy (source: NRC, 1981).

TABLE IV. 1
Opportunities for conserving energy in agricultural production (source: NRC, 1981)

Preparing the land	—Reduced or zero tillage (assuming no increased chemical usage) —Chisel versus moldboard versus disk plowing
Irrigating the land	—More efficient pumps —More efficient water use on farms —Reducing water loss in courses and canals —Salinity control
Applying nutrients	—Efficient application of fertilizers —More energy-efficient formulations of fertilizers, including organic fertilizers from human and animal wastes —Energy-efficient manufacture of fertilizers —Proper soil testing —Biological nitrogen fixation
Controlling pests and disease	—Biological methods —Integrated system —Ultra low-volume sprayers —Resistant varieties of seeds/plants —Intercropping
Harvesting	—Increased efficiency of equipment use, including reducing losses

the following general approaches may be applicable and should be considered in developing national food strategies:

—Grow low-volume, high-value export crops and import food, taking into consideration normal market fluctuations.
—Increase production (acreage) of crops requiring low energy input.
—Decrease feeding of intensive grain supplies to livestock because they are inefficient grain converters.
—As appropriate, emphasize crop production of high-protein, high oil-content crops—for food and fuel.
—Increase production, if possible, of food and fuel in areas where people are concentrated, such as in and around urban areas.
—Shift crop production to the most favorable areas and exchange production between regions or countries.
—Grow only crops that are the most adaptable to particular regions within countries.

GERM PLASM AND ITS IMPACT ON ENERGY USE†

Available seed or other planting material for each crop constitutes the germ plasm on which production of that crop is based. Plant breeding programs, which increase

†Adapted from NRC (1981).

the number of crop varieties with desirable characteristics, depend on the worldwide pool of genetic material, including wild varieties, for the diversity that is necessary to support plant breeding.

Most germ plasm research has been directed towards increasing crop yields per unit area per day and maximizing economic return to the producer. Some research has focused on other crop attributes—short and strong stems to support heavier seedheads without lodging; protein quality; and resistance to disease, insects, and storage damage—which have implications for other uses. Almost no research has been targeted specifically toward energy considerations. For example, in forest genetics, although the selection of fast-growing, high quality tree species also may implicitly support higher energy productivity of biomass, most forestry research has been geared to lumber quality, stand uniformity, and disease resistance, rather than energy content.

The following are examples of the need for developing countries to assess the potential of a number of aspects of germ plasm selection directly related to energy use in agriculture:

—Multi-purpose crops might provide higher proportions of residues to be used for energy, even at the expense of reduced food content. (The risk of losing topsoil and nutrients should be carefully weighed in comparison to potential energy benefits.)

—Fuel quality of crop residues would be significantly enhanced by lowering the silica content in rice husks, straw, and peanut hulls, if this could be achieved without affecting productivity, lodging, or pest resistance.

—Other opportunities may result from higher energy costs and would make selection for energy factors feasible, such as response to lower levels of fertilizer use (i.e. the opposite of efforts to increase biological fixation), improved water efficiency, and greater pest resistance, which might eliminate the need for a particularly expensive pesticide. Although market forces presumably will encourage these potential economies, developing countries are frequently the last to derive the benefit from new technologies unless governments and technical assistance agencies provide support.

FERTILIZER AND BIOLOGICAL NITROGEN FIXATION†

Mineral fertilizers are only one of several nutrient supply sources (Fig. IV.2). The productivity of mineral fertilizers depends largely on the nature and amount of other complementary farm inputs and management practices. Thus, while this section discusses fertilizers in isolation from other inputs, in reality, agricultural production is a complex integrated system. An improvement in fertilizer use efficiency promises significant energy savings and increases economies of its use. Increased efficiency in fertilizer use will also mean less energy consumed in fertilizer manufacturing and distribution.

† Adapted from Mudahar and Hignett (1982).

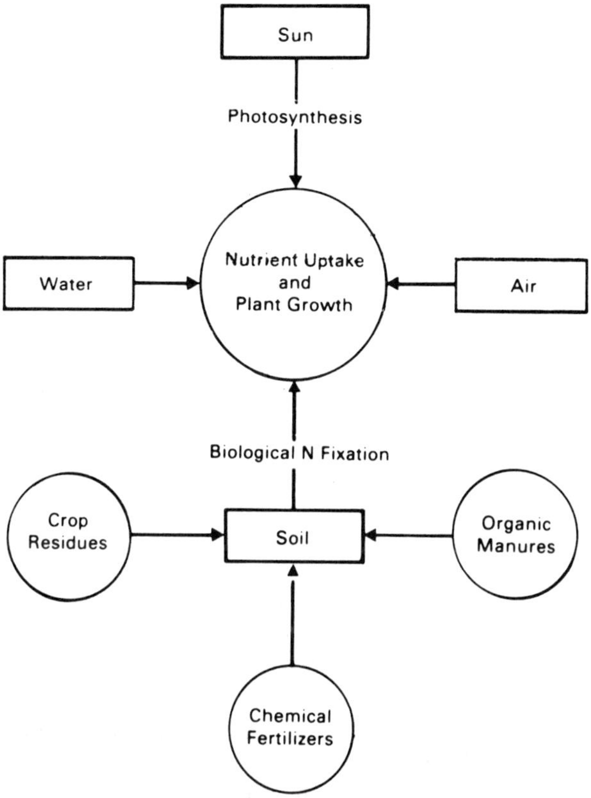

Fig. IV.2. Sources of nutrient supply and plant growth (source: Mudahar & Hignett, 1982).

If any one of the 16 essential plant nutrients is lacking, a plant cannot produce to its potential. Of these 16 elements, three (carbon, hydrogen, and oxygen) are obtained from air and water. The six so-called 'macronutrients' (nitrogen, phosphorus, potassium, calcium, magnesium, and sulfur) must be supplied by commercial fertilizers, soil amendments, or recycled organic materials when soil supplies are inadequate.

Nitrogen, phosphorus, and potassium have traditionally been the main concern of the fertilizer industry and are known as primary nutrients, but increasing frequency of deficiencies of calcium, magnesium, and sulfur has forced attention to supplying one or more of these nutrients in fertilizers. The other seven essential elements—boron, chlorine, copper, iron, manganese, molybdenum, and zinc—are called 'micronutrients' because plants require them in small quantities. Supplies of these are usually adequate, but increasing incidences of deficiencies of some elements, such as zinc, have been noted in various developing countries.

Nitrogen is regarded primarily as the most important fertilizer nutrient because its supply more often limits crop yields, although phosphorus is the limiting factor in many tropical soils. Figures IV.3 and IV.4 illustrate typical nitrogen response curves

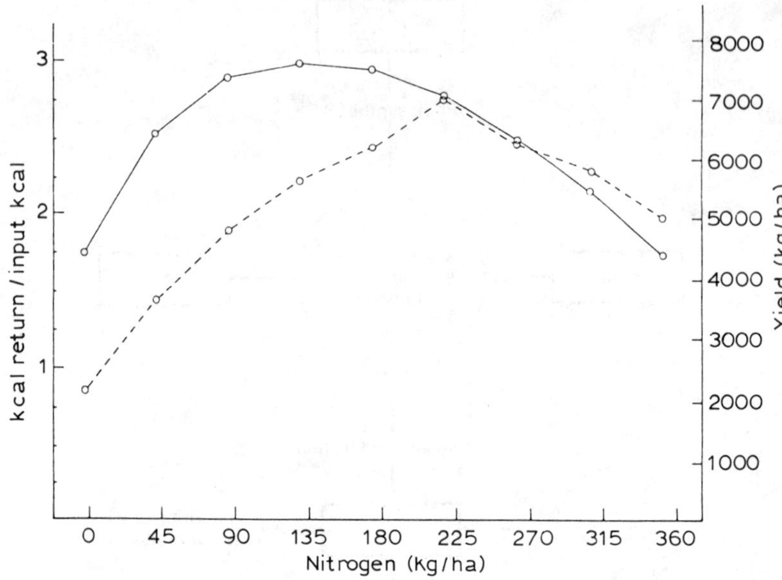

Fig. IV.3. Corn yields (kg/ha, ○---○) with varying amounts of nitrogen applied per hectare (phosphorus = 37 kg/ha), kcal return per input kcal (○——○) (source: Pimentel, 1980; © 1980 CRC Press, Inc., Boca Raton, FL).

for maize, rice, and wheat. Nitrogen fixation requires energy, whether by natural or man-made processes. Nitrogen is also fixed in nature by biological processes through microorganisms in the soil. This process also requires energy which is supplied indirectly by the sun for the synthesis of the organic matter that nourishes microorganisms or by the plant itself in the case of symbiotic bacteria. Biological fixation of nitrogen is most efficient in soils that are adequately supplied with other

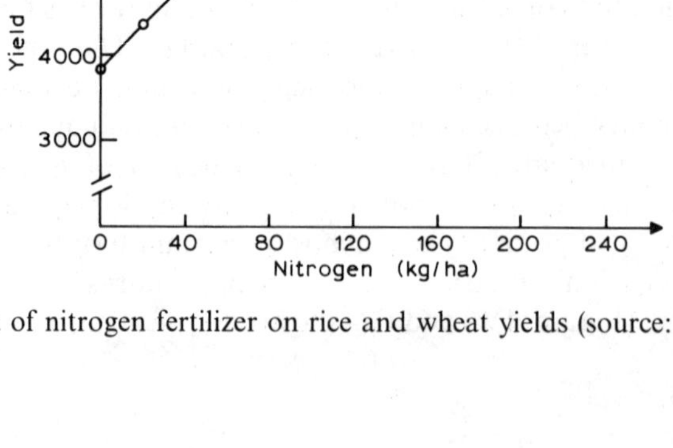

Fig. IV.4. Effect of nitrogen fertilizer on rice and wheat yields (source: Pimentel, 1974).

nutrients, particularly phosphorus and calcium. However, application of mineral nitrogen fertilizer tends to inhibit biological fixation with legume crops depending on quantities applied and timing and placement.

The natural sources of nitrogen, when properly managed and conserved by crop rotations and recycling crop residues and manures, have previously proven adequate for a modest level of cereal yields in many developed countries. However, they are not adequate for the intensive agriculture (high yields and multiple crops) that is needed to feed the present and future world population. The present challenge is to provide enough nitrogen, industrially or naturally fixed, to meet these requirements.

Table IV.2 lists a number of economic and technological characteristics of mineral and organic fertilizers. The raw materials for organic fertilizers are byproducts of other biological and/or economic activities. The conversion of these raw materials into organic fertilizers not only increases their economic value but also solves the disposal problem.

It is well documented that no more than 25–40% of the applied nitrogen is recovered by plants, especially by rice under Asian conditions. The rest is lost to the crop root zone of the soil, which can constitute potential pollution problems with high application rates. On the other hand, under general farming conditions, approximately 15–20% of the applied phosphate fertilizer is depleted by the crop to which it is applied. Unlike nitrogen, however, the rest of the phosphate becomes mixed into the soil and continues to be available to succeeding crops for several years. The behavior of potash appears to be somewhat between nitrogen and phosphate.

Ideally, nutrient losses can be substantially reduced, by perfectly matching nutrient availability with nutrient requirements of the crop throughout various stages of its growth.

There are several known mechanisms of nitrogen loss. These include volatilization, nitrification–dentrification, leaching, and erosion. Regardless of the source of fertilizer nutrient, fertilizer use efficiency can be significantly improved through better fertilizer management. Four major aspects of better fertilizer management include:

1. Adequate type and dosage of fertilizer.
2. Proper balance among nutrients.
3. Correct time of application.
4. Accurate method of application.

Fertilizer use efficiency can also be increased by modifying fertilizer materials in order to attain proper timing and placement. Nitrogen efficiency for rice can be improved through: (1) split application; (2) deep placement of fertilizer nitrogen in the active root zone of the plant; (3) pesticide site and use of various coatings and inhibitors to control the release of nitrogen; and (4) use of nitrogen compounds that have low water solubility. Table IV.3 shows the potential energy savings through improved nitrogen use efficiency.

TABLE IV.2
Economic and technological characteristics for chemical and organic fertilizers (source: Mudahar & Hignett, 1982)

Characteristics	*Chemical fertilizers*	*Organic fertilizers*
1. Primary nutrient content	Relatively high.	Relatively low.
2. Other nutrients	Only when prescribed in the grade, with some exceptions.[a]	Both secondary and micronutrients are normally present.
3. Nutrient composition	Predetermined.	Highly variable.
4. Nutrient availability	Immediately after application, phosphorus and potassium are fixed in soil and become available over time.	Slowly released through mineralization over time.
5. Organic matter	Absent.	Normally high.
6. Direct manufacturing cost	Relatively high and increasing.	Relatively small and stable.
7. Transportation cost	Relatively small but transported across long distances.	Relatively high but transported only across short distances.
8. Ease in handling	Relatively easy to handle and clean.	Relatively bulky, messy, and difficult to handle.
9. Direct energy use—manufacturing	Highly energy-intensive, especially nitrogen.	Almost negligible. Rather a good source of energy through bioconversion.
10. Direct energy use—distribution	Relatively low.	Relatively high.
11. Raw material exhaustability	Supply depends largely on nonrenewable, finite raw materials.	Supply depends largely on renewable, infinite resources.
12. Spatial availability	Highly localized.	Highly diffused and decentralized.
13. Fodder land requirement	None.	Large, but needs to be accounted for only when livestock is kept mainly for fertilizer production.
14. Environmental pollution	Known to create water and air pollution.	Solves the disposal problem. Pollution depends on the efficiency in collection, storage, distribution, and application.
15. International trade	Largely traded in the international market.	Does not enter the international market.
16. Amenability to economic analysis	Relatively easy and straightforward.	Relatively difficult to account for the costs and benefits.

[a] Some chemical fertilizers contain sulfur, calcium, or micronutrients derived from phosphate rock or basic slag that are not identified in the grade.

TABLE IV.3

Estimated potential savings in energy and nitrogen through improved nitrogen use efficiency during 1978–9 (source: Mudahar & Hignett, 1982)

Region	*Nitrogen consumed* (10^6 *Mt*)	*Nitrogen uptake*[a] (10^6 *Mt*)	*10% Efficiency improvement*[b]		*25% Efficiency improvement*[b]		*50% Efficiency improvement*[b]		*100% Efficiency improvement*[b]	
			Nitrogen savings (10^6 *Mt*)	*Energy savings*[c] (10^6 *bbl*)	*Nitrogen savings* (10^6 *Mt*)	*Energy savings*[c] (10^6 *bbl*)	*Nitrogen savings* (10^6 *Mt*)	*Energy savings*[c] (10^6 *bbl*)	*Nitrogen savings* (10^6 *Mt*)	*Energy savings*[c] (10^6 *bbl*)
Developed market economies	21·3	8·52	0·85	10·85	2·13	27·20	4·26	54·40	8·52	108·80
North America	10·5	4·20	0·42	5·36	1·05	13·41	2·10	26·82	4·20	53·63
Western Europe	9·4	3·76	0·38	4·85	0·94	12·00	1·88	24·01	3·76	48·02
Other developed market economies	1·4	0·56	0·06	0·77	0·14	1·79	0·28	3·58	0·56	7·15
Developing market economies	10·2	4·08	0·41	5·24	1·02	13·03	2·04	26·05	4·08	52·10
Africa	0·5	0·20	0·02	0·26	0·05	0·64	0·10	1·28	0·20	2·55
Latin America	2·5	1·00	0·10	1·28	0·25	3·19	0·50	6·39	1·00	12·77
Near East	1·7	0·68	0·07	0·89	0·17	2·17	0·34	4·34	0·68	8·68
Far East	5·5	2·20	0·22	2·81	0·55	7·02	1·10	14·05	2·20	28·09
Centrally planned economies	20·0	8·00	0·80	10·22	2·00	25·54	4·00	51·08	8·00	102·16
Asian centrally planned economies	7·8	3·12	0·31	3·96	0·78	9·96	1·56	19·92	3·12	39·84
Eastern Europe and USSR	12·2	4·88	0·49	6·26	1·22	15·58	2·44	31·16	4·88	62·32
Developed countries	33·5	13·40	1·34	17·11	3·35	42·78	6·70	85·56	13·40	171·12
Developing countries	17·9	7·16	0·72	9·19	1·79	22·86	3·58	45·72	7·16	91·43
World total	51·4	20·56	2·06	26·31	5·14	65·64	10·28	131·28	20·56	262·55
Value equivalent[d] (billion $)	28·27	11·31	1·13	0·79	2·83	1·97	5·65	3·94	11·31	7·88
Grain output (10^6 Mt) from saved nitrogen at:										
1 kg N = 10 kg grain			21		51		103		206	
1 kg N = 15 kg grain			31		77		154		308	

[a] Assuming that the average nitrogen uptake by plant is about 40%. This can be a weighted average across agroclimatic regions, crops, and N sources.
[b] Refers to improvements in nitrogen use efficiency in excess of 40%, with no reductions in crop yields.
[c] In the form of crude oil by assuming that, on the average, 1 Mt of N is equivalent to 12·77 bbl of crude oil.
[d] The prices assumed are $550/Mt of N and $30/bbl of crude oil.
Mt = million tonnes; bbl = barrels.

Mineral Fertilizers (NRC, 1981)

Figure IV.5 classifies countries into four categories depending on their degree of energy and nitrogen self-sufficiency. Table IV.4 provides further details on a number of food, fertilizer, and energy indicators. Two points should be emphasized: (1) the self-sufficiency in Fig. IV.5 is represented by a ratio rather than a percentage, and (2) a ratio of 2·00 indicates that 200% of the country's needs are available. Countries that have a surplus of nitrogen but imported feedstocks, such as Japan, may consider phasing out part of their nitrogen industry, whereas countries with plentiful domestic energy feedstocks but a limited nitrogen industry, such as Mexico, may emerge as the major nitrogen producers of the world.

The first effort by human beings to fix nitrogen imitated nature's lightning flashes; high-energy electric arc generators associated with hydroelectric generators fixed the nitrogen of the air to form nitrates. However, the process wasted energy and was soon abandoned. In areas where cheap, abundant hydroelectric power is available, electrolytic production of hydrogen to make ammonia by the Haber–Bosch process is preferable. With a given amount of electricity, four times as much nitrogen could be fixed as compared to the arc process, and higher analysis of nitrogen fertilizers could be made. Synthesis of ammonia from nitrogen and hydrogen requires very little energy, but the production of hydrogen is energy-intensive.

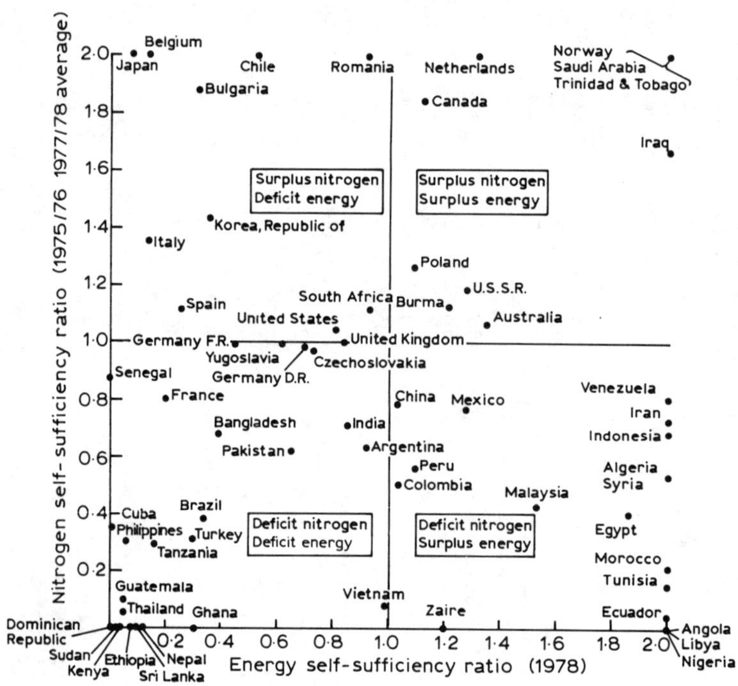

Fig. IV.5. Energy and nitrogen self-sufficiency: two-way country classification (source: Mudahar & Hignett, 1982).

Low-cost hydroelectric power is available only in a few countries, and this premium form of energy is often so valuable for industrial purposes that other sources of energy and feedstock are now used to produce nitrogen fertilizer. Only 1% of the world's nitrogen fertilizer is produced with electricity as a major input. Natural gas is the most efficient source of both energy and chemical feedstock and supplies more than 70% of the world's nitrogen fertilizer. Natural gas-based plants are the most energy-efficient and the least capital-intensive of all known methods for production of synthetic nitrogen fertilizer. Also, natural gas is usually a less expensive form of energy than fossil-derived fuels or electricity.

While the value of industrially fixed nitrogen is well established and cannot be underestimated, it has been increasingly difficult for some nations to meet their nitrogen fertilizer requirements, mainly because of problems related to their balance of payments. Their cost can be a constraint for adequate fertilizer use, especially in countries where foodgrain prices are controlled at a low level.

Fortunately, natural gas deposits are widely, although unevenly, distributed and known reserves are increasing as discoveries continue. Several countries are building or planning new nitrogen plants based on recently discovered natural gas deposits, although financing of new projects is difficult because of prevailing underuse of existing production capacities. Countries that do not have natural gas face more difficult problems. Naphtha has become too expensive for nitrogen fertilizer production. Heavy fuel oil is a possibility, but costs are likely to be higher than costs for imported nitrogen. Use of coal as fuel and feedstock to make nitrogen fertilizer is technically feasible, but highly capital-intensive, and, with present technology, the process has a low level of reliability.

Eventually, the nitrogen fixation process may use renewable resources. It is technically feasible to use methane derived from biogas or alcohol generated from biomass. Direct gasification of biomass is another possibility. Various other schemes, based directly or indirectly on solar power, are undergoing study but currently none seem both technically and economically promising. For the remainder of this century, we must rely mainly on nitrogen fertilizers made from natural gas. Countries that do not have natural gas face alternatives of importing ammonia or finished nitrogen fertilizers or subsidizing domestic production of nitrogen fertilizer using more expensive feedstocks or processes. Higher prices may be partially offset by efficient use of fertilizer nitrogen, increased conservation by recycling, more reliance on crops and methods using natural fixation processes, and increased dependence on less energy-intensive fertilizers and soil amendments.

In developing countries as a whole, only 2·7% of their total commercial energy use is in the form of fertilizer. With imported fertilizers, most of the energy is expended outside of the importing country. The energy feedstock may well be natural gas that would otherwise be wasted by flaring. Agricultural energy use in developing countries, including fertilizer, averages only 4% of total commercial energy use. Considering the importance of the agricultural sector in most countries, particularly in food production, a higher priority must be assigned to energy allocation in agriculture.

TABLE IV.4

Food, fertilizer, and energy indicators in selected countries (source: Mudahar & Hignett, 1982)

Region/country	*Self-sufficiency, % in:*				*Population 1979 (millions)*	*Arable land, % of potential in use*	*Fertilizer use, kg/ha of arable land and permanent crops, 1979*	*Per capita consumption of commercial energy, 1978 (kgoe)*
	Food 1974–76 average	*Fertilizer 1977/78–1979/80 average*		*Commercial energy 1978*				
		N	*NPK*					
Developed market economies								
North America								
Canada	119	210	515	112	24	48	41	6 755
United States	111	106	104	81	220	49	111	7 737
Western Europe								
Belgium	98	363	286	14	10	72	540	4 134
France	110	85	88	20	53	67	312	2 971
Germany, Federal Republic	81	98	133	45	61	80	479	4 092
Italy	83	135	95	14	57	65	189	2 197
The Netherlands	121	333	289	132	14	73	805	3 624
Norway	79	381	215	228	4	42	320	3 790
Spain	96	108	130	26	36	58	82	1 636
United Kingdom	72	99	88	84	56	63	324	3 546
Yugoslavia	99	95	76	62	22	73	110	1 384
Oceania								
Australia	170	95	88	135	14	29	29	4 505
Other developed market economies								
Japan	73	198	98	8	116	81	478	2 602
South Africa	112	116	94	93	28	66	64	2 162
Developing market economies								
Africa								
Algeria	70	54	64	779	18	83	23	468

Angola	95	0	0	935	7	8	4	131
Ethiopia	104	0	0	8	32	57	6	13
Ghana	139	0	0	31	11	46	7	112
Kenya	102	0	0	4	16	59	17	95
Morocco	94	25	211	19	20	87	29	194
Nigeria	102	0	6	1 811	75	80	3	72
Senegal	109	80	92	0	6	58	6	123
Tanzania	96	33	47	16	17	25	6	44
Tunisia	92	27	511	202	6	95	12	369
Zaïre	97	0	0	120	28	6	1	47
Latin America								
Cuba	139	30	19	1	10	99	156	794
Dominican Republic	124	0	0	1	6	89	59	316
Guatemala	117	9	9	5	7	72	59	177
Mexico	98	78	79	128	68	67	49	941
Trinidad and Tobago	68	951	550	372	1	63	54	3 377
Argentina	124	54	27	92	27	66	4	1 274
Brazil	108	36	44	34	123	19	58	540
Chile	84	198	89	53	11	96	25	678
Colombia	104	43	39	103	29	17	53	476
Ecuador	114	4	10	371	8	61	36	344
Peru	89	71	57	109	17	24	32	442
Venezuela	77	87	47	471	14	17	60	2 033
Near East								
Egypt	87	47	55	186	41	99	212	315
Libya	52	0	0	3 113	3	83	24	1 285
Sudan	104	0	0	2	18	27	4	117
Iran	84	16	47	672	37	99	24	1 230
Iraq	76	203	314	2 458	13	101	19	430
Saudi Arabia	50	1 290	836	4 818	8	87	19	888
Syria	89	36	23	200	8	86	21	659
Turkey	100	37	38	35	44	98	53	539

(continued)

TABLE IV.4—*contd.*

Region/country	*Self-sufficiency, % in:*				*Population 1979 (millions)*	*Arable land, % of potential in use*	*Fertilizer use, kg/ha of arable land and permanent crops 1979*	*Per capita consumption of commercial energy 1978 (kgoe)*
	Food 1974–76 average	*Fertilizer 1977/78–1979/80 average*		*Commercial energy 1978*				
		N	*NPK*					
Developing market economies—*contd.*								
Far East								
Bangladesh	93	57	44	39	86	100	45	29
Burma	104	89	67	121	34	24	11	43
India	98	69	64	85	678	93	30	121
Indonesia	98	120	90	344	148	55	44	189
Korea, Republic	87	173	138	36	37	107	384	925
Malaysia	137	34	18	153	13	54	103	487
Nepal	100	0	0	12	14	85	9	7
Pakistan	98	51	43	65	80	72	52	117
Philippines	122	20	24	6	49	66	35	231
Sri Lanka	82	0	0	10	15	83	72	74
Thailand	129	3	1	5	46	57	17	222
Other developing market economies								
Centrally planned economies								
Asian centrally planned economies								
China	102	85	84	103	945	131	129	569
Vietnam	92	19	41	99	51	77	30	85
Eastern Europe and USSR								
Bulgaria	106	178	126	32	8	73	193	3 415
Czechoslovakia	94	101	58	73	15	65	335	5 123
GDR	91	113	277	70	17	73	325	4 844
Poland	98	77	70	109	35	79	218	3 806
Romania	105	22	167	92	22	88	137	2 749
USSR	97	121	126	128	264	66	75	3 742

Fertilizer Use
The most urgent fertilizer-related objective should be the more efficient use of fertilizers at the farm level. This applies particularly to nitrogen fertilizers, which are by far the most popular and energy-intensive. Methods for increasing efficiency in fertilizer use include more balanced use of fertilizers, more accurate timing and placement, and correction of recognized deficiencies in secondary nutrients and micronutrients in the soil. Long-term attention must be devoted to improved forms of fertilizers that are easier to use efficiently, as well as to more precise determination of crop nutrient requirements. Greater use of compound fertilizers may facilitate a more balanced use.

Continued research is warranted to identify farm management practices that will fully use biological nitrogen fixation and organic waste materials.

Fertilizer Transportation
Relatively minor energy savings in fertilizer transportation and packaging can be expected in the short term. In fact, improved packaging (smaller and stronger bags) and package handling procedures, as well as more prompt delivery, are needed even at the expense of greater energy use. Prolonged savings can be achieved through increased use of rail and water transportation.

Fertilizer Policy
In response to financial procurement problems and to ensure increased crop production, further information must be generated on the role, options, and implications of various economic incentive programs, including fertilizer subsidies. This information should be communicated in the context of fertilizer-related national goals, including (1) improving the efficiency of fertilizer use, (2) achieving balanced fertilization, and (3) saving nonrenewable energy.

Biological Nitrogen Fixation (Mudahar & Hignett, 1982)

World agriculture would be significantly enhanced if energy-intensive mineral fertilizers could be further displaced by biological sources of nitrogen. Biological nitrogen fixation refers to the fixation of atmospheric nitrogen, which cannot be used directly by plants, into ammonia or other usable forms of nitrogen. The biological process of converting atmospheric nitrogen into a usable form is rather complex and not yet easily understood.

Two biological processes are responsible for fixing atmospheric nitrogen: (1) symbiotic fixation of nitrogen, and (2) asymbiotic fixation of nitrogen. Symbiotic fixation refers to the fixing of nitrogen by bacteria, known as rhizobium, in a symbiotic relationship with a host plant, primarily legumes. Asymbiotic fixation, on the other hand, refers to the fixing of nitrogen by free-living bacteria and blue-green algae, which have no direct relationship to plants.

The available literature on biological nitrogen fixation agrees on four empirical observations. First, nitrogen fixation is influenced by the growth of bacteria and/or host plants and their nitrogen requirements. Secondly, nitrogen fixation is

Fig. IV.6. A farmer spraying rice seedlings with pesticide (source: FAO Photo 9858/C/35).

determined essentially by the host plant's degree of photosynthetic activity or by the availability of carbon dioxide. Thirdly, a negative correlation seems to exist between the amount of nitrogen fixed biologically and the presence of nitrogen supplied by mineral fertilizers. Finally, large numbers of bacteria are known to fix nitrogen, but their fixing ability varies tremendously.

Biological nitrogen fixation has significant potential for expanded application in developing countries that are energy-poor and need to offset the costs of imported food, fiber, and fertilizers.

PEST CONTROL†

Control of weeds, insects, and plant diseases is a basic requirement of food production throughout the world (Fig. IV.6). Pest control in developing countries is critical because of inadequate funds, energy, and pesticides. Worldwide basic crop losses to all pests, including insects, diseases, and weeds, are estimated to be almost 48%. This includes 35% pre-harvest and 20% post-harvest losses.

Protection of crops requires a considerable expenditure of energy, especially in controlling weeds by plowing and cultivating the soil. Table IV.5 estimates the yield

†Adapted from NRC (1981).

TABLE IV.5
Yield increase needed to justify application of agricultural chemicals (source: Robinson & Mollan, 1982)

Chemical	*Energy content (MJ/kg)*	*Application rate (kg/ha)*	*Energy input from chemicals (MJ/ha)*	*% Yield increase needed*[a]
Diuron	274	0·5–1·6	140–440	0·5–1·5
Atrazine	188	1–3	190–560	0·6–1·9
Trifluralin	151	0·5–1·25	75–190	0·3–0·6
Paraquat	459	0·8	370	1·2
2,4-D	87	0·5–1·25	45–110	0·2–0·4
Chloramben	170	2–3	340–510	1·1–1·7
Propanil	220	3–6	660–1 320	2·2–4·4
Propachlor	290	4–6	1 160–1 740	3·9–5·8
Glyphosate	454	0·5–2·0	230–910	0·8–3·0
Methyl parathion	160	0·5–1·0	80–160	0·3–0·5
Carbofuran	454	1–2	450–910	1·5–3·0
Carbaryl	153	1–2	155–310	0·5–1·0

[a] Total energy input to cereals from all sources is about 30 GJ/ha.

increase necessary to justify the use of a few sample pest control chemicals. Most developing countries are in the tropics and subtropics, where pest control problems are enormously more complex than in temperate zones, and the range and severity of pests are greater. Pests also tend to be more persistent because of the absence of winter cold. In some areas, year-round and multiple cropping provides continuous host plants to sustain pests.

IRRIGATION†

The world contains approximately 13·4 billion hectares of land surface with crops occupying about 1·4 billion hectares, or about 11% of the land area. Of the cropped land, about 40% is in humid regions, 40% in subhumid regions, 15% in semi-arid regions, and only 5% in arid regions. The irrigated lands of the world exceed 0·2 billion hectares. This area amounts to approximately 15% of the cropland of the world. There are more than 0·5 billion additional hectares of potentially irrigable land in the world if water can be provided.

Importance in Developing Countries

Of the world's irrigated land, more than 0·15 billion hectares are in developing countries. The inhabitants of these countries are highly dependent on irrigation for their food. Figure IV.7 illustrates the plight of a peasant farmer in India suffering from drought. Underground water is available, but requires tubewells and reliable

† Adapted from Smerdon and Hiler (1985).

Fig. IV.7. A farmer in Rajasthan, India, an area suffering from severe drought. Tubewells are needed to tap underground water, but reliable energy supplies will be needed (source: FAO Photo 5786/D/7,11).

energy supplies. The annual energy requirement for irrigation in developing countries, excluding the energy necessary to provide water, was $161{\cdot}6 \times 10^{15}$ J in 1986. This is a 55% increase in the irrigation energy over the energy used in 1973 and is equivalent to 27 million barrels of oil per year (1 barrel = $0{\cdot}159\ m^3$). Most irrigation energy is used for pumping.

On a global basis, only 15% of the cropland is irrigated, but it still produces 30% of the world's food. Also, it provides food security against droughts similar to those that have occurred in recent years in parts of Africa. Therefore, the importance of irrigation in producing food is clearly evident.

Energy Use in Irrigation

Energy in irrigation includes the energy required for constructing the water supply source, providing the conveyance works, installing the field irrigation systems, and operating and maintaining the system. Both the direct and indirect energy uses must be considered. The indirect energy requirements include the inputs necessary for manufacturing the materials in dams, canals, pumps, pipe and equipment, as well as the energy for constructing the works and building the farm irrigation systems. Direct energy uses, which are recurring energy expenditures, are for pumping and operating the farm irrigation systems.

Energy for the Water Supply

When surface water can be used, dams usually must be built and canals constructed to deliver water to the fields. Based on analyses of the cost and energy requirements for constructing typical irrigation supply reservoirs and associated canal systems, it is possible to determine the total energy that must be committed to providing a surface water supply for irrigation. Expressed as the annual energy required per hectare of land to be supplied with irrigation water from surface sources, the energy requirement for developing the surface water supply is 748 MJ/ha per year.

When groundwater is to be used, wells must be drilled and pumping units installed. Considering the energy required to drill wells, as well as to manufacture and install pumps and pump motors, a corresponding energy requirement to provide irrigation water from groundwater sources is determined. The figure for groundwater sources, again expressed as energy required annually per hectare served, is 1727 MJ/ha per year.

These statistics represent typical irrigation conditions in the world. They consider the expected life of the various components in the irrigation water supply works, and the energy requirements are, therefore, recurring annual energy needs. Therefore, for typical water supply systems, an annual energy investment of 748 and 1727 MJ/ha per year is necessary to provide the water supply from surface water and groundwater sources, respectively. (These values are equivalent to 0·13 and 0·29 barrels of oil/ha per year.)

Energy to Provide Field Irrigation

Farm irrigation systems include surface irrigation of various types, sprinkler systems, and, in some cases, trickle systems. To calculate the energy needed to provide a surface irrigation system, the energy required to construct ditches, level fields, manufacture and install the necessary equipment, and provide other necessary inputs needs to be determined. For a sprinkler or trickle irrigation system, the energy to manufacture all the components and install them must also be determined. These calculations have been completed for nine typical types of irrigation systems. The results for five typical field irrigation systems include: surface irrigation, 466 MJ/ha per year; surface irrigation with irrigation run-off recovery system (IRRS), 1219 MJ/ha per year; hand-moved sprinkler, 808 MJ/ha per year; side-roll sprinkler, 1155 MJ/ha per year; and trickle, 4215 MJ/ha per year.

Energy to Operate Irrigation Systems

When the water supply is groundwater, pumping is always required and the energy increases proportionally to the depth of the supply. Also, when pressurized systems such as sprinklers or tricklers are used, additional pumping energy is expended. This pumping energy is in addition to the amount required to provide the water supply and construct the field irrigation system discussed in the preceding sections.

It is not possible to operate an irrigation system without some water losses. These losses occur in seepage and leakage, evaporation, and percolation in the fields below the crop root zone and beyond the crop. The greater the amount of water lost, depicted by a lower irrigation efficiency, the greater the wasted energy. Therefore,

TABLE IV.6

Annual energy required for irrigation (source: Smerdon & Hiler, 1985)

System type (irrigation efficiency in parentheses, %)	*Installation energy*	*Surface water supply*			*Groundwater—50-m lift*			*Groundwater—100-m lift*		
		Energy for supply	*Pumping energy*	*Total energy*	*Energy for supply*[a]	*Pumping energy*	*Total energy*	*Energy for supply*	*Pumping energy*	*Total energy*
Surface (50)	0·47	0·75	4·7[b]	5·9	1·29	83·3	85·0	1·72	161·8	164·0
Surface (70)	0·47	0·75	3·4[b]	4·6	1·29	59·5	61·2	1·72	115·6	117·8
Surface (85)[c]	1·22	0·75	4·6[b]	6·6	1·29	50·8	53·3	1·72	97·0	99·9
Sprinkler (75)[d]	0·81	0·75	55·5	57·1	1·29	107·9	110·0	1·72	160·1	162·6
Trickle (90)[e]	4·22	0·75	30·6	35·6	1·29	74·3	79·8	1·72	117·8	123·7
LEPA (90)[f]	1·20	0·75	14·3	16·2	1·29	58·0	60·5	1·72	101·6	104·5

Data in this table (except for the LEPA system) are adapted from research at Utah State University by J. Keller. Energy figures are in GJ/ha per year (1 GJ = 10^3 MJ = 10^9 J). Systems are designed to meet a peak water use rate of 8·4 mm/day and provide 1 m net irrigation. Pump efficiency is assumed to be 50%, and pump power unit is assumed to have a net efficiency (thermal energy to mechanical energy) of 25%.

[a] Energy to provide supply (drill and equip the well) for a well with a 50-m pumping lift was estimated to be 75% as much as for a deeper well with a 100-m pumping lift.

[b] Some pumping energy is assumed even for surface irrigation with open ditch to account for friction head loss in pipe and a slight elevation of the water to the level of the ditches. For systems in which canal water is supplied at sufficient elevation to permit gravity flow, pumping energy is zero except for the modest amount of energy required for the system with an IRRS.

[c] This is a gated pipe surface system with a pumped IRRS. These data can also be used for a surge system.

[d] This is a hand-moved sprinkler system.

[e] Trickle system is designed for orchard crops.

[f] Operating pressure of 140 kPa at command platform.

systems designed to have high irrigation efficiency, if properly operated, save energy as well as water.

Similarly, efficient pumping systems save both money and energy. Irrigation pumps theoretically may have pump efficiencies of more than 70%, but field tests of irrigation pumps in the United States show typical pump efficiencies to be between 50 and 55%, and often less. Irrigation pump efficiencies in developing countries are unlikely to be higher than in the United States and probably do not exceed 50%. This efficiency could be increased with properly designed, operated, and maintained pumps. The efficiency of converting fuel energy to mechanical energy is about 25% for diesel engines or electric systems when the efficiency of power generation (33%), electricity transmission (85%), and electric motors (88%) are all considered. This gives a typical combined efficiency (pump and motor) of 12·5% for pumping units in converting fuel (thermal) energy to lifted water for irrigation (mechanical energy).

Using the data on total energy to provide the water supply, install the farm irrigation system, and operate the system allows the formation of curves relating the total annual energy required per hectare to irrigate using the water provided with different pumping lifts for various kinds of irrigation systems. Pumping lifts for surface water supplied systems would be small, and in some cases zero. Table IV.6 and Fig. IV.8 provide the results for typical irrigation systems designed to provide 1 m of net irrigation annually and satisfy a peak water use rate of 8·4 mm/day. The six irrigation systems illustrated, with assumed irrigation efficiencies in parentheses, include: hand-moved sprinkler (75%); surface (50%); surface (70%); surface with IRRS (85%); trickle (90%); and low energy precision application system (LEPA) (90%). The pump efficiency is 50% and pump engine efficiency is 25%.

The LEPA was introduced in 1981 by Texas A&M University (Lyle & Bordovsky, 1981, 1983). Basically, it is a continuously moving sprinkler system (side-roll or center-pivot) modified to distribute water at low pressure from drop tubes instead of high pressure sprinklers. The water drops into small basins created by miniature furrow dams established with a special machine in the cultivation process. The pressure required is about 140 kilopascals (kPa) for a 57 liters/s system. The installation energy is estimated to be 1·2 MJ, similar to that for a side-roll sprinkler system. The energy saving in experimental work with the LEPA was reported to have ranged from 43 to 86% in relation to conventional sprinkler systems. In comparison, the figures from Table IV.6 provide a total energy saving of 36–72% compared to sprinkler systems supplied from groundwater with 100-m lift and surface water, respectively.

Surge irrigation is a recently developed surface irrigation method with considerable potential for saving both water and energy. It is appropriate for conditions in developed countries, such as the United States, where fields tend to be quite large with long irrigation runs. Its applicability is estimated to be somewhat less in developing countries where farms are very small. Nonetheless, an estimate of the energy requirement for surge irrigation is presented. Surge irrigation uses gated pipes and requires extra valves to control the on/off pulsing of flow to the furrows. However, the installation energy and pumping energy would be similar, or perhaps slightly less, than the gated pipe surface system with irrigation return system used by

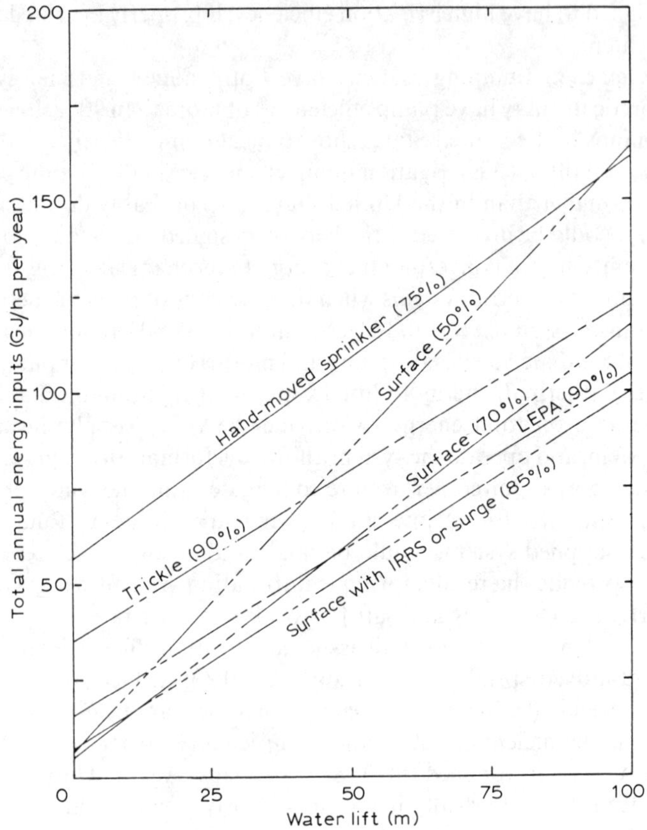

Fig. IV.8. Total annual energy requirement to provide 1 m net irrigation for different irrigation systems with irrigation efficiencies as indicated in parentheses. Pump efficiency is assumed to be 50% and net power unit efficiency (thermal energy to mechanical energy) to be 25% (source: Smerdon & Hiler, 1985).

Batty *et al.* (1975). The expected irrigation efficiencies of 85% for the two systems are also similar. Therefore, the same energy use data for surface irrigation with irrigation run-off recovery system is appropriate for surge irrigation.

It is obvious from Fig. IV.8 that the total annual energy required increases markedly with increased water lifts. This illustrates the importance of pumping lifts, including the operating pressure of pressurized systems in irrigation energy considerations. Also, by comparing the surface systems, the curves dramatically show the importance of providing well-designed systems with high irrigation efficiency.

Potential Energy Savings in Irrigation

Several field tests on irrigation pumping plants have been conducted during the last few years and are summarized in Fig. IV.9. A good design and installation results in

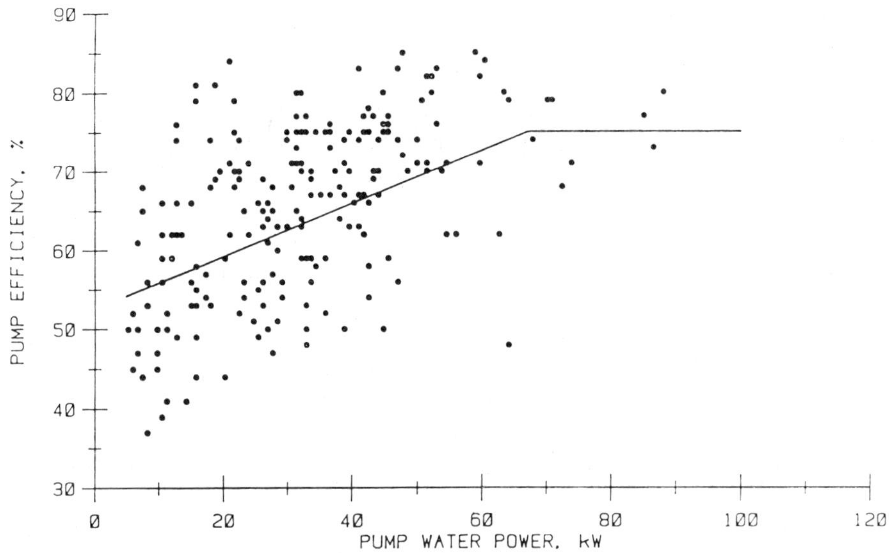

Fig. IV.9. Irrigation pump field efficiencies (source: Gilley & Supalla, 1985).

a pump efficiency approaching 75%. Field data in Fig. IV.9 indicate that pump efficiencies often fall far below 75%. Proper system design and pump repair can significantly improve efficiency.

A general equation can be developed showing the energy savings impact of improvements in individual components of an irrigation system. This equation for potential energy savings from improvements in the irrigation system in percentage, PES, is as follows:

$$\text{PES} = \left\{1 - \left(\frac{D_a}{D_b}\right)\left(\frac{H_a}{H_b}\right)\left(\frac{E_b}{E_a}\right)_1\left(\frac{E_b}{E_a}\right)_2 \cdots \left(\frac{E_b}{E_a}\right)_n\right\}100\%$$

D is the depth of net irrigation required by the crop; H is the total head (pumping lift) required of the irrigation pump; and E represents the efficiency of the various components of the irrigation system, including the efficiency of the pump, the efficiency of the watercourse in conveying water, and the efficiency of the irrigation application system on the farm. The subscript b indicates conditions before improvements and the subscript a shows the conditions afterwards. The subscript numbers outside the parentheses denote the individual components in which the efficiency is improved, and any number of components, n, may be included.

As an example, assume the following: the net irrigation application is reduced from 800 to 700 mm, which results in the ratio $D_a/D_b = 0{\cdot}875$; the total head is reduced from 50 to 30 m, which results in the ratio $H_a/H_b = 0{\cdot}60$; the efficiency of the pump is increased from 0·55 to 0·67, which results in the ratio $(E_b/E_a)_1 = 0{\cdot}821$; the watercourse conveyance efficiency is increased from 0·50 to 0·70, which results in the ratio $(E_b/E_a)_2 = 0{\cdot}714$; and the efficiency of water application in the fields is increased from 0·65 to 0·80, which results in the ratio $(E_b/E_a)_3 = 0{\cdot}812$.

The combined effect results in the following potential energy savings:

$$PES = \{1 - (0{\cdot}875)(0{\cdot}60)(0{\cdot}821)(0{\cdot}714)(0{\cdot}812)\}100\% = (1 - 0{\cdot}25)100\% = 75\%$$

Therefore, in this hypothetical case, 75% of the energy originally used can be saved through this combination of improvements in the irrigation practices. Stated another way, only one-fourth as much energy would be required to provide irrigation as before. Although it would seldom, if ever, be possible to make all these improvements in a single system, the equation can be used to assess the magnitude of energy savings resulting from each improvement, either individually or collectively.

How might these savings occur in developing countries? Through the use of improved crop varieties, the water required by crops, *D*, can be reduced. The total head required, *H*, can be reduced by providing low-pressure irrigation systems where high-pressure systems existed before. Irrigation pumps in the field often have low efficiencies and can be repaired or replaced to markedly improve their efficiency. Watercourses can be renovated and, thereby, improve water delivery efficiency. And, finally, fields can be leveled and better on-farm irrigation practices installed to decrease water losses which result from poor on-farm irrigation practices. Some of these improvements resulting in energy savings are usually feasible in developing countries.

MECHANIZATION OF AGRICULTURAL PRODUCTION†

Mechanization includes the use of tractors and machines for land preparation, planting, cultivation, and harvesting. The purposes of mechanization are:

- —To increase the productivity of the farm family or to multiply its ability to do work.
- —To improve the quality of work or to make it less laborious.
- —To perform tasks that cannot be completed manually or to overcome time constraints of critical operations. For example, land clearing is difficult to accomplish without specialized equipment such as root rakes, root plows, and heavy-duty offset disk plows.

Mechanization may improve yields through better land preparation, more precise placement of seed and fertilizer, and more efficient harvesting. Tillage operations illustrated in Figs IV.10 and IV.11 require reliable supplies of liquid fuel. Mechanization also encourages the production of difficult crops, such as soybeans. Worldwide soybean production probably would not have reached its current level without development of the combine (combined harvester–thresher).

Tractor Selection, Operation, and Maintenance

Many factors should be considered before selecting a tractor. The engine type, accessories, transmission design, tire type, and power capacity all affect the fuel

†Adapted from NRC (1981).

Fig. IV.10. Tractor operations require large amounts of liquid fuel. This ridger is preparing fields for planting sugar cane in the Philippines (source: FAO Photo 11054/3).

Fig. IV.11. Paddy field preparation in Malaysia (source: FAO Photo 4194/M/1).

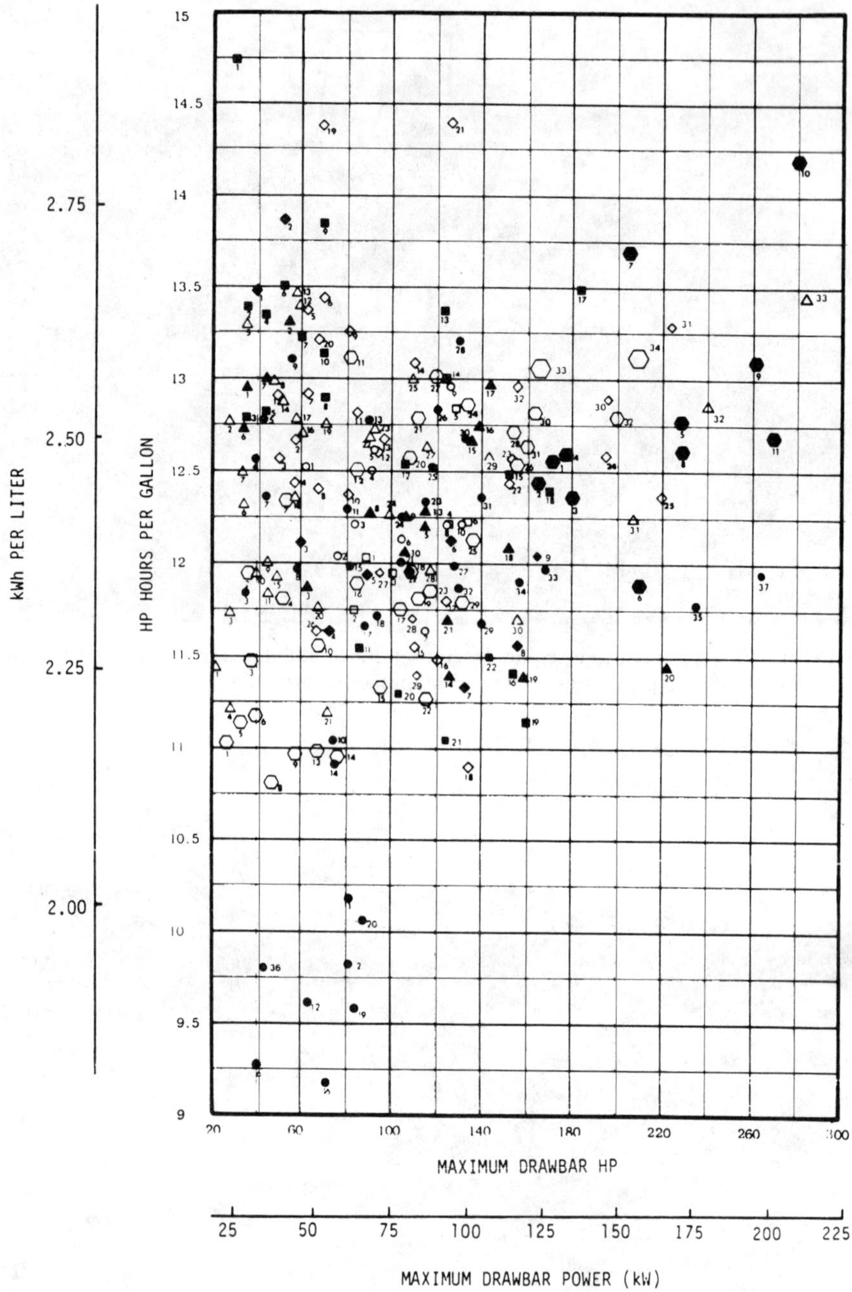

Fig. IV.12. Fuel consumption of diesel tractors at 75% of maximum drawbar power, 1967–80 (source: Stout, 1984; see Table IV.7 for key to symbols).

consumption rate. Before making a purchase, farmers should obtain performance data from one of the recognized testing agencies such as the Organization for Economic Cooperation and Development (OECD) or the Nebraska Tractor Testing Laboratory. An important index of efficiency is the specific fuel consumption (sfc) expressed in kilowatt-hours of work produced per liter of fuel consumed. Figure IV.12 indicates that a range of about 2·2–2·6 kWh/liter is common. Table IV.7 identifies the manufacturer and specific model of some of the tractors represented in Fig. IV.12. All other factors being equal, it is best to select a tractor that produces more work from each unit of fuel consumed.

Engines operate most efficiently within a relatively narrow range of engine loads and speeds, as depicted on the engine performance map shown in Fig. IV.13. The broken lines on the map represent percentage of power produced. Another method to measure sfc is in terms of the weight of fuel consumed to produce a unit of work (kg/kWh). When expressed in these terms, small values of sfc are desirable. This engine operates most efficiently (0·220 kg/kWh) at about 90% load and 70% engine speed. When tractors are operated at part load, a common practice on farms, significant fuel savings can be achieved by the practice of 'shift up and throttle back'. Nebraska Tractor Test data for 50% of pull at maximum power show fuel savings potential of 17–28% for randomly selected large diesel tractors. Table IV.8 provides this fuel consumption data.

Suppose a particular operation requires only 50% of the drawbar pull that a tractor is capable of producing. The John Deere 4440, when operated in fourth gear at 2305 rpm engine speed, produces 46·5 kW at a forward travel speed of 7·94 km/h (Table IV.8). The work output is 2·09 kWh/liter.

The same job can be performed in the same time by shifting to a higher gear and

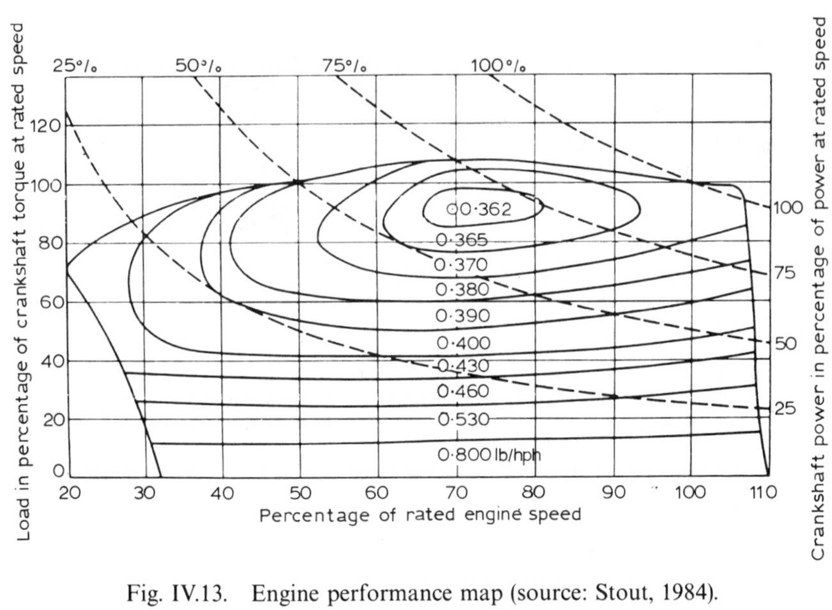

Fig. IV.13. Engine performance map (source: Stout, 1984).

TABLE IV.7
Fuel consumption of selected diesel tractors, 1967–80 (source: Stout, 1984)

	Tractor make and model	*Maximum drawbar (kW)*	*kWh per liter*
Allis–Chalmers		▲	
2	175	41·23	2·63
3	185	48·54	2·34
6	5040	26·75	2·51
7	5050 Diesel	33·73	2·56
9	7020 Power Shift	78·60	2·41
13	7040PS Diesel	87·84	2·42
14	7045 Power Director	95·37	2·25
15	7050 Diesel	100·79	2·50
18	7080 Diesel	114·72	2·39
20	8550 Diesel	166·65	2·26
Case		◇	
1	770 Manual	36·68	2·55
2	770 Manual	42·61	2·50
3	770 Power Shift	38·83	2·48
4	770 Power Shift	42·95	2·45
8	870 Power Shift	50·33	2·45
—	970 Power Shift	55·60	2·53
11	1070, 1090 Manual	63·54	2·53
13	1070 Power Shift	71·15	2·50
15	1270	81·89	2·28
18	1370 Diesel	100·15	2·15
24	2670 Diesel	145·17	2·48
25	2870 Diesel	163·83	2·44
International Harvester		●	
3	Int'l 454	26·73	2·34
5	Farmall 544	33·90	2·52
8	Int'l 666 Diesel	43·52	2·36
10	Farmall 766	55·61	2·18
13	Farmall 856	67·21	2·52
22	Int'l 1066 Turbo Diesel	81·22	2·36
28	Farmall 1468	96·47	2·61
29	Int'l 1566 Turbo Diesel	104·70	2·30
33	Int'l 4366 Turbo Diesel	126·00	2·36
35	4568	175·85	2·32
37	4786 Diesel	198·10	2·35
John Deere		⬡	
1	830 Diesel	21·23	2·17
3	1530 Diesel	28·99	2·26
4	2030	39·04	2·33
10	2840 Diesel	50·33	2·28
11	4000	61·61	2·59
15	4240 Diesel	71·00	2·24
18	4430 Quad-Range	81·66	2·36
John Deere—*contd.*		⬡	
23	4620 Syncro-Range	87·57	2·34
25	4640 Diesel	102·50	2·39
26	4840 Diesel	117·19	2·47
34	8640 Diesel	155·45	2·59
Steiger		⬢	
2	Bearcat	123·57	2·46
—	Bearcat II PT-225	134·92	2·13
6	Cougar III ST-250	156·92	2·35
8	Cougar III ST-270	215·56	2·49
10	Panther III ST-325	210·33	2·80
—	Panther III PT-350	227·89	2·71
Ford		△	
1	1600 Diesel	13·82	2·26
2	2000 6-speed	20·51	2·52
5	3000 6-speed	26·77	2·62
7	3600 8-speed	25·44	2·47
8	4000 8-speed	35·76	2·56
12	5000 8-speed	44·88	2·64
21	7700 16-speed	53·39	2·21
24	8700 16-speed	74·69	2·42
27	9600 Dual Power	86·83	2·49
29	TW-30 Diesel	107·55	2·48
31	Fw 30	158·17	2·41
33	Fw 60	213·04	2·64
Massey–Ferguson		■	
1	MF230 Diesel	19·89	2·91
2	MF235 Diesel	27·07	2·64
5	255 12-speed	33·38	2·53
7	MF275 Diesel	44·88	2·61
8	MF285 Diesel	52·62	2·54
11	MF1105 Diesel	64·35	2·28
12	MF1135 Diesel	79·46	2·48
13	MF1150	92·47	2·64
15	MF1500	113·97	2·46
18	MF1805 Diesel	126·95	2·45
White		◆	
1	2-50 Field Boss	31·24	2·66
3	2-70 Diesel	44·36	2·39
4	2-85 Diesel	53·66	2·29
5	2-105 Field Boss	67·74	2·35
6	2-150 Field Boss	95·11	2·39
8	4-180 Field Boss	116·78	2·28

TABLE IV.8

Tractor fuel consumption at rated engine speed and reduced engine speed (50% of pull at maximum power) (source: Stout, 1984)

Tractor make and model	*Engine speed*	*kW*	*km/h*	*rpm*	*Gear*	*Fuel consumption*		*Fuel savings (%)*
						(kWh/liter)	*(liters/h)*	
Allis–Chalmers 7020 Diesel	Rated	44·4	8·52	2 510	5 (1F)	2·04	21·80	
	Reduced	44·1	8·50	1 815	7 (2F)	2·51	17·56	19
Case 2200 Diesel	Rated	44·7	11·04	2 252	8 (3I)	1·93	23·16	
	Reduced	44·9	11·01	1 473	10 (4L)	2·56	17·56	24
Deutz DX-110 Diesel	Rated	35·3	9·13	2 371	9 (1S)	2·48	14·19	
	Reduced	35·1	9·11	1 492	12 (5Z)	2·97	11·85	17
Ford TW-20 Diesel	Rated	50·4	8·98	2 412	8 (5PD)	1·99	25·32	
	Reduced	50·4	8·99	1 561	11 (6PD)	2·77	18·21	28
International 1086 Diesel	Rated	46·3	9·11	2 576	8 (1-HiTA)	2·07	22·37	
	Reduced	46·3	9·14	1 513	12 (2-HiDD)	2·76	16·77	25
John Deere 4440 Diesel Power	Rated	46·5	7·94	2 305	4	2·09	22·26	
	Reduced	46·3	7·94	1 364	6	2·79	16·62	25
Massey–Ferguson 2745 Diesel	Rated	51·5	9·76	2 770	14 (5I)	1·85	27·82	
	Reduced	51·5	9·76	1 998	17 (6I)	2·43	21·23	24
White Field Boss 2-135 Diesel	Rated	48·1	10·29	2 342	11 (4-D)	1·92	25·09	
	Reduced	48·4	10·27	1 561	14 (5-D)	2·51	19·27	23

throttling back, and less fuel is required. In sixth gear at 1364 rpm, the tractor produces 46·3 kW at 7·94 km/h, but the work output increases to 2·79 kWh/liter. However, the fuel consumption rate decreases from 22·26 to 16·62 liters/h—a saving of 25%.

The correct combination of tractors and implements, use of tractor mounted equipment, and power-take-off driven implements all influence energy use.

Tire size and type also affect energy use efficiency. Suppose a 45-kW tractor is fitted with 12·4/11·36 tires and a three-furrow mounted plow. Suppose the weight on the rear axle of the tractor with the plow in operation is 1180 kg, and it is intended to plow land with a resistance of 3·5 kilonewtons (kN) per furrow in poor traction conditions. The total pull is 10·5 or 5·25 kN per driving wheel. The slip would be more than 20% for a 12·4/11·36 tire, and the pull would be significantly above the value of 4·2 kN required for the maximum tractive efficiency (D. Hunt, 1983).

Following are three methods to increase tractive efficiency (J. Matthews, 1982):

(1) Increase the weight per driving wheel to 1600 kg by adding ballast. This procedure necessitates increasing the inflation pressure from 0·8 to 1·5 bar. The tractor will then be able to pull the plow comfortably and the maximum tractive efficiency will be 61%.
(2) Reduce the plow to two furrows. The pull per driving wheel will then be reduced to 3·5 kN, which will be within the pull available with a load of 1180 kg per driving wheel. The inflation pressure can be maintained at 0·8 bar and the tractive efficiency will be 66%. Therefore, provided the tractor can be operated at a sufficiently high speed to use all of the engine power available, the rate of work will be 8% higher than with method (1).
(3) Assemble 16·4/14·30 tires and ballast to 1600 kg load per driving wheel. A pull of 5·25 kN is within the capability of a 16·9/14·30 tire, even in poor traction conditions, and the inflation pressure need only be 0·8 bar. The maximum tractive efficiency will be 64% so that, provided the full engine

TABLE IV.9
Comparison of energy inputs for tilling 1 ha of soil by humans, oxen, 6-hp tractor, and 50-hp tractor (source: Pimentel & Pimentel, 1979)

Tilling unit	*Required hours*	*Machinery input (kcal)*	*Petroleum input (kcal)*	*Human power input (kcal)*	*Oxen power input (kcal)*	*Total input (kcal)*
Human	400	6 000	0	194 000	—	200 000
Oxen (pair)	65	6 000	0	31 525	260 000[a]	297 525
6-hp tractor	25	191 631[d]	237 562[b]	12 125	—	441 318
50-hp tractor	4	245 288[e]	306 303[c]	2 400	—	553 991

[a] Each ox is assumed to consume 20 000 kcal of feed per day.
[b] An estimated 23·5 liters of gasoline was used.
[c] An estimated 30·3 liters of gasoline was used.
[d] An estimated 191 631 kcal machinery was used in the tillage operation.
[e] An estimated 245 288 kcal machinery was used in the tillage operation.

power is used, the rate of work will be 5% higher than with method (1) above. This performance is not quite as good as with method (2), but may be easier to attain because it does not necessitate an increase in speed to use total engine power. Radial tires are also more energy efficient than bias ply tires.

Extending the working life of a machine helps justify the energy cost associated with its manufacture. Therefore, proper maintenance and adjustments save energy as well as repair costs.

Tillage

Tilling the soil requires a considerable amount of energy, whether accomplished by human, animal, or tractor power (Table IV.9). Figure IV.14 shows a typical digging hoe widely used for tillage operations in developing countries. Figure IV.15 shows a buffalo-drawn plow. In developing countries, tillage accounts for most of the commercial energy used in agricultural production since seeding, weeding, and harvesting is frequently performed by hand, and crop drying is usually achieved through solar energy. Savings in energy inputs for tillage could significantly alleviate the dependence of crop production on fossil fuels. Tables IV.10, IV.11, and IV.12 provide results of various tillage energy studies.

Fig. IV.14. Farmer with digging hoe in the Upper Solo River basin of Indonesia (source: FAO Photo 9178/P/27a).

Fig. IV.15. Buffalo-drawn plow being used by a Thai farmer (source: FAO Photo 7156/H/20,25).

Conservation tillage is a term that includes the following energy and soil saving practices (Sims *et al.*, 1983).

—*Direct drilling:* the planting of seeds (and placement of fertilizer) into uncultivated or fallow ground where the existing cover has been eliminated by chemicals.

—*Over drilling:* the planting of seeds into uncultivated or fallow ground without use of herbicides to remove competition. Zero tillage is another term used to describe this practice.

—*Reduced or minimum tillage:* conventional cultivation practices, but with the fewest realistic number of tillage operations and passes to establish an acceptable seedbed. This may include the use of specialized combinations of tillage equipment and/or chemicals to control previous cover or weeds.

The advantages of these limited tillage systems include (Fluck & Baird, 1980):

—Less fuel because of fewer and less energy-intensive field operations.

—Higher yields, particularly in dryland farming and on well-drained land.

—Less time and labor.

—Intensified land use. Double cropping potential has been extended several hundred miles north.

—Ability to farm lower quality land. Row crops may be grown on slopes up to 20%.

TABLE IV.10

Approximate analysis of energy used in tillage for all crops in the United Kingdom (source: Matthews, 1982)

Item	*Energy (GJ)*	*Efficiency*	*Proportion (%)*
Primary tillage	3·7	—	—
Secondary tillage and seedbed preparation	1·8		
Planting	1·0		
Total for tillage (T)	6·5		
Traction efficiency (E_1)		0·6	
Energy at axles (T/E_1)	11·0		
Transmission efficiency (E_2)		0·9	
Energy from engines (T/E_1E_2)	12·0		
Engine efficiency (E_3)		0·24	
Fuel energy ($T/E_1E_2E_3$)	50		
Petroleum fuels used on farms	85		
Proportion on tillage			59
Energy available from all tractors in UK[a]	72		
Proportion on tillage			17

[a] Calculated from 400 000 tractors × 50 kW mean × 1000 h per annum × 3600 J/kW.

TABLE IV.11

Energy and labor requirements for cultivation of clay loam soil (source: Lewis, in Robinson & Mollan, 1982)

Cultivation system	*Number of passes*	*Worker (h/ha)*	*Energy at implement (MJ/ha)*	*Tractor fuel (GJ/ha)*	*Energy for manufacture and mtce (GJ/ha)*	*Energy for labor (GJ/ha)*	*Total (GJ/ha)*
Plow (20 cm)							
Disk-harrow: drill	3	3	320	2·07	0·62	0·21	2·90
Shallow plow (10 cm)							
Combined cultivator and drill	2	1·5	180	1·17	0·25	0·10	1·52
NIAE rotary digger							
Combined cultivator and drill	2	1·5	180	1·17	0·27	0·10	1·54
Spray							
Direct drill ('chemical plowing')	2	0·5	30	0·19	0·10	0·03	0·58[a]
Tractor efficiencies							
Field efficiency					75%	overall 15·4%	
Wheel slip and rolling resistance					70%		
Fuel efficiency					33%		
Factor for fuel oil 1·12 GJ input/GJ of calorific value							

[a] Including herbicide—0·56 kg/ha paraquat, 0·26 GJ/ha.

Source: NIAE data (average 1971–74) for winter wheat.

TABLE IV.12

Comparative energy inputs, excluding traction losses, to NIAE experimental plots during 6 years, MJ/ha (primary tillage only in parentheses) (source: Matthews, 1982)

	Soil and crop		
	Clay loam winter wheat	*Silty loam winter wheat*	*Clay loam spring barley*
Traditional plow system	320 (245)	180 (118)	324 (307)
Shallow (100 mm) plow, combined seedbed preparation, and drill	187 (115)	108 (68)	203 (133)
Chisel plow (125 mm) twice, combined seedbed preparation, and drill	286 (203)	194 (147)	308 (213)
Rotary digger (tines 200 mm, rotor 100 mm), combined seed preparation, and drill	176 (117)	133 (88)	201 (156)
Direct drill, preceded by herbicide	38 (—)	43 (—)	54 (—)

TABLE IV.13

Comparison of energy inputs for spraying herbicide on 1 ha by manual labor and for using a 50-hp tractor (source: Pimentel & Pimentel, 1979)

Spraying unit	*Required hours*	*Machinery input (kcal)*	*Petroleum input (kcal)*	*Human input (kcal)*	*Total input (kcal)*
Human	3·0	8[a]	0	900	908
50-hp tractor	0·7	21 463[b]	30 327[c]	210	52 000

[a] An estimated 8 kcal of machinery was used in the spraying operation.
[b] An estimated 21 463 kcal of machinery was used in the spraying operation.
[c] An estimated 3 liters of gasoline was used.

Other Operations

Table IV.13 illustrates the relative efficiencies of applying herbicides manually and with a tractor-powered sprayer. While the tractor operation is much faster, it requires significantly more energy.

ENERGY MANAGEMENT IN THE POST-HARVEST SYSTEM (NRC, 1981)

As in agricultural production, the use of energy is critical to the efficiency and modernization of the post-harvest system (Fig. IV.16). Because the major portion of food crops is produced in the rural sector of developing countries and remains predominantly outside the cash economy, comparatively little fossil fuel is used in harvesting, drying, processing, and transporting these basic staples, with notable exceptions, such as rice milling. The greatest amount of energy consumed in the

POST-HARVEST SYSTEM

Primary processing and storage

- Drying
- Threshing

Products—Markets—Consumers

Secondary processing

Cereals and pulses

—Milling/hulling
—Parboiling
—Grinding
—Oil expelling and crushing

Perishables

- Potatoes, cassava, and other tubers
  —Drying/grinding
  —Fermenting
  —Refrigeration
- Fruit, vegetables, fish, and meat
  —Refrigeration
  —Salting, drying, and smoking
  —Pickling
  —Canning and bottling

Commercial crops

- Tobacco
  —Drying
- Coffee
  —Drying and roasting
- Sugar cane
  —Crushing
  —Refining

Markets

Consumers

Fig. IV.16. Energy use in post-harvest operations (source: NRC, 1981).

rural sector is used to cook, and, because of the important agricultural effects of deforestation, cooking—the final and most critical step in the food chain—must be considered along with energy use in agriculture as a whole. The production of fuelwood on marginal lands, roadsides, hedgerows, and other areas around fields and settlements, in addition to more traditional forest conservation efforts on watersheds, is becoming increasingly important to support sustained agricultural productivity. Demand is increasing for rural wood supplies to produce cooking fuel, particularly charcoal, as a replacement for electricity, kerosene, and bottled gas in urban areas.

Energy use in the post-harvest system must increase to improve the quality and quantity of food and food products and to increase their value to the consumer. However, this may be accomplished by increasing the efficiency of energy use and using different sources of energy more efficiently to carry out post-harvest operations.

For example, increased food processing can lead to energy conservation by prolonging shelf life, reducing transport bulk or weight, decreasing home cooking, and producing more byproducts.

Reduction of post-harvest losses could result in the utilization of energy that has already been expended in the production of the food; this is obviously of particular importance in areas where energy-intensive fertilizer and other inputs have been used in its production.

Transportation

The demand for transportation intensifies as urban populations grow and agricultural production increases. The transportation sector in developing countries currently uses 50% or more of the liquid fuel supplies, most (80%) of which is allocated to road transportation. Estimates indicate that 60% or more of this supply is used in trucks, and about half of this 60% is for transporting food and other agricultural commodities or agricultural inputs. In total, between 12 and 15% of the liquid fuel supplies in developing countries are consumed in agricultural transportation, although these percentages vary according to the degree of industrialization and the geography of particular countries. In addition, large amounts of animal and human energy are used in transportation at the local village level.

Significant substitutions for liquid fuels in the transportation sector are unlikely in the near future. Therefore, the overriding question is the following: What is the least expensive system of assembling and transporting agricultural products, foodstuffs, and inputs between geographically dispersed production units and consumers?

In the following examination of liquid fuel conservation, the agricultural transportation system is divided into three segments:

1. *Local transportation:* the movement of agricultural products from the farm to the local market, usually a distance of less than 25 km. Estimates show that this segment of the system accounts for a relatively small portion (2%) of total consumption of liquid fuels, because most local transportation is by traditional methods such as bullock carts, handcarts, and bicycles.
2. *Inner-city transportation:* the movement of goods from the local market to the gates of the town or city where they will be consumed. This is long-distance transportation (200–300 km) by trucks of various sizes, or by trains or waterways. This segment of the agricultural transportation system comprises the largest share, perhaps 10%, of total liquid fuel consumption.
3. *Urban transportation:* the distribution of agricultural products within the city or town. The distances are relatively short: 25 km on the average. Although both commercial and traditional transportation is available, the more extensive use of commercial transportation in the interurban segment and the

prevalence of congested traffic imply that the share of liquid fuel consumption is about 4%. This percentage is higher than the share consumed by the local sector but much lower than that of the inner-city transportation sector.

For each of these sectors, three types of conservation possibilities are examined (Table IV.14):

1. Optimizing or minimizing the demand for agricultural product transportation.
2. Shifting to more energy-efficient forms of transportation.
3. Improving road transportation efficiency.

Optimizing Transportation Services

Liquid fuel supplies can be conserved by taking the following steps:

—*Local assembly.* Product bulking at local market centers and temporary storage facilities is necessary to accumulate sufficient quantities to ship to more distant markets in larger and more efficient vehicles.

—*Processing.* Processing products in the assembly areas is a method of reducing the physical bulk and net tonnage of products before long-distance shipment to

TABLE IV.14

Opportunities to conserve energy in agricultural transportation (source: NRC, 1981)

	Local	*Inner-city*	*Urban*
Optimizing demand for transportation	Minimize commercial traffic through relocation of storage and collection facilities	Storage facilities, both local and city Processing and drying Improved load factors: —information communication —coordination of agricultural input and output transportation —more common carriers rather than capture trucks	Break-bulk facilities and organized wholesaling services
Shifting modes of transportation	Incorporate mechanization, if available	Rail—improvement of existing facilities Water—improvement of port facilities	
Increasing efficiency	Utilize appropriate technology, maintenance energy, and energy-efficient vehicles	Maintenance Highways Speed limits Improved driving Energy-efficient trucks Larger trucks Route selection	Efficient trucks Rational break-bulk facilities
Implementing fuel substitution	Producer gas	Ethanol Methanol Vegetable oils	LPG Electricity Producer gas from charcoal

urban centers. It can also be important in the conversion of perishable products that require high-cost, specialized transport and handling services (i.e. milk, fruits, and vegetables) into more durable commodities. Feasibility studies can determine the ideal number, size, and location of processing facilities to minimize assembly and shipping costs.

—*Urban distribution.* The wholesaling function requires market terminal facilities and institutional arrangements for dividing shipments into sizes suitable for movement to retail outlets. Here, optimization in the location and layout of the wholesale market and terminal facility can also reduce transportation and energy costs.

—*Distribution of agricultural inputs and foodstuffs to rural areas.* The backflow of these commodities can be less costly if shipments are coordinated with the outbound movement of agricultural products to urban centers. Information systems that link trucking companies with potential shippers are essential for the effective use of trucking equipment.

The seasonality of agricultural production and farm demand for commercial inputs, such as fertilizers, complicates the coordination of transportation services, Temporary storage in rural assembly and distribution centers is required to meet market demands.

National, regional, and urban planning units can influence the overall costs of agricultural transportation services through careful projections of demand and analyses of investment decisions involving market facilities, transportation equipment, and institutional development assistance.

Shifting to More Energy-Efficient Forms of Transportation

Within the food sector, the transportation industry functions within three distinct settings, each of which has particular operational and technical requirements and, consequently, unique opportunities to shift to more energy-efficient methods.

In the *local* setting, products are transported from the farm to storage facilities or local markets by mechanized vehicle or by human or animal power. Although human- and animal-powered transportation is still prevalent in many traditional societies, mechanized vehicles have become increasingly important as a method of local transportation. This mechanized trend is expected to continue, although perhaps at a somewhat slower rate than in the recent past.

In local settings, transportation cost is the only value-added increment to farmgate price. Hence, at this level, transportation represents a fairly large share of product price because of short distances (typically an average of 25 km one way). However, fuel is not a particularly large component of this price, thus fuel price is not a major factor in local transportation.

Because produce must be transported by the most available means, the key factors in the choice of transportation methods at the local level are their availability and cost. If mechanized vehicles and the fuel needed for their operation are available, they will be used regardless of energy efficiency. Subsidized fuel prices may artificially constrain fuel supplies, thereby limiting the availability of mechanized vehicles, but generally speaking, the existence of such vehicles virtually ensures their use for local transportation.

These facts have important implications for energy use in local transportation and suggest a possible role for appropriate new technologies, such as producer gas generators now being introduced for jeepneys in the Philippines.

In *long-distance* transportation, agricultural products are moved from local storage facilities to regional terminals or markets by rail, water, and truck travel. In some cases, particularly with export crops, long-distance transportation may supersede local transportation as produce is moved directly from production center to regional market or storage facility. Rail and water systems are comparable in energy efficiency. When a rail system is in place and the railhead or siding is sufficiently accessible to permit its use, rail transportation is a viable, energy-efficient option.

If rail lines do not exist, traffic density must be of the order of 1 million tonnes/year to justify investment in railroad infrastructure. Since this level of traffic density is quite unlikely in most cases, increased rail transportation is generally not a feasible energy conservation option. Water transportation, on the other hand, has far more potential, particularly because many developing countries have relatively undeveloped water transportation systems. In general, river transportation is three to four times more energy efficient than long-haul trucking and, if feasible, its use should be encouraged.

In the *urban* setting, agricultural products are transported by truck from regional terminals, typically wholesale markets in congested core areas, to retail facilities. In rationalizing the transportation system, these terminals could be located on the periphery of the urban area and local distribution could be handled by small liquid petroleum gas (LPG), or electric-powered trucks and operations (i.e. load matching) could be coordinated through regional terminals. This method assumes, of course, that the urban electricity is not generated by fossil fuels.

Improving the Energy Efficiency of Road Transportation

One important way to increase the energy efficiency of the transportation sector is to increase the efficiency of a particular transportation mode through improved maintenance, more effective use of existing vehicles, and incorporation of more efficient vehicles.

By using turbochargers, electronic injection systems, improved gear-axle ratios, etc., new truck models are 10–15% more energy-efficient than the vehicles they replace. Use of lightweight materials also decreases tare weight, thereby permitting increased payload capacity. However, these new vehicles are more difficult to maintain, require better lubricants, and need to be properly tuned. In particular, tuning requires somewhat sophisticated and often scarce equipment. As a result, even when capital is available, these new trucks may not yet be feasible for use in developing countries with a shortage of skilled mechanics and adequate repair facilities.

Important improvements in energy efficiency can also be achieved by improved use of existing equipment. Experience has shown that energy savings of between 3 and 4% can be achieved through proper tuning, appropriate lubrication, and the timely replacement of air filters. Correct tire pressure is also important; one tire manufacturer estimates that energy consumption could increase by 7–8% when tire pressures are 20% below optimal.

Driving behavior is also important in energy conservation. Experience in Europe has shown energy savings of as much as 9–10% by better driving habits. Special devices now available can indicate the instantaneous fuel consumption of a motor vehicle. These devices can be used to train truck drivers. In addition, salary bonuses related to energy conservation can be used to motivate better and more efficient driving.

Energy consumption is directly related to road gradient and curvature and to traffic congestion. All infrastructure improvements that will reduce one or more of these conditions will improve the energy efficiency of the truck fleet. Reduction in road roughness through improved road maintenance, better road rehabilitation, and more road paving should also result in improvements in energy efficiency; however, these improvements often result in increases in average vehicle speed. Because speeds beyond 60–80 km/h are extremely costly in relation to energy, the net result of road improvements may be increased energy consumption. However, road improvements are justified by other resulted savings, such as reduced vehicle maintenance costs and reduced travel time. Many countries have introduced speed limits on their main highway networks as a conservation measure and have achieved satisfactory results.

More careful load management techniques could make the use of much larger trucks possible. Currently, the average size of interurban trucks in developing countries is about 8 t, consuming an average of about 0·04 liters/t km. Significantly larger trucks (15–20 t) use only about 0·025 liters/t km, but in order for operation of the larger trucks to be profitable, larger loads must be available on a predictable basis.

Table IV.14 summarizes opportunities for conserving, or more efficiently utilizing energy in the agricultural transportation system.

Food Processing (NRC, 1981; Syed, 1982)

A strategy for increasing the output and efficiency of energy use in food processing systems should include the following goals (NRC, 1981):

—To increase the use of energy for food processing and preservation to improve the quantity and quality of foods available, emphasize the use of renewable fuels, and decrease the use of fossil fuels when possible.

—To increase fuel conservation in the food systems of developing countries by promoting the recycling of waste heat and the adoption of efficient food processing equipment.

—To use food processing to reduce the weight and volume of food for transportation, thus reducing the energy cost for distribution of processed food.

—To promote food processing to extend the shelf life and availability of foods, thereby reducing the wastage of energy inputs for agricultural production.
—To increase the use of byproducts, such as residues and wastes from food processing, for energy generation, including production of biogas; to adapt or retrofit renewable energy power sources in areas where they are economically viable.
—To increase training at the professional and technical level to attain more energy-efficient food processing; to increase the trained cadre for assessing energy needs and alternatives for food processing and disseminating appropriate technologies in developing countries.
—To increase access to food technology databases and distribute relevant literature on energy-efficient food processing operations, projects, and technologies.
—To substitute alternative food commodities that require less energy for processing and consumption (cooking) when culturally feasible; to promote food processing for commodities that are locally abundant and economically suitable as processed export goods.

The overall goal is not to reduce the total energy expended in processing food; instead, the goal is to increase the amount of processed foods when yield losses in the post-harvest system reflect both energy and food waste. This goal actually requires more energy. However, the most desirable energy system for a particular area, the efficiency of each system, and the economic and cultural impact of using alternative energy systems for food processing and consumption vary significantly. Therefore, training and expertise are crucial in making wise choices among alternative energy sources. Training deserves strong support in all development programs, especially in choosing appropriate measures for energy use in food processing.

ENERGY AND THE FOOD CONSUMPTION SYSTEM

Fuel Use in the Food Consumption System (NRC, 1981)

The food consumption system is defined to include homes, restaurants, food stands, hotels, and other places where food is prepared for final consumption. This system is a large consumer of energy in developing countries, generally accounting for at least 20% of the total energy use and more than 50% of the rural energy use. Energy is used primarily for cooking, but also for drying, washing, and refrigeration. A diversity of fuels are commonly used, such as agricultural residues, dung, kerosene, bottled gas, and electricity. Table IV.15 presents a matrix of energy uses by different types of establishments, and Table IV.16 indicates the general nature of the current energy delivery for food consumption systems.

Traditional energy sources are the principal forms of biomass energy, particularly in the form of charcoal and wood. The Middle East is an exception because biomass is not readily available. The mixture of wood, twigs, agricultural residues, dung, and

TABLE IV.15
Fuel mix in the food consumption system (source: NRC, 1981)

	Charcoal	*Wood*	*Agricultural residues (including dung)*	*Coal briquettes*	*Kerosene*	*LPG*	*Electricity*
Urban households	×	×			×	×	×
Rural households		×	×	×	×		
Hotels	×				×	×	×
Restaurants	×	×			×	×	×
Food stands	×	×			×		

charcoal fuels varies according to the country. In rural areas, fuels for household use are generally collected free of charge, while in urban areas, fuels are primarily purchased. Energy used in cooking is estimated at 0·5–0·7 t of air-dried wood per capita annually. As incomes increase, the use of commercial fuels for cooking expands in both rural and urban areas. There is increasing usage of LPG as a substitute for kerosene by upper-income urban families in areas where it is readily available, such as Southeast Asia and North Africa. The LPG distribution network, however, does not generally extend beyond major towns.

The current food consumption system is characterized by low efficiency of energy conversion. Household cooking processes generally average between 5 and 10% efficiency compared with 30% for a modern stove. Fire in a developing country often serves multiple functions: cooking; preventing insects; preserving food; and providing security, social interaction, lighting, and heating. Therefore, energy efficiency considerations alone may not determine what is socially desirable and acceptable.

TABLE IV.16
Current energy delivery for the food consumption system (source: NRC, 1981)

Source	*Conversion device/process*	*Product*
Petroleum	Refineries	Kerosene
		LPG
	Generators	Electricity
Biomass	Cutting	Fuelwood
	Digesters	Agricultural residues
		Methane gas
	Kiln	Charcoal
	Caking	Dung cakes
Coal	Carbonization plants	Briquettes
Hydroelectricity		Electricity

TABLE IV.17
Characteristics of cooking heat required for different foods (source: NRC, 1981)

Low heat	*High heat*
Wet	
Stews: meat, fish	
Stews: vegetable	Porridge/foufou
Beans	Rice
Hominy: maize, sorghum	Maize on cob (boiled)
Boiled sweet potatoes	Stir fry
Boiled potatoes	Soups
Boiled squash, pumpkin, marrow, papaya	Boiled eggs
Couscous	Tea, coffee
Bulgur wheat	Candies
Curries	
Soybean curd	
Custard	
Tamales	
Beer	
Dry	
Roast meat	Scones and bread
Roast corn	Chapatis
Roast peanuts	Tortillas
Roast squash	Sweet breads and cakes
Baked potatoes	
Baked sweet potatoes	
Preparation of semolina, beef jerkey, fish jerkey, parched corn	

Cooking requirements vary according to the type of food and quantity cooked (Table IV.17). Fuel use may vary depending on the degree of moisture, temperature, and heat duration. The availability and moisture content of particular crop residue fuels, such as jute sticks or corn cobs, is seasonal. Use of residues may result in additional costs to the production system through reduced soil nutrients and organic matter and increased erosion.

Impact of Oil Prices and Traditional Fuel Availability on Cooking Patterns and Nutrition (NRC, 1981)

The demographic revolution of the late 1950s and 1960s, and the explosive rise in oil prices in the 1970s, led to a crisis in the availability of cooking fuels in developing countries. Deforestation pervades the Third World because of the increased demand for biomass fuels. Major shifts in fuel consumption patterns are occurring in urban areas where the middle-income and poor families can no longer afford to use kerosene, butane, and LPG. These families are being forced to substitute fuelwood and charcoal for fossil fuels. This phenomenon further aggravates the deforestation problem.

If the increasing urban demand for firewood and charcoal can be met through increased supply and development of alternative heating technologies, rural areas can better supply their own needs. However, problems arise when rural areas must meet the energy requirements for both urban and rural cooking.

Even though world oil prices dropped in the mid-1980s, energy prices in many developing countries remained high. High fuel costs and convenience have an impact on urban consumption patterns and cooking processes. For example, urban residents increasingly use bread, sandwiches, tortillas, chapatis, ready-to-eat cereal products, and other convenience meals that are pre-cooked by commercial vendors. These products shift cooking from the individual household and allow for economies of scale. At the same time, the demand for wheat products increases, thus entailing foreign exchange costs.

Technological Alternatives (NRC, 1981)

Charcoal—Improved Kilns, Stoves, and Other Devices

Improved charcoal kilns of many types and sizes—metal drum, retort, brick beehive, and dome—have been tested in a variety of circumstances. Few have been adopted, and most charcoal production in developing countries is maintained by itinerant entrepreneurs, using traditional pit or mound methods. In areas where wood is gathered from public land for only the cost of labor, there is little incentive to acquire a costly, difficult-to-transport device to improve the efficiency of charcoal recovery.

In many countries, particularly in arid areas where trees are scarce, a remarkably sophisticated and efficient system of contractual relationships exists among charcoal makers, transporters, and consumers, which provides employment and income to numerous individuals at the expense of the countries' dwindling forest resources. This system has proven difficult to alter, even in extreme situations in which the charcoal must be transported hundreds of kilometers. The capital investment and political desire to achieve a balance of use with regeneration that includes, rather than displaces, present charcoal producers have eluded most countries.

Successful fuelwood plantations for production of charcoal are operating in a limited number of locations, notably in Brazil and Argentina. More serious attention by governments, communities, and their technical assistance donors needs to be given to this aspect of the national energy supply. Charcoal use is likely to increase even more with the adoption of other technologies, such as producer gas generators, that will generate electricity and power fishing boats and trucks. Technologies for increasing the efficiency of production (kilns and retorts) and use (stoves, briquettes, and char gas) should be promoted when appropriate.

Biogas

Biogas generation requires considerable capital investment; a plentiful and regular source of substrate, preferably animal dung; and a fair level of technical competence to obtain sufficient quantities of gas to justify the effort. With notable exceptions, successfully operating biogas generators are typically associated with fairly sophisticated, integrated systems of waste management based on cattle, swine, or poultry production, in which the gas generated is a valuable byproduct rather than the main product. The Chinese experience, in which public health and nutrient conservation needs are the main objectives of the extensive biogas program, and the Indian experience, with community and household gobar gas generators, are consistent with this viewpoint.

Improved Cooking Stoves

Cooking foods using traditional methods is inefficient (Table IV.18). Development of low-cost, more efficient cooking stoves is a high priority (Foley & Moss, 1983). Despite the amount of effort concentrated on the design and diffusion of improved cooking stoves, based primarily on fuelwood, success has been sporadic. Some communities and governments have adopted new technologies with enthusiasm, but design, cultural, and cost factors have hindered more widespread adoption.

As of yet, no demonstrable connection between the introduction of 'efficient' cooking stoves and a reduced rate of fuelwood consumption exists, although reduced fuelwood consumption is usually the justification for the effort. The ways in which the efficiency of cooking stoves is measured and reported vary considerably. Indeed, improvement in efficiency and reduction of fuel consumption attributed to devices are often assumed, not tested or measured; and so far, no published evidence exists that indicates a reduction in fuelwood consumption resulting from the large-scale introduction of improved stoves. Much of the information on diffusion of improved stoves concerns the relatively affluent who have sufficient resources to acquire the technology and for whom access to fuelwood is not a critical limitation.

Thus, even in areas where stoves are constructed and used as their designers intended, they may use available fuel more efficiently for additional purposes, rather than reduce overall fuel use. Only a small amount of critical information is available about the desperately poor and the impact of technology on reducing the use of animal dung for fuel. In addition, scientists have demonstrated that when traditional (three-stone) cooking is performed carefully, efficiencies equal to those of well-designed chula or lorena stoves can be obtained. Countries suffering from fuelwood shortages obviously use their supplies carefully. Improved stoves should

TABLE IV.18

Range of efficiency results for different cooking methods (source: Manibog, 1984; reproduced, with permission, from the *Annual Review of Energy*, 9; © 1984 by Annual Reviews Inc.)

Method	*Range (%)*
Open fire	2–30
Indian chula	5–14
Brick chula	13–16
Closed fire	10–43

be introduced to rural and urban areas, but should not be expected to solve the problems of fuel shortages.

Recent advances in solar cookers (coupled with higher fuel costs) have made some types of solar cooking more attractive. Previous attempts to introduce solar cookers had limited success, but the time has come when some forms of solar heating can be used more extensively.

Solar Heating

In the short term, solar-heated hot water introduced into the cooking cycle will have a major impact on energy use in food preparation. A large portion of food preparation now relies on boiling water, steam, or hot water. If households can begin wet cooking processes with solar-preheated water, savings in fuel consumption would be in the magnitude of 40–60%, especially if water for tea and coffee is included. If the current estimates indicating that 60% of the energy required for agriculture is consumed in cooking are accepted, the domestic adoption of solar-heated hot water could result in an energy saving of as much as 30% for the agricultural sector. Solar-heated hot water would also have health benefits for bathing and washing.

Hot Pots

Solar heating can be used to boil stews, as well as other dishes that depend on wet cooking. The cooking process can then be completed by placing the heated dish in a fireless, insulated cooker in which cooking continues as long as the heat is retained.

EFFICIENT ENERGY MANAGEMENT IN AGRICULTURE IS GOOD BUSINESS

Energy is used in agriculture in a multitude of processes as discussed in this chapter. Sometimes increased energy input improves yields and profits. Examples include energy invested in fertilizers and pesticides which, up to a point, increases yields. Energy invested in irrigation also increases yields if rainfall is low.

In many agricultural operations, energy conservation is important and helps reduce input costs. Proper selection and management of tractors and machinery can also conserve fuel costs.

Post-harvest operations such as food processing, transportation, and cooking use large quantities of energy and require diligent conservation efforts. Pre-harvest and post-harvest food losses indirectly reflect energy losses. Thus, a reduction in food losses not only provides more food for consumption, but also decreases wasted energy.

CHAPTER V

Energy from Biomass

INTRODUCTION

The sun is the source, either directly or indirectly, of virtually all forms of renewable energy. Photosynthesis, the process by which the sun fixes carbon and produces all forms of plant life, is the principal energy conversion process in agriculture. The photosynthetic process, as expressed by plant productivity, is substantially enhanced by the judicious use of cultural energy in the form of fertilizer and fuels for pumping water and operating machines, and by energy subsidies in other forms. Photosynthesis is discussed in more detail in the next section of this chapter.

While petroleum-based fuels are used widely in agriculture, agriculturalists who are either unable to obtain petroleum or to afford it are interested in new and renewable energy forms for use in the food system. This chapter, as well as Chapters VI, VII, and VIII, discuss in detail direct solar energy applications in agriculture, indirect solar energy in the form of biomass fuel, small-scale wind energy conversion, and small-scale hydropower systems.

Biomass is defined as all organic matter except fossil fuels. The concept of renewability on an annual or short-term basis differentiates between biomass and fossil fuels. Biomass comes in many forms: wood, annual crops, crop residues, animal manure, marine plants, algae, and others. Biomass may be burned directly to produce heat or converted to more useful solid, liquid, or gaseous forms. Then, it can be stored and subsequently used for fuel (Fig. V.1). Conversion processes are generally classified as biological, extraction, or thermochemical as illustrated in Fig. V.2. A more detailed breakdown of thermochemical conversion options is provided in Fig. V.3.

A generalized scheme for recycling and utilizing organic materials for fuel and other purposes is shown in Fig. V.4. Some advantages and disadvantages of using biomass for fuel are listed in Table V.1. A major advantage is that biomass fuels are available and widely distributed in all countries. Although not necessarily free or

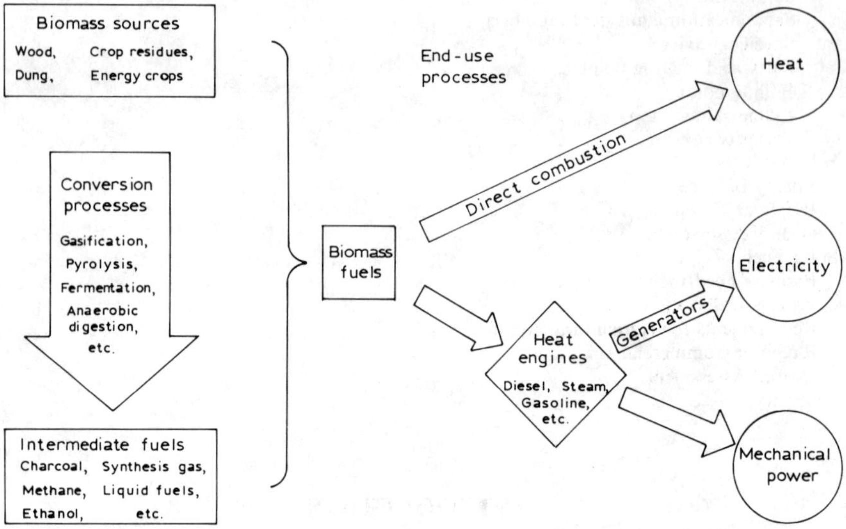

Fig. V.1. Methods of using biomass for energy (source: D. O. Hall *et al.*, 1982; © 1982 Pergamon Press plc).

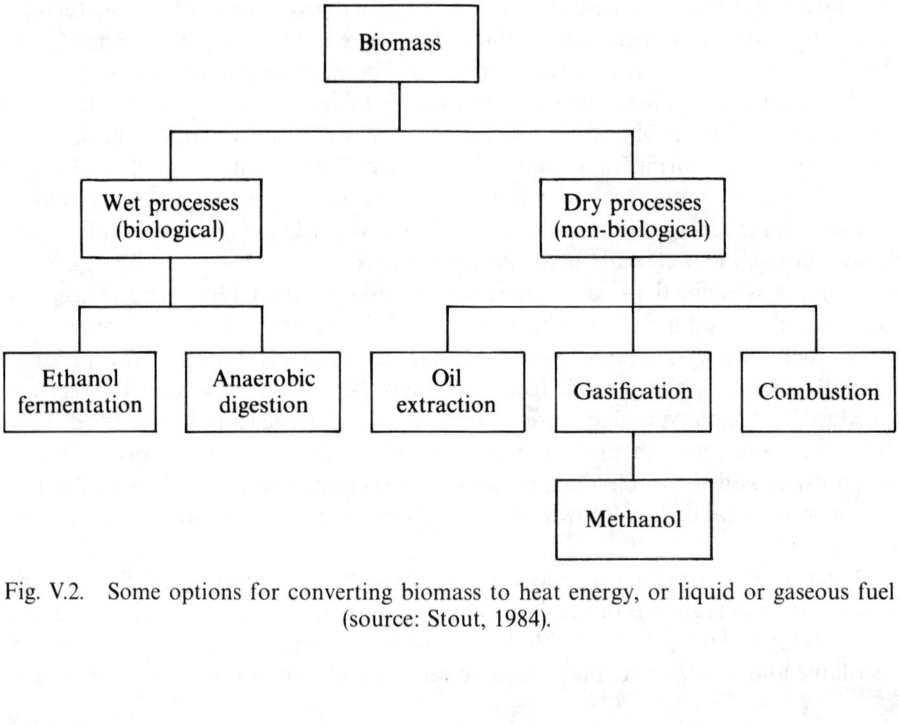

Fig. V.2. Some options for converting biomass to heat energy, or liquid or gaseous fuel (source: Stout, 1984).

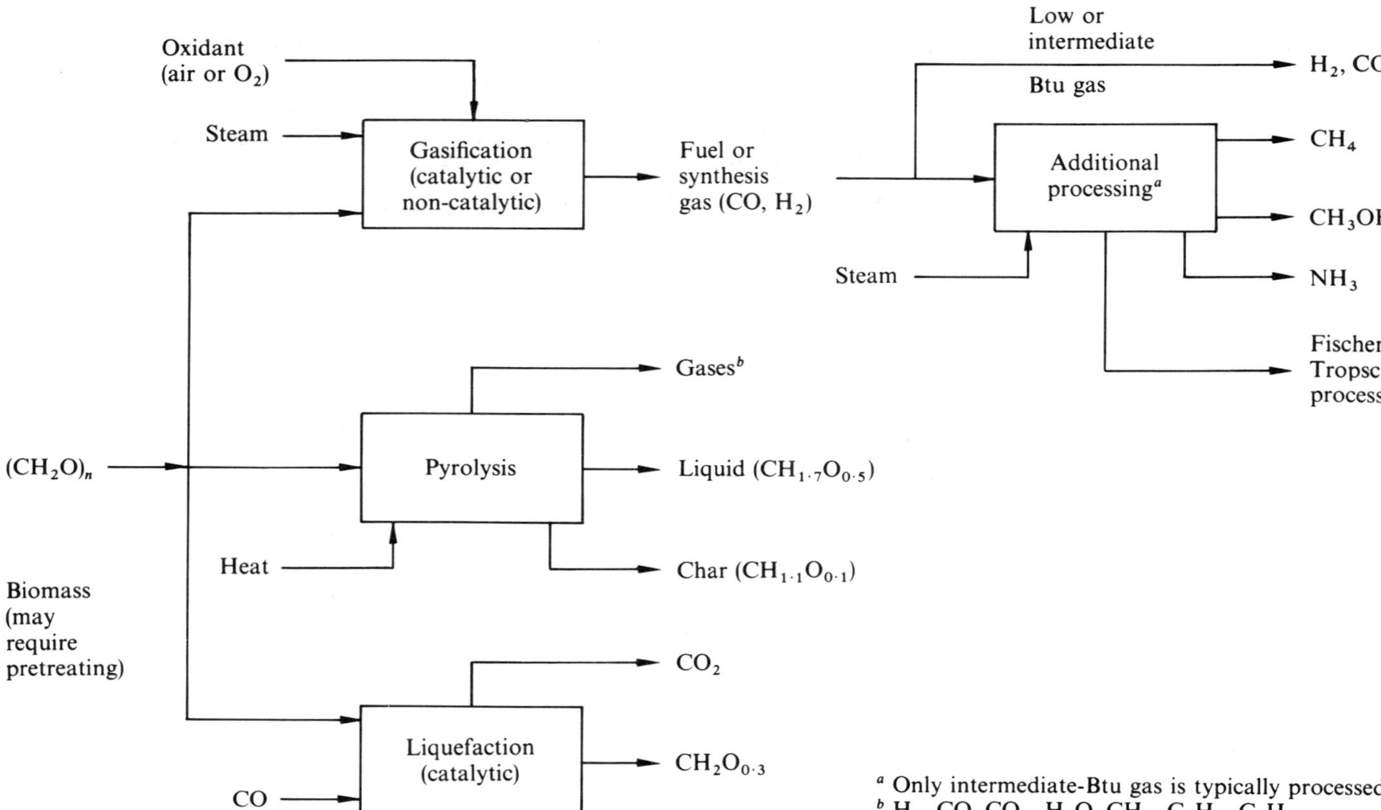

[a] Only intermediate-Btu gas is typically processed
[b] H_2, CO, CO_2, H_2O, CH_4, C_2H_4, C_3H_6

Fig. V.3. Generalized block flow diagram for the thermochemical conversion of biomass (source: Sofer & Zaborsky, 1981).

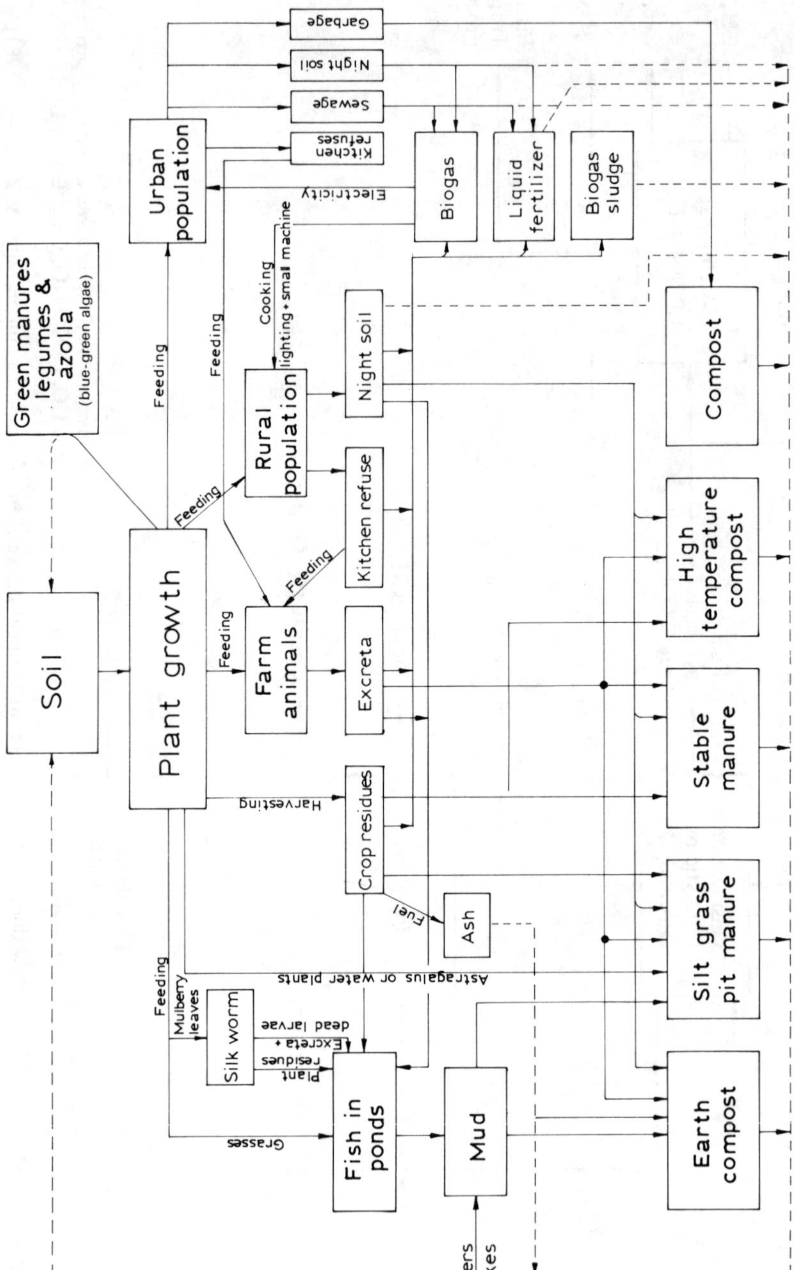

Fig. V.4. Recycling of organic wastes in the People's Republic of China (source: FAO, 1977).

TABLE V.1
Some advantages and disadvantages of using biomass for fuel (source: Mislin & Bachofen, 1982)

Advantages	*Disadvantages*
1 Stores energy	1 Land use competition
2 Renewable	2 Land areas required
3 Versatile conversion and products; some products with high energy content	3 Supply uncertainty in initial phase
4 Dependent on technology already available with minimum capital input; available to all income levels	4 Costs often uncertain
5 Can be developed with present manpower and material resources	5 Fertilizer, soil, and water requirements
6 Large biological and engineering development potential	6 Existing agricultural, forestry, and social practices
7 Creates employment and develops skills	7 Bulky resource; transport and storage possible problems
8 Reasonably priced in many instances	8 Subject to climatic variability
9 Ecologically inoffensive and safe	
10 Does not increase atmospheric CO_2	

inexpensive, these fuels are renewable and dispersed. But are they appropriate? Does it make sense to use biomass for fuel? The answer is complex and depends on many factors discussed in this chapter.

Rask (FAO, 1981) categorized various countries according to their energy and agricultural self-sufficiency (Fig. V.5). For countries with agricultural surplus and energy deficits (upper-left quadrant), biomass fuels probably make sense and should be explored. For countries with both agricultural deficits and energy deficits (lower-left quadrant), biomass residues for fuel may be logical if they do not detract from food production, and if they are economically viable. For countries with both agricultural and energy surpluses (upper-right quadrant), biomass for fuel might be appropriate if it is economically advantageous. Countries with agricultural deficits and energy surpluses (lower-right quadrant) may want to sell energy and buy food. Whether or not agricultural residues should be used for energy depends on local economic factors.

Many aspects of biomass energy systems are highly location-specific; for example, a gasification plant is most appropriate located near a rice mill or cotton gin. In both cases, a supply of biomass feedstock is collected and may even have a negative value because of high disposal costs.

Liquid fuels are highly desirable for mobile vehicles as well as for other

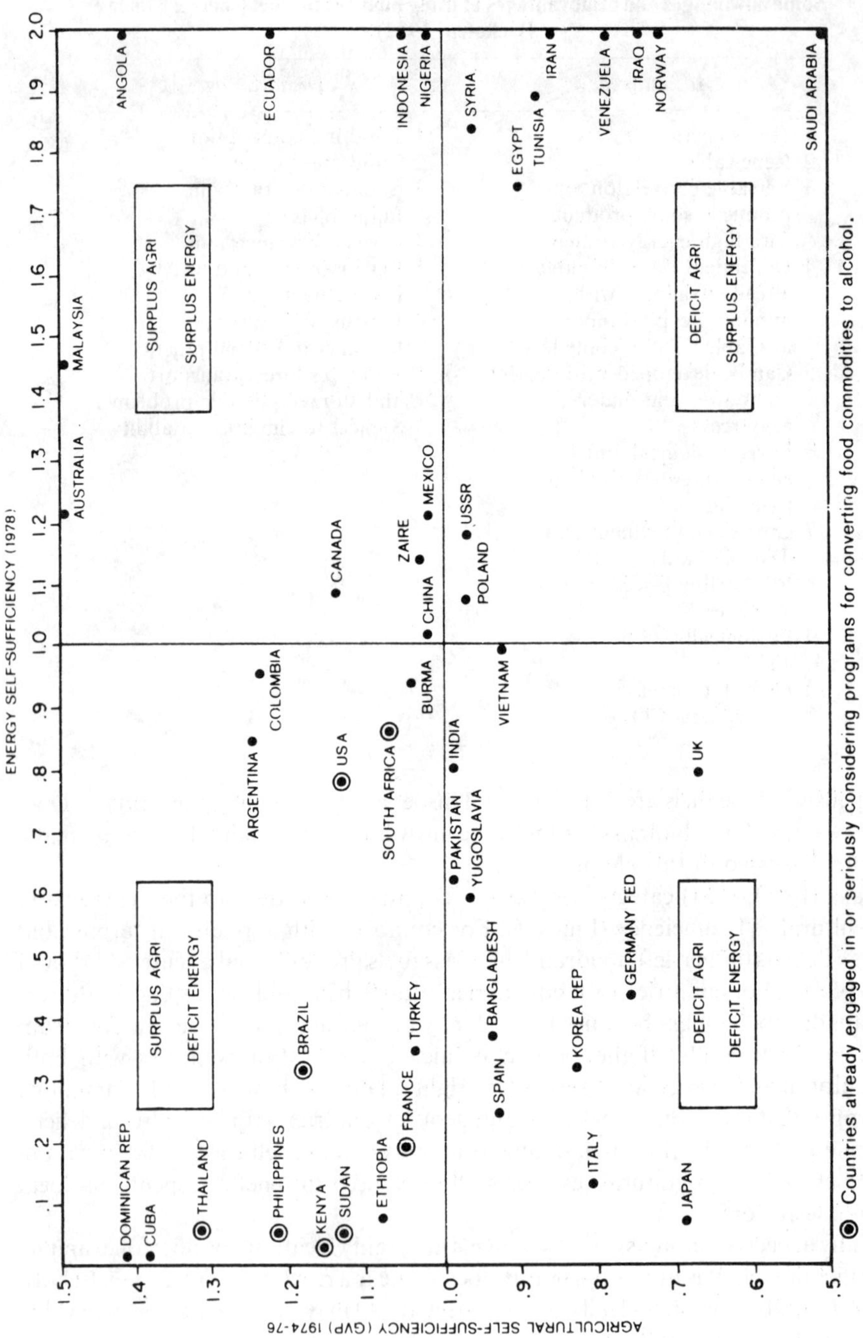

Fig. V.5. Energy and agricultural self-sufficiency (source: Rask, in FAO, 1981).

TABLE V.2
System versatility of liquid fuel from biomass options (source: D. O. Hall *et al.*, 1982)

Fuel	*Feedstock*	*Production system*			*Conversion system*		
		Agronomic[a] requirements for deriving high yields	*Range[b] of soil types tolerated*	*Drought tolerance*	*Relative capital cost*	*Relative technical sophistication*	*Suitability for small-scale applications*
Ethanol	Sugar cane	Strict	Narrow	Low	Moderate to high	Moderately complex	Yes[d]
	Cassava	Moderate	Wide	High			
Vegetable oils	Oil palm	Strict	Wide	Low	Low	Simple	Yes
	Sesame/castor groundnut, etc.	Strict	Wide[c]	High			
Methanol[e]	Wood	Moderate	Wide[c]	High	High	Complex	No
Synthetic[e] petroleum (Fischer–Tropsch) process)	Wood	Moderate	Wide[c]	High	Very high	Very complex	No

[a] Refers to fertilizer requirements and the need for irrigation, pest control, etc.
[b] Refers to the ability of crops to grow on poor soils.
[c] Different species are suitable for different soil types.
[d] Small-scale production of ethanol for fuel is technically feasible but not yet fully proven.
[e] Technically feasible but unproven on a large scale.

TABLE V.3

Limitations on the availability of biomass materials for use in energy (source: D. O. Hall *et al.*, 1982)

Biomass resource	*Nature of limitation*			
	Theoretical limitations	*Practical limitations*	*Environmental limitations*	*Socio-cultural limitations*
Wood	Amount of forest and woodland remaining and the levels of sustainable wood production	Distribution relative to population centers, collection problems, and competition for other uses	Need for good collection practices to prevent forest damage	Ownership of forests, poor forestry knowledge, and inappropriate forest-use traditions
Crop residues	Amount of residues produced—a function of crop types, yield, and farmland area	Competition for other uses and seasonality of supply	Soil degradation due to excessive removal of residues and inadequate nutrient recycling	Inequalities in land ownership and rights to use crop residues
Animal manure	Amount of dung produced—a function of the number and type of animals and levels of health	Collection problems with free-ranging animals	Removal of nutrients from the soil, unless measures are taken to recycle them	Ownership of animals and cultural objections to collecting and using dung
Food processing wastes	Amounts of wastes generated—depends on the quantities of food processed and the processing efficiency	Competition for other uses and seasonality of supply	May be an advantage in some cases in which wastes present a disposal problem	Ownership of wastes

TABLE V.4

Generic socioeconomic impacts of biomass energy systems (source: Braunstein *et al.*, 1981; © 1981 Pergamon Press plc)

Area of interest	*Impact or issue description*
Land use	Land requirements for biomass energy production are orders of magnitude larger than for other energy technologies. This raises significant aesthetic, practical impact, and land-availability questions. Severe conflicts among competing land-use interests can be expected.
Market linkage	Linkages among biomass product markets can be complex and lead to either increasing or decreasing overall market stability. Diversity in feedstock and product mix and vertical or horizontal process integration should probably be encouraged.
Dynamics of market expansion	Excessive subsidization, stored resources, and probable expansion of more profitable biomass markets could imply wasteful and destructive overexpansion of biomass energy production.
Food	Competition between food and energy from biomass could affect domestic and international food supplies and prices.
Employment and demographic changes	Location of biomass facilities in rural areas would generally produce significant employment and community impacts. Regional demographic shifts may also occur as a result of variations in biomass availability.
Institutional issues	Relationships among biomass energy producers, existing utilities, and potential customers for cogenerated steam need to be developed. Environmental regulations, tax treatment and government revenue implications, and capital availability are also major areas of uncertainty.
Political issues	Decentralization and subsidies among competing energy programs may become even more prominent political issues.
Transportation	Road upgrading and maintenance requirements, as well as traffic pattern changes may be significant. Indirect impacts of transportation system alterations should also be considered.

applications. Thus, biomass-derived liquid fuels have received considerable attention. Table V.2 tabulates and evaluates a number of biomass production and liquid-fuel conversion options.

Biomass fuels are no panacea; many unresolved problems still exist. Some limitations on the availability of biomass fuels are noted in Table V.3. The socioeconomic impacts of biomass energy systems are complex and not always positive as indicated in Table V.4. Before implementing a biomass energy program, the technical, economic, and social aspects of such a program should be carefully analyzed.

PHOTOSYNTHESIS†

Photosynthesis is the process by which plants convert solar energy into biomass, or the photochemical reduction of carbon dioxide with electrons from water. Total biomass production is total photosynthesis less respiration occurring during growth. An estimated 40×10^{12} W of solar power is captured by the green leaves of plants each year. Once the inorganic compounds, water, and carbon dioxide are synthesized into carbohydrates, solar energy is stored chemically. However, efficiency of photosynthesis is very low—perhaps only 0·1 or 0·2%. Nevertheless, the world annual energy consumption is only one-tenth the amount of photosynthetic energy stored annually in biomass.

Photosynthesis is considered to be the most important physiological process in the world. In its simplest form, the product of photosynthesis is sugar (glucose). The basic equation is as follows:

$$6CO_2 + 6H_2O \xrightarrow{\text{light}} \underset{\text{(glucose)}}{C_6H_{12}O_6} + 6O_2$$

Both glucose and oxygen contain chemical energy in excess of the amount found in carbon dioxide and water. Two basic aspects of photosynthesis are (1) conversion of light energy into chemical energy and (2) conversion of carbon dioxide (fixation) into organic compounds (biomass). In reality, photosynthesis is an extemely complex process which leads to the formation of all biomass constituents, e.g. sugar, starch, cellulose, and lignin.

Most of the energy that civilization transforms into heat, light, electricity, and mechanical work was originally converted through photosynthesis to fossil fuels, e.g. petroleum, coal, natural gas, and biomass (e.g. wood, plant residue, and animal manure). Another important function of photosynthesis is the ability to return oxygen to the earth's atmosphere, thereby ensuring that a nearly constant oxygen supply is maintained.

On an annual basis, the overall efficiency of the utilization of light energy through photosynthesis in field conditions is only a fraction of a per cent, but under ideal conditions, many plant species convert 25% of the absorbed light into chemical

† Adapted from R. E. Anderson (1979); Stout (1983*a*); and Beadle *et al.* (1985).

energy. Photosynthetic efficiency is significantly less in field conditions because environmental factors such as light intensity, temperature, water availability, and carbon dioxide concentration may vary.

WOOD FOR FUEL

Wood is by far the most common form of biomass (Figs V.6 and V.7). This resource is so important to the well-being of people in developing countries that it has been the subject of numerous studies and publications (Eckholm, 1976; Openshaw, 1978). Its use and availability has been analyzed thoroughly. Openshaw (1978) noted that the problem of inadequate wood supplies may be even greater than statistics indicate because the present consumption of fuelwood and charcoal may be nearly three times the recorded production. Household consumption per capita is approximately 1 t of wood/year. Cooking represents 50% of the household use, while heating, the second most important use, accounts for 30%. As the world's population becomes more prosperous, many will substitute charcoal and other more convenient non-wood fuels for fuelwood use. However, the subsistence sector will continue to depend heavily on fuelwood as long as local supplies are available.

The consequences of deforestation, resulting primarily from the removal of firewood for cooking, are considered serious. Numerous reforestation projects are

Fig. V.6. Women carrying heavy loads of firewood in the Himalayan foothills (source: FAO Photo 10395/D/26).

Fig. V.7. Cooking on an open fire in a rural home. In Burma, an average family of five burns 5 t of fuelwood per year for cooking alone (source: FAO Photo 10831/C/26).

underway as indicated in Table V.5. The World Bank has stated that the economic, environmental, and human consequences of firewood shortages are severe. Fuelwood gathering is one of the contributory causes of deforestation, which is already claiming 10 million hectares/year of forests in the developing world, leading to soil erosion, reduced agricultural productivity, and siltation of reservoirs, riverbeds, and irrigation canals (World Bank, 1983).

TABLE V.5
Fuelwood projects financed by the World Bank (1977–80) (source: NRC, 1982)

Country	*Reforestation area (ha)*	*Approximate cost per hectare ($)*	*Economic rate of return (%)*
Burundi	8 500	388	17
India	52 000	884	13
Korea	120 000	250	18
Malawi	28 000	500	14
Mali	3 400	1 900	11
Niger	400	6 000[a]	11
Philippines	28 000	300	22
Thailand	11 000	227	13
Upper Volta	3 500	2 000	16

[a] Irrigated.

SOURCES AND AVAILABILITY OF BIOMASS FEEDSTOCKS

The principal bioenergy feedstocks are residues, energy crops, marine plants, and algae. While information is fragmentary and inconclusive, the following section illustrates that total biomass resources are extensive and should be considered as a serious supplement to petroleum and other fuels.

Residues

Numerous studies have been conducted to determine the quantity of residues available in various regions (Smil, 1981; D. O. Hall *et al.*, 1982; FAO, 1982*a*). FAO, in its publications on residue utilization (1982*a*), also devotes a section to compiling information from individuals via a questionnaire on quantitative availabilities of residues. Information on the various materials includes: annual amount produced, actual utilization, present disposal practices, proposed utilization, and constraints that hamper proper utilization.

The easily accessible residues—straws, stalks, leaves, vines, and roots—have always been important sources of energy in the rural areas of developing countries. Residue yields can be estimated by applying appropriate multipliers to the known crop harvest. Smil (1981) compiled a number of residue coefficients for various crops (Table V.6). The range is wide, reflecting various harvesting and handling practices, as well as possible inaccuracies in data collection, but the obvious indication is that huge quantities of residues are produced. Smil also multiplied the coefficients by the land areas of various crops to obtain estimates of the total residues produced in various countries and around the world (Table V.7).

All residues are not necessarily available for fuel. Residues have many uses, including animal feed, fertilizers and soil organic matter, industrial materials, and

TABLE V.6
Residue coefficients for major crops (source: Smil, 1981; © 1981 Pergamon Press plc)

Crop	*Residue/crop ratio*
Rice (paddy)	0·75–2·51
Wheat	1·10–2·57
Corn	0·55–1·30
Barley	0·82–1·50
Oats	0·95–1·75
Rye	1·20–1·75
Sorghum	0·85–1·99
Potatoes	0·20–0·30
Soybeans	1·10–2·60
Other legumes	1·20–1·50
Sugar cane	0·20–0·25
Cotton	1·40–3·00
Rape	1·85–2·00

TABLE V.7
World production of major crop residues in 1975 (source: Smil, 1981; © 1981 Pergamon Press plc)

	Cereal straws (Mt)	*Tuber vines (Mt)*	*Legume vines (Mt)*	*Sugar cane, stalks, and leaves*	*Total (Mt)*	*Total residues (10^6 TJ)*	%
North America	430	3	2	5	440	7·38	19
Europe	330	22	4	—	356	5·96	15
USSR	203	18	8	—	229	3·83	10
Latin America	118	9	7	58	192	3·21	8
Africa	99	15	8	10	132	2·21	6
Asia	836	44	38	54	972	16·29	41
Oceania	29	—	—	5	34	0·57	1
Developed	1 035	46	14	13	1 108	18·56	47
Developing	1 009	66	53	119	1 247	20·89	53
World	2 044	112	67	132	2 355	39·45	100

feedstocks. Even if the crop residues used for feed and manufacture could be completely replaced by fodder crops and substitute materials, complete removal of residues from fields is undesirable. Posselius and Stout (1983) developed a model that considered the relevant factors affecting wind and water erosion, soil compaction, nutrient maintenance, and soil physical properties. The model provided estimates of the amount of residue that must be returned to the soil to avoid detrimental side-effects and, thereby, determined whether excess biomass residue was available for fuel or other purposes.

It is important to distinguish between total residues and residues that are actually available. Many residues have a variety of uses and a value associated with each use; therefore, they cannot be removed from the soil indiscriminately. For example, crop residues may be used as follows: to provide soil surface protection; to act as a storehouse of nutrients; to stabilize soil structure and improve the tilth; to reduce soil bulk density; to enhance infiltration and moisture retention; to provide energy for microorganism activity; to increase cation exchange capacity; and to release carbon dioxide. The amount of energy available from crop residues can be estimated from the data in Tables V.8 and V.9.

Animal dung is another important form of residue in developing countries (Tables V.10 and V.11). Similar to plant residues, animal dung has many uses, including direct application to the soil, composting, or as a feedstock for an anaerobic digester (Fig. V.8).

Growing Crops for Fuel

Energy plantations have been proposed to produce biomass for fuel, but the concept is controversial (Fig. V.9). Critics question: (1) the use of potential food or feed for fuel; (2) whether or not the energy balance could be favorable; (3) the economics; and (4) the appropriateness of the technology.

TABLE V.8

Per capita production of cereal crop residues in selected developing countries (source: D. O. Hall *et al.*, 1982; © 1982 Pergamon Press plc)

Country	*Energy content of residues from cereal crops (GJ/capita per year)*
Argentina	25·0
Thailand	9·3
Malawi	8·6
Brazil	8·1
Nepal	7·1
China	6·8
India	5·5
Upper Volta	5·1
Bangladesh	4·3
Ethiopia	3·3
Peru	2·9
Somalia	2·1
Congo	1·1
Zaire	0·9
Developing countries (average)	5·6

Note: Cereal residues are not the only farm agricultural residues. Significant amounts of residues from roots and tubers, vegetables, nuts, cotton, sugar cane, etc., are produced in some countries.

TABLE V.9

Production of residues from cereal crops in developing countries (per hectare) (source: D. O. Hall *et al.*, 1982; © 1982 Pergamon Press plc)

Crop	*Crop yield (t/ha per year)*		*Crop/residue ratio*	*Residue production*[a] *(t/ha per year)*	
	Range	*Average*		*Range*	*Average*
Rice	0·7–5·7	(2·5)	1:2	1·4–11·4	(5·0)
Wheat	0·6–3·6	(1·5)	1:1·75	1·1–6·1	(2·6)
Maize	0·5–3·7	(1·7)	1:2·5	1·3–9·3	(4·3)
Sorghum	0·3–3·2	(1·0)	1:2·5	0·8–8	(2·5)
Barley	0·4–3·1	(2·0)	1:1·75	0·7–5·4	(3·5)
Millet	0·5–3·7	(0·6)	1:2	1·0–7·4	(1·2)

[a] Calculated from crop yield and residue ratio. These figures are, at best, approximate since crop/residue ratios are likely to show regional as well as varietal variations. The highest values for residue production are likely to be somewhat exaggerated because high yielding crop varieties tend to have lower crop/residue ratios.

TABLE V.10
Manure production by domesticated animals (source: D. O. Hall *et al.*, 1982; © 1982 Pergamon Press plc)

Animal	*Manure production (t/head per year)*
Cattle, buffalos, camels	1·00
Horses, donkeys, asses	0·75
Pigs	0·3
Sheep, goats	0·15
Chickens, poultry	0·005

Annual Crops
Energy ratios ranging from 1 to almost 17 can be achieved for different crops in various agricultural systems. (Energy ratio is the energy content of biomass produced divided by cultural energy input. Cultural energy includes fertilizer and fuel input, as well as indirect energy invested in machinery and buildings.) Maize production typically has an energy ratio ranging from 2·5 to 1, to as much as 5 or 6 to 1 (Merrikin & Ward, 1981). Additional examples of energy ratios for various crops and farming systems are shown in Fig. V.14.

TABLE V.11
Per capita production of animal dung in selected developing countries (source: D. O. Hall *et al.*, 1982; © 1982 Pergamon Press plc)

Country	*Energy content of animal dung (GJ/capita per year)*
Botswana	55
Argentina	40
Somalia	28
Sudan	22
Bolivia	19
Brazil	16
Nepal	13
Peru	7·6
Upper Volta	7·3
India	6·3
Bangladesh	5·1
Thailand	3·9
China	3·4
Malawi	2·8
Zaire	1·1
Congo	1·0
Developing countries (average)	5·7

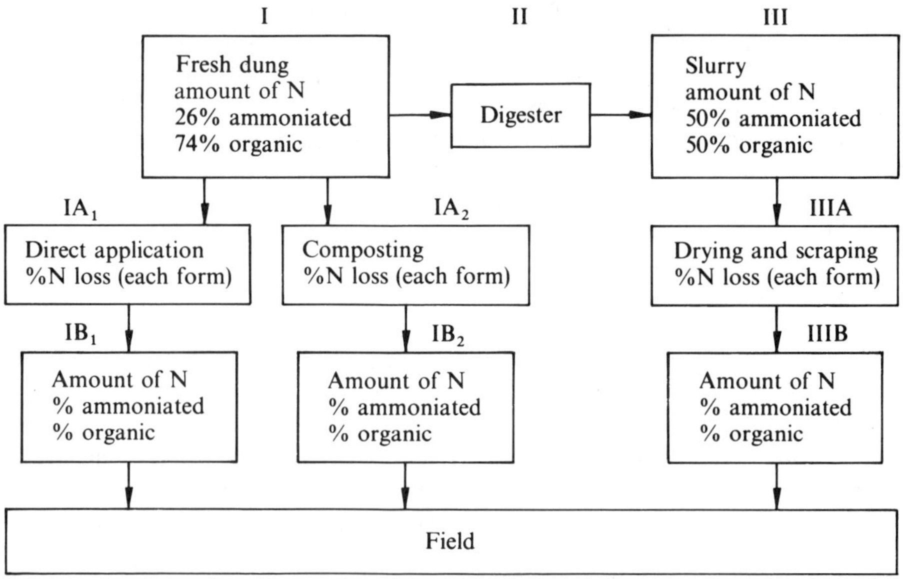

Fig. V.8. Alternative methods for handling dung (source: Fluck & Baird, 1980).

Fig. V.9. Sweet sorghum can be grown as a renewable feedstock for the production of methane or ethanol (source: DOE Photo 216 001 002).

Appraising the economics of annual crop production for fuel is more difficult. Although diverse opinions exist, many economists and engineers believe producing maize for conversion to ethanol or any fuel feedstock is uneconomical. Generalizations, however, can be misleading or incorrect. When assessing the economic potential of biomass fuel projects, analyzing each specific proposal is safer and more realistic. The cost of feedstocks, local technological and managerial capabilities, local markets for fuel and byproducts, and other site-specific considerations must all be taken into account.

Forests and Other Perennials

Although difficult to measure, world forest resources are enormous. Estimates of the annual above-ground wood resources are provided in Table V.12. Note that more than 9 billion m^3 or 51% of the world's total forest increment is produced in the tropics. Of this total, an estimated 11% is currently being used—2% for industry and 9% for fuel. In terms of coal equivalent (CE), the unused increment in the developing world is 3·2 billion tonnes annually (Bene *et al.*, 1979).

Although forest resources are extensive, they are not distributed evenly. Deforestation is a serious problem in many developing countries, and plenty of disaster areas exist where the use of fuelwood is much higher than the annual growth increment.

Oil/Hydrocarbon Crops

Euphorbia and other species which produce low molecular weight hydrocarbons have been proposed for fuel crops. A recent development is interest in vegetable oil from sunflowers, soybeans, rapeseed, groundnuts, palm, and other oilseeds. Research on production, extraction, and refinement of vegetable oils for diesel fuels is underway in South Africa, Brazil, Australia, Germany, the United States, and other countries. Vegetable oils for fuel are discussed in detail later in this chapter.

TABLE V.12

World's annual wood production (source: Bene *et al.*, 1979)

Forest type	*Area ($ha \times 10^6$)*	*Annual increment of wood[a] per hectare*		*Total increment wood[a]*		
		(m^3)	*(t)*	*($m^3 \times 10^9$)*	*($t \times 10^9$)*	*($tCE \times 10^9$)*
Cool coniferous	800	4·1	3·0	3·3	2·4	1·4
Temperate mixed	800	5·5	4·0	4·4	3·2	1·9
Warm temperate	200	5·5	4·0	1·1	0·8	0·5
Equatorial rain	500	8·3	6·0	4·1	3·0	1·8
Tropical moist deciduous	500	6·9	5·0	3·5	2·5	1·5
Dry	1 000	1·4	1·0	1·4	1·0	0·6
Totals and means	3 800	4·7	3·4	17·9	12·9	7·7

[a] Estimated to include all above-ground wood.

Composition of Feedstock

The physical and chemical composition of a feedstock will determine its most practical use as a fuel. If the feedstock is dry and the end-use is heat, direct combustion may be the best process. If, however, a high energy-density liquid fuel is needed for a mobile engine, ethanol fermented from sugar or starch crops may be preferred for use in spark-ignition engines. Oil extracted from oilseeds is generally preferred for diesel engines. Feedstocks must be matched to conversion processes and end-uses. Ethanol from cellulosic feedstocks is desirable but not generally considered a practical commercial technology today.

Food/Fuel Conflicts

To generalize on a global scale about possible conflicts between biomass production for food and fuel is impossible. A country's approach to energy cropping is likely to depend particularly on the degree to which it is self-sufficient in food, agricultural products, and energy. Other aspects being equal, countries with surplus agricultural production and deficit energy production are most likely to follow energy incentive policies. The countries in this category are influenced by many other factors, and include not only Brazil and the United States, which thus far have the most comprehensive energy cropping programs, but also most of the other countries actively considering energy cropping plans (New Zealand, the Philippines, South Africa, and Thailand).

Large-scale energy cropping could, in the long run, have significant effects on world food supplies that can only be approximately assessed at this early stage. Some of the possible consequences of energy cropping were examined by the FAO expert consultation, Energy Cropping Versus Food Production, in June 1980 (FAO, 1981). The findings emphasized the great diversity among countries, which makes generalizing on this issue dangerous. Timing is also critically important because of the rapid technological development of many biomass conversion processes, which, together with changes in energy prices, alters their relative economic efficiency. The consequences of some of the conversion processes, the extent of the biomass resource base, and the effects of energy cropping are all uncertain. Flexibility is a key ingredient to national policy development.

Although in the short term, only small food price impacts seem likely, in the long term, energy cropping on a substantial scale, such as in food exporting and in potentially exporting countries, could limit food availability, increase its price, and result in detrimental effects on nutrition of the poor. Energy cropping could also have implications on policies regarding food aid and national and international reserve stocks.

Freshwater and Saltwater Plants (OTA, 1980)

Although aquaculture and mariculture are practiced in various countries, the science of farming in water is still relatively unexplored and undeveloped. Because water plants have a continuous supply of water, they are never water-stressed. To achieve maximum production, nutrients and carbon dioxide for submerged plants should be added.

Aquaculture
Aquatic plant types are diverse—from micro-algae to large marsh plants such as cattails, water hyacinths, and even some trees, such as mangroves. Such plants can achieve very high rates of biomass production, often exceeding 45 t/ha per year. Micro-algae can be a source of biomass, but because they are completely submerged, they are subject to significant light losses. Data on aquatic plant productivity are still too limited to allow confident extrapolation to large-scale systems.

Mariculture
Ocean farms might produce giant kelp or other macro-algae. About 2 million wet tonnes of macro-algae are now cut annually around the world, but experts estimate that the total potential is 10 times this amount. Large ocean kelp farms could theoretically supply significant quantities of natural gas (methane), but formidable ocean farming problems would have to be resolved.

CLASSIFICATION OF BIOMASS FUELS

Energy is defined as the capacity to do useful work. Heat is one form of energy; mechanical work produces shaft or draft power. Sugars and starches, cellulose and lignins, and other constituents of biomass may be burned directly to produce heat, and to perform mechanical work in an external combustion engine, such as a steam engine. Or biomass may be converted to liquid or gaseous fuels which may be stored and subsequently burned to produce heat or used to fuel internal combustion engines, such as spark ignition or diesel engines, thereby producing mechanical work (Fig. V.10).

Gaseous Fuels

Gasification of solid biomass to produce a variety of gaseous fuels may be accomplished by several processes. The chemical changes that take place as solid biomass is converted into gaseous fuels are presented in Fig. V.11.

Gasification is defined as any process that converts solid fuels to the gaseous fuel stability region. Raw biomass, a complex mixture of cellulose, lignin, or other materials is made up of carbon, hydrogen, and oxygen. The corners of the triangle represent pure carbon, hydrogen, and oxygen. Heating or combustion of biomass releases gaseous products. The combustion of any fuel is represented by a vector toward the oxygen corner.

Combustion is complete and flame temperatures maximum on the water/carbon dioxide line. Incomplete combustion terminates to the left of the water/carbon dioxide line and contains some hydrogen and carbon monoxide.

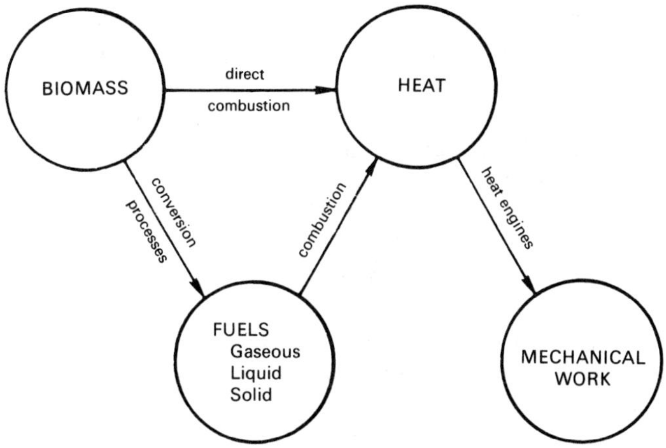

Fig. V.10. Biomass conversion to heat or mechanical work (source: Stout, 1984).

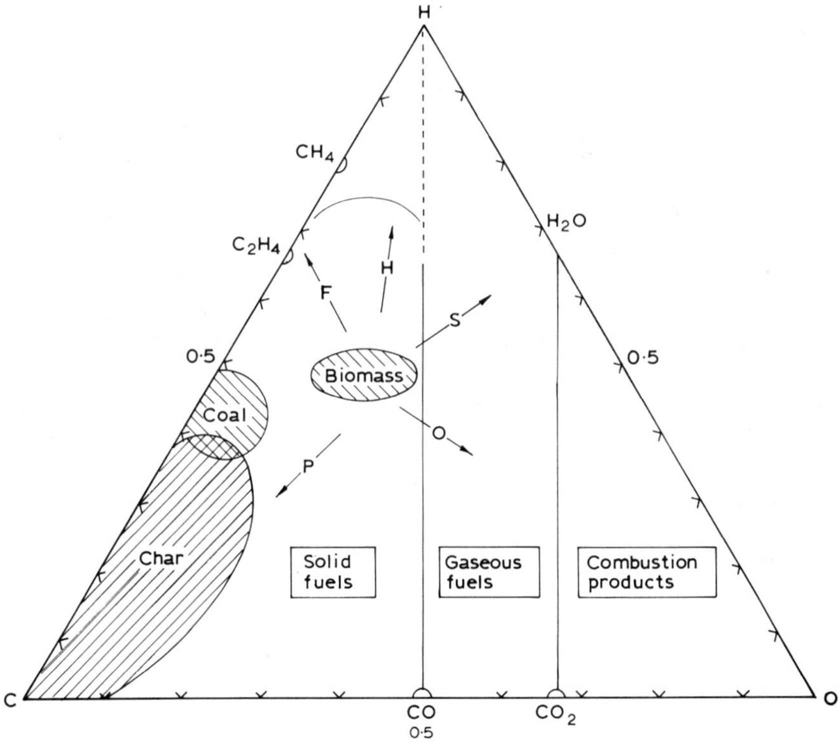

Fig. V.11. Chemical changes during gasification (source: Reed, 1979). O = Oxygen process; S = steam process; H = hydrogen process; F = fast pyrolysis process; P = slow pyrolysis process.

Above 1000°C, the only stable gas molecules under normal conditions are carbon monoxide and hydrogen, so high-temperature gasification tends to terminate on the line connecting carbon monoxide and hydrogen. Compositions on the lower portion of the left half of the triangle are solid or liquid fuels.

The arrows on Fig. V.11 represent a number of biomass gasification options. Pyrolysis (P) is the simplest form of gasification. When biomass is heated to about 500°C, it breaks down into gases, liquids, and char. The energy involved in this process is only a small fraction of the heat of combusion of the fuel.

Fast pyrolysis (F) can minimize the initial production of char and liquid and produces large amounts of ethylene.

Oxygen/air gasification (O) is the simplest means of obtaining gaseous fuels from solids. High reaction temperatures (about 1000°C) result in gases rich in carbon monoxide and hydrogen which are useful for synthesis to methanol or other chemicals if oxygen is used. If air is used, nitrogen dilution produces a low-energy gas (5–10 MJ/m^3) suitable for combustion, but not for synthesis or pipeline use.

Hydrogen gasification (H) can be attractive if an inexpensive source of hydrogen exists. Steam gasification (S) is widely used with coal and may have application with biomass.

Liquid Fuels

Most internal combustion engines in today's vehicles—tractors, trucks, automobiles, etc.—were designed to use liquid fuels, such as gasoline or the equivalent. Engines of the future could be designed to burn a wider range of liquid fuels such as ethanol, methanol, and vegetable oil, but the transition to a new fuel with different chemical and/or physical characteristics is complex, especially if supplies of the new fuel are sporadic and unpredictable, or if quality control is uncertain. Manufacturers require years of lead time and an ensured supply of the new fuel before they will make a significant investment in a new type of engine. Brazil is an example of a country where a number of manufacturers have introduced engines that burn wet ethanol.

Three types of liquid fuel from biomass are considered in this report—methanol produced by synthesis of producer gas, ethanol from fermentation of sugar, and starch or cellulosic materials and vegetable oil from a variety of oilseeds for use as diesel engine fuel.

Solid Fuels

Solid fuels are the predominant fuels used for cooking in developing countries and are important for many other heat applications as well. Raw biomass, such as wood and agricultural residues, along with charcoal, fall into this category.

Energy Balance

The laws of thermodynamics do not allow more energy output from any process than input. Energy is neither created nor destroyed. Thus, if all input energy is considered, the output energy will be exactly the same, but with reduced availability (increased entropy). Standard energy accounting procedure, however, neglects the solar energy input that produces biomass through photosynthesis. A debate then results about whether or not there can be a net energy gain from the entire biomass production and conversion system.

This may be a more important question than the overall energy balance. Premium fuel gains can often be achieved if processing energy is restricted to non-premium forms, such as biomass, coal, etc. Detailed energy balance studies have been conducted by many researchers.

The photosynthetic process is a method of storing solar energy in the form of biomass. Some representative energy crop yields are shown in Fig. V.12. In each case, the energy value for the whole crop is provided, followed by a reduced amount of energy for crop components or after various conversion processes to liquid or gaseous fuels.

A more specific comparison of the energy output per hectare per year for ethanol and vegetable oil from various species is illustrated in Fig. V.13.

The energy ratio for several biomass fuels from various crops is shown in Fig. V.14. Sunflower oil gives an energy ratio of 6·1 in this analysis conducted in Australia, whereas ethanol produced energy ratios ranging from 0·56 to 2·75. An energy ratio of 1 indicates that the energy content of the fuel produced is exactly equal to the cultural energy input to grow the crop and convert the biomass to a liquid fuel. Many other studies of energy output have been conducted with similar results.

Utilization of Biomass Fuels

This publication would not be complete without mentioning the use of biomass fuels in internal combustion engines. Figures V.15 and V.16 show the energy content of various liquid fuels on a weight and volume basis. Most engines are designed to operate on a fairly narrow range of petroleum-based fuels. Manufacturers specify fuel characteristics that will result in satisfactory long-term engine performance. Usually, biomass fuels deviate from manufacturers' fuel specifications, so either the engine or the fuel must be modified to be compatible (Fig. V.17). Numerous experiments have been conducted in both areas.

Liquid fuels are generally preferred for mobile vehicles because of their high energy density. Several approaches may be taken to produce liquid fuels from biomass, including:

—Production of ethanol by (yeast) fermentation of sugar crops, starch crops, or surplus molasses.
—Production of methanol from wood, through gasification to produce synthesis gas followed by catalytic conversion to methanol.
—Production of gasoline from wood-using gasification as above, coupled with Fischer–Tropsch conversion to gasoline, or, alternatively, wood-derived methanol conversion to gasoline using the Mobil process.
—Cultivation of oil producing crops such as oil palm, sunflowers, coconuts, sesame, and groundnuts.

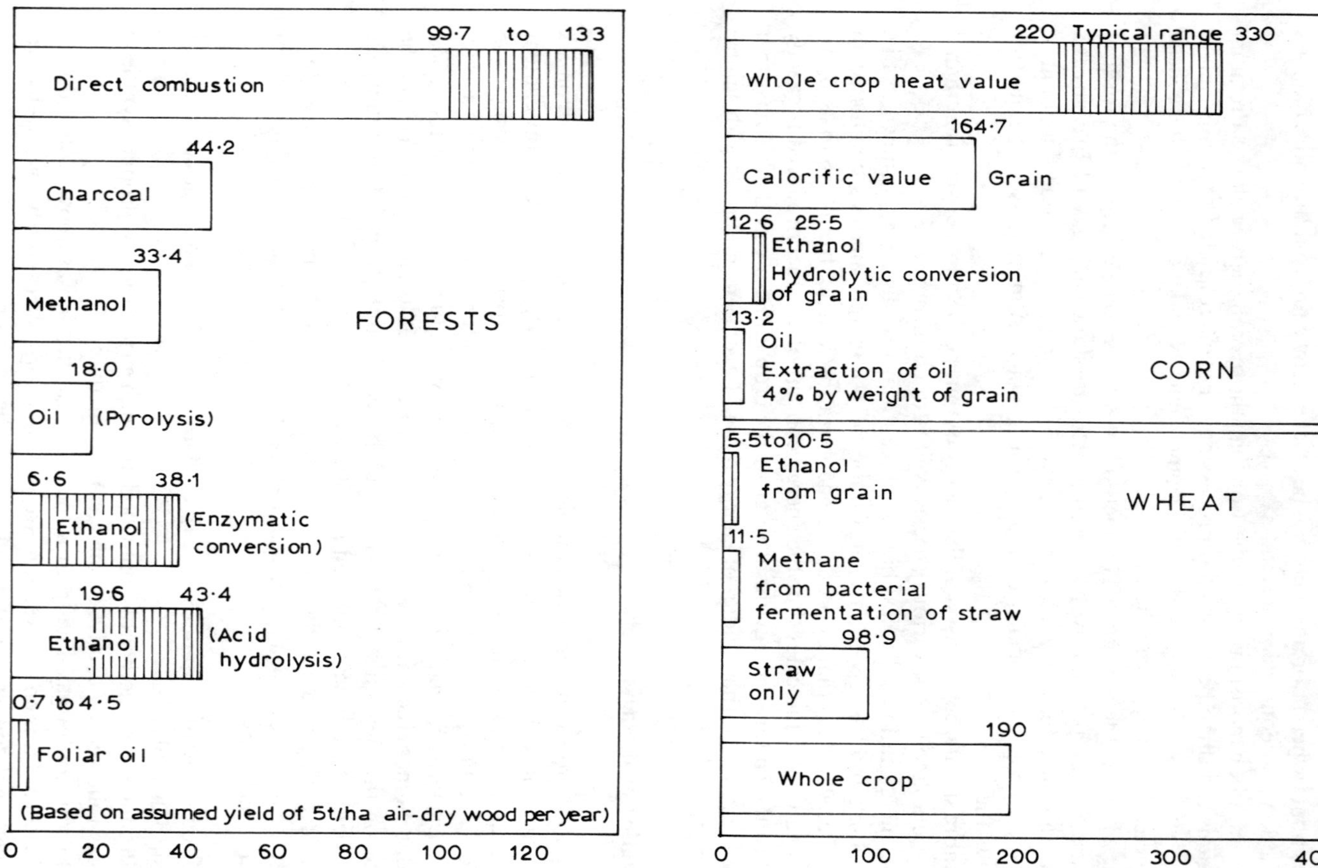

Fig. V.12. Representative energy crop yields, in terms of GJ/ha per year. Note that gaseous and solid fuel possibilities are included as well as liquid fuels (source: Quick, 1980).

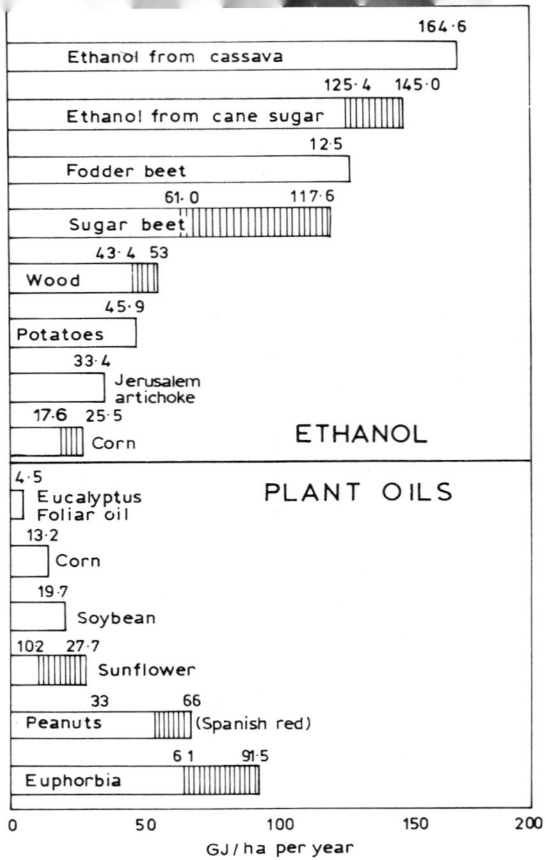

Fig. V.13. Comparison of some representative crops in terms of ethanol or plant oil production, respectively, based on fuel energy yield per hectare per year. The data for *Euphorbia*, a latex-producing arid land plant, are tentative and based only on results interpolated by Nobel laureate Melvin Calvin from small-scale field trials and reduction to hydrocarbon fuel in the laboratory (source: Quick, 1980).

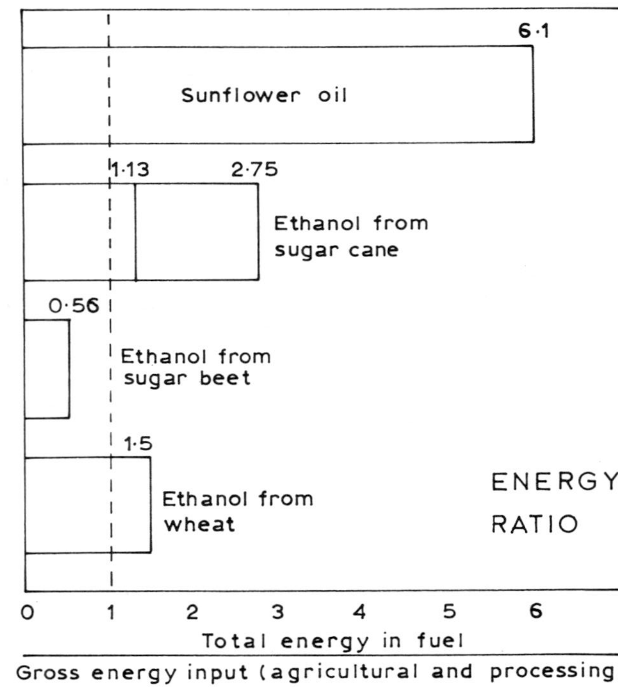

Fig. V.14. Ratios of fuel energy output to total energy inputs to produce and process crops. A ratio of 1·0 means that the output energy is exactly the same as the input (source: Quick, 1980).

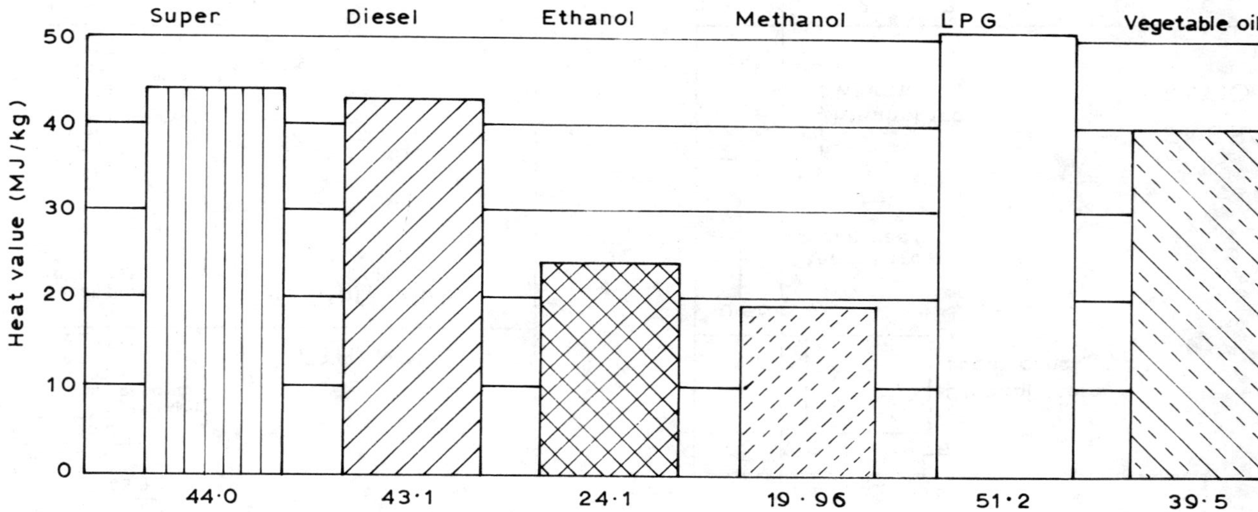

Fig. V.15. Fuel energy density spectrum on a weight basis; heat values are low calorific values (source: Quick, 1980).

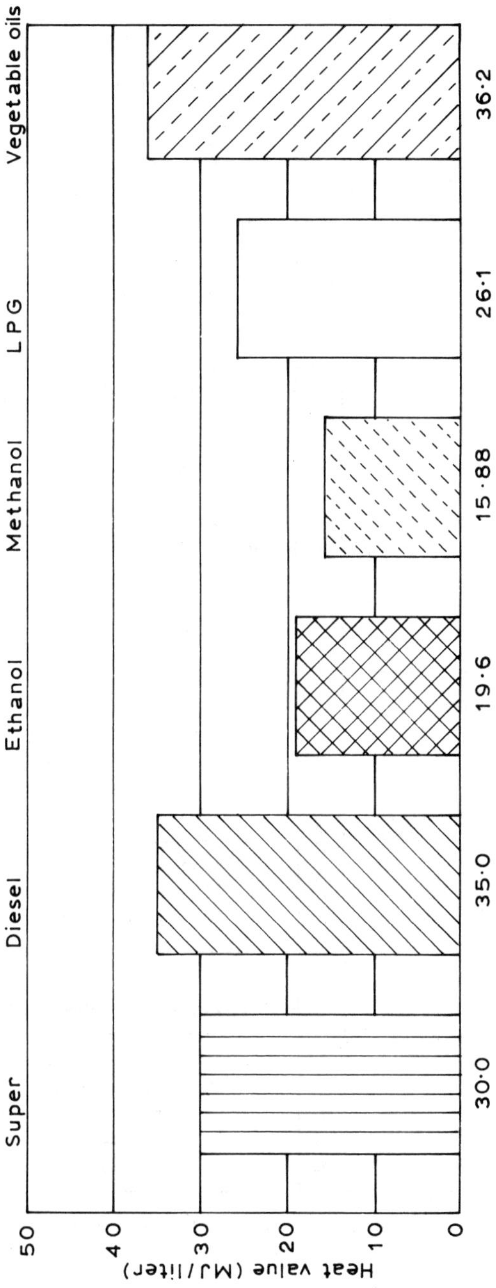

Fig. V.16. Fuel energy density spectrum on a volume basis (source: Quick, 1980).

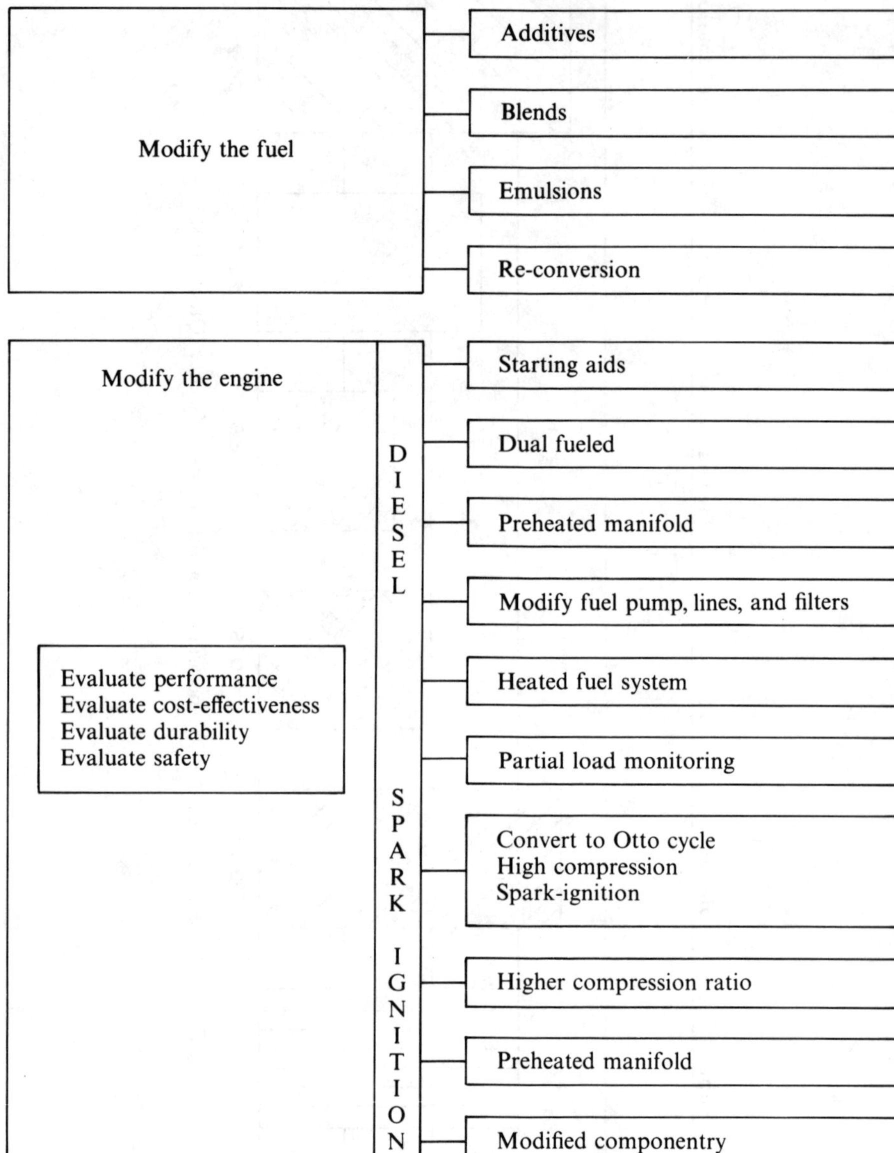

Fig. V.17. Fuel alternatives—strategies for adopting nonconventional fuels (source: Quick, 1980).

TABLE V.13

Liquid fuel yields from biomass under optimum conditions (source: D. O. Hall *et al.*, 1982; © 1982 Pergamon Press plc)

Fuel	*Feedstock*	*Feedstock yield (t/ha per year)*	*Conversion yield (% by weight)*	*Fuel yield (t/ha per year)*	*Heat content (GJ/t)*	*Fuel yield (GJ/ha per year)*
Ethanol	Sugar cane	50	6	2·8	27	76
	Cassava	12	14	1·7	27	46
Vegetable oils	Oil palm	25	20	5·0	40	200
	Groundnut	1·3	45	0·6	40	24
Methanol	Wood	20	30	6·0	20	120
Synthetic petroleum (Fischer-Tropsch process)	Wood	20	14	2·7	40	108

TABLE V.14

Fuel alternatives for gasoline and diesel engines (source: D. O. Hall *et al.*, 1982; © 1982 Pergamon Press plc)

Type of engine	*Alternative fuel*	*Percentage substitution of gasoline/diesel*	*Engine modifications required*
Diesel engines	Methane Gasifier gases	80	Minor[a]
	Ethanol Methanol	20–45	Minor[b]
	Vegetable oils	20–100[e]	Minor
Gasoline engines	Methane Gasifier gases	100	Major[c]
	Methanol Ethanol	20	None
	Methanol Ethanol	100	Major[d]
	Synthetic petroleum	100	None

[a] Gases are aspirated into the cylinder and ignited by the simultaneous injection of diesel fuel.

[b] Alcohols are not sufficiently soluble in diesel fuel to be blended directly. Instead, they can be injected simultaneously or fumigated into the air intake. Alcohol containing up to 20% water can be used.

[c] Major engine redesign is needed to increase compression ratios.

[d] Engines require higher compression ratios, modified carburetors, and measures to avoid corrosion.

[e] Purification and esterification are required if vegetable oils are to be used pure.

The relative suitability of these various liquid fuel options to a particular set of conditions is determined by a wide variety of site-specific factors as well as policy decisions. Technical feasibility obviously relies on growing biomass efficiently, which, in turn, depends on the type of land available, the climate, and the cultivation methods used. Theoretical liquid fuel production levels can be very high under optimum conditions. A comparison of optimistic production figures (on a gigajoule per hectare basis) for the main liquid fuel alternatives are shown in Table V.13.

The suitability of various liquid fuels, as well as methane gas for spark-ignition and diesel engines is tabulated in Table V.14.

ENERGY PROFILES

Biomass may be converted to convenient fuel forms through a number of processes, some of which are considered in this chapter:

Gaseous fuels
- —Gasification to produce CO and H_2 (producer gas)
- —Anaerobic digestion to produce methane (biogas)
- —Biophotolysis to produce hydrogen

Liquid fuels
- —Fermentation of ethanol
- —Synthesis of producer gas to produce methanol
- —Extraction of oil from seeds

Solid fuels
- —Direct combustion to produce heat

In general, the following areas are discussed for each process presented in this chapter:

biomass feedstock
process description
primary products
byproducts
environmental and safety aspects
net energy balance
R & D or commercial use
overall assessment

Figure V.18 presents an overview of the fuel uses discussed in this section.

Producer Gas

Gasification is the conversion of the carbonaceous solids in biomass into a combustible gas by controlling or limiting the rate of oxygen (air) admitted to the fuel bed. Only gasification, in which air is the source of oxygen, is considered for biomass conversion on the farm. The combustible components in the gas are

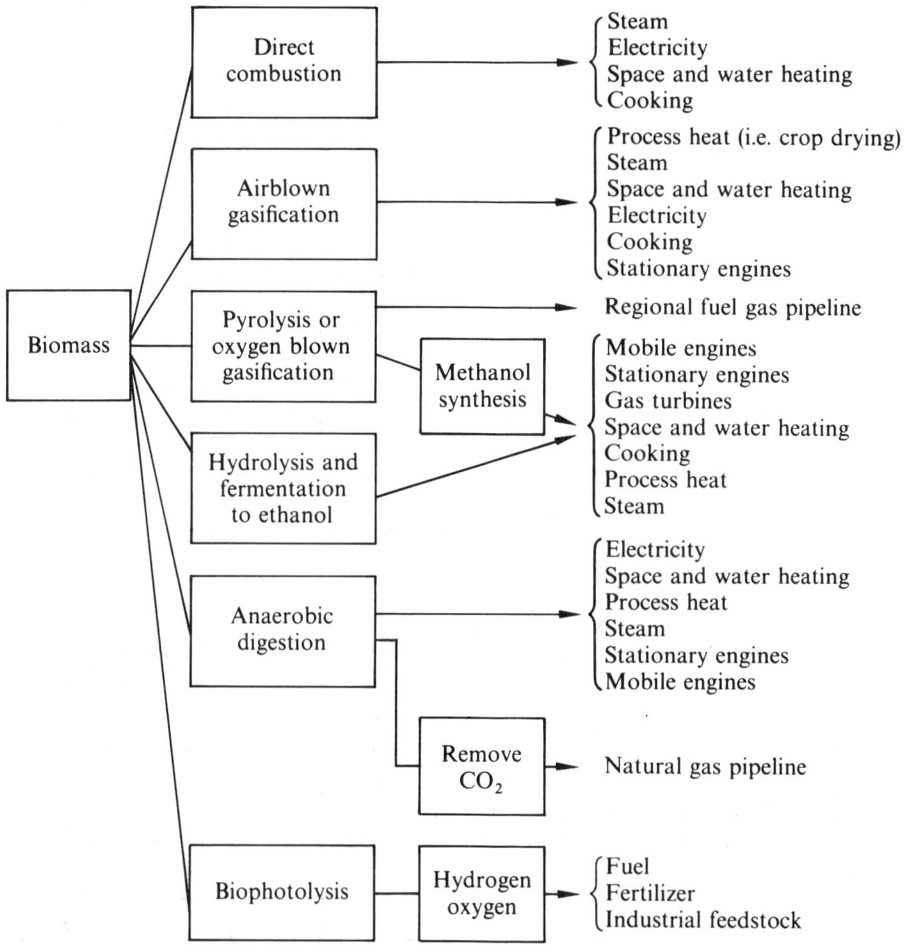

Fig. V.18. Fuel uses for biomass (source: OTA, 1980).

primarily carbon monoxide, hydrogen, and traces of methane. The resulting product of biomass gasification is referred to as producer gas, gengas, or woodgas. Often considered a low-energy gas, it contains about one-fifth as much energy per unit volume as natural gas. This product also contains a large portion of non-combustible gases, including nitrogen and carbon dioxide.

Many routes to gasification are possible as shown in Fig. V.19. Pyrolysis is the breakdown of biomass by heat into gases, tars, and char. The heating rate, temperature, and vapor residence time determine the final products (Fig. V.20). Air gasification, while requiring a pyrolytic step, uses a minimum amount of air and steam to convert char to gas in a single unit. Air gasification of biomass is particularly simple and approximately 1 million units were built to operate mobile vehicles during World War II. Low energy gas is produced because it is diluted with nitrogen from the air.

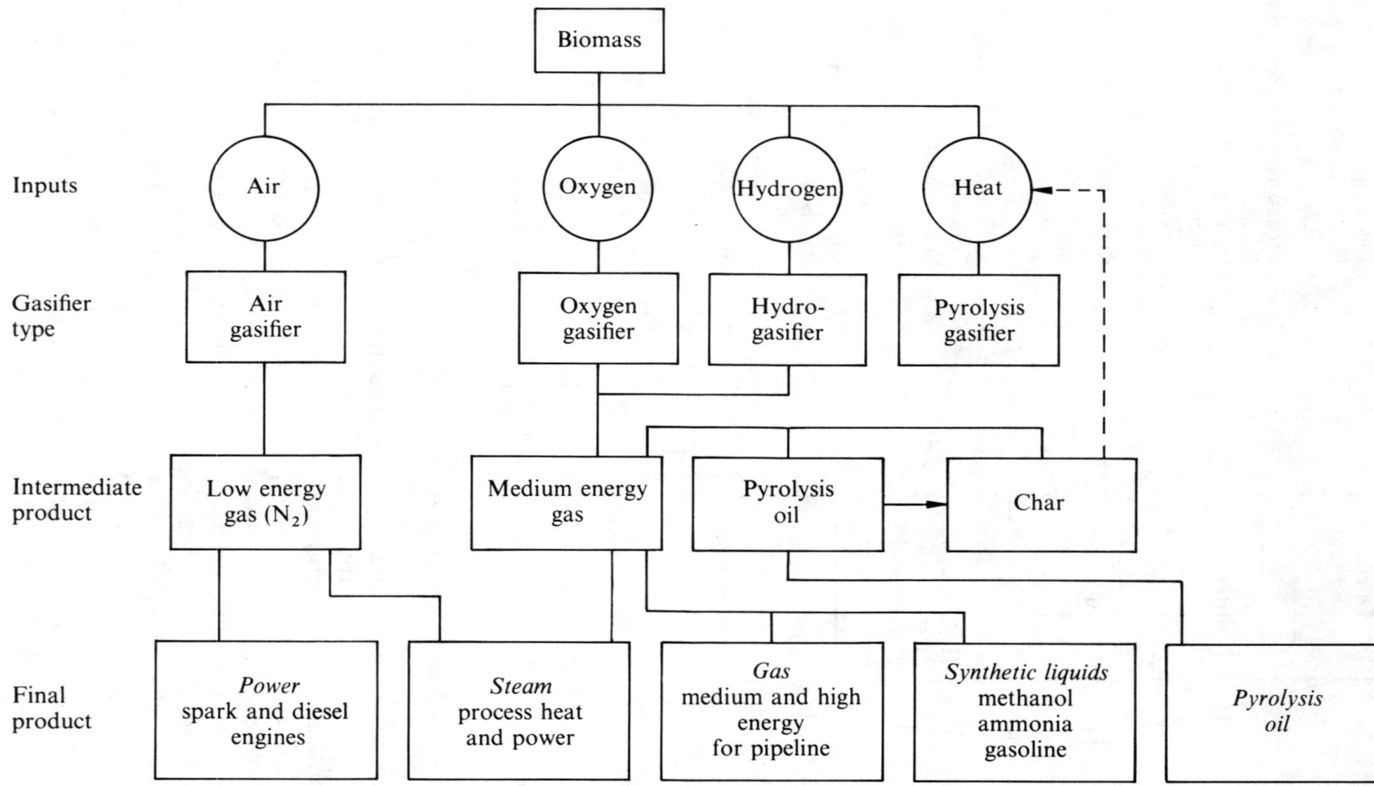

Fig. V.19. Gasification processes and their products (source: Reed, 1980).

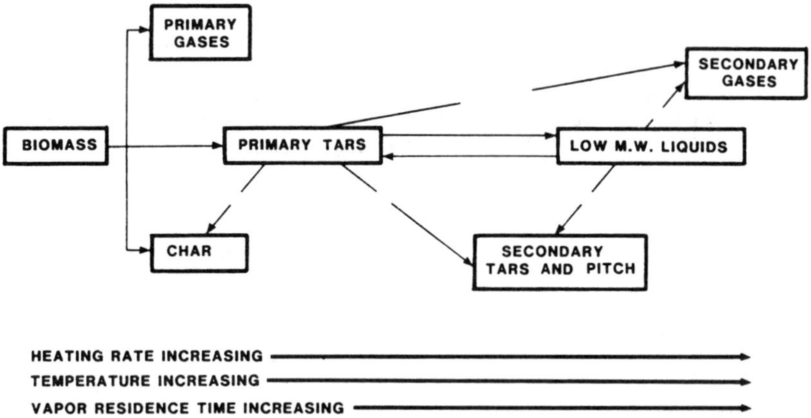

Fig. V.20. Effects of heating rate, temperature, and vapor residence time at temperature on products formed in wood pyrolysis (source: Hiler & Stout, 1985). M.W. = molecular weight.

Oxygen gasification is also a relatively simple process that produces a medium energy gas composed primarily of carbon monoxide and hydrogen. While satisfactory for burning, it can also be used for chemical synthesis to make methanol and other products.

Biomass Feedstock

Table V.15 lists some desirable properties of biomass for gasification. The feedstock should be relatively dry and low in ash, particularly the silica which tends to form slag deposits under certain conditions. The carbon, sulfur, and ash content of selected fuels is provided in Table V.16.

Biomass feedstocks for gasification fall within three categories: (1) agricultural crop residues (i.e. corn cobs, corn stover, wheat and rice straw, and sunflower stalks); (2) crop processing residues (i.e. walnut shells; cherry, plum, and prune pits; rice hulls; peanut shells; and cotton gin waste); and (3) forestry products and residues (i.e. wood, woodbark, logging, and mill residues). Research is underway to raise rapidly

TABLE V.15

Desirable properties of biomass for gasification (source: Sofer & Zaborsky, 1981)

Average moisture content of less than 50%
Average heating value (HHV) of not less than 9·8 MJ/kg
Average feedstock size range greater than 1·27 and less than 7·62 cm
Ash fusion temperature of not less than 1149°C
Low ash content (6–10%)
Easy ignition characteristics
Uniform chemical composition
Can form structurally strong char
Relatively simple to collect, store, and handle
Economically justifiable transportation cost
Available in adequate quantities to consistently meet the gasifier load demand

TABLE V.16

Approximate carbon, sulfur, and ash content of selected fuels, particularly biomass, in percentage (Source: C. W. Hall, 1981. Reproduced with permission. The entire publication *Biomass as an Alternative Fuel* is available from Government Institutes, Inc., 966 Hungerford Drive, #24, Rockville, Maryland 20850.)

Fuel	*Carbon*	*Sulfur*	*Ash*
Bagasse	31	0	7·3
Bagasse, dry	43		2
Coal	70–80	0·6–5	4–14
Cotton, ginning wastes	42	1·7	6·5–10·2
Manure, feedlot	21·4	0·2	8·9
Peat (dry)	60·5		3·3
Rice hulls, straw	38–40	0–1	19
Sawdust pellets (dry)	47·2	0	0·97
Sewage	14	0·7	71
Solid waste, municipal	27·5	0·3	38
Wood	50–60	0·013	1·12

growing trees for harvest every 3–5 years as biomass to convert energy into producer gas or some other form of fuel.

Because the energy density of biomass is relatively low, about 12–16 MJ/kg, investment of large amounts of energy to collect, compact, transport, or process biomass could cause the net energy output to approach zero or even to become negative.

Gasifier Designs

Two gasifiers offer potential for use with the fixed-bed, moving-bed, or fluidized-bed designs. Fixed feedstock and moving-bed gasifiers include updraft, downdraft, and cross-flow types. Fixed-bed gasifiers permit only batch operation. The feedstock for a moving-bed gasifier can be fed through a slide valve for batch feeding or through an airlock for continuous feeding.

In a downdraft gasifier (Fig. V.21), air is injected through nozzles into the combustion zone of the chamber and drawn down through the charcoal bed along with the tars and moisture. This causes the oil vapors to crack into low tar gases suitable for use with internal combustion engines.

In an updraft gasifier, air contacts the base of the burning bed, generating hot carbon monoxide and carbon dioxide gases, which then pass through the gasifier, first volatilizing the biomass and then drying it (Fig. V.22). The gases cool as they move upward through the feedstock bed, and convert volatile matter to tar if not burned immediately. For this reason, updraft gasifiers are more suitable for close-coupled steam boilers or crop dryers.

In the cross-flow gasifier design, air is introduced on one side of the reactor while producer gas is removed on the other side. It is used mostly for close-coupled boilers.

For fine particle feedstock, a fluidized-bed gasifier is most suitable (Fig. V.23). Usually, an inert material such as sand is the fluidization medium. The sand and

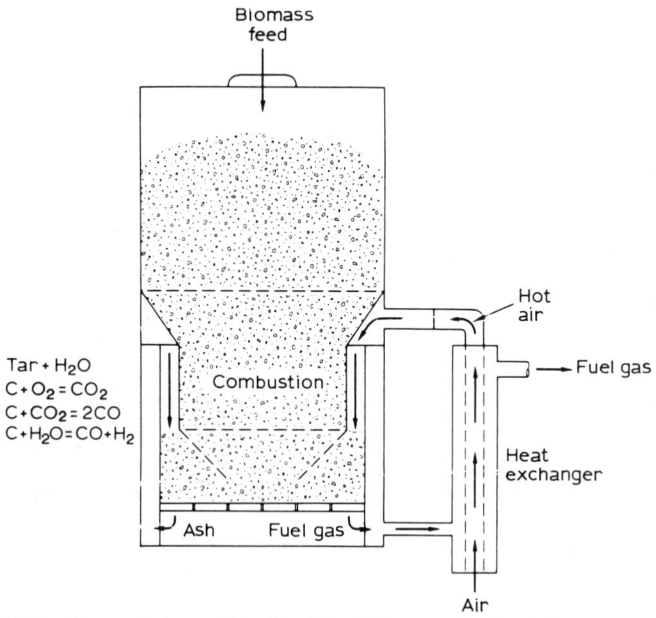

Fig. V.21. Downdraft gasifier (fixed-bed) (source: Sofer & Zaborsky, 1981).

feedstock mixture is subjected to an upward flow of air supplied through a distributor plate. The air velocity is such that the mixture becomes fluidized. Partial combustion of a small portion of the feedstock causes the sand to heat uniformly from 650°C to 850°C. As the gas resulting from this thermochemical conversion of biomass rises, it is directed into a cyclone separator to remove solid particulate material and then scrubbed to help eliminate unwanted tars.

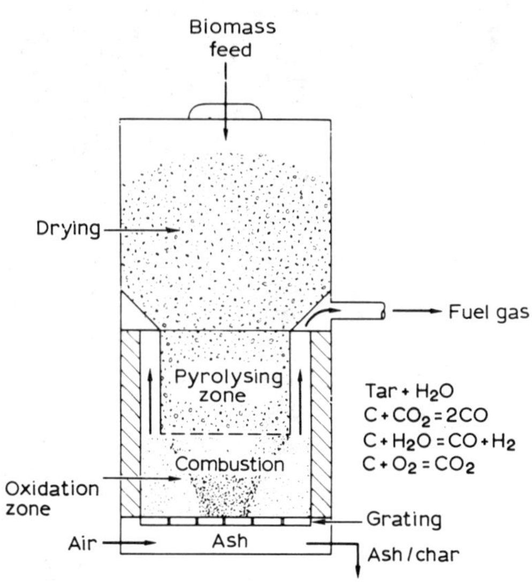

Fig. V.22. Updraft gasifier (fixed-bed) (source: Sofer & Zaborsky, 1981).

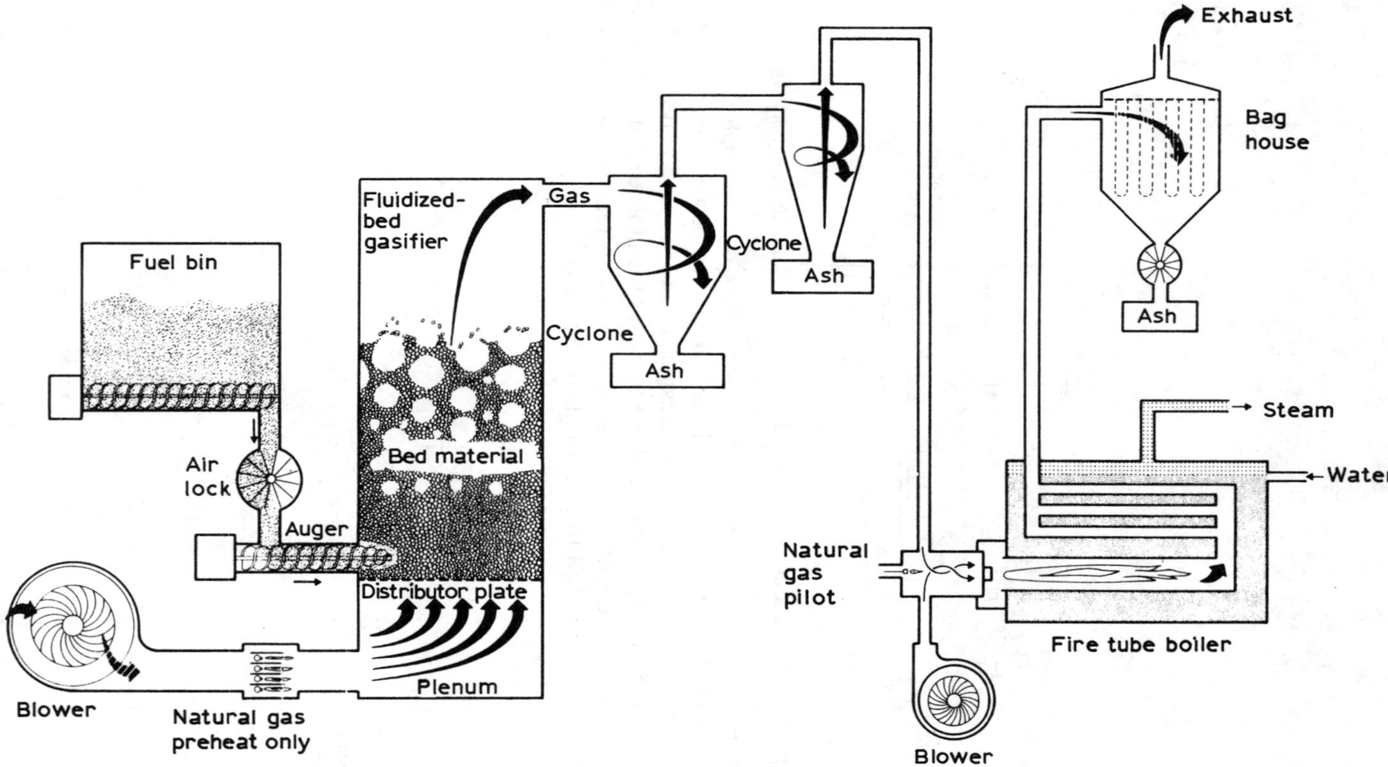

Fig. V.23. Texas A&M University fluidized-bed gasifier Boiler II (source: Hiler & Stout, 1985).

The fluidized-bed design offers certain advantages: rapid heat transfer maximizes gas formation, the gas production rate is easily controlled, and a variety of feedstocks can be used. Slag formation can be a problem with certain high ash content feedstocks, but it can be controlled with the fluidized-bed.

Technology Status

Close-coupled gasifiers for new boilers or to convert existing gas/oil boilers to biomass feedstock are available commercially. These relatively simple units operate at 85–90% conversion efficiencies. The gas has a heating value of 4–11 MJ/m^3 when burned without scrubbing. Using close-coupled gasifiers may result in a 5–10% boiler derating unless the feedstock is dried to 15% moisture content. Many private manufacturers have installed commercial units ranging from 0·8 to almost 16 GJ/h.

Retrofitting an existing oil or gas boiler with a gasifier costs about two-thirds as much as a new wood-fired boiler. Gas produced from these units costs about \$1·42–\$2·56/GJ for biomass originally purchased at \$11·02–\$22·05/dry tonne.

Several companies produce downdraft gasifiers that use forest byproducts as feedstock. Downdraft gasifers usually produce cleaner gas and are, therefore, more appropriate for use with internal combustion engines. These units are used for applications such as operation of a pickup truck, boiler firing, and engine generation sets. The size of these units ranges from 0·5 to 6·3 GJ/h.

Bailie (1976) presented cost estimates of producer gas-fueled electrical generating systems. A 1-MW installation costs about \$800 000. Approximately 30% of this amount is for the gasifier and clean-up plant, 40% for the modified diesel engine, and 30% for the electrical plant. The cost breakdown is about \$800/kilowatt of electrical energy (kWe), or \$240 for the gasifier, \$320 for the engine, and \$240 for the generator. Smaller units (10 kW or less) would cost from \$1000/kWe to \$1200/kWe. However, generator units below 8 kWe may not be competitive. An estimated \$10 billion in capital investment would be required to use 22% of the crop residues from 1·3 million United States farms.

Primary Products

The gasification process produces a combination of hot, odorless gases. The quality and condition of the gases depend on the gasifier design. Tables V.17 and V.18 present typical gas analyses from both updraft and downdraft producer gas generators.

The heat of combustion of producer gas is approximately 4–11 MJ/m^3. This constitutes a characteristically low-energy gas, which is high in nitrogen.

Byproducts

The byproduct of gasification is ash. In batch-fed gasifiers, ash is removed after each batch; in a continuous operation, provisions are made for its continuous removal. Depending on the feedstock used, gasification residues weigh 1–20% of the original raw material. A walnut shell substrate produces only few residues, while a corn cob feedstock produces the highest amount. Residues have no known use except to replenish land by returning nutrients to the soil.

TABLE V.17
Gas analysis from an updraft gasifier (charcoal feedstock) (source: Williams & Horsfield, 1977)

Gas component	*Percentage by volume*
CO	25–30
H_2	10–14
CO_2	5–8
O_2	0·5–1·5
CH_4	0–2·5
N_2 and others	50–53

Environmental and Safety Aspects
Gasification of biomass does not result in any serious environmental hazards. The sulphur content of biomass is usually minimal; consequently, the gas products of combustion are environmentally safe. Fluidized-bed gasification produces large amounts of particulate matter which must be removed by a cyclone separator before combustion (LePori *et al.*, 1981*a*).

Excessive removal of agricultural and forestry biomass may cause excessive erosion and water run-off. Several factors, such as soil type, crop, field slope, and weather, should be considered before biomass is removed from the land for energy conversion. Biomass is essential to maintain soil fertility.

The main combustible component of producer gas is fatal if inhaled in large quantities and for a long period of time. Adequate precautions must be exercised to prevent gas leakage when operating indoors under positive pressure. Some manufacturers design gasifiers to operate at negative pressure to prevent leakage.

The environmental effects of returning large quantities of ash to the soil on a regular basis are unknown. Additional research is needed to assess these effects.

Net Energy Balance
The conversion efficiency of a producer gas generator depends on the type of feedstock and its moisture content. Williams and Horsfield (1977) report that typical conversion efficiencies of 70–80% can be obtained. In calculating the efficiency, the

TABLE V.18
Gas analysis from a downdraft gasifier (wood feedstock (source: Williams & Horsfield, 1977)

Gas component	*Percentage by volume*
CO	20–22·5
H_2	12·5–15
CO_2	9·5–10
O_2	0·5–1·5
CH_4	2·5–3·4
N_2 and others	50–54

potential heat in wet feedstock and sensible heat from input air were included as input heat. Heat loss occurred in solid char and in condensate from the steam generated by producer gas.

The net energy balance which encompasses the production of biomass until its final use as a producer gas depends on many factors. The following sample energy balance is presented as one scenario:

> Corn is grown on 16·0 ha with an expected yield of 7500 kg/ha. Both cobs and stover are removed for gasification purposes. Moisture content of corn is 15·5% at harvest. Material is collected and transported at 8 km to the gasifier. A total of 960 kg/ha of residue is returned to the land and gasifier efficiency is 55·2% (a conservative figure). To calculate the net energy balance, the energy of production of residue is prorated according to the grain-to-residue ratio, which, for this example, is 1. Energy equivalent of fertilizer removed in residue is also included in the analysis. In this scenario, the energy ratio is 2·78. That is, the gasifier returns 2·78 times as much energy as is required to produce, collect, and transport the residue to the gasifier.

R & D or Commercial Use

Although producer gas technology is more than 40 years old and was used extensively during World War II, more research is needed to perfect it as a viable energy alternative. The gasification characteristics of various feedstocks, operation of the gasifier, effects of the final product, and improvement of engine efficiency while operating on producer gas all need further study. Figure V.24 illustrates how a gasifier may be used to power a stationary or mobile engine.

Currently, research is also underway to develop producer gas generators for agricultural applications. In the United States, Battelle Laboratories, Texas A & M University, Texas Tech University, and Kansas State University have conducted downdraft gasifier research and development. The Agricultural Engineering Department at the University of Kentucky has developed a cross-flow gasifier for crop drying. At Texas A & M University, cotton gin trash was used to produce steam as shown in Fig. V.23. The overall efficiency of converting cotton gin trash to steam ranged from 42 to 56% (Table V.19). A systems analysis indicated that 41·6 GJ/h of cotton gin trash could produce 1 MW of energy (Fig. V.25).

Several manufacturers are marketing gasifiers for commercial installation. The most common commercial applications have been to generate electricity and retrofit steam boilers for heating purposes.

Overall Assessment

Gasifiers appear promising for converting relatively dry biomass to a low-energy gas suitable for thermostatically controlled burners or internal combustion engines. Intensive research has resulted in the development of techniques for maintaining the high temperatures necessary for controlling particulate emissions and producing clean gas suitable for internal combustion engines. Gasifiers are now available for commercial use.

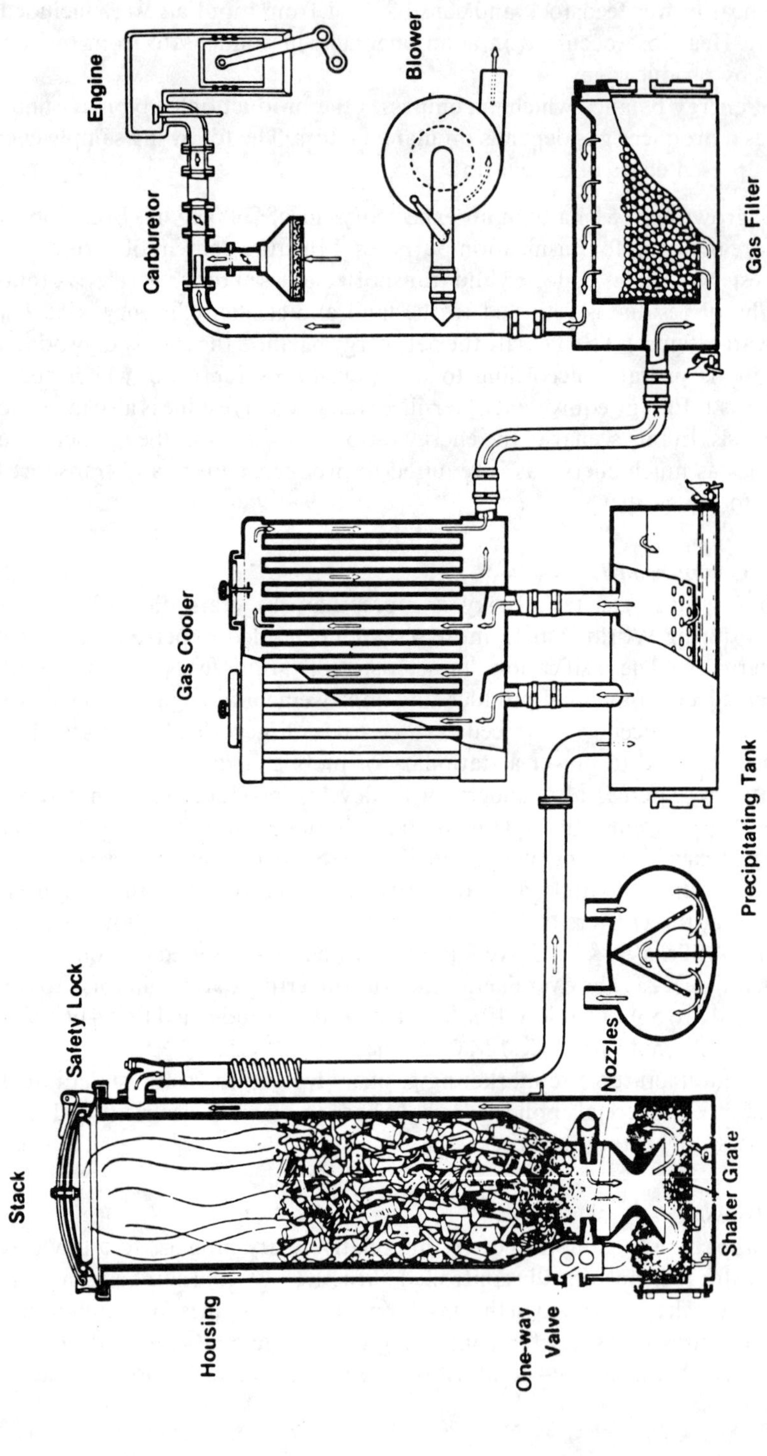

Fig. V.24. Basic layout of gas producer system, Svedlund design (source: NRC, 1983*a*).

TABLE V.19
Overall energy conversion efficiency for steam production (source: Hiler & Stout, 1985)

Cotton gin trash		*Steam production rate*		*Overall efficiency*	
Fuel rate (kg/h)	*Standard deviation*	*(kg/h)*	*Standard deviation*	*%*	*Standard deviation*
169	10	436	92	42	9·1
122	6	420	19	53	2·4
118	6	426	23	56	0·9
123	1	382	37	47	4·6
123	1	426	19	53	2·4

Note: Fluidized-bed gasifier, 610-mm diameter, flared steam at 121°C.

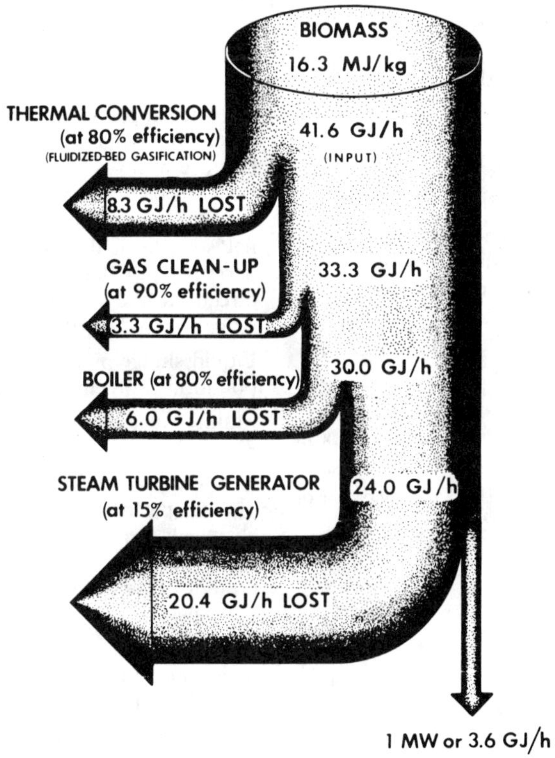

Fig. V.25. Energy flow in proposed small power plant (source: Hiler & Stout, 1985).

Biogas

Biogas is a product of the anaerobic digestion of biomass. Table V.20 lists some advantages and disadvantages of anaerobic digestion (anaerobic means 'without air'; aerobic means 'with air').

In China, more than 4 million single digesters are producing biogas. Figures V.26 and V.27 illustrate how local workers construct family-sized biogas plants. This type of plant is also shown in Fig. V.28.

The methane concentration of biogas is normally in the range of 50–70%. In a field situation, the composition of biogas may vary from practically no methane to nearly 80% methane. Because of the nature and kinetics of anaerobic digestion, this variation may occur in a short time.

TABLE V.20
Advantages and disadvantages of anaerobic digestion (source: Smil & Knowland, 1980)

Advantages	*Disadvantages*
Produces large amount of methane gas	Possibility of explosion
Methane can be stored at ambient temperature	High capital cost (However, if operated and maintained properly, the system may pay for itself.)
Produces free-flowing, thick, liquid sludge	May develop a volume of waste material much larger than the original material, because water is added to substrate. (This may not be a disadvantage in the rural areas of developing countries where farm fields are located close to the village, thus permitting the liquid sludge to be applied directly to the land, serving both as irrigation and fertilizer.)
Sludges are almost odorless; odor not disagreeable	Liquid sludge presents a potential water pollution problem if handled incorrectly
Sludge has good fertilizer value and can be used as a soil conditioner	Maintenance and control are required
Reduces organic content of waste materials by 30–50% and produces a stabilized sludge for ultimate disposal	Certain chemicals in the waste, if excessive, have the potential to interfere with digester performance. (However, these chemicals are encountered only in sludges from industrial waste waters and therefore are not likely to be a problem in a rural village system.)
Weed seeds are destroyed and pathogens are either destroyed or greatly reduced in number	Proper operating conditions must be maintained in the digester for maximum gas production
Rodents and flies are not attracted to the end product of the process. Access of pests and vermin to wastes is limited	Most efficient use of methane as a fuel requires removal of impurities such as CO_2 and H_2S, particularly when the gas is to be used in internal combustion engines
Provides a sanitary method for disposal of human and animal wastes	
Helps conserve scarce local energy resources such as wood	

Fig. V.26. Initial stage of construction of an underground biogas plant in rural Sichuan, China (source: FAO Photo 10428/J/9a).

The amount of gas produced depends not only on the digester size and the volatile solids loading rate, but also on the temperature within the unit and the biodegradability of the feedstock. If the temperature is not controlled and fluctuates with the seasons, the gas production rate will also vary. If the temperature is controlled, a certain portion of the energy provided is used to heat the digester and is not available for beneficial use. Therefore, gas utilization plans must allow for lower production amounts during cold weather periods.

Types of Digesters

A typical anaerobic digester system using animal manure as a feedstock is illustrated in Fig. V.29. A settling pond is used to store manure prior to digestion. After several weeks in the digester, the manure exits and is stored in an effluent lagoon. Gas exits from the top of the digester tank.

Fig. V.27. Village neighbors working together to complete construction of an underground biogas plant in rural Sichuan, China (source: FAO Photo 10428/G/11).

The small hydrogen sulfide content is removed if necessary and the biogas may be used to fuel an internal combustion engine that drives an electric generator. The system may also supply electricity for on-site use, and heat from the engine may be used on-site as well. Table V.21 compares various types of digester systems. Figures V.26, V.27, V.28, V.30, and V.31 show different biogas plant designs.

Biomass Feedstock

Animal manure is the most common feedstock for anaerobic digestion, but human wastes, food processing wastes, by-products of ethanol fermentation, sweet sorghum juice, and other materials may be used. Loading of manure into a small biogas plant is illustrated in Fig. V.32.

Table V.22 summarizes the results of anaerobic digestion of various types of biomass. Almost any biomass material can serve as a substrate for a digester.

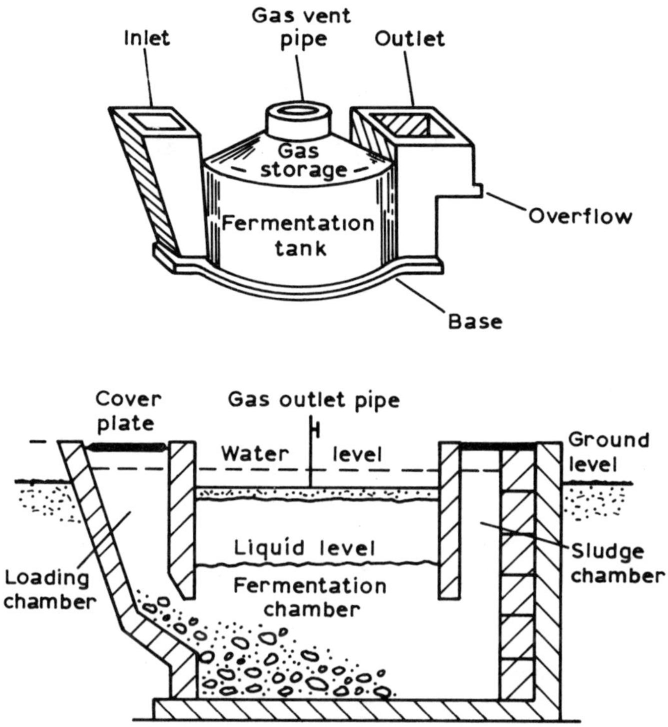

Fig. V.28. Chinese design of a biogas plant (source: OTA, 1980).

Normally, for optimum methanogenic performance, a 30:1 carbon to nitrogen (C/N) ratio is preferred. However, values ranging from 15:1 to 45:1 have been satisfactory. Although plant wastes (which are high in carbon) are occasionally added to animal manure, animal manure is generally used alone because it falls within the most desired C/N range.

The major use of manure throughout the world is for fertilizer or energy through

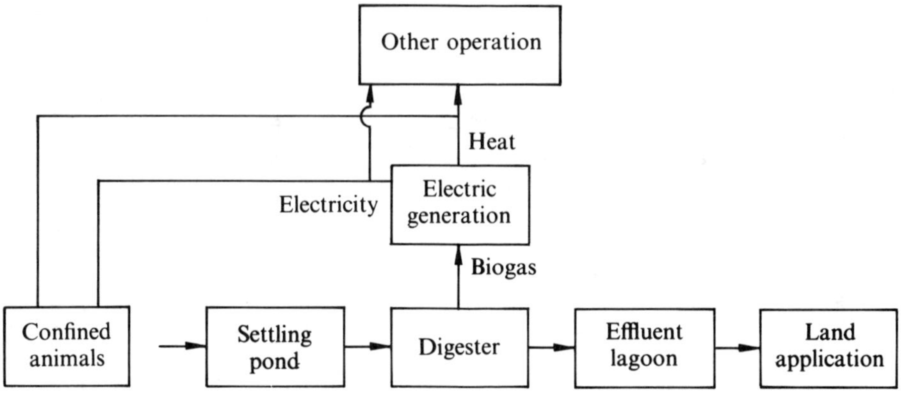

Fig. V.29. Anaerobic digester system (source: OTA, 1980).

TABLE V.21

Comparison of different types of digester systems (source: D. W. Robinson & Mollan, 1982)

Type of digester	*Suitable wastes*	*Volumes, solids content*	*Typical retention times*	*Degree of mixing*	*Operating temperatures*	*Gas production*	*Degree of control required*	*Comments*
1. Batch	Agricultural, irregular, or seasonal. Fibrous or difficult to digest.	Low volumes up to 25% solid	60 days or more	Little needed	Usually 30–35°C	Irregular and discontinuous	Little once started	Messy and time consuming to start
2. Plug-flow, horizontal, vertical	Agricultural, continuous, or regular flows. Less fiber content	Larger volumes 5–15% solids	30–60 days	Occasional	30–35°C	Continuous	Simple	Loading and scum removal can be messy
3. Conventional sewage works	Continuous sewage sludge	Less than 5% solids	30–60 days	Occasional	30–35°C or unheated	Continuous	Simple	Not very effective
4. High rate sewage digestion								
Primary	Sewage sludge	4–10% solids	10–30 days	Regular	30–35°C	Continuous	More sophisticated	Automatic
Secondary	From primary digesters	4–10% solids	20–60 days	None	Unheated	None collected	Simple	
5. High Rate	Agricultural Industrial	4–15% solids	5–20 days	Continuous	30–35°C	Continuous	More sophisticated	Can be automated
6. Anaerobic contact	Industrial (Agricultural)	Low solids	0·5–5 days[a]	Continuous	30–35°C	Continuous	Sophisticated	Automatic
7. Anerobic filter	Industrial	Low solids (low organic content)	0·5–5 days[a]	None needed	30–35°C	Continuous	Sophisticated	Automatic
8. Upflow anaerobic sludge blanket process	Industrial	High volumes Low solids	0·5–1 day	None	30–35°C	Continuous	Little once started	Automatic

[a] Liquid retention time.

TABLE V.22
Anaerobic treatment of agricultural wastes (source: Regan, 1975)

Waste	*Temperature (°C)*	*Process loading (kg VS/m³ per day)*[a]	*Hydraulic retention time (days)*	*% VS reduced*	*Gas produced (m³/kg VS added)*	*Methane %*	*Process used*
Slaughter house	33	1·8	—	—	—	—	Two-stage with solids return
Pear	35	1·6–7·4	0·5–1·0	—	—	—	Two-stage with solids return
Whey	37	2·1	6	—	—	65	Single-stage
Beef cattle	23	1·6–6·4	10–20	—	—	—	Lagoon
	35	1·4	10	71·0	0·56	58	Single-stage
	35	2·9	10	53·0	0·55	57	Single-stage
	35	4·3	10	44·3	0·74	52	Single-stage
	35	5·8	10	55·5	0·71	53	Single-stage
Dairy manure	—	2·1–3·5		10–15			Single-stage
		0·6					Unmixed
		1·2					Mixed
	35	1·6	12	45·3	0·06	77	Single-stage
	35	1·6	20	44·7	0·05	79	Single-stage
	35	2·9	12	37·8	0·04	77	Single-stage
	35	2·9	20	53·3	0·07	74	Single-stage
Dairy bull	—	3·8	10	18–26		61–66	Single-stage
Swine	—	3·8	10–15	49–61		58–61	Single-stage
	35	4·0–8·0	10–20				Single-stage
Poultry manure	—	2·7–5·0		50			Single-stage
	35	1·4					Single-stage
	35	2·9	10–15	57–68		52–58	Single-stage

VS = volatile solids.
[a] Loading calculated based on ratio biological oxygen demand to volatile solids equal to 1·0.

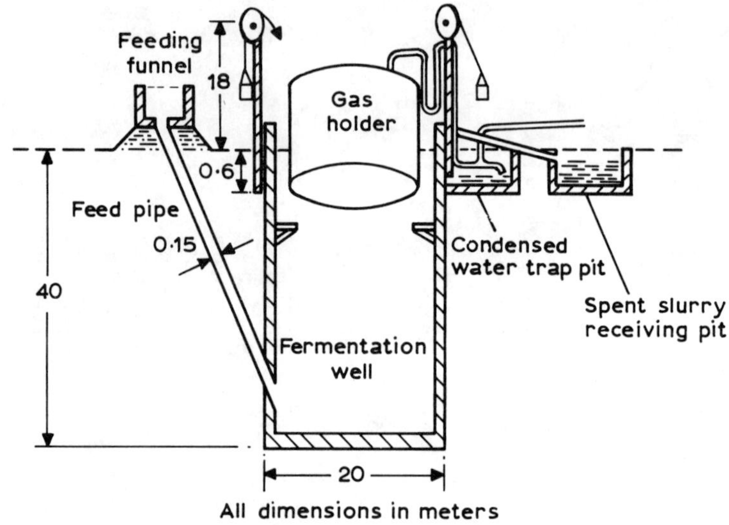

Fig. V.30. Indian Gobar gas plant (source: OTA, 1980).

direct burning. Anaerobic digestion stabilizes the nutrients and nearly all the nitrogen, phosphorus, and potassium value remains in the digester effluent. Thus, the process provides for both energy extraction and maximum nutrient retention for use as fertilizer.

The use of crop residue as a substrate for anaerobic digestion has not been researched as extensively as manure. The C/N ratio of crop residue, or even mixtures of crop residue and manure, is normally too high for satisfactory biogas production. The ratio can be reduced by adding human excrement, including urine or chemical nitrogen.

Table V.23 summarizes the number of livestock in various parts of the world. Table V.24 tabulates manure production of the various species. Although some manure is used productively as fertilizer or fuel in developing countries, some remains unused. A considerable amount of energy can be extracted from manure, and it will still retain much of its fertilizer value.

TABLE V.23
Livestock numbers, 1983 (source: FAO, 1984*d*)

	Horses	*Mules*	*Asses*	*Cattle*	*Buffalo*	*Camels*	*Total*
Africa	3 752	2 245	12 053	174 333	2 393	12 557	207 333
North & Central America	18 606	4 042	3 236	183 138	8	—	209 030
South America	13 122	2 979	4 225	213 597	570	—	234 493
Asia	17 440	5 207	18 810	372 678	120 402	4 155	538 692
Europe	5 050	520	1 203	133 317	433	—	140 523
Oceania	620	—	5	31 134	—	—	31 759
USSR	5 601	2	344	117 186	320	240	123 693
	64 191	14 995	39 876	1 225 383	124 126	16 952	1 485 523

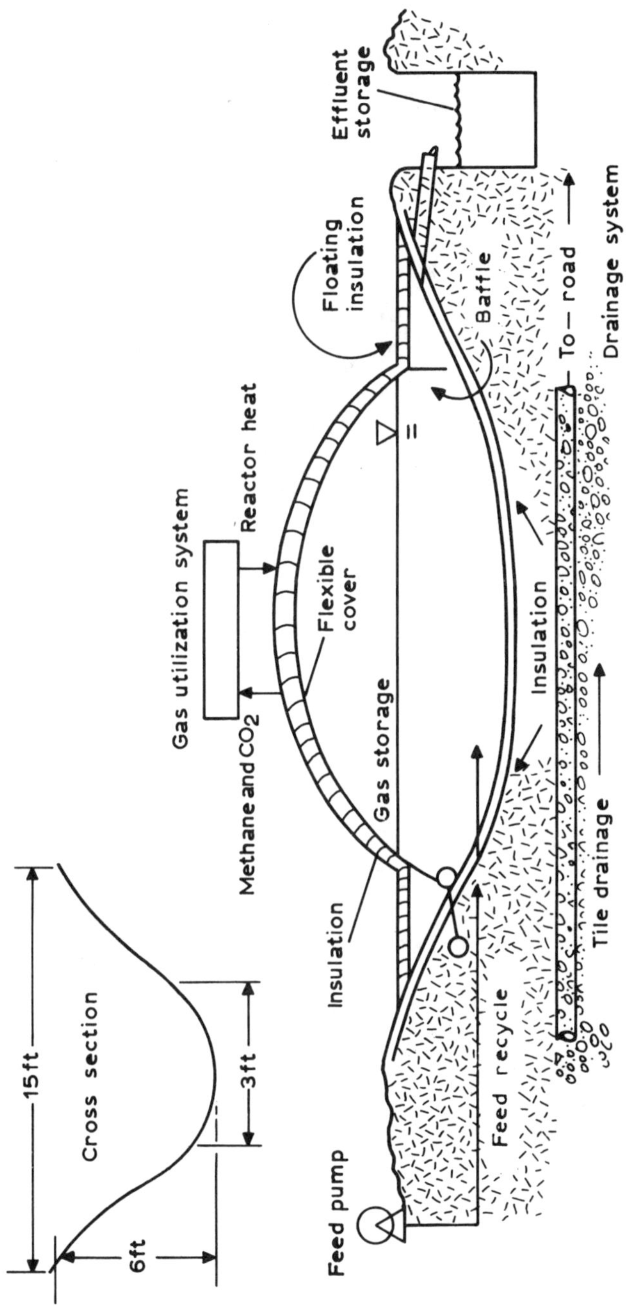

Fig. V.31. Plug-flow digestion system (source: OTA, 1980).

Fig. V.32. A biogas unit in Sichuan, China. The owner is filling the digester with effluent from another biogas plant, to accelerate bacterial action (source: FAO Photo 10428/M/23a).

TABLE V.24
Manure production data for various species (source: Makhijani & Poole, 1975)

Animal	*Fresh manure per 1 000 kg liveweight (kg/year)*	*Assumed average liveweight (kg)*	*Fresh manure production assumed per head (kg/year)*	*Assumed moisture content of fresh manure (%)*	*Nitrogen content percentage of dry matter*	
					Solid & liquid wastes	*Solid wastes only*
Cattle	27 000	200	5 400	80	2·4	1·2
Horses, mules donkeys	18 000	150	2 700	80	1·7	1·1
Pigs	30 000	50	1 500	80	3·75	1·8
Sheep & goats	13 000	40	500	70	4·1	2·0
Poultry	9 000	1·5	13	60	6·3	6·3
Human feces without urine	—	40–80	50–100	66–80	—	5–7
Human urine	—	40–80	18–25 kg dry solids/year	—	15–19 (urine only)	—

Anaerobic Digestion Process

Anaerobic digestion is a biological process performed by living microorganisms. Expressed as a simple equation, anaerobic digestion is defined as:

$$\text{organic matter} + \text{bacteria} + \text{water} \rightarrow \underset{(CH_4)}{\text{methane}} + \underset{(CO_2)}{\text{carbon dioxide}} + \underset{(H_2S)}{\text{hydrogen sulfide}} + \text{stabilized effluent} + \text{bacteria}$$

This process occurs only in strict absence of free oxygen.

Anaerobic fermentation is a three-step process (Fig. V.33). Step 1 involves hydrolysis of high molecular weight carbohydrates, fats, and proteins into low molecular weight fatty acids and amino acids. This process is known as acid fermentation. The bacterial extracellular enzymes convert the solid material into a soluble form. During step 2, volatile fatty acids (acetic and propionic); lower alcohols; and aldehydes, carbon dioxide, hydrogen, ammonia, and sulfuric acid are produced. This phase is known as acid regression. In step 3, methane and carbon dioxide are generated from fatty acids, alcohols, and aldehydes; methane and water are synthesized from carbon dioxide and hydrogen. This process is known as alkaline fermentation. The bacteria present in steps 1 and 2 are collectively called acidogens or acid formers, and the bacteria evident in step 3 are called methanogens or methane formers (Fig. V.34).

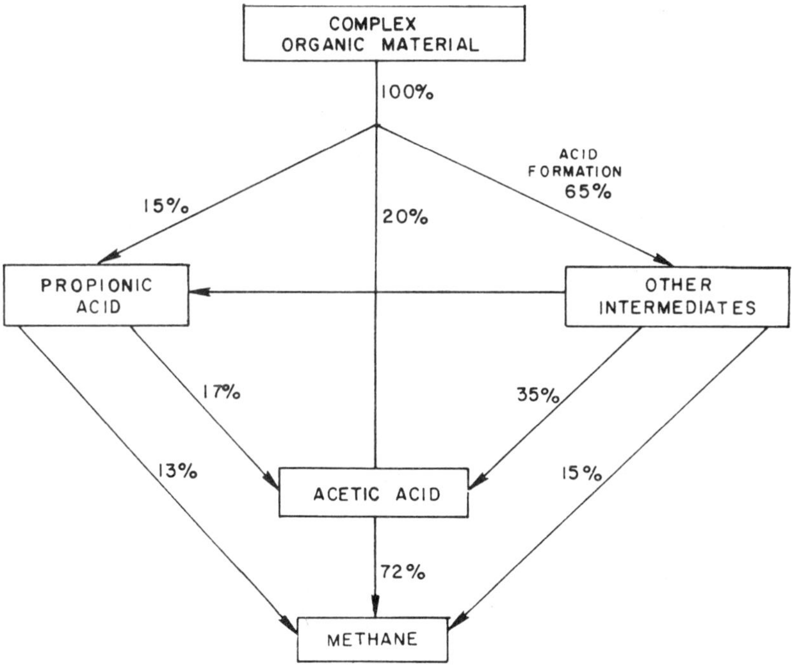

Fig. V.33. Methane formation from a complex organic material. Percentages refer to the conversion of chemical oxygen demand, or COD (source: Hiler & Stout, 1985).

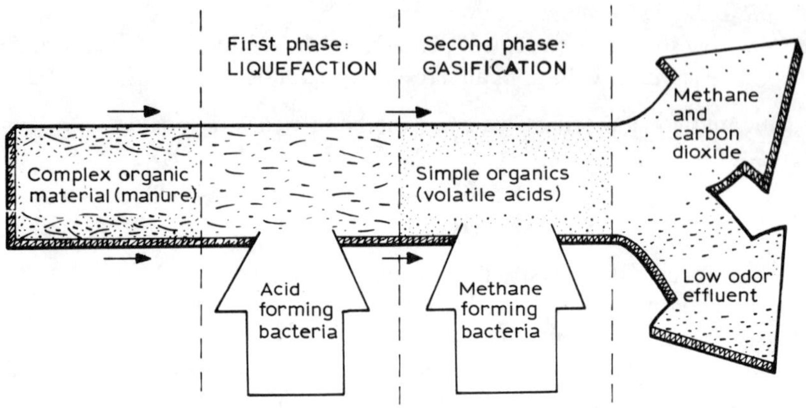

Fig. V.34. Two phases of methane production by anaerobic digestion (source: Parsons, 1984).

Methane-forming bacteria are more environmentally sensitive than the acid formers. Methane-formers are strict anaerobes—slow-growing, pH-sensitive, and temperature-dependent. Acid-formers, however, are hardy and grow rapidly under a variety of environmental conditions. Any environmental constraints, therefore, are due to requirements of the methane formers.

pH. The pH is an important indicator of methanogen performance. Gas is produced if the pH is from 6·6 to 7·6, but gas production is highest when the pH is from 7·0 to 7·2. In the absence of other indicators, pH values alone have been used to infer the proper digester environment.

Temperature. The two most common kinds of methanogenic bacteria are those that prefer temperatures near 35°C (mesophilic) and those that prefer temperatures near 54°C (thermophilic) (Fig. V.35). The thermophilic bacteria produce methane at a faster rate, but are more sensitive to temperature fluctations than the mesophilic bacteria. A third kind of bacteria, psychrophilic, functions at temperatures below 20°C, but little is known about it.

Toxicity. Many compounds can be toxic to methanogens in sufficient concentration. These include heavy metals such as copper, zinc, and nickel; antibiotics in livestock feed; ammonia in excess of 3000 mg/liter and various salts. (Some salts are needed, but they quickly become inhibitory.)

Conversion efficiency. Anaerobic digestion is usually around 60–70% efficient. Efficiency is measured as:

$$\frac{\text{Weight of volatile solids converted to biogas}}{\text{Weight of volatile solids available}}$$

The efficiency of a particular digester depends on operational factors, such as

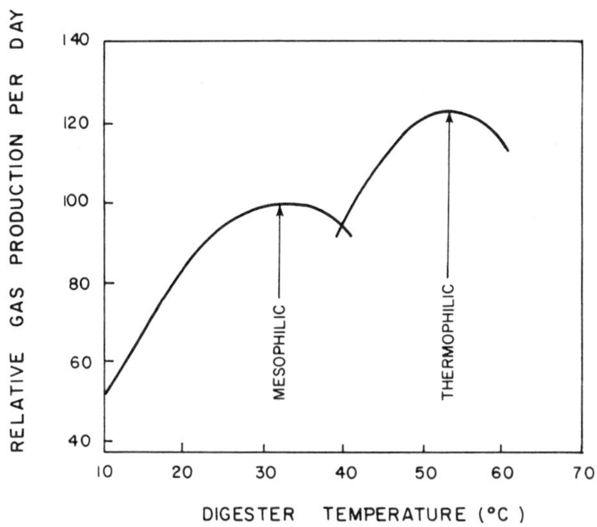

Fig. V.35. Effect of temperature on gas production rate (source: Hiler & Stout, 1985).

temperature and loading rate. A high loading rate implies a lower efficiency, although a high efficiency may not necessarily be the most economically attractive alternative.

Digester Designs

The controlled anaerobic fermentation process takes place in a digester, which is basically a container to keep the waste isolated from the air. There are many different digester designs, which are often classified according to the waste loading schedule. Using this classification, two digester types emerge: batch and continuous flow.

In a batch design, the digester is loaded once and then sealed until fermentation is nearly complete. This system offers some advantages: (1) raw material availability is sometimes sporadic or comes in batches, such as crop residues; (2) daily management is minimal; and (3) construction is relatively inexpensive. Disadvantages include: (1) significant labor is required to load and empty the digester; (2) gas production is not constant; and (3) batch digesters are not as efficient as continuous digesters. (Most early digesters were the batch type.)

The continuous flow digester, an improvement in digester design, is loaded daily or even hourly. An equal volume of digester slurry is removed with each new loading. The new substrate must be introduced anaerobically. There are two types of continuous flow digesters: completely mixed and plug flow.

In a completely mixed digester, each new load is mixed thoroughly with the old or existing slurry in the digester. Therefore, the removed effluent will contain material that has been retained in the digester for varying amounts of time. However, material that has been in the digester for only a fraction of the total retention time can be removed.

The completely mixed digester (Fig. V.36) has either a single effluent outlet at the bottom or an outlet at the top to remove liquids, as well as an outlet at the bottom to remove the more concentrated sludge. Tops can be fixed or floating; floating tops pressurize the biogas. Allowances must be made in either the digester design or in a separate gas storage area for increasing gas volumes.

In a plug-flow digester, each daily feeding performs like a separate plug, eventually flowing to the opposite end of the digester where it is removed. Each new load of substrate displaces the previous load along the digester, passing a point of maximum fermentation along the way. As a result, all of the effluent material is retained in the digester for the same length of time—the specified retention time. Plug-flow digesters are not mixed or stirred. Unfortunately, this may lead to solid separation problems under higher loading rates.

Plug-flow digesters are sometimes accompanied by two or more completely

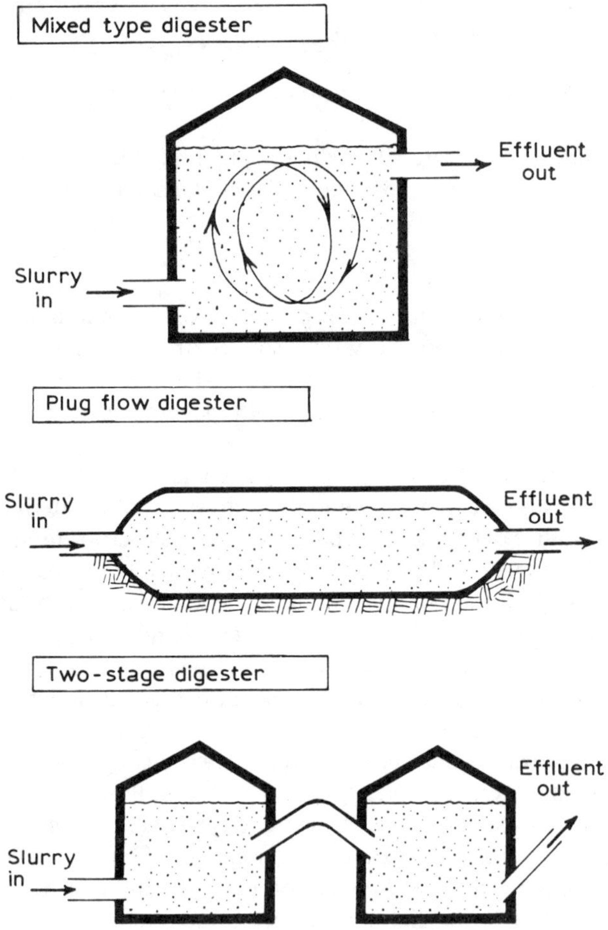

Fig. V.36. Typical digester designs (source: Parsons, 1984).

mixed digesters connected in a series. Using this arrangement, higher loading rates are possible and more uniformly digested effluent can be produced. However, digester cost increases per unit of output.

To maximize the conversion of solids to methane (or efficiency), a plug-flow system is recommended. To maximize methane production per unit volume of digester, the single-stage, completely mixed digester is best. The multi-stage digesters are between these two extremes.

Primary Products

The gas produced in anaerobic digestion of organic material is a mixture of mostly methane and carbon dioxide with small amounts of other gases present. Fermentation of animal waste or other substrates with a C/N ratio of around 30 will generally produce a mixture containing about 60% methane. Table V.25 lists ranges of the composition expected in biogas produced from farm wastes.

In a digester operating at maximum capacity, the biogas produced can generally be expected to have a heat value of about 22 MJ/m^3 at atmospheric pressure. In comparison, natural gas has a heat value of 37 MJ/m^3, and a liter of gasoline contains about 33 MJ/m^3.

Table V.26 compares the energy in biogas produced from various manures with the quantity of other fuels required to yield the same amount of energy.

Biogas Utilization

The success of a biogas production unit depends on using the resulting methane gas to its maximum benefit. Gas utilization may involve direct use of the total mixture including impurities, or it may involve purification steps to remove the following: (1) carbon dioxide to increase the energy value of gas per unit volume; and (2) hydrogen sulfide to decrease corrosiveness of the mixture.

The carbon dioxide content can be reduced by bubbling the gas through a water scrubber with pH controlled to prevent acid conditions from developing. Lime may be added to control pH, but in some cases, ammonia is sufficient in the solution to maintain a neutral pH. Hydrogen sulfide can be removed by transferring the gas through iron-impregnated wood chips or iron filings.

TABLE V.25
General composition of biogas produced from farm wastes (source: Fry, 1973)

Gas		%
CH_4	Methane	54–70
CO_2	Carbon dioxide	27–45
N_2	Nitrogen	0·5–3
H_2	Hydrogen	1–10
CO	Carbon monoxide	0·1
O_2	Oxygen	0·1
H_2S	Hydrogen sulfide	Trace

TABLE V.26
Approximate daily production and heat values for biogas (source: J. S. Converse & Graves, 1974)

Livestock 454 kg body weight	*Approximate biogas production (m^3/day) (20°C)*[a]	*Approximate heat value (kcal)*	*Approximate equivalents*		
			Liters[b] *gasoline*	*Liters diesel fuel*[b]	*m^3 natural gas*[b]
Beef	0·9	4 500	0·57	0·53	0·51
Dairy	1·3	6 800	0·87	0·76	0·77
Poultry broilers	2·6	13 900	1·74	1·59	1·56
Poultry layers	2·0	10 900	1·36	1·21	1·22
Swine	0·8	4 400	0·57	0·49	·48

[a] Assumes biogas containing about 70% methane or a heating value of 5 339 kcal/m^3.
[b] Heating values: gasoline, 7 990 kcal/liter; diesel fuel, 8 860 kcal/liter; natural gas, 8 900 kcal/m^3.

Researchers evaluating biogas or purified methane storage conclude that compressing and storing gas produced by small digesters on farms is not practical. In addition, significant quantities of energy cannot be compressed into small enough tanks to make efficient use of biogas in mobile engines, tractors, and other vehicles. However, there are some exceptions as shown in Fig. V.37.

Major biogas usage must be at the production site in cook stoves, space heaters, stationary engines, or crop dryers (Figs V.38, V.39, V.40, and V.41). Other possible uses include electric generators or low-grade (low-temperature) heat production.

Stationary engines operated on methane should have high compression chambers designed for this application. Full engine power can be achieved only if carbon dioxide is removed from the biogas mixture to increase its energy content. Longer engine life can be attained if hydrogen sulfide is also scrubbed from the gas before use.

Sludge as a Fertilizer

Digester waste or sludge is an excellent fertilizer (Fig. V.42). Although some disagreement exists, waste from the anaerobic digester is believed to be a better fertilizer than undigested manure or effluent from an aerobic waste treatment operation. Compared to undigested manure, spent slurry from the anaerobic digester contains all the original potassium and phosphorus and up to 99% of the original nitrogen. In addition, the amount of boron, calcium, iron, magnesium, sulfur, and antimony is unchanged.

The primary difference between the fertilizer value of anaerobically digested waste and either undigested or aerobically digested waste is in the form the nitrogen takes. In anaerobically digested sludge, nitrogen is in the form of ammonia, or more precisely, the ammonium ion; in aerobic oxidation, nitrogen has been oxidized to form nitrates and nitrites. For fertilizer application, ammonium may be a more

Fig. V.37. Although biogas is more suitable for stationary applications because of its low energy density, mobile vehicles such as this bus in Sichuan, China, can be operated with biogas (source: FAO Photo 10428/B/36a).

Fig. V.38. Woman lighting the burner of her stove in a peasant kitchen in Sichuan, China. She produces her own biogas for cooking from the manure from the pig sty and the family latrine (source: FAO Photo 10428/E/3).

Fig. V.39. Biogas can also serve as a fuel for gas lamps (source: FAO Photo 10428/I/14).

appropriate source of nitrogen than either the nitrates or nitrites. Ammonium is less likely to leach away and more apt to become fixed to the negatively charged particles (clay and humus). Nitrogen as ammonium must be reconverted to oxidized nitrogen in the soil for plant fertilization. However, differences in soil type and crops grown may be the decisive factors in determining the advantages or disadvantages of ammonium versus nitrate and nitrite as fertilizer.

A potential exists for ammonium volatilization or nutrient loss, the extent of which is uncertain. Effluent may need to be stored in a sealed container and then injected into the soil to obtain the maximum nutrient retention and utilization.

The digested slurry is a liquid, soupy mixture composed of solids suspended in the liquid. The waste can be used as a liquid (agitating may be necessary), or the solids and liquid can be separated for simplified handling. Digested slurry is easily dewatered and many have speculated about the potential of dried fertilizer marketing.

Fig. V.40. With proper cleanup, biogas can be used as a fuel for internal combustion engines. A small rice mill is being driven by a biogas-fueled engine (source: FAO Photo 10428/F/09).

Fig. V.41. A biogas-fueled engine driving a small electric generator. Pig manure or other wastes can be converted to a more useful form of energy—electricity (source: FAO Photo 10428/G/35).

Fig. V.42. Effluent is used as fertilizer. It increases yields and improves the soil (source: FAO Photo 10428/C/33a).

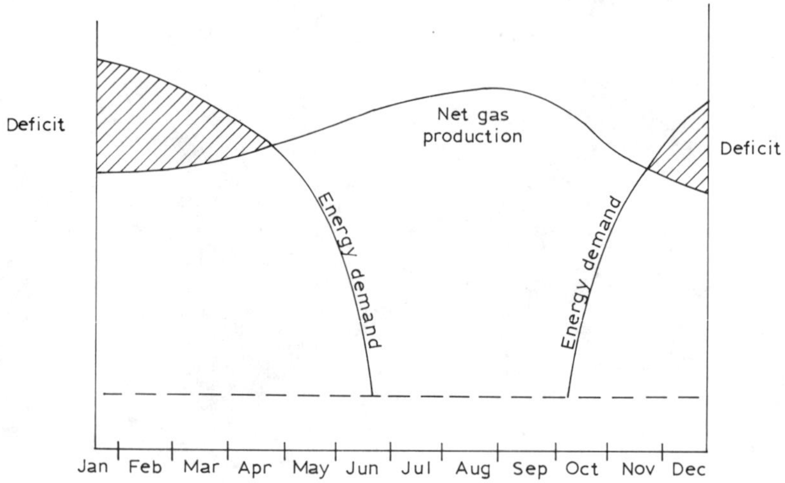

Fig. V.43. Production of biogas in relation to farm energy demand (source: Būttner, 1981; © 1981 Pergamon Press plc).

Sludge as a Livestock Feed

Stabilized effluent obtained from the anaerobic digester can be used as feed for livestock if mixed with molasses, grains, and straws. When the economic value of the protein in effluent is compared with biogas on a per animal basis, the effluent may be more than 10 times as valuable. Thus, a significant amount of interest exists in 'refeeding'. Effluent is high in protein and has an essential amino acid concentration similar to soybean meal. Unfortunately, effluent is also high in moisture, and, if used directly, may cause handling problems, such as freezing during the winter and a shorter bunk life in the summer.

Centrifuges have been used to extract solids; however, much of the protein is in the finer microbial solids which escape during the extraction process. The protein concentration of the remaining centrifuged cake is similar to that of alfalfa.

Environmental and Safety Aspects

One important positive aspect of anaerobic digestion is its role as a waste treatment or waste disposal system. As a waste disposal system, the anaerobic digester has the following characteristics:

—Digested waste is stabilized and virtually odor free.

—Waste is handled as a liquid and readily integrated with the increasingly popular liquid handling systems.

—Although there is usually no net reduction in liquid volume, and may even be a net increase, the anaerobic digester experiences a solid volume reduction of about 50%.

—Nutrients are retained in the waste.

—Anaerobic digesters minimize nitrogen loss during the treatment process in spite of potential ammonia volatilization.

Net Energy Balance

Since methane-forming bacteria require digesters to be maintained at elevated temperatures, they must be heated in cooler climates. Thus, net biogas output is reduced in the winter. An example of the contribution of a 25-cow biogas plant to the energy needs of a farm (for heating water and buildings, and for cooking) is shown in Fig. V.43.

R&D or Commercial Use

Millions of small biogas plants are in operation in many countries around the world. Most are simple plants constructed of local materials. An exception is the large biogas plant located at Bartow, Florida, that uses cow manure to replace more than 20 000 liters of fuel oil each week in a processing plant (Fig. V.44). The performance of smaller biogas production plants is illustrated in Table V.27.

Overall Assessment

The relative economy of anerobic digesters is probably the single most important factor in determining their feasibility. However, an accurate economic analysis of

Fig. V.44. A large biogas plant located in Bartow, Florida, provides heat for a nearby processing plant (source: DOE Photo 207 001 003).

the usage of anaerobic digesters is difficult to obtain because many unknowns and uncertainties exist:

—Cost of energy. Increases in conventional fuel prices make biogas more attractive as an energy source. However, the cost of constructing biogas plants may rise along with increased energy costs.

—Sludge use. Feed is clearly more expensive than fertilizer, and plans for refeeding wastes have resulted in favorable economic analyses. Since sludge may be used for many purposes, the relative supply and demand of wastes will be influential in the future.

—Type of system. The anaerobic digester has not yet been perfected. Additional study and research are needed both in design and operational parameters. Design breakthroughs exist which may significantly lower construction costs. A relatively efficient and reliable design could be mass-produced at a lower per-unit cost.

—Waste handling system. The digester compares favorably in efficiency and handling ease with other waste handling systems, such as aerobic lagoons. In addition to the potential economic justifiers of energy or feedstock production, individuals investing capital in a new liquid handling system could write off some of the anaerobic system costs to waste handling.

TABLE V.27

Biogas production from various types of manure (source: D. W. Robinson & Mollan, 1982)

A.	*1 000 fattening pigs at 90 kg finishing weight*	
	Manure production/day	4 600 kg
	Volatile solids kg/day	540
	Biogas production liters/day = 204 m^3/day.	204 000
	204 m^3 of biogas ≡	85·5 liters of diesel equiv/day
B.	*25 000 poultry (1·4 kg finishing live weight)*	
	Manure production/day	2 025 kg
	Volatile solids (kg/day)	435
	Biogas production (liters/day) = 214 m^3/day	214 500
	214 m^3 of biogas ≡	108·9 liters of diesel equiv/day
C.	*100 dairy cows (500 kg live weight)*	
	Manure production/day	2 832 kg
	Volatile solids (kg/day)	294
	Biogas production (liters/day) = 96 m^3/day	95 555·6
	96 m^3 biogas/day ≡	48·4 liters of diesel equiv/day
D.	*200 fattening cattle (435 kg average weight)*	
	Manure produced/day	5 200 kg
	Volatile solids/day	540 kg
	Biogas production (liters/day) = 175 m^3/day	175 500
	175 m^3 biogas/day ≡	89·1 liters of diesel equiv/day

—Use of salvaged material. Currently, builders could use salvage or used material and tanks to lower costs substantially. A long-term supply of salvageable material may not be available, however.

—Nature of farm operation. The economics of anaerobic digestion may be site-dependent. For example, feasibility may depend on the amount and form of energy currently being used, the nature of the waste disposal system, the availability of existing equipment that can be coupled with the digester, and the size of the farm operation.

Increased fuel and fertilizer costs have resulted in increased farm interest in anaerobic digesters. Anaerobic digestion may become a viable alternative as farmers develop the appropriate combination of digester design, farm operation, and fuel and fertilizer savings.

Hydrogen (Hoffman, 1981; Parker, 1981)

With the atomic symbol H, atomic weight 1·007 97, and atomic number 1, hydrogen is the simplest and lightest element. The colorless gas has a density at 0°C and

101 kPa [1 atmosphere (atm)] of 0·089 87 g/liter. Its specific gravity compared with air is 0·0695.

Process Description
Hydrogen may be produced using several methods (Fig. V.45), Current industrial production, about 20 million tonnes/year, is mostly from the steam reforming of natural gas or petroleum. Hydrogen can readily be produced from coal. Steam reacts with coal (at about 950°C) and high pressure to produce carbon monoxide, carbon dioxide, and hydrogen.

Combustion of biomass in a limited air or oxygen supply results in producer gas byproducts that include hydrogen and carbon monoxide. A detailed discussion of producer gas appears earlier in this chapter in the section on producer gas.

Electrolysis, or water splitting, is another way to produce hydrogen. Electricity from any source, including solar photovoltaic cells or nuclear energy, could be used to produce hydrogen.

Biophotolysis
Through biophotolysis, certain microscopic algae can produce hydrogen and oxygen gas from water and sunlight. This process, in effect, is the photosynthetic splitting of water.

Biomass Feedstock
Green and blue-green algae have the ability to free hydrogen gas photosynthetically. Biophotolysis is attractive as a concept because the substrate (feedstock) is water and extremely plentiful, the driving force is unlimited (the sun), and the product (hydrogen) is storable and an excellent nonpolluting fuel.

Primary Products
Hydrogen gas is the primary product. Its properties are given in Table V.28.

An excellent fuel, hydrogen may be burned directly as a gas with a high heat of combustion of 118 MJ/kg or as a compressed gas at 12 000–13 000 kPa. Or, it may be

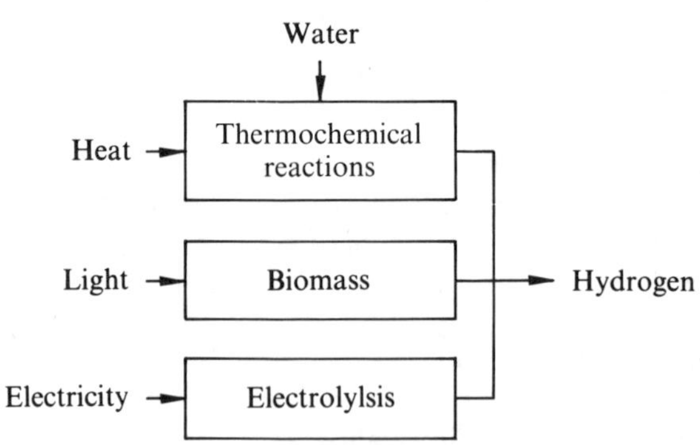

Fig. V.45. Methods of producing hydrogen (source: Parker, 1981; © 1981 McGraw-Hill).

TABLE V.28
Properties of hydrogen gas (source: Considine, 1977)

Property	*Value*
Melting point	−259·2°C
Boiling point at 1 atm	−252·8°C
Density of solid at −259·2°C	0·0866 g/cm^3
Density of liquid at −252·8°C	
Critical temperature	−240·0°C
Critical pressure	13·0 atm
Critical density	0·030 1 g/cm^3
Specific heat at constant pressure	
Gas at 25°C	3·42 cal/(g)(°C)
Liquid at −256°C	1·93 cal/(g) (°C)
Solid at −259·8°C	0·63 cal/(g) (°C)
Heat of fusion at −259·2°C	14·0 cal/g
Heat of vaporization at −252·8°C	107 cal/g
Viscosity at 25°C	0·008 92 centipoise

liquefied and stored in insulated tanks. Properties of liquid hydrogen are shown in Table V.29.

Another way to store hydrogen is to combine it with nitrogen to form hydrozine, a liquid with a melting point of 1·4°C and a boiling point of 113·5°C. Hydrogen can also be combined with nitrogen to form metal hydrides.

Hydrogen has the greatest specific energy per unit mass of any chemical fuel (Fig. V.46); therefore, it may become an important fuel in the future. Hydrogen has several fuel advantages: it has a large amount of energy per unit mass; the only by-product of its combustion is water; and, it may be used in every application in which hydrocarbon fuels are now used.

The primary disadvantage of hydrogen is its low boiling point (1252°C), which permits handling only in cryogenic containers. As a liquid, its energy density per unit volume is about one-third the density of gasoline (Fig. V.46). Another disadvantage is the risk of fire or explosion. Liquid hydrogen is frequently used in the United States space program, however, so it can be handled safely.

TABLE V.29
Properties of liquid hydrogen (source: Parker, 1981; © 1981 McGraw-Hill)

Property	*Value*
Boiling point	20·4 K
Liquid density	0·070 8 g/cm^3
Latent heat of vaporization	108 cal/g or 450 J/g
Energy release upon combustion	20 000 cal/g or 2 050 cal/cm^3 or $1{\cdot}21 \times 10^5$ J/g
Flame temperature	2 483 K
Auto-ignition temperature	858 K

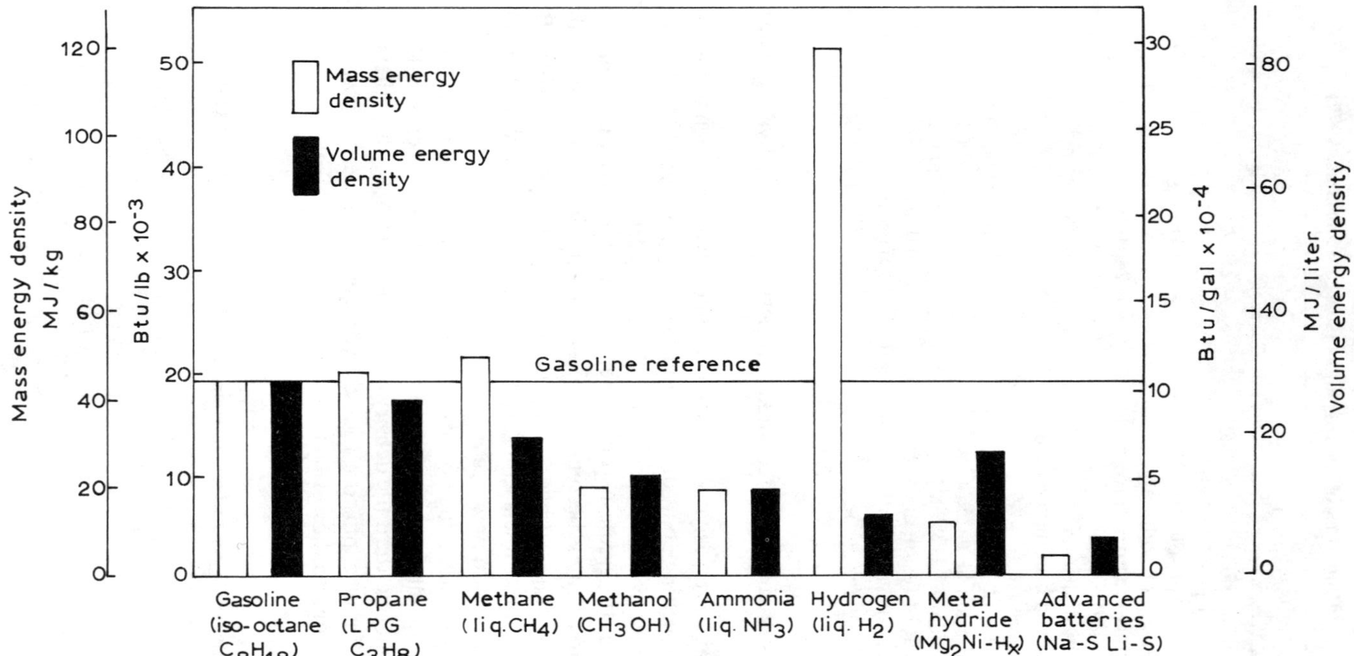

Fig. V.46. Energy density of selected fuels (1 MJ/kg = 430 Btu/lb; 1 MJ/liter = 3587 Btu/gallon) (source: Parker, 1981; © 1981 McGraw-Hill).

Byproducts
No significant byproducts result from hydrogen production except in the gasification process in which carbon monoxide is emitted.

Environmental and Safety Aspects
Algae production can have a negative effect on water quality and plug filters as well. A notable advantage of hydrogen as a fuel is that it has no polluting effect from combustion because the only byproduct is water.

Net Energy Balance
Production of pure hydrogen from biomass is in the very early stages of development. No energy balance data are available for practical applications.

R&D or Commercial Use
The largest single use of hydrogen is for ammonia synthesis. Natural gas or petroleum refinery byproducts are used as a hydrogen source for combination with atmospheric nitrogen to form ammonia, the foundation of most nitrogeneous fertilizers.

Hydrogen is also required for catalytic hydrogenation of unsaturated liquid vegetable oils to make solid fats. Hydrogenation is commonly used to manufacture organic chemicals such as alcohols from esters and glycerides. Methanol is produced by synthesis of hydrogen and carbon monoxide. Reaction of hydrogen with chlorine produces hydrochloric acid.

Research is underway to implement biophotolysis as a method of producing hydrogen. Current publications contain numerous reports on this research (Hall, 1979).

Overall Assessment
Hydrogen does not appear to have potential as a fuel for agriculture, at least for the next decade. For many years to come, its primary importance in agriculture will be as a feedstock for ammonia.

Ethanol (Ofoli & Stout, 1980)

Ethanol, (C_2H_5OH), also known as ethyl alcohol or grain alcohol, is a good fuel for spark ignition engines. It has recently been the subject of intensive study and development in Brazil, the United States, and other countries (Table V.33).

Ethanol and methanol are produced by completely different processes than a variety of sometimes similar feedstocks. Figure V.47 illustrates the various feedstocks for ethanol and methanol.

In making ethyl alcohol, the distiller first produces a sugar solution from the feedstock, then ferments the sugar to ethanol, and finally separates the ethanol from water through distillation (Fig. V.48). Simple distillation concentrates the ethanol up to 95%. Higher concentration requires addition of another chemical, such as benzene, followed by further distillation. Nearly 100% (200 proof) concentration or

TABLE V.30
World methanol supply demand (source: Anon., 1983)

Area	*1983 (million gallons/year)*		
	Supply	*Demand*	*(Deficit) Surplus*
Canada	531	80	451
Japan	286	425	(139)
USA	1 710	1 300	210
Western Europe	1 264	1 307	(43)
Rest of World	1 621	1 011	430
Total	5 412	4 123	909

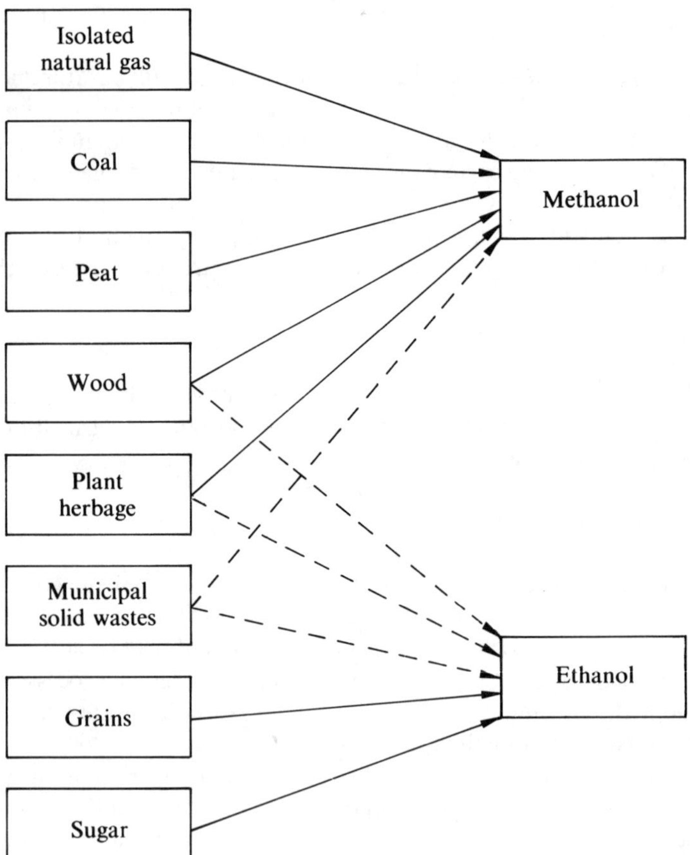

Fig. V.47. Feedstocks for fuel alcohols (source: OTA, 1980).

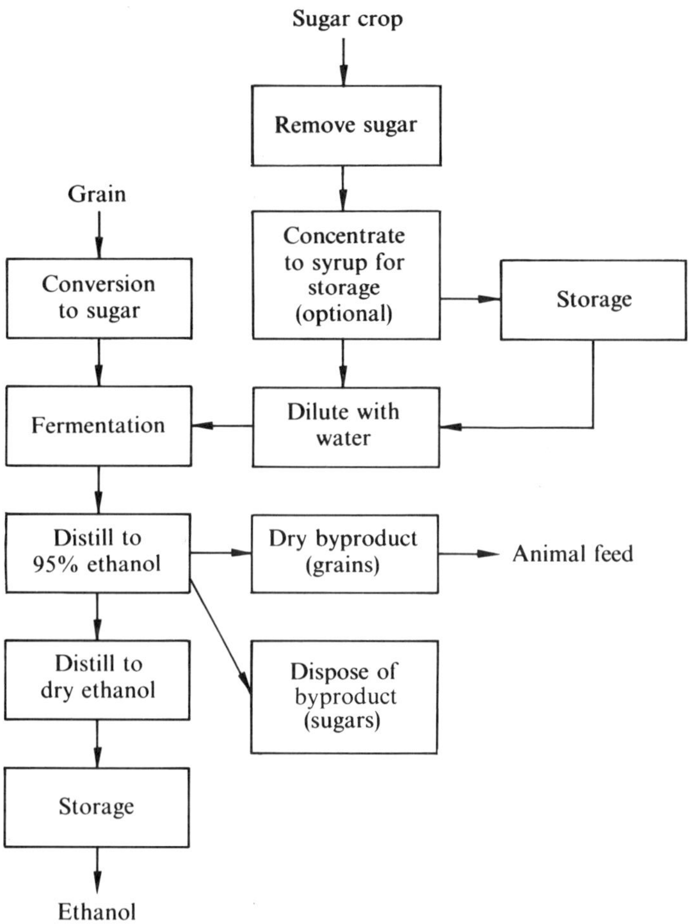

Fig. V.48. Synthesis of ethanol from grains and sugar crops (source: OTA, 1980).

anhydrous alcohol is required for blending with gasoline. (Alcohol concentration is commonly expressed as 'proof'. To obtain proof, simply double the percentage concentration, i.e. 50% = 100 proof and 100% = 200 proof. One-hundred per cent alcohol concentration is also called 'neat' or 'anhydrous'.)

Originally produced from wood, methanol is sometimes referred to as 'wood alcohol'. Methanol synthesis consists of gasifying the feedstock to make a carbon monoxide mixture. The ratio of these two components is adjusted and the mixture is cleaned and pressurized in the presence of a catalyst to produce methanol (Fig. V.49). A more detailed discussion of methanol is presented later in this chapter.

Biomass Feedstock

Ethyl alcohol can be produced from a wide range of raw materials: sacchariferous materials or raw materials that contain sugar naturally, i.e. sugar cane, sugar beet, or

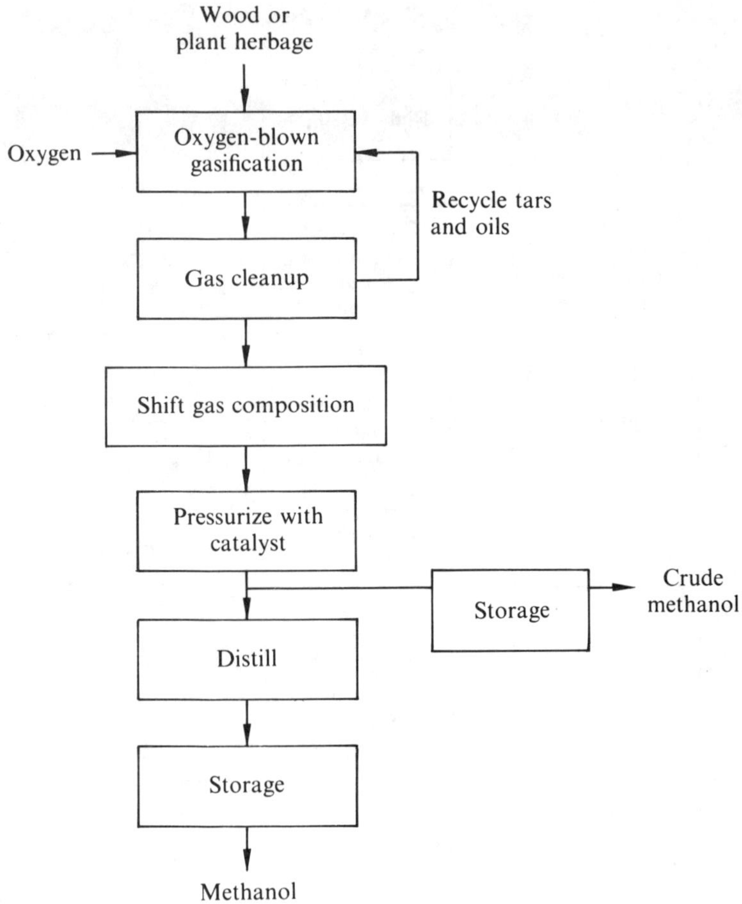

Fig. V.49. Methanol synthesis (source: OTA, 1980).

sweet sorghum; amylaceous or starchy materials, including cereal grains, cassava, potatoes, and Jerusalem artichokes; and cellulosic materials.

While sugars can be directly fermented to alcohol, starches, hemicelluloses, and celluloses must first undergo saccharification. Starches are normally saccharified using enzymes, but celluloses and hemicelluloses are usually broken down by acid hydrolysis (Fig. V.50), although enzymatic splitting (cellulase) is gaining importance. Table V.31 lists selected materials and their alcohol yields.

The choice of raw material to use should be based on the following: yield of alcohol per tonne of material (consistent units should be used when comparing two materials); percentage of fermentable sugars obtainable; alcohol yield per hectare of crop; byproduct yield; and cost of feedstock.

Generally, the feedstock material does not affect the quality or fuel value of the ethanol produced.

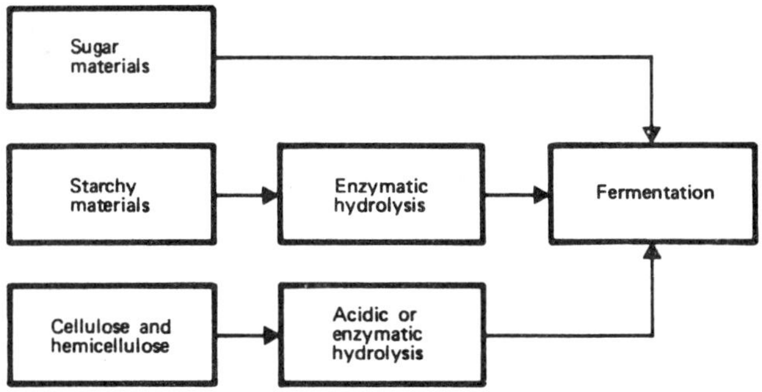

Fig. V.50. Routes to fermentation (source: Ofoli & Stout, 1980).

Sacchariferous materials. These materials are directly fermentable to ethanol and usually have high yields per unit of land area. Molasses contains about 50% sugar and, normally, no pretreatment is needed before fermentation. Sugar cane, a perennial grown in many developing countries, has high biomass yield and is a common feedstock for ethanol. Sweet sorghum is also being considered. It is resistant to most diseases and insects, needs little fertilizer and water, and can be harvested with conventional farm machinery.

Amylaceous raw materials. Starches must be broken down into simple sugars (i.e. glucose) before fermentation.

Yield of alcohol from grains depends on the starch content of the grain. The theoretical yield is 0·568 g of ethanol/g of starch. Actually, the yield is about 90% of the theoretical value, or 0·5 g of ethanol/g of starch.

Sprout-damaged grain can be used for alcohol production if the grain does not lose too much starch from weathering or germination. However, losses may be as high as 30–40% of the starch content.

TABLE V.31
Alcohol yields from selected raw materials (source: Ofoli & Stout, 1980)

Material	*Yield (liters/t)*
Corn	372
Grain sorghum	372
Mandioc (cassava)	367
Molasses, beet	267
Molasses, cane	313
Potatoes	96
Potatoes, dried	346
Rice	354
Sugar beet	92
Sugar cane	62
Wheat	346

The high gluten content of wheat causes excessive foaming during fermentation. Special processing equipment and techniques are needed to implement this process.

Cellulosic materials. Principal sources of biomass residues are agricultural crop refuse, logging residues, and collectible manure waste. However, sewage, urban wastes, and paper pulp also can be used to make ethanol.

Many species of timber can be used for alcohol production. The eucalyptus tree, with its low water requirement and high energy conversion rate, is a good energy source. In addition to live timber, wood wastes from milling and from the pulp and paper industries provide raw materials for alcohol production.

Process Description

The process of making ethanol from corn is illustrated in Fig. V.51. First, the grain is ground, mashed, and cooked. Enzymes added during the mashing stage convert the starch to sugar. The resulting mash is then fermented, using yeast. The process yields a beer, usually of 8–12% alcoholic content, as well as carbon dioxide.

The beer is then distilled to 190 or 191 proof alcohol. A dehydration process will yield essentially 100 proof ethanol if anhydrous alcohol is required. Some countries require denaturing by adding a chemical to make the alcohol unfit for human consumption.

Distillers' grains, a major byproduct of alcohol production can be extracted at three different points (Fig. V.51)—just before fermentation if saccharification is complete, just before distillation, or after distillation. If this byproduct is

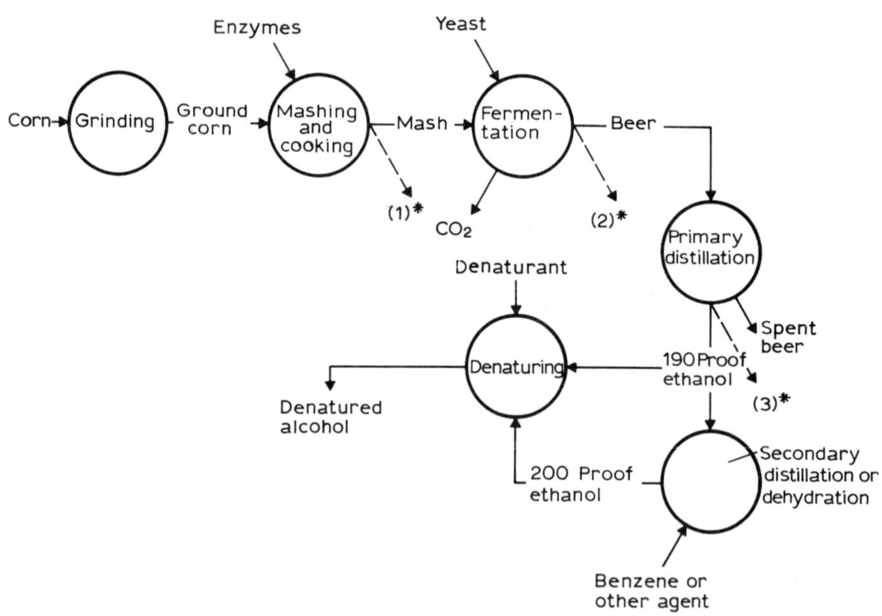

Fig. V.51. A flow diagram for ethanol production. * Distillers grains may be taken out at (1), (2), (3) (source: Ofoli & Stout, 1980).

withdrawn before distillation, the corn grind need only be fine enough to allow effective conversion of starches to sugars. However, when distillers' grains are removed after distillation, corn must be finely ground so that the distillation column is not plugged.

The three major products—ethanol, distillers' grains, and carbon dioxide—are produced in nearly equal quantities.

Factors most affecting alcohol yields are: grain mash preparation methods; how adequately pH and temperature are controlled during the process (Fig. V.51); and the time allowed, including the efficiency and completeness, for saccharification.

The ratio of alcohol yield to available fermentable sugars is 0·5 under optimum conditions.

Grinding, mashing, and cooking. While grains should be ground well enough to enable proper cooking and saccharification, grinding is not crucial for efficient ethanol production.

Mashing the grains solubilizes the starchy material and prepares the starch content for conversion to sugar. Mashing is a hydrolysis reaction during which enzymes convert starch to sugar.

Cooking the grain helps burst the starch cells, making them more accessible to enzymes. It also further prepares the grain for conversion to sugars by gelatinizing the starches. Cooking units need agitators or stirrers because the mash should be stirred continuously during the cooking process to keep solids in suspension.

Ethanol can be produced in many ways. Figure V.52 illustrates one recipe. Enzymes used for this scheme were brand names obtained from the Miles Laboratory in Elkhart, Indiana, and are suited to operations at the temperatures shown in the figure. Enzymes from other manufacturers may require other operating temperatures. Manufacturers will specify these temperatures and the quantities to be used.

Mashing is achieved by mixing ground corn with water, at a ratio of approximately 75 liters of water to 25 kg of corn, to form a slurry. (Water temperature is not specified in Fig. V.52 because it varies with locality, ranging from 7 to 12°C.) Next, an alpha-amylase, such as Taka-Therm, is added to the mixture to convert starch, amylase, and amylo-pectin to soluble dextrins and small quantities of glucose and maltose (Ofoli & Stout, 1980). Then the slurry must be heated to about 90°C and maintained at that temperature for 30 min.

The temperature of the mixture should then be raised to near its boiling point, usually 100°C. (The boiling point will depend on the elevation of a given locality.) After the mixture has boiled for 10 min, the slurry is cooled to 93°C. Then, more alpha-amylase should be added and the mixture should be cooled for more than 30 min to 54°C.

At this point, a gluco-amylase such as Diazyme must be added to convert starch, amylase, amylo-pectin and soluble dextrins to glucose. The slurry temperature should be held at 54°C for 30 min, and then cooled to about 30°C before the yeast is added for fermentation (baker's yeast is the most appropriate).

For best results the temperature should be maintained between 27 and 32°C. The

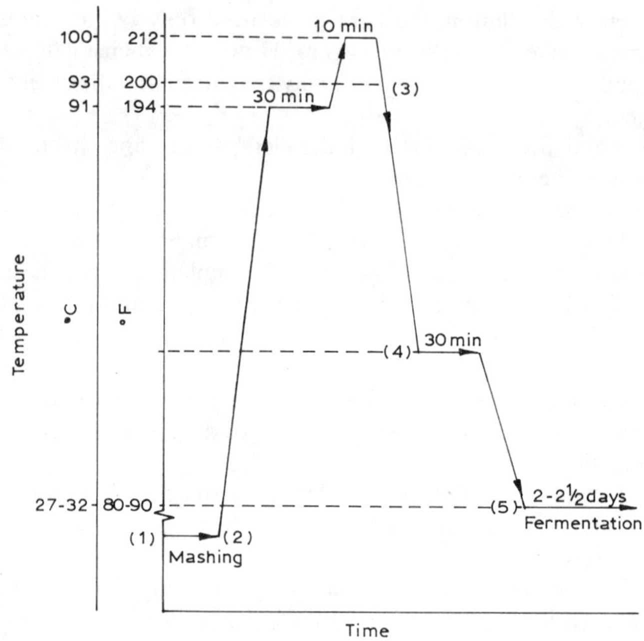

Fig. V.52. A scheme for cooking corn for ethanol production (source: Ofoli & Stout, 1980). Key: (1), initial temperature of process water; (2), add Taka-Therm (pH 6–6·5); (3), add Taka-Therm (pH 6–6·5); (4), add Diazyme (pH 4–4·5); (5), add yeast.

higher end of this range reduces fermentor residence time for the mash, but if temperatures exceed 32°C, the yeast cells will die. Lower temperatures ensure yeast survival but prolong fermentor residence time. The complete fermentation process requires 2–2½ days.

Saccharification. Saccharification is the process of converting starch (a polysaccharide) to simple sugars (monosaccharides) by enzymatic action. An important factor in this process is pH control.

Fermentation. Fermentation is the conversion of sugar to alcohol and carbon dioxide through yeast activity. This process must be monitored and adjusted for pH (4·0–4·5), nutrient availability to yeasts, and temperature (27–32°C). Although fermentation is an anaerobic process, small amounts of oxygen promote yeast cell growth.

The beer produced after fermentation is usually 8–12% alcohol. Most baker's yeasts currently used in fuel alcohol production will not survive in an alcoholic content of more than 12%. Therefore, if the sugar content of the mash causes fermentation to yield a beer of more than 12% alcohol, the mash must be diluted before adding yeast for fermentation. Some wine yeasts will tolerate alcoholic concentrations of as much as 17%.

Fermentation efficiency depends on the process duration, pH, temperature, and completeness of the prior hydrolysis, as well as the level of yeast activity.

Distillation

Distillation is the separation of a liquid mixture of two or more substances into individual or groups of components. Separation is achieved through repeated vaporization and condensation of the liquid mixture until the desired concentration or composition is attained.

Distillation is possible in a packed column or a plate column. The plate column is more commonly used for distilling alcohol. A plate in a column is shown in Fig. V.53. Beer or aqueous alcohol flows down the column through a downcomer, accumulates on the plate and in the cup, and overflows down the next downcomer to the plate below. Vapor, on the other hand, flows up the column through the perforations in the plate and strips the liquid of alcohol. This method forces intimate contact between the vapor and the liquid until equilibrium is established.

Each plate constitutes a stage. Once equilibrium is reached after each stage, the vapor becomes richer in alcohol and the liquid becomes richer in water. At the top of the still, the vapors are collected and condensed to a liquid which may be any composition of ethanol up to 191 proof.

Major parts of a distillation column are shown in Fig. V.54. The stripping and absorption sections can be separate columns, or combined (as shown). The stripping section depletes the incoming beer of alcohol, while the absorption section extracts water from the vapor and sends it down the column. The absorption section is also called the rectifier.

Vapors cool to liquid form in the condenser. Condensed alcohol is stored in the accumulator and can be withdrawn and returned to the top of the tower as a reflux. The reflux helps maintain the desired temperature at the top of the tower. For example, at atmospheric pressure and sea level, the temperature at the top of the tower should be 78·2°C for 190 proof ethanol to be withdrawn from the column.

Under atmospheric pressure and at sea level, the highest proof of ethanol obtained by ordinary distillation is 190–191. At this concentration, alcohol and

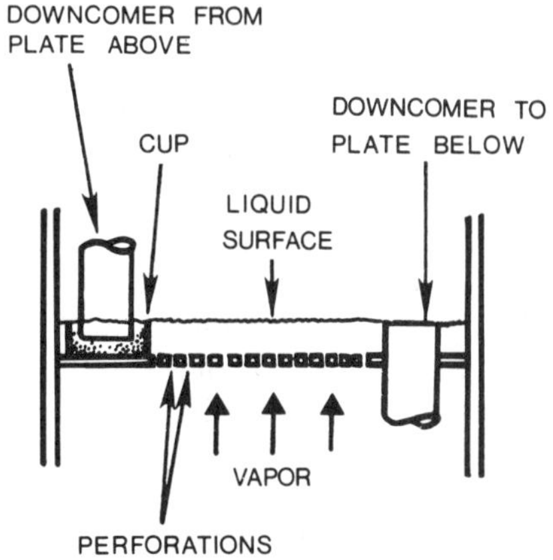

Fig. V.53. Arrangement of a plate, downcomer, and cup in a distillation column (source: Ofoli & Stout, 1980).

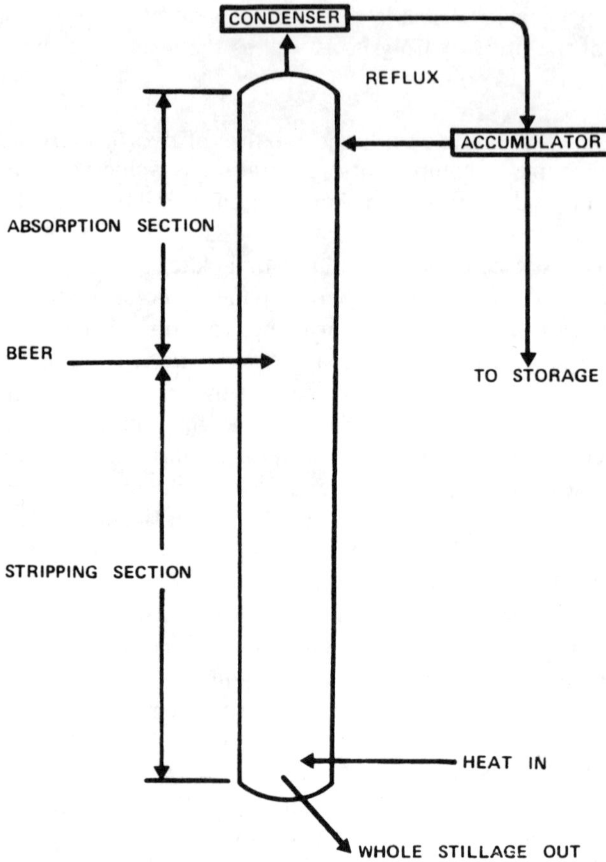

Fig. V.54. A distillation column with combined stripping and absorption sections (source: Ofoli & Stout, 1980).

water form a constant boiling mixture and can no longer be separated. Further separation to produce anhydrous alcohol can be accomplished by azeotropic distillation, in which benzene or another agent is added to the mixture of alcohol and water to change its composition. The new compound can then be separated under atmospheric pressure. Azeotropic distillation involves hazardous chemicals and is not recommended for farm-level operations.

Primary Products

Ethanol is the primary product. The concentration varies from 8 to 12% immediately following fermentation, up to 100% anhydrous if all the water is removed. Other characteristics of ethanol are provided in Table V.32.

Energy content. Ethanol contains about two-thirds as much energy as gasoline and three-fifths as much as diesel fuel. This low energy content of ethanol results in

TABLE V.32
Properties of ethanol, methanol, and gasoline (source: Ofoli & Stout, 1980)

	Ethanol	*Methanol*	*Gasoline*
Chemical formula	C_2H_5OH	CH_3OH	C_4–C_{12} hydrocarbons
Molecular weight (g)	46·0	32·0	100–105
Composition, percentage by weight			
Carbon	52·2	37·5	85–88
Hydrogen	13·1	12·6	12–15
Oxygen	34·7	49·9	0
Specific gravity	0·794	0·796	0·72–0·78
Density (kg/liter)	0·794	0·796	0·72–0·78
Boiling point (°C)	78	65	27–225
Flash point (°C)	12·8	11·1	−42·8
Latent heat of vaporation (kJ/kg)	845	1 100	353
Heat of combustion (MJ/kg)	29·7	22·3	47·0
Stoichiometric ratio	9·0	6·5	14·7
Auto-ignition temperature (°C)	423	464	257
Research octane number	107	109	93
Road blending octane value	126 (average)	130	

increased specific fuel consumption. Thus, a larger quantity of ethanol fuel must be burned when compared to petroleum-based fuels to deliver the same amount of work.

Octane rating. Octane is a measure of the ability of a fuel to resist detonation (knocking) during combustion. Ethanol's significantly higher octane rating makes it a better fuel than gasoline for spark ignition engines. Because of the higher octane rating, ethanol can be used in more efficient, high compression engines. If an engine is properly designed or modified, the gain in fuel efficiency through higher compression offsets a portion of the increase in fuel consumption due to lower energy content. Figure V.55 illustrates the octane number of gasoline produced by various methods and shows the octane-boosting capability of ethanol blends.

Cetane rating. An important property for fuels used in compression-ignition (diesel) engines is the cetane rating. Cetane is a measure of the ignition delay for fuel. Ethanol has a very low cetane rating (Fig. V.56). The low cetane number makes ethanol inappropriate for direct fueling of diesel engines.

Vaporization. Alcohols have a higher heat of vaporization and a lower vapor pressure than gasoline. These two properties combined create problems in starting cold engines on alcohol. Because more heat is required to evaporate alcohol, the intake air must be relatively warm to provide heat for vaporization. To start a carbureted engine on alcohol, the intake air must be 10°C or above. For smooth, efficient operation in cold weather, the intake air should be elevated to nearly 90°C.

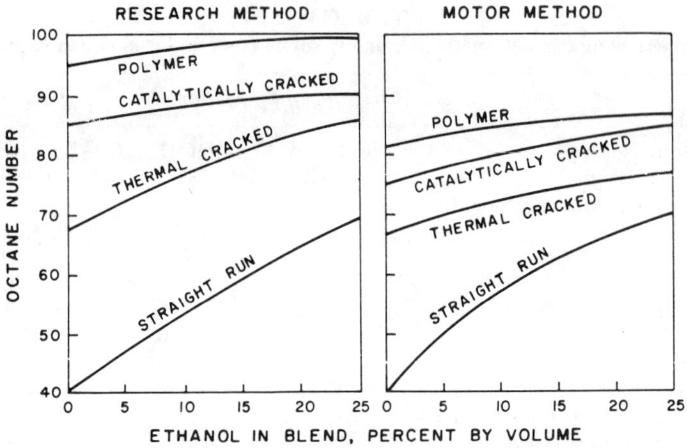

Fig. V.55. Increase of octane rating in several gasolines with alcohol addition (source: Hiler & Stout, 1985).

Corrosion. Alcohols react with certain materials that are compatible with petroleum fuels. Alcohol is a solvent which breaks down some types of plastics and rubbers. Certain metals also corrode more readily when exposed to alcohol. Generally, ethanol is not as corrosive as methanol. Neither form is as corrosive when maintained relatively free of additives, because some additives chemically react with alcohol and increase corrosion.

Farm-Scale Distillation Units

A large farm-scale distillation unit is shown in Fig. V.57. Removing the final 5% water component from the ethanol solution in small, farm-level distillation units is not considered to be technically feasible. Thus, the maximum proof from farm-level stills is 190 or 191. In climates where temperature fluctuates significantly, alcohol containing more than 1 or 2% water cannot be blended with gasoline without risking phase separation. Thus, ethanol from farm-level stills is not a candidate for blending with gasoline, and, therefore, the engine must be modified to burn straight alcohol.

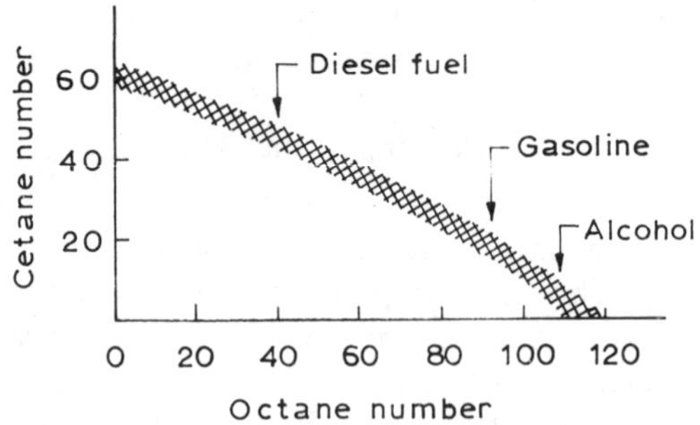

Fig. V.56. Octane–cetane relationships for various fuels (source: ASAE, 1981*a*).

Fig. V.57. Alcohol still at Texas A&M University for converting starchy grains to ethanol. Cooker–fermenter tanks (upper). Distillation, ethanol, and stillage storage tanks (lower) (source: Photo by Henry O'Neal, Texas A&M University).

Burning Straight Ethanol in Spark Ignition Engines

Straight ethanol of 140–200 proof can be burned in gasoline engines with minor modifications as shown in Fig. V.58. Because alcohol has a lower energy content, it requires a lower air-to-fuel ratio for the intake mixture. When burning ethanol, the cross-sectional areas of the jets should be increased to about 1·5 times their original size to meter proportionally more fuel into the intake air. Thus, all jets must be replaced or drilled to a new diameter of 1·27 times the original diameter. After this modification is completed, the engine no longer operates efficiently on gasoline (Rotz *et al.*, 1980).

All incompatible parts (plastics and rubbers) must be removed from the fuel system. Alcohol will slowly dissolve these materials, causing fuel contamination and damaged parts.

Ignition timing may be shifted slightly when ethanol is burned. The change depends on the engine used and previous adjustments. The higher octane fuel should tolerate an advance in the timing, which will allow a small increase in engine thermal efficiency.

With these changes, an engine should start and perform satisfactorily on a warm day. Benefits obtained from the conversion include slightly increased engine thermal efficiency and, frequently, increased torque and power. The increase in thermal efficiency means that more energy or work is obtained from the engine for each weight unit of fuel energy consumed. But the fuel has a lower volumetric energy content, resulting in higher fuel consumption.

This form of conversion for a gasoline engine is relatively simple and inexpensive, but it presents problems with cold engine starts. As stated previously, the intake air must be about 10°C or above. Special equipment is needed to start engines at lower temperatures. A torch or electric-resistant heating element may be used to heat

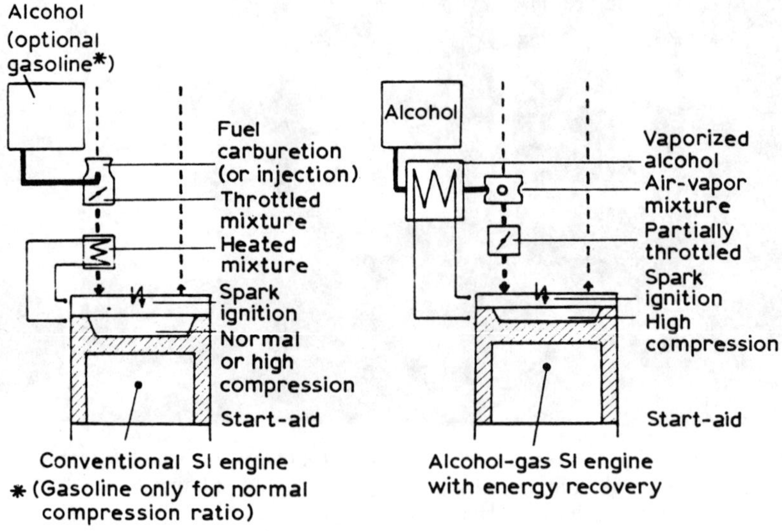

Fig. V.58. Alcohol spark ignition engines (source: Bandel, 1981).

incoming air, but is not recommended because ethanol is highly volatile and fire may occur when alcohol is exposed to an open flame or high temperatures. A more appropriate solution may be to use a special starting fuel such as gasoline or propane. By adding a small propane tank or gasoline injection device, starter fuel can be released into the intake manifold to aid cold weather starting.

Because ethanol has a higher heat of vaporization and a higher boiling point than gasoline, the intake air must be heated after the engine is started for smooth, efficient operation. To provide this heat, a heat exchanger is needed to transfer heat from the exhaust gases or the engine coolant to the intake air. Specially designed intake manifolds are available with a water jacket to allow circulation of the engine coolant through the manifold. With any of these methods, the incoming air can be warmed to the necessary 90°C after engine warm-up.

Heating the intake air may cause another problem. The expanded hot air entering the engine is less dense, reducing the volumetric efficiency of the engine. The result is a slight decrease in the maximum available power.

When a gasoline engine is converted, as described previously, one major problem still exists. Ethanol is a high quality, high octane fuel which permits high compression. When used in an engine designed for gasoline (compression ratio of 6:1–8:1), the thermal efficiency is lower than optimum for alcohol fuel; therefore, fuel consumption is higher than necessary. Fuel consumption may be as much as 1·5 times higher than the normal volumetric consumption of gasoline. Along with the high fuel consumption, the exhaust emission of hydrocarbons (unburned fuel) will be high.

A solution to this problem is to increase the engine compression ratio to 12:1. An engine could be modified to provide this compression but gasoline engines are not designed to withstand the increased pressures that would occur in the cylinder. A blown head gasket or major breakdown may result under heavy engine loads.

Yet, some benefits are obtained when a gasoline engine is converted to run on straight alcohol. Ethanol burns relatively cool, so the engine is not likely to overheat, exhaust valves are less likely to burn, and lubricating oil may last longer. Lower exhaust temperatures also provide lower emissions of nitrogen oxides, a major pollutant from internal combustion engines.

Alcohol is a clean-burning fuel. No soot or carbon build-up should appear on engine parts. Alcohol may clean old carbon deposits and dissolve varnish and gums which have formed with gasoline usage.

Industrial-Scale Distillation Systems

Large industrial-scale stills produce anhydrous ethanol. A blend of 10% anhydrous ethanol with 90% unleaded gasoline (gasohol) is common in the United States, whereas a 20–80 blend is used in Brazil. The advantage of these blends is that they can be used in unmodified spark ignition engines.

Ethanol Use in Diesel Engines

Many problems exist in using ethanol as a fuel for diesel engines. Ethanol has a low cetane number and poor lubricity; therefore, it is not a good diesel fuel. Blending ethanol and diesel fuel is possible, but appears impractical. Ethanol can be fumigated into the intake manifold of a diesel engine, and can displace some diesel fuel, but the process is in the experimental stage and significant problems must be resolved (Fig. V.59). A spark ignition system or a second injector system to utilize the ethanol could be added to a diesel engine, but these major modifications would be extremely expensive and are not currently recommended.

Byproducts

The byproducts depend on the feedstock, but always include carbon dioxide. Carbon dioxide recovery is generally not feasible. Distillers' grains are a major by-product when corn is used as a feedstock.

Other than carbon dioxide and distillers' grains, fusel oil forms 0·1–0·7% of the crude distilled spirit and is a mixture of amyl and isoamyl alcohols. Corn oil, another possible byproduct, can be separated by wet-milling. Corn gluten, corn germ, and fodder yeast can also be separated as byproducts but, in most cases, are not.

Environmental and Safety Aspects

If sugar crops or grains are the primary feedstocks, surpluses and processing wastes would be considered first. New cropland, devoted to feedstock production, would have soil erosion implications and subsequent impact on land and water quality.

Much of the energy used to operate an ethanol plant is generated on-site in conventional boilers. Maintenance of air quality requires appropriate emission controls. Handling and disposal of distilling slop also poses formidable problems because of the large quantities involved—7–10 times the volume of ethanol produced.

The effects of alcohol–gasoline blends on automotive emissions depends on how the engine is tuned. The use of ethanol–gasoline blends would normally result in the following:

increased evaporative emissions;
decreased carbon monoxide emissions;
increased aldehyde emissions;
increased nitrogen oxide emissions with decreased exhaust hydrocarbons (or vice versa depending on engine adjustment).

Pure alcohol as a fuel for spark ignition engines generally has a positive effect on emissions.

Energy Balance

A study by Ofoli and Stout (1980*a*) considered corn as a feedstock and included energy inputs for: (a) producing corn on the farm, (b) transporting corn from the farm to the alcohol plant, and (c) implementing the commercial production process.

By total energy accounting procedure, alcohol production yields 4600 kJ/liter and 2400 kJ/liter more energy than required to produce alcohol using dryland corn and irrigated corn, respectively. A premium energy analysis indicates that 4900 kJ/liter more premium energy is required to manufacture alcohol from dryland corn, and

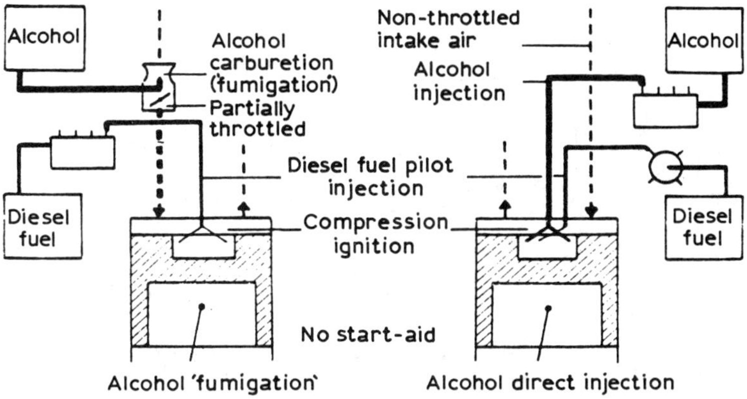

Fig. V.59. Alcohol/diesel dual-fuel compression ignition engine with diesel fuel pilot injection (source: Bandel, 1981).

7100 kJ/liter more is required using irrigated corn rather than the premium energy yielded.

A third method assumes that of the energy inputs required for on-farm production, off-farm transportation, and the alcohol plant, 35, 0, and 85%, respectively, could conceivably be replaced by non-premium sources of energy. With this analysis, ethanol is a net premium energy producer; 15 200 and 13 800 kJ/liter more premium fuel is produced than is required for dryland corn and irrigated corn, respectively. This production results in a net premium fuel gain of 2·8 for dryland corn and 2·4 for irrigated corn.

Numerous other energy balance studies have been conducted; for example, Da Silva *et al.* (1978) analyzed energy balances for making ethanol from sugar cane, cassava, and sweet sorghum. Figures were tabulated for energy input into agricultural production and industrial conversion to ethanol. A net energy gain of 90·8 GJ/ha per year was calculated for sweet sorghum (stems and grain from original planting, plus ratoon), 89·2 GJ/ha per year for sugar cane, and 7·5 GJ/ha per year for cassava.

Ethanol from corn provides little or no energy gain, whereas ethanol from sorghum or sugar cane has a potential energy ratio of more than 2:1.

Other examples of energy balance studies are provided in Figs V.60 and V.61.

R&D or Commercial Use

Industrial-scale ethanol plants are common in many countries, but farm-scale plants have not been viable for sustained commercial operation.

Overall Assessment

Ethanol is an excellent fuel for spark ignition engines and is valuable as an octane booster. Millions of liters of ethanol are being produced and used for fuel each year (Table V.33). The technology is well-established and current research promises to increase the process efficiency and lower the cost. Inadequate feedstock supplies and high costs are the greatest areas of uncertainty for most countries.

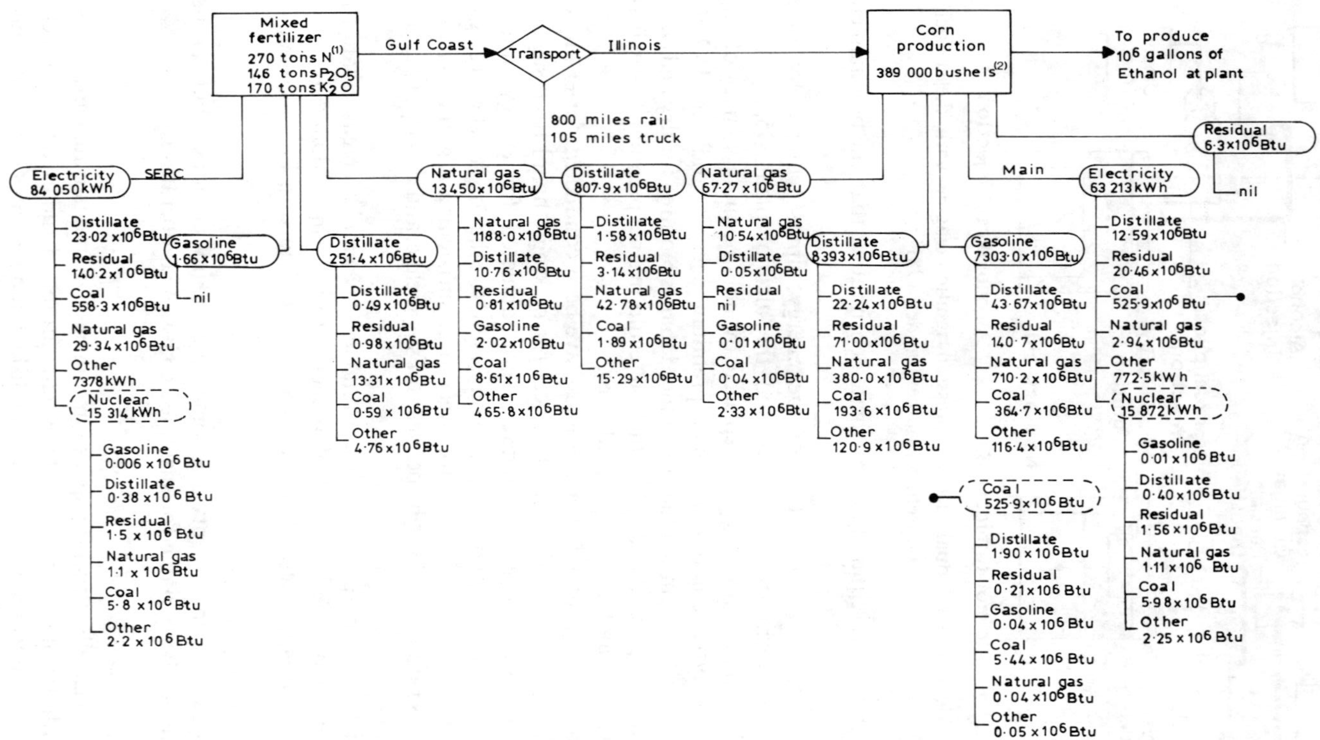

Fig. V.60. Fuel investments in corn crop production (source: TRW, 1980). (1), 2000 lb; (2) 389 000 bushels ≡ 9 900 tonnes.

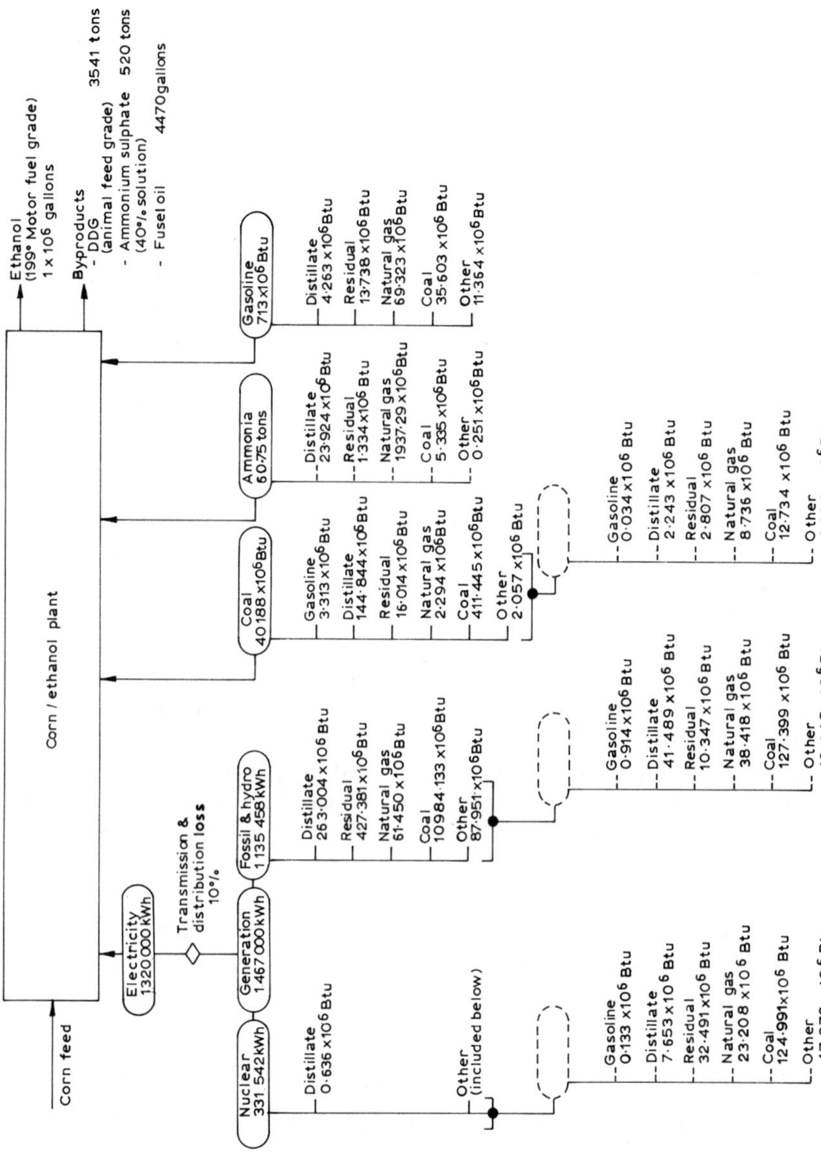

Fig. V.61. Fuel investments in corn fermentation to ethanol (source: TRW, 1980).

TABLE V.33

Alcohol-fuel production and use, Brazil and the United States, 1976–84 (source: L. R. Brown *et al.*, 1985)

Year	*United States*			*Brazil*		
	Production (*million barrels*)	*Use*	*Share of auto fuel* (%)	*Production* (*million barrels*)	*Use*	*Share of auto fuel* (%)
1976	—	—	—	2·0	1·1	1·0
1977	—	—	—	6·2	4·0	3·8
1978	0·3	0·6	0·02	12·5	9·5	8·4
1979	0·5	1·2	0·05	19·6	14·1	12·2
1980	0·9	1·9	0·08	20·1	16·9	16·3
1981	1·8	2·0	0·09	23·9	16·0	16·1
1982	5·0	5·5	0·24	32·9	23·3	22·5
1983	8·9	10·3	0·45	47·7	39·0	35·7
1984[a]	10·0	12·0	0·52	57·2	47·0	43·0

[a] Estimated.

Methanol (Ofoli, 1980)

Methanol, CH_3OH, also known as methyl alcohol or wood alcohol, is a colorless, odorless, water-soluble liquid. At a molecular weight of 32, it is the lightest of the alcohols. Because it has one carbon atom less than ethanol, its heat of combustion is about 22% less.

Methanol will corrode lead, zinc, and magnesium parts and could lead to deterioration of plastic and rubber components in automobile engines. It is miscible with water in all proportions, and with most organic liquids. It is also an effective and widely used antifreeze for automobiles. Prolonged breathing of the poisonous vapors can cause blindness, and it must be used only in well-ventilated places. Methanol evaporates quickly. It is a good solvent for dyes, gums, and some vegetable waxes. It contains no trace metals, sulfur, or aromatics.

Biomass Feedstock

Methanol can be made from wood, crop residues, and many other biomass feedstocks as well as from coal, petroleum, natural gas, and other nonrenewable energy sources (Fig. V.62).

Process Description

Crude methanol is obtained when 2 mol hydrogen and 1 mol carbon monoxide react in a catalyst-filled converter under pressure ranging from 2·8 to 10 Pa. The crude gas is then refined. The reaction is:

$$CO + 2H_2 \rightarrow CH_3OH$$

In practice, synthesis gas contains both carbon monoxide and carbon dioxide; therefore, both the carbon monoxide and part of the carbon dioxide are converted

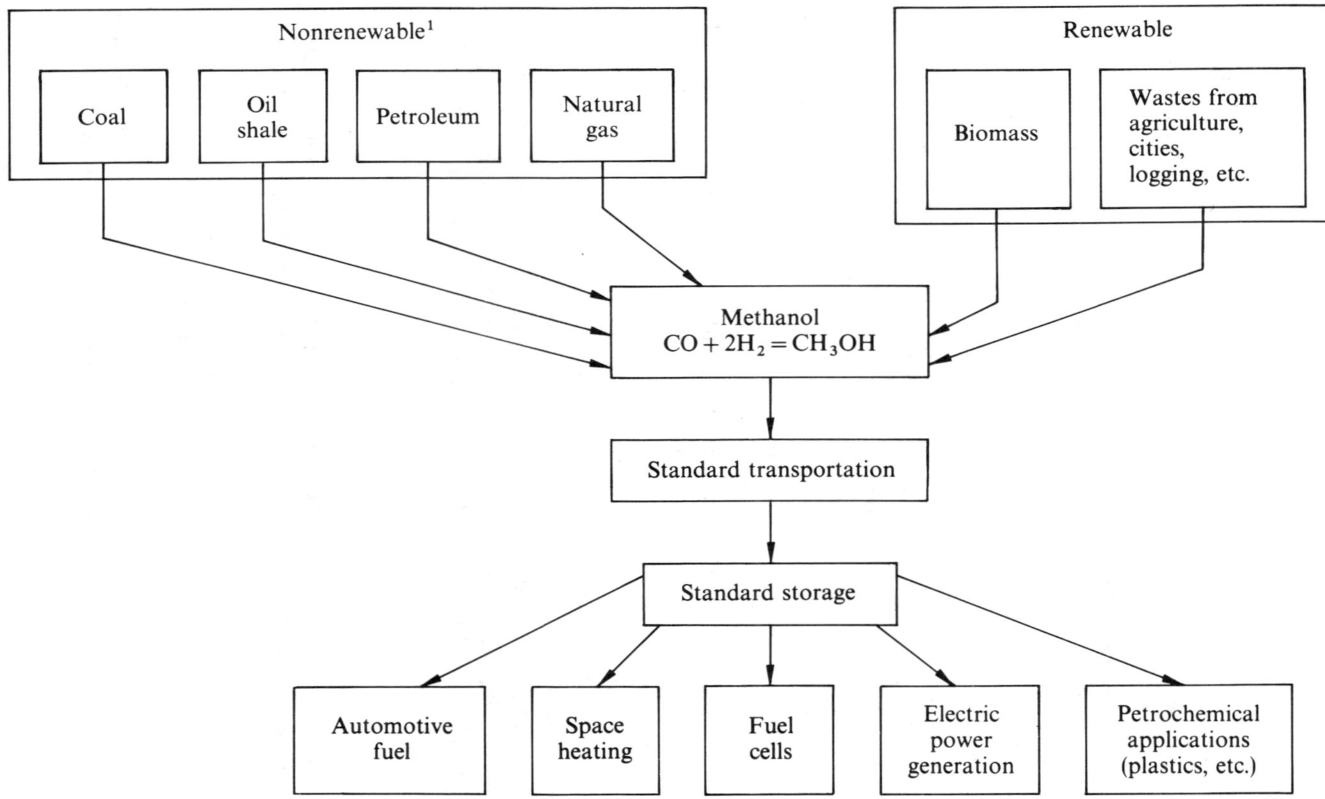

[1] Technically, these fossil fuels are renewable, but renewal requires hundreds of thousands of years, so, in reality, they are practically nonrenewable.

Fig. V.62. Pathways to and from methanol (source: Ofoli, 1980).

to methanol. The reaction with carbon dioxide is either:

$$CO_2 + 3H_2 \rightarrow CH_3OH + H_2O \qquad \text{(1 step)}$$

or

$$CO_2 + H_2 \rightarrow CO + H_2O$$

$$CO + 2H_2 \rightarrow CH_3OH \qquad \text{(2 steps)}$$

Any carbon-containing material can be used to synthesize methanol. Compared to synthesis from natural gas, using coal, lignite, wood waste, agricultural residue, or garbage, requires several additional steps; yields are also lower.

Hokanson and Rowell (1977) provide a 12-step process for converting wood waste into methanol:

1. Use partially oxidized wood waste.
2. Clean and cool the resulting crude gas. After partial oxidation, the crude gas contains water vapor, tars, organics, hydrocarbons, nitrogen, and carbon dioxide. After scrubbing these materials, the gas (primarily hydrogen and carbon monoxide) must be cooled from about 80 to 30°C.
3. Compress the gas to about 700 kPa.
4. Remove carbon dioxide from the gas. During this phase, use a potassium carbonate system to reduce the carbon dioxide content to about 300 ppm.
5. Remove any residual carbon dioxide. Here, the carbon dioxide content must be further reduced to 50 ppm by scrubbing with monoethanolamine (MEA).
6. Remove nitrogen and hydrocarbons from the gas. This is accomplished in a low temperature (cryogenic) system.
7. Further compress the gas to 2800 kPa.
8. Shift the gas to 2 parts hydrogen and 1 part carbon monoxide. This step is required because after step 6, the synthesis gas (syngas) is not yet in the ratio (2 moles of hydrogen and 1 mole of carbon monoxide) required for methanol synthesis. The shifting enables a portion of the carbon monoxide to react with water and form additional hydrogen (in the presence of an iron catalyst). This reaction is:

$$CO + H_2O \xrightarrow{\text{iron catalyst}} H_2 + CO_2$$

9. Remove the carbon dioxide formed during the shift. The hot potassium carbonate must be used again to remove carbon dioxide from the gas.
10. Compress the gas to 10–28 kPa.
11. Convert the hydrogen and carbon monoxide formed during step 8 to methanol. This process occurs in a zinc chromium catalyst reactor in which 95% of the syngas is converted to methanol.
12. Refine the crude methanol obtained in step 11. A distillation column will separate the methanol from higher alcohols (primarily iso-butanol, ethanol, and propanol).

Several commercial processes are available for making methanol, including Purox, Lurgi, Moore–Canada, and others.

Primary Products

An important difference between methanol and gasoline is the stoichiometric air/fuel ratio requirement of 6·5 for methanol and about 14·7 for gasoline (Table V.32). Therefore, the carburetor main jet must be enlarged to permit a greater flow of liquid methanol into the air stream to provide a richer mixture. Another important difference is the increased heat of vaporization of methanol. Methanol is an important chemical feedstock and potentially a major fuel of the future. Table V.30 provides world supply and demand statistics.

Utilization of Methanol in Spark Ignition Engines

Methanol, both pure and in blends, has been tested for use in automotive engines. Engine modifications include: enlarging the carburetor jet to accommodate methanol's lower air-to-fuel ratio and recycling exhaust heat to the intake manifold to compensate for methanol's higher heat of vaporization (11 MJ/kg as compared to 3·5 MJ/kg for gasoline).

Modifications for Methanol Use in Automobiles and Tractors

The following modifications are needed to operate current engines on pure methanol:

—The stoichiometric ratio (ratio of air-to-fuel consumed) should be decreased from 14 for gasoline to 6 for methanol.
—Heat from the exhaust can be recycled to help vaporize the fuel.
—Some provision must be made for cold starting. (The engine could be started on another fuel until it warms up, and then switched to operate on methanol.)
—Plastics and other parts of the engine that could deteriorate from exposure to methanol must be replaced by corrosion-resistant materials. This includes magnesium fuel-system parts, electric fuel pumps, and resin-coated paper fuel filters. High concentrations of methanol in methanol-gasoline blends may also lead to this corrosive problem.
—The carburetor must be rejetted to allow for the low stoichiometric air-to-fuel ratio for methanol.
—A larger tank is needed in automobiles to provide the same driving range as a tankful of gasoline.

Advantages of Methanol as Automotive Fuel

The following are advantages of using methanol as an automotive fuel:

—Blending gasoline with methanol improves the octane rating of the blend. When straight methanol rather than gasoline is used, the engine can be operated at a higher compression ratio, thus increasing the overall power output. An increase of one point in the fuel octane rating increases the compression ratio by about 4%.
—Because it emits less pollutants, methanol burns clearer than gasoline. Straight methanol also reduces oxides of nitrogen emissions, compared to gasoline. In addition, exhaust emissions of carbon monoxide and hydrocarbons may be lower for methanol, compared to gasoline.

—With its 50% oxygen content—12 g of carbon, 4 g of hydrogen, and 16 g of oxygen—and relatively low air-to-fuel ratio, methanol burns more efficiently than gasoline.
—Methanol has a higher thermal efficiency than gasoline, but a lower volumetric efficiency.

Disadvantages of Methanol as Automotive Fuel

Following are the disadvantages of incorporating methanol as an automotive fuel:

—Methanol blends, if water-contaminated, are susceptible to phase separation, which among other consequences, could result in engine failure.
—The use of methanol can lead to vapor lock problems. This occurs when fuel vapor is trapped in the fuel pump, preventing it from operating normally. Methanol–gasoline blends are more volatile than pure gasoline and more susceptible to vapor lock problems.
—Gasoline and methanol are not interchangeable in a given carburetor design. Therefore, a switch from the pure state of one fuel to the other requires a carburetor modification.

Methanol can also be used as a utility fuel to fire conventional boilers, for home heating, and in power plants to generate electricity without pollution. It is also suitable for power generation by fuel cells.

Byproducts

No commercially significant byproducts are produced.

Environmental and Safety Aspects

In a test of methyl-fuel, a mixture of methanol and small amounts of ethanol, propanol, and isobutanol, the following were observed:

—Particulates were not released from the stack.
—Nitrogen oxide emissions were less than either natural gas or fuel oil emissions.
—Carbon monoxide concentrations were less than oil and natural gas concentrations.
—No sulfur compounds were emitted.
—Methanol produced negligible amounts of acids, aldehydes, and unburned hydrocarbons.
—Previous deposits of soot (from other fuels) were burned off with the use of methanol, permitting higher heat transfer rates.

R&D or Commercial Use

Except for racing cars, methanol is not commonly used for fuel today. Numerous research projects on methanol fuel development have been conducted in many countries.

Overall Assessment

Because of its versatility and potential as a liquid fuel, methanol provides an alternative source of liquid fuel that warrants attention. Economy of scale limits methanol production to large-scale, capital-intensive plants. Widespread use of methanol for fuel is likely in the future, but production in small-scale, farm-size plants is unlikely.

Vegetable Oil

Vegetable or plant oils are mixtures of triglycerides, i.e. glyceride esters of fatty acids. Common saturated fatty acids in vegetable oils include: lactic (12:0), palmitic (16:0), and stearic (18:0). (Symbolic notation for fatty acids: the first number refers to the number of carbon atoms; the second number, to the number of double bonds.)

Common unsaturated fatty acids in vegetable oils include: oleic found in olive and safflower oils (18:1); linoleic found in safflower, cottonseed, soybean, and sunflower oils (18:2); erucic found in rapeseed (22:1); and ricinoleic found in castor oil (18:1). Monounsaturated fatty acids have one double bond in the hydrocarbon chain; whereas, polyunsaturated fatty acids have more than one double bond.

From the earliest experiments with compression ignition engines, their ability to operate with a wide variety of fuels has been recognized. Reports of diesel engines operating on coconut oil, peanut oil, palm oil, and other vegetable oils or animal fats are not uncommon. K. S. Fang outlined the characteristics of vegetable oils in relation to the fuel requirements of diesel engines (Fang, 1949).

As noted previously in this chapter, alcohols do not possess the basic fuel characteristics required by diesel engines. The current concern about liquid fuel supplies has renewed interest in vegetable oils for diesel fuel. Extensive research and testing of vegetable oils for diesel fuel is underway, although the current cost of vegetable oil is at least twice the cost of petroleum.

Production of vegetable oil for diesel fuel is a relatively simple process which involves extracting the oil from oilseed, filtering, degumming, and possibly reducing the viscosity of the liquid through transesterification (Fig. V.63).

TABLE V.34
Wrold production of selected vegetable oils (source: Pryde, 1981)

Primarily edible fat or oil	*Production*, 1 000 *t*	
	1971/1972	*1978/1979*
Soybean	7 156	12 440
Sunflower	3 488	4 335
Groundnut	2 958	2 745
Coconut	2 668	2 710
Rapeseed	2 343	3 225
Cottonseed	2 731	2 950
Palm	1 682	3 545
Olive	1 655	1 640
Sesame	681	655
Palm kernel	467	545

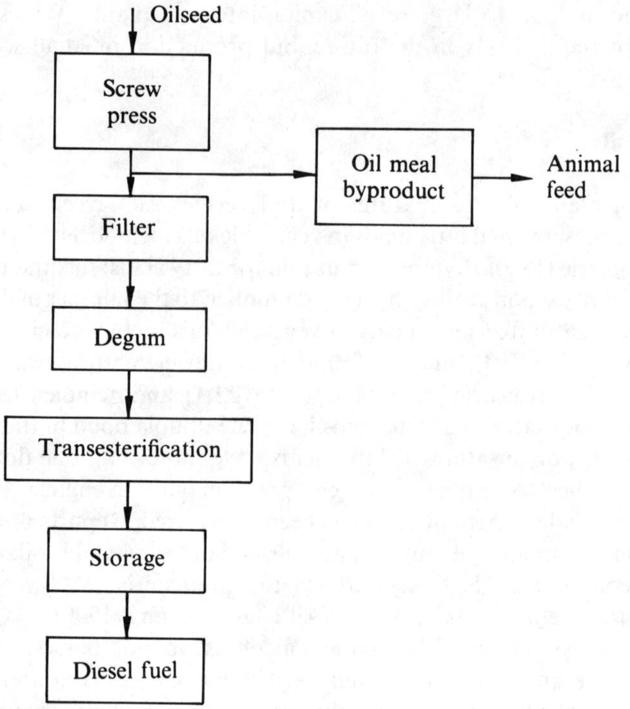

Fig. V.63. Production of vegetable oil for diesel fuel (source: Hiler & Stout, 1985).

TABLE V.35
Oil content and average yield of selected oilseeds (source: Pryde, 1981)

Crop	*Oil content (wt%)*	*Average yield (kg/ha)*	
		Oil	*Total*
Babassu	60–65	—	—
Copra			
(Phillipines)	65–68	669	1 045
(World)		1 045	
Corn kernel	4·8	254	5 282
Corn germ	34·5	—	—
Cottonseed	18–20	140	863
Flaxseed	35–42	230	650
Oil palm	45–50	3 895	4 455
Peanuts	45–50	790	2 494
Rapeseed	40–45	409	1 166
Safflower	30–35	762	2 119
Sesame	50	260	—
Soybean	18–20	319	1 788
Sunflower	35–45	589	1 469

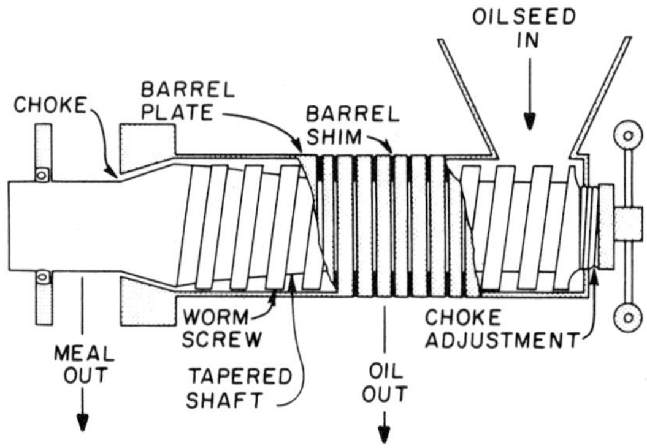

Fig. V.64. Typical screw press for oilseed extraction (source: Hiler & Stout, 1985).

Biomass Feedstock

A dozen or more vegetable oils are candidates for diesel fuel. World production of several vegetable oil crops is shown in Table V.34. The oil content and average yield of selected oilseeds are given in Table V.35.

The Oil Extraction Process

Vegetable oil extraction is relatively simple on a small scale; involves no legal complexities (compared to ethanol production); leaves a dry, high protein meal; and requires little energy or water. Figure V.64 shows a typical screw press for oilseed extraction. The specific oil extraction process varies with each oilseed. Examples are provided for sunflower and oil palm.

Sunflower oil extraction. Commercial oilseed processing plants use a solvent (hexane) extraction process to remove virtually all oil from the seed. For farm-level extraction, a screw press is used to recover 60–80% of the oil from the meal. The expeller operation requires the following processes:

—Cleaning is necessary to remove stones or metal from the whole seed.
—Dehulling is often desirable to reduce the fiber content and increase the protein in the meal.
—Rolling ruptures the seed and increases oil recovery efficiency.
—Cooking maximizes oil recovery.
—Pressing separates the oil from the meal.
—Filtering removes small seed particles after the meal is pressed.

The final product, crude oil, may need further processing to remove phosphatides and waxes, a process known as degumming, as illustrated in Fig. V.65.

Palm oil extraction. The expeller operation requires the following processes:

—Fruit sterilization sterilizes fruits and bunches by boiling them for about an hour.

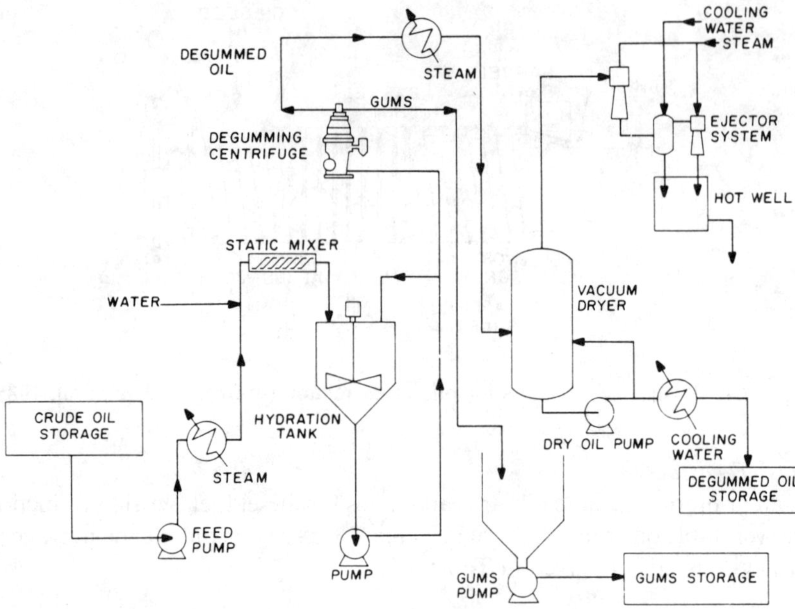

Fig. V.65. Process flow diagram for degumming plant oils (source: Hiler & Stout, 1985).

—Fruit stripping separates fruits from the stalk.
—Maceration pulps and separates the mesocarp from the nut.
—Pressing squeezes the oil out of the fibers.
—Fiber and nuts are separated.
—Fiber is dried to about 15% moisture.
—Nuts are dried.
—Nuts are cracked to obtain the kernel.
—Kernel oil extraction squeezes the oil out under high pressure and temperatures ranging from 70 to 100°C after the kernel is crushed.

The final output is palm oil and palm kernel oil. Byproducts include fibers, nut shell, palm kernel cake, bunch stalks, and palm fronds.

Primary Products

Properties of selected vegetable oils are given in Table V.36.

Several physical and chemical properties must be considered when using vegetable oils for fuel, including the following:

Energy content. Energy content is the heat of combustion, a general indicator of the work output that can be expected from the fuel.

Cetane number. The cetane number is a measure of the ability of a diesel fuel to ignite quickly once injected into the combustion chamber. In high speed engines, the longer the ignition delay period, the greater the accumulation of unburned fuel, and the more severe the diesel ignition knock. A high cetane number, such as 45–50 for #2 diesel fuel, is desirable.

TABLE V.36
Properties of vegetable oils used in diesel engine tests (source: Pryde, 1981)

Oil	*Density (g/cm^3 at 21°C) (70°F)*	*Dynamic viscosity, centipoise at 21°C (70°C)*	*Cetane number*	*Energy content higher heating value (MJ/kg)*	*(MJ/liter)*
Diesel fuel (#2)	0·845	3·8	45	45·2	39·0
Rice bran	0·915	67·4	—	39·2	—
Soybean	0·918	57·2	—	37·1	34·3
Sunflower	0·918	60·0	40	39·4	36·5
Olive	0·915	74·0	—	39·3	36·0
Coconut	0·915	51·9	—	37·5	—
Peanut	0·914	67·1	—	39·4	36·2
Palm	0·898	88·6	—	39·5	36·5
Palm kernel	0·904	66·3	—	39·6	—

Solidification point. This is the temperature at which a vegetable oil or other liquid turns to a solid and ceases to flow (Table V.37).

Saponification equivalent (number). This is the number of milligrams of potassium hydroxide required to saponify 1 g of fat or oil. The term 'saponify' signifies the hydrolysis of an ester into an acid or salt (Table V.37).

Specific gravity. Specific gravity is the relative weight compared to water (Table V.37).

Fatty acid melting point. This temperature must be low for a liquid fuel (Table V.37).

Boiling point. The boiling point for stearic acid is 244°C; for glycerides (soybeans and olive), 308°C. Decomposition becomes significant around 300°C.

Iodine value. This value indicates the degree of unsaturation and is the number of grams of iodine absorbed by 100 g of unsaturated fatty material. It describes the

TABLE V.37
Physical properties of selected vegetable oils (source: Pryde, 1981)

Vegetable oil	*Solidification point (°C)*	*Saponification value*	*Iodine value*	*Specific gravity*
Castor	−17 to −18	175–183	84	0·960–0·967
Coconut	14–22	253–262	6–10	0·926
Cottonseed	+12 to −13	194–196	103–111	0·917
Linseed	−19 to −27	188–195	175–202	0·930–0·938
Palm	35–42	200–205	49–59	0·958
Peanut	3	186–194	88–98	0·917–0·926
Rapeseed	−10	168–179	94–105	0·913–0·917
Safflower	−13 to −18	188–203	122–141	0·925–0·928
Soybean	−10 to −16	189–194	122–134	0·924–0·927
Sunflower	−17	188–193	129–136	0·924–0·926

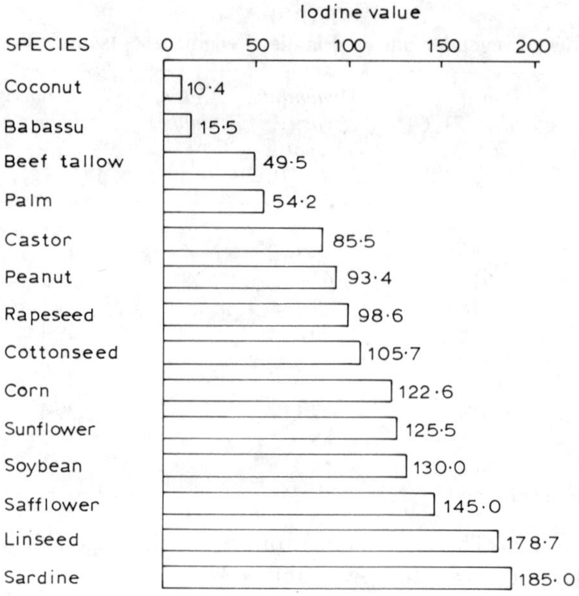

Fig. V.66. Iodine number ranking for a range of vegetable and other oils (source: Quick, 1980*a*).

tendency of a thin film of oil to form a skin (Fig. V.66). Quick (1980) reported that linseed oil, high in iodine value, was also prone to injector coking. Coconut and palm oils, at the low end of the iodine scale, show promise for use as diesel fuel (Table V.37).

Viscosity. Viscosity indicates the consistency of a liquid. A high viscosity number implies a thick liquid. Figure V.67 lists the viscosity of a number of vegetable oils, lubricating oils, and other fuels. At 37·8°C, the viscosity of most vegetable oils is in the range of 30–40 mm^2/s, whereas diesel fuels range from 2 to 4 mm^2/s. Note that the viscosity of most vegetable oils is only slightly less than SAE 10 lubricating oil. The characteristic high viscosity of vegetable oils could result in a pressure buildup in the injector pump that exceeds its design limit. Also, a high viscosity liquid does not atomize as easily as a lower viscosity liquid. Poor fuel atomization in the combustion chamber could lead to faulty combustion, cold starting problems, and crankcase sludge formation.

Blending vegetable oils and diesel fuels together lowers the viscosity of the resulting oil mixture (Fig. V.68).

Another way to reduce the viscosity of a vegetable oil is through transesterification, a relatively simple process (Fig. V.69). Ethyl or methyl alcohol is mixed with vegetable oil. Glycerol is formed along with an ester of vegetable oil which has a much lower viscosity and appears to be a suitable diesel fuel. Note that the safflower methyl ester has a viscosity of 7·6 mm^2/s (Fig. V.67). The glycerol can be removed and sold as a separate commercial product.

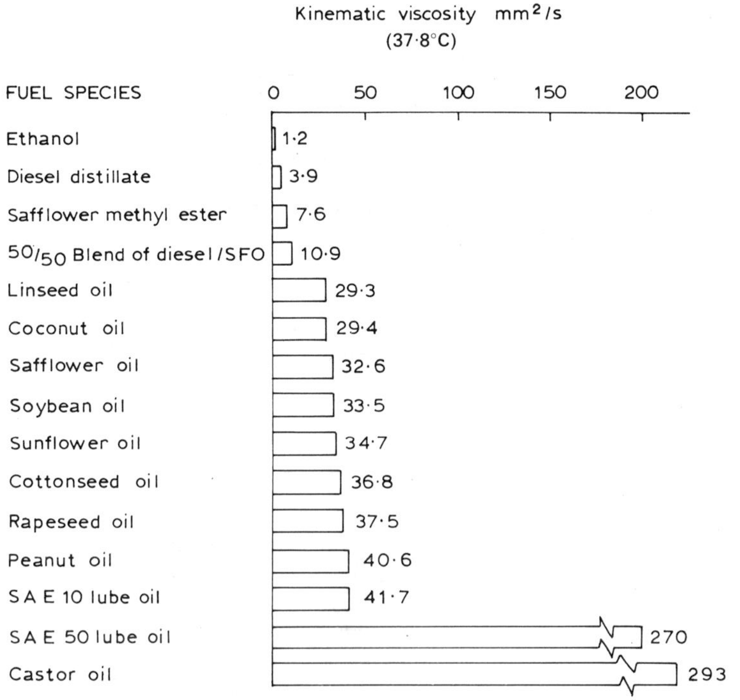

Fig. V.67. Kinematic viscosity ranking of a range of oilseed fuel candidates at 37·8°C. Data for diesel fuel, ethanol, a diesel sunflower oil blend, safflower methyl esters, and certain lubricating oils (representative values) are presented for comparison (source: Quick, 1980*a*). SFO = Sunflower oil.

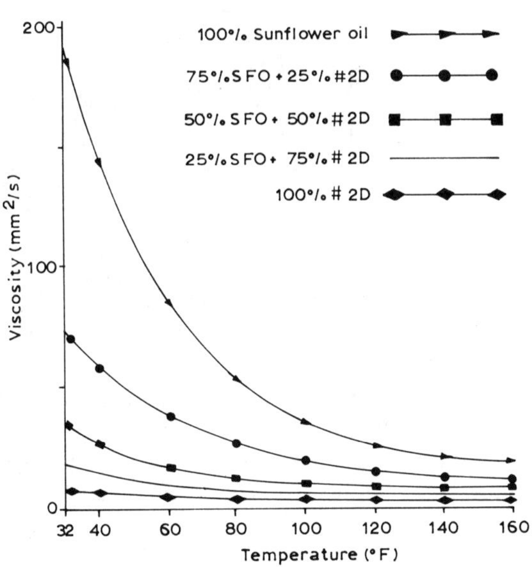

Fig. V.68. Viscosity of sunflower oil, diesel fuels, and blends (source: Hofman *et al.*, 1981).

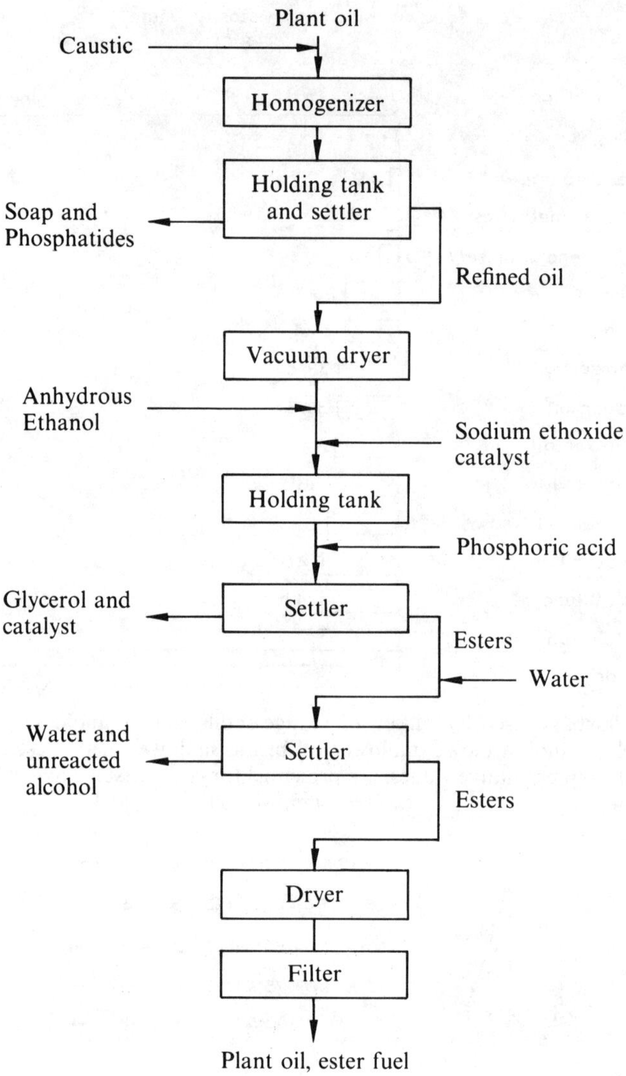

Fig. V.69. Process flow diagram for processing crude sunflower oil into ethyl ester fuels (source: Hiler & Stout, 1985).

Byproducts

Byproducts vary for each oilseed. Soybean oil meal and cottonseed oil meal have long been used as a high protein feed supplement.

Consider sunflower oil meal as a lesser known feed supplement. Three types of sunflower oil meal are being studied—a 42% protein meal from dehulled seed, a 34–45% protein meal from partially dehulled seed, and a 28% protein meal from seeds with hulls.

In experiments with cattle and sheep, sunflower oil meals yield results similar to soybean oil meal when substituted in animal feed on an equal protein basis. The energy value of sunflower oil meals is less than soybean oil meal because of the added fiber content when some or all of the hulls are included.

In swine rations, the limiting factor of sunflower oil meal is the lack of the amino acid, lysine. In high protein meals, those with 42% or more crude protein from dehulled seeds, the lysine content is only 1·5% compared to about 3% in soybean oil meal. In the 34% protein meal, the lysine is about 1·2%, whereas the 28% meal contains about 1·0% lysine.

Substituting sunflower oil meal for one-third or two-thirds of the soybean oil meal and adding synthetic lysine to equal the lysine provided by the soybean oil meal produced similar results in barley-based rations for swine. Substituting all the soybean oil meal with sunflower oil meal, even with lysine additions, provided more variable results. However, the differences in the results were minimal.

In corn-based rations for swine, one-third, two-thirds, or all of the soybean oil meal could be replaced if lysine was added to make up the difference between the sunflower oil meal and soybean oil meal rations.

Lysine supplement is expensive; for example, when all the soybean oil meal was replaced in corn-based rations, the cost of the lysine was equal to or greater than the cost of the sunflower oil meal supplied to furnish the necessary protein. In using sunflower oil meal in swine rations, costs must be calculated to determine how much sunflower oil meal can be used economically.

Energy Balance

Energy ratios as high as 5 to 1 or 6 to 1 have been projected for production and extraction of sunflower oil (Table V.38). Other studies have resulted in more modest energy ratios between 2 to 1 and 4 to 1 (Table V.39).

TABLE V.38

Energy inputs and returns from sunflower, North Dakota, United States (source: Hofman *et al.*, 1979)

	MJ/ha
Energy imput	
Sunflower production—175 liters/ha at 39·0 MJ/liter	6 825
Sunflower oil extraction—495 liters/ha requiring 1·1 MJ/liter	545
Total	7 370
Energy returns	
Sunflower oil—495 liters/ha at 36·2 kJ/liter	17 900
Sunflower meal—1 057 kg/ha at 20 650 kJ/kg	21 800
Total	39 700

$$\frac{\text{Output}}{\text{Input}} = \frac{39\,700}{7\,370} = 5{\cdot}4{:}1$$

TABLE V.39
Estimated energy inputs and outputs (MJ/ha) for on-farm extracted vegetable oils in Idaho, USA dryland production (source: McIntosh *et al.*, 1984)

Item	*Sunflower*		*Safflower*		*Winter rape*
	North ID	*South ID*	*North ID*	*South ID*	*North ID*
Seed	1 373	1 373	2 043	2 043	277
Fertilizer	4 148	2 602	4 363	2 981	5 044
Herbicides	352	352	352	352	
Pesticides	204				204
Machinery	543	623	600	611	638
Diesel	2 951	3 057	2 541	2 940	2 644
Gasoline	1 987	2 137	2 295	1 769	2 118
Aircraft	13				25
Gasoline[a]	268				536
Labor	13	13	11	12	11
Transportation	336	329	437	356	429
Total input	12 188	10 486	12 642	11 064	11 946
Oil recovery	1 259	1 049	1 173	957	1 753
Total	13 447	11 535	13 815	12 021	13 699
Yield (oil)	16 391	13 659	13 346	10 918	29 567
Percentage charged to oil	45·35	45·35	41·99	41·99	51·29
Ratio of output/input	2·69	2·61	2·30	2·16	4·21

[a] Gasoline for aircraft only.

A farmer could produce enough sunflower oil on 10% of the land to provide mobile power to produce a crop of corn. Many farmers find this concept of energy independence attractive.

R&D or Commercial Use
Extensive research has been conducted in the past decade on extracting vegetable oils, processing them more suitably for fuel use, and utilizing them in diesel engines.

Overall Assessment
Currently, vegetable oil costs at least twice as much as diesel fuel. As long as petroleum-based diesel fuel is available, no great shift to vegetable oils is expected, unless relative price changes make vegetable oil more competitive.

From a technical viewpoint, vegetable oil is a good diesel fuel. Several manufacturers, including Caterpillar, Perkins, and Deutz now extend their normal factory guarantees, under certain conditions, to include plant-oil fuels. Problems with gum and high viscosity should be resolved. Many research projects are underway to learn to use vegetable oils for diesel fuel and the technical prospects for this resource are encouraging.

Solid fuels (Claar *et al.*, 1981)

Production of most solid biomass fuels requires no special or advanced technologies. Firewood harvesting or residue collection is often performed primitively in developing countries (Fig. V.70).

Charcoal is an important form of bioenergy in most developing countries (Fig. V.71). Charcoal production techniques vary widely, but the following basic steps are involved: partial combustion of the feedstock to initiate carbonization, further heating of the charge in the absence of air, and cooling.

Biomass fuels, such as forest and/or crop residues, could be a viable alternative if appropriate technology and hardware become available to release biomass energy. Direct combustion is one of the oldest known methods of releasing biomass energy. Some direct combustion furnaces under current development have excellent potential in agricultural applications.

Fig. V.70. A woman drying cow dung in her yard. Manure is precious in fuel-hungry Ethiopia, where very little wood remains for cooking (source: FAO Photo 10486/A/16).

Fig. V.71. A charcoal kiln operating in Uganda. This kiln converts 100 m^3 of low quality wood into approximately 6000 kg of charcoal/month (source: FAO Photo 7389/11).

The characteristics of four main types of biomass combination systems are provided in Table V.40. Each system has its own advantages and disadvantages, and choosing the best system for a particular application obviously depends on the relative importance of a variety of system features. These include factors such as biomass cost and the resulting importance of combustion efficiency in determining overall costs; the type of biomass feedstock available and the need for drying and reduction of particle size; and the facility size requirement and capital investment limits.

Biomass Feedstock

Any relatively dry organic matter can be used to fire a furnace. The higher the moisture content, the lower the output heat. To characterize a fuel, the chemical–physical characteristics are expressed as the proximate analysis, and the chemical composition of a fuel, without reference to the physical form in which the compounds appear, is expressed as the ultimate analysis.

The proximate analysis involves determining the percentage of: (a) moisture; (b) ash; and (c) volatile matter, as well as calculating the percentage of fixed carbon by difference. Table V.41 gives the proximate analysis for a number of biomass fuels on a dry-matter basis. The ultimate analysis of a dried fuel sample involves determining the percentage of (a) carbon, (b) hydrogen, (c) sulfur, (d) nitrogen, and (e) ash, as well

TABLE V.40
Characteristics of biomass combustion systems (source: D. O. Hall *et al.*, 1982; © 1982 Pergamon Press plc)

Type of combustion system	*Main variants*	*Main characteristics*
Pile burning	Dutch ovens Cyclonic furnaces	Biomass is burned in a pile within the furnace. Dutch ovens are the oldest and simplest type of furnace and can take large pieces of biomass but are difficult to control. Cyclonic furnaces use a jet of circulation air to improve conversion efficiency. Both can handle fairly wet biomass and are generally suited to small- to medium-scale applications.
Grate burning	Fixed grate Moving grate	Grates are used to separate ash from incoming biomass. Various grate designs exist and are designed to improve air supply, uniformity of biomass feed, and ash removal
Suspension burning	—	Biomass is ground into fine particles and suspended in preheated air before entering the furnace where it is burned under turbulent conditions. No grate is needed. Biomass must contain no more than 14% moisture
Fluidized-bed combustion	—	Biomass is ground and burned in a floating, agitated 'bed' of non-combustible mineral. This is the most sophisticated and expensive approach but it gives high efficiency and low pollution levels

as calculating the percentage of oxygen by difference. Tables V.42 and V.43 present proximate and ultimate analyses for corn cobs and corn stover, respectively. Table V.44 presents an ultimate analysis for rice hulls.

Process Description
Combustion is defined as the rapid chemical combination of oxygen with the combustible elements of the fuel. In agricultural crop residues, such as corn stover or cobs, the combustible elements are carbon, hydrogen, and metallic elements such as potassium and sodium. The chemical products formed during the combustion are carbon dioxide, water, and metallic oxides. These reactions are exothermic, and the heat released is approximately 33 MJ/kg of carbon burned and 14 MJ/kg of hydrogen burned.

The objective of good combustion is the release of heat from the fuel while minimizing the losses from imperfections and excess air. The combination of the combustible compounds of a fuel with oxygen requires (1) temperatures high enough to ignite the fuel in oxygen (air), (2) mixing or turbulence in the airstream,

TABLE V.41

Proximate analysis (%, weight-dry basis) ASTM-standard D-3172-73 (1979) (source: Kaupp & Goss, 1983)

Fuel	*Volatile matter*	*Ash*	*Fixed carbon*	*Total*
Alfalfa seed straw	68·2	6·0	25·8	100·0
Almond shell	69·2	4·8	26·0	100·0
Barley straw (high)	64·9	10·3	24·8	100·0
Barley straw (low)	64·9	10·3	24·8	100·0
Bean straw	68·0	10·2	21·8	100·0
Corn cobs	74·2	1·5	24·2	99·9
Corn stalks	73·3	6·4	20·3	100·0
Cotton gin trash	63·6	17·6	18·8	100·0
Cotton stalks	62·9	17·2	19·9	100·0
Olive pits	75·6	3·2	21·2	100·0
Peach pits	73·9	0·9	25·2	100·0
Prune pits	78·8	0·5	25·7	100·0
Refuse derived fuel (RDF)	72·0	10·4	17·6	100·0
Rice hulls	60·2–62·1	16·9–17·2	20·7–21·8	100·0
Safflower straw	68·6	6·0	25·4	100·0
Walnut shell (cracked)	79·6	1·1	19·3	100·0
Walnut shell (1/4 in pellets)	70·2	5·8	24·0	100·0
Wheat straw (50/50 corn stalks)	68·4	7·4	24·2	100·0
Wood blocks	75·0	0·2	24·8	100·0
Wood chips (city tree prunings)	71·7	3·0	25·3	100·0
Wood chips (mfg waste, hogged)	78·0	0·3	21·7	100·0
Wood chips (Manzanita)	72·8	0·4	26·8	100·0
Wood chips (prune tree)	76·4	1·1	22·5	100·0
Wood chips (barked whole log)	76·4	0·1	23·5	100·0

TABLE V.42

Corn-cob analyses (source: Pioneer Hi-Bred Seed Company, Des Moines, IA, 1981, pers. comm.)

Proximate analysis		*Ultimate analysis*	
Component	*Weight (%)*	*Component*	*Weight (%)*
Moisture	15·0	Moisture	0·0
Volatile matter	76·6	Carbon	48·4
Fixed carbon	7·0	Hydrogen	5·6
Ash	1·4	Sulfur	—
Total	100·0	Nitrogen	0·3
		Oxygen	44·3
		Total	100·0

Higher heating value (15% moisture content wet basis): 18·6 MJ/kg
Net heating value (15% moisture content wet basis): 14·4 MJ/kg
Ash fusion temperature: 790–815°C

TABLE V.43
Corn stover analyses (source: Claar *et al.*, 1981)

Proximate analysis		*Ultimate analysis*		*Ash analysis*	
Component	*Weight (%)*	*Component*	*Weight (%)*	*Component*	*Weight (%)*
Moisture	35·00	Moisture	4·94	SiO_2	72·15
Volatile matter	54·60	Carbon	42·48	Al_2O_3	4·30
Fixed carbon	7·15	Hydrogen	5·04	Fe_2O_3	4·69
	3·25	Sulfur	0·18	TiO_2	0·22
		Nitrogen	0·75	P_2O_5	1·05
Total	100·00	Oxygen	42·65	CaO	5·15
		Ash	3·96	MgO	3·92
				Na_2O	1·29
		Total	100·00	K_2O	4·41
				SO_3	0·15

Higher heating value (35% moisture content wet basis): 10·7 MJ/kg.
Ash fusion temperature: 840–954°C.

and (3) sufficient reaction time for the combustion process. These factors are referred to as the 'Three Ts' of combustion: temperature, turbulence, and time.

The actual combustion of a crop residue or wood occurs in three consecutive overlapping phases: (1) evaporation of moisture, (2) volatilization and burning of volatile matter, and (3) combustion of fixed carbon. Initially, the moisture maintained by the residue is evaporated. Then, the heat is absorbed by the residue, raising its temperature and depleting the volatile gases. Hence, these gases are burned to sustain combustion. When the majority of the volatiles are distilled out of

TABLE V.44
Ultimate analysis of rice hulls (source: Kaupp & Goss, 1983)

Element	*Mass fraction (%, dry basis)*	*True density, δ_t (g/cm^3)*
C	41·44	1·8–2·26
O	37·32	—
Si	14·66	2·0–2·5
H	4·94	—
K	0·59	0·86
N	0·57	—
S	0·30	2·0–2·9
P	0·07	1·82–2·69
Ca	0·06	1·55
Na	0·035	0·97
Fe	0·006	7·86
Mg	0·003	1·74
Zn	0·006	7·14

the residue, the highly reactive surface of the remaining fixed carbon is burned in the presence of oxygen.

Any air supplied in excess of the required quantity of air to burn a specific quantity of fuel is known as 'excess air'. All furnaces are designed to operate at excess air levels. The excess air supplied for agricultural residue, wood, and bagasse should be in the range of 35–50% and not more than 100%. Further, the excess air requirements are related to the moisture content of the fuel. For example, high moisture content fuels require high air levels for satisfactory combustion while low moisture content fuels may burn with low levels of excess air.

The level of excess air in the combustion chamber must be limited because it: (1) cools the combustion reaction and slows the combustion reaction rate; (2) reduces the overall combustion system efficiency; (3) increases the flue gas velocities in the furnace; and (4) requires the fan system to move unnecessary air through the combustion chamber, which, in turn, uses more power.

Combustion calculations are performed to determine heat losses and efficiency of the furnace. The values are obtained by assuming various amounts of excess air flue gas exit temperatures. The calculated values should be verified against actual flue gas analysis tests to determine the furnace efficiency. The usual losses associated with burning wood or crop residues include: (1) dry flue gases (carbon dioxide, oxygen, and nitrogen) exiting at temperatures higher than ambient; (2) energy required to vaporize the moisture in the fuel, and heat it to the flue gas temperature; (3) loss from unburned carbon monoxide; (4) loss resulting from hydrogen in the fuel; (5) carbon remaining in the ash; (6) carbon discharged up the stack; and (7) other undetermined losses, such as radiation and variations in load changes.

Requirements for Agricultural Applications

A furnace system for agricultural applications should meet several design criteria:

—The system should provide heat for several farmstead applications.
—The furnace should handle a variety of fuel sources.
—Its construction should incorporate simple fabrication techniques.
—The system should be operated with minimum labor requirements.
—It should be easy to control and maintain.
—The furnace should provide good combustion characteristics.

Combustion systems are classified by the fuel burning method. Pile burners, semi-pile burners, and semi-suspension burners have been, and are currently in use with high moisture fuels, including wood and bagasse.

Based on the design criteria for agricultural application, the following pile burning systems merit consideration: (a) the moist-fuel cell furnace, (b) the vortex furnace, (c) the furnace with a gasification chamber, and (d) the sloping-grate furnace.

French researchers described the design features and operation of these four pile-burning furnaces (INRA, 1983). These systems have a high thermal inertia to carry them through minor changes in the rate of fuel supply and fluctuations in the moisture content. In addition, they demand minimal operator attention.

R&D or Commercial Use

Since the early 1970s, several public, private, and industry supported investigations have been conducted to study the use of direct combustion systems for converting crop residues and processing wastes into thermal energy.

A direct combustion, fire chamber, incineration-type furnace that burned cobs was developed by Pioneer Hi-Bred Seed Company (Capital Square, 400 Locust St, Des Moines, IA 50309 USA) to dry whole-ear seed corn. Several problems arose during operation: slag formation on the grate, corrosion on the metallic parts of the dryer, and deposits of particulate material on the corn on the drying floor.

A concentric vortex, mobile field burner was developed and tested for incinerating grass straw and stubble for the Oregon grass seed industry (Van Vaerenbergh, 1985).

Researchers at Oregon State University developed and tested several conventional square-bale burners for heating farm buildings and crop drying. These units feed two bales into the combustion chamber from opposite sides and allow the fire to burn on the opposing faces (Page, G., 1979, pers. comm.).

Crop residue burners have been developed and are being marketed by two American manufacturers, Middle States Manufacturing and Stormor, Inc. (Reddy, 1985). These units burn large round bales of straw and stover for drying grain. Middle State Manufacturing also markets a unit to burn only corn.

In Germany, Orth *et al.* (1976) and Strehler (1981) have investigated the burning of straw in furnaces (Figs V.72, V.73, and V.74). Orth has reported the combustion capabilities of a two-stage furnace that burns and gasifies large round bales of straw. The straw is volatilized in the primary combustion chamber, and the gases are burned in the secondary chamber. This furnace is being marketed by PSW—Gesellschaft zur Herstellung von Verbrennungsanlagen GmbH. One of these furnaces has been operated at Lynnhurst Farms, in Canning, Novia Scotia, supplying heat for a farmstead grain drying system (Peill, 1980).

In Europe, several straw-fired furnaces have been developed and are being marketed for heating farmstead buildings and drying grain. One such manufacturer is Orum Passat Aps of Orum Sonderlying, Denmark (Fig. V.74).

Many commercially available heat and power plants have been designed to use wood as a fuel. Examples from EKONO Company in Finland are provided in Table V.45.

Overall Assessment

Direct combustion furnaces have a significant number of potential applications in agricultural and processing systems. They allow the biomass fuel to be stored and used in almost its original form. The furnaces are relatively simple to construct and operate, and capable of efficiently releasing high levels of heat energy which can be used readily for space heating and drying operations.

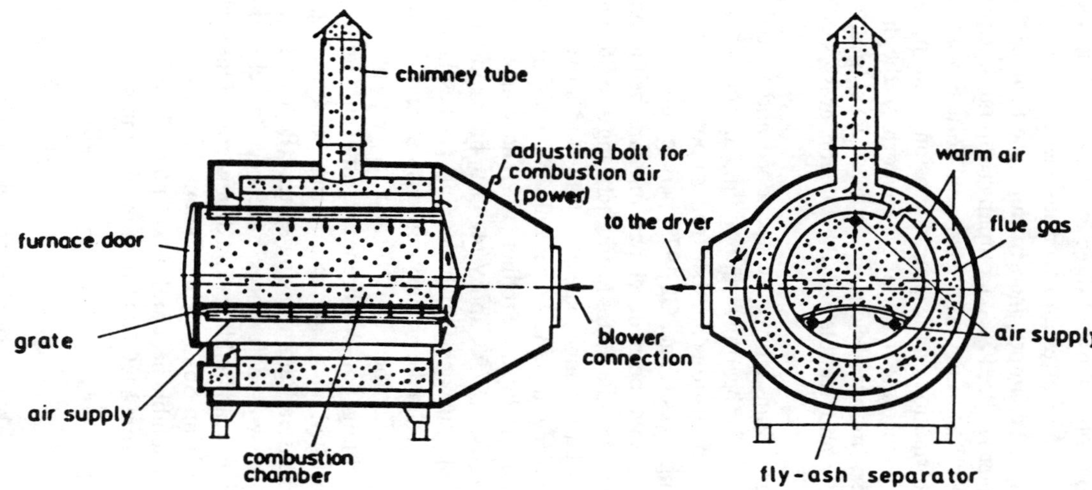

Fig. V.72. Warm-air-generator (250 kW) with combustion of 1 m of woodlogs (source: Strehler, 1981. © 1981 Pergamon Press plc.)

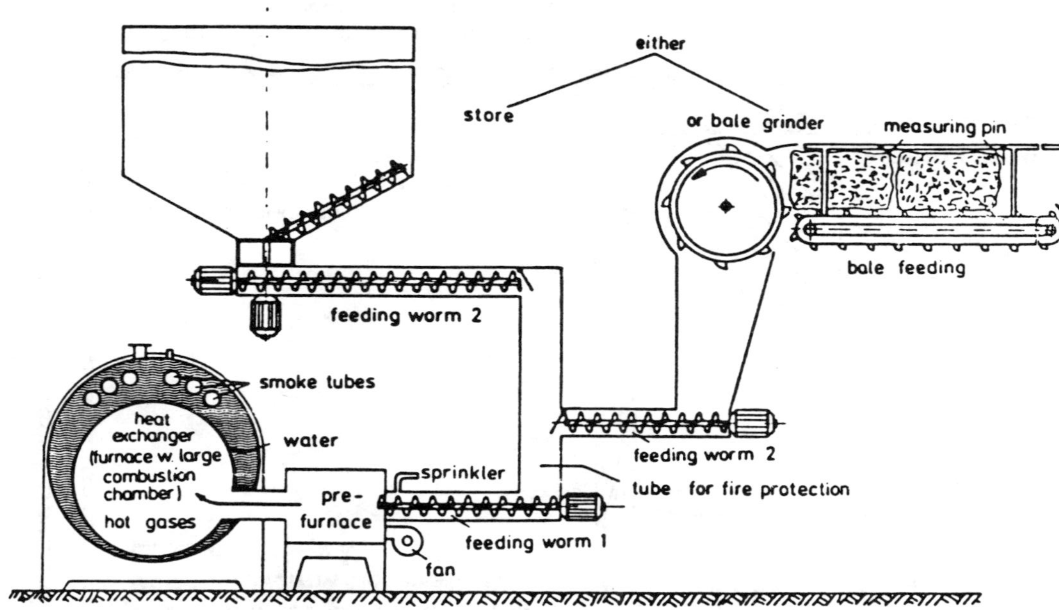

Fig. V.73. Energy from straw and wood (source: Strehler, 1981. © 1981 Pergamon Press plc.)

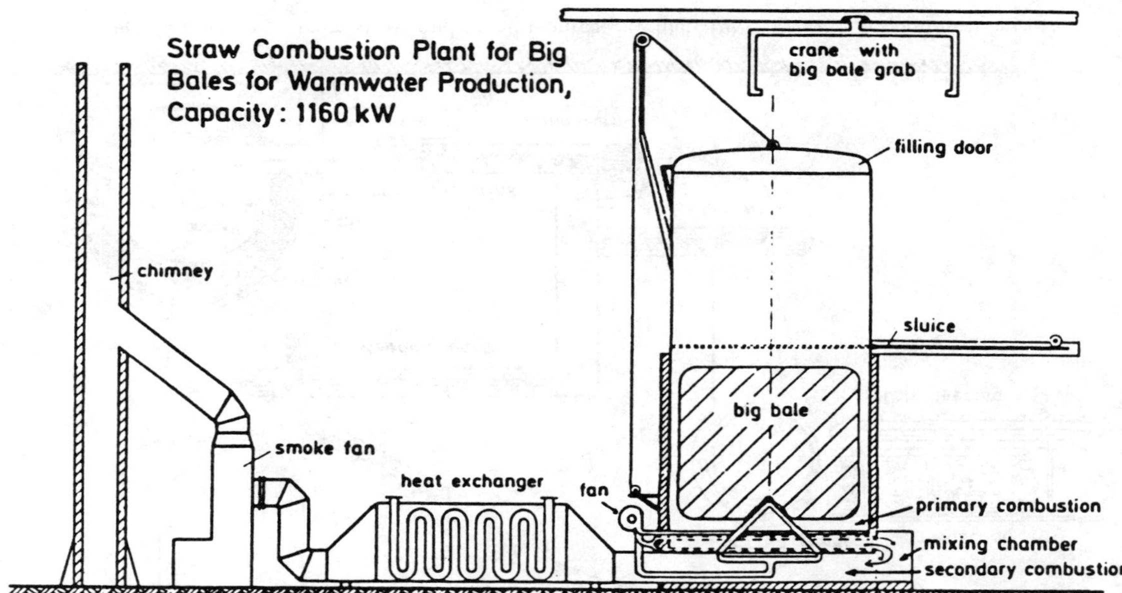

Fig. V.74. Furnace for large bales. This furnace is favorable for high capacity (more than 300 kW). Prices are between DM 100 and DM 300/kW heating capacity (source: Strehler, 1981. © 1981 Pergamon Press plc).

TABLE V.45
Technical and economic features of selected heat and power plants (source: FAO, 1980)

	Heat plant		*Power plant*	
	A	*B*	*C*	*D*
Output —thermal —power	 1 MW —	 2·5 MW —	 2 MW 250 kW	 8 MW 1 MW
Form of energy	Hot water	Medium pressure steam	Low pressure steam power	Low and medium pressure steam power
Fuel	Hogged sawmill waste Fuel oil	Hogged wood waste and fuel wood Fuel oil	Hogged sawmill waste Fuel oil	Hogged wood waste and fuel wood Fuel oil
Boiler design	Warm water	Flame-fire tube, three-pass	Two-drum water tube	One-drum, water tube
Prime mover	—	—	Steam engine	Steam turbine
Boiler output conditions	10 bar, 120°C	1·1 kg/s, 15 bar, saturated (200°C)	1·1 kg/s, 16 bar, 350°C	3·9 kg/s, 40 bar, 360°C
Fuel consumption (chips) at full load	0·7 m^3/h	1·5 m^3/h	2·0 m^3/h	6·3 m^3/h
Plant building volume	220 m^3	1 300 m^3	900 m^3	4 500 m^3
Time of delivery and assembly	11 months	12 months	12 months	18 months
Approximate transport weight	20 t	40 t	50 t	210 t
Total cost (US $)	320 000	588 000	1 452 000	2 511 000
Pay-back time of investment compared with oil fired equipment	3·3 years	4·3 years	7·7 years	2·7 years

SUMMARY

Biomass is a significant energy source. Wood, charcoal, and crop residues are currently important sources of energy.

Many conversion processes are available to convert biomass to more useful liquid or gaseous forms. These technologies are in various stages of development and commercial use. This chapter provides information to help readers understand the conversion processes and select the most appropriate process for a given application.

CHAPTER VI

Solar Energy

FUNDAMENTALS OF SOLAR ENERGY

The sun converts mass into energy at the rate of millions of tonnes per second. Each year the solar radiation passing through the earth's atmosphere is about 700×10^{12} MWh. This is 13 000 times the current world energy use (Halacy, 1980), but it represents only $0{\cdot}5 \times 10^{-9}$ of the total energy radiated by the sun (Androsky, 1973).

About 30% of the radiation reaching the earth's atmosphere is reflected back into space (Fig. VI.1). About 47% is absorbed as heat by the atmosphere, land, and water. But about 30% of the solar radiation supplied to the ground is reradiated into the atmosphere. Evaporation and precipitation use 23% of the atmospheric solar radiation. Solar energy also produces wind and ocean currents, but they use less than 1% of the energy reaching the outer atmosphere. Surprisingly, less than 0·03% of solar energy is used for photosynthesis.

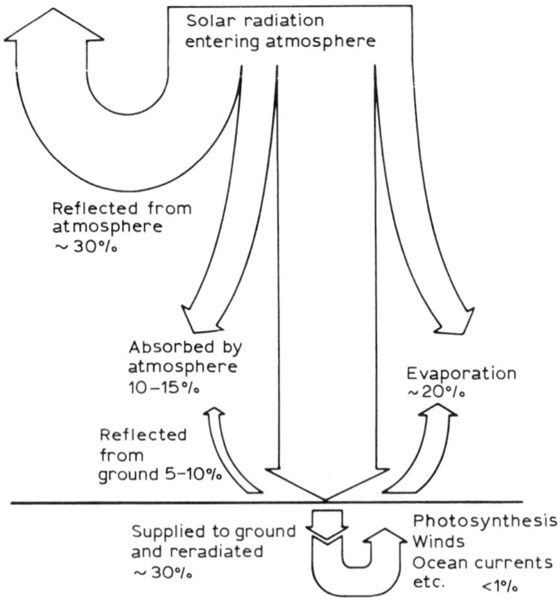

Fig. VI.1. Solar energy intercepted by the earth (source: Steinhart & Steinhart, 1974).

Solar Intensity

The average intensity of solar energy, measured on a plane perpendicular to its path at 160 km above the earth's surface, is about 1·35 kW/m^2 (Halacy, 1980). This number is known as the solar constant. The maximum intensity of solar radiation at the earth's surface is about 1·2 kW/m^2, but is encountered only near the equator on clear days at noon. Under these ideal conditions, the total energy received is from 6 to 8 kWh/m^2 per day (Spillman *et al.*, 1979). In an urban location, the maximum intensity normally exceeds 0·8 kW/m^2 and may reach about 1·0 kW/m^2 under standard atmospheric conditions at sea level (Meinel & Meinel, 1976).

Figure VI.2 shows the intensity of solar energy in different settings: the desert, an urban location, and under standard sea level atmospheric conditions. Both direct (D) and scattered (S) radiation are shown. The latter, also called diffuse radiation, is the solar energy that has been scattered or diffused by atmospheric dust and clouds. Direct radiation is the solar energy that reaches the earth's surface without being deflected by clouds or dust. On clear days the radiation might be 90% direct, and on cloudy days it could be 100% diffuse. The solar radiation striking a surface will always be composed of these two components.

When the wind speed is low and minimal airborne dust exists, the scattered component for a desert location in Fig. VI.2 is only about 8% of the total, but for an urban location the amount rises to 22%. For some urban locations and for dusty desert conditions, the amount of scattered radiation could be significantly higher (Meinel & Meinel, 1976).

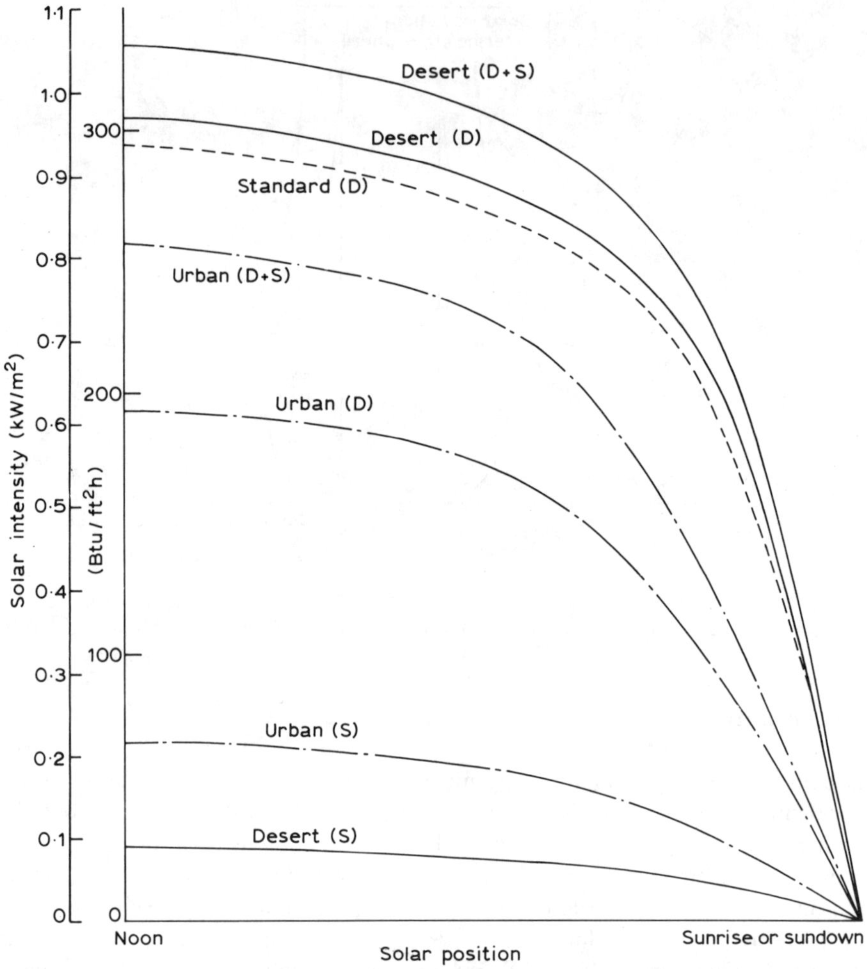

Fig. VI.2. Solar intensity of direct (D) and scattered (S) radiation for urban and desert locations (source: Thekaekara, 1976).

Geographical Distribution

The sunniest regions of the earth are between approximately 20–30° N latitude and 20–30° S latitude (Meinel & Meinel, 1976). In these regions, the dry air results in few clouds and sparse rainfall most of the year. The great deserts of the earth are located in these areas. Cloudiness increases and total insolation decreases both north and south of these two bands.

Daily total and monthly average measurements of solar energy received are available from many weather stations around the world. The measurements are coordinated through the World Meteorological Organization. Figure VI.3 shows the global distribution of solar energy recording and estimating stations. A large

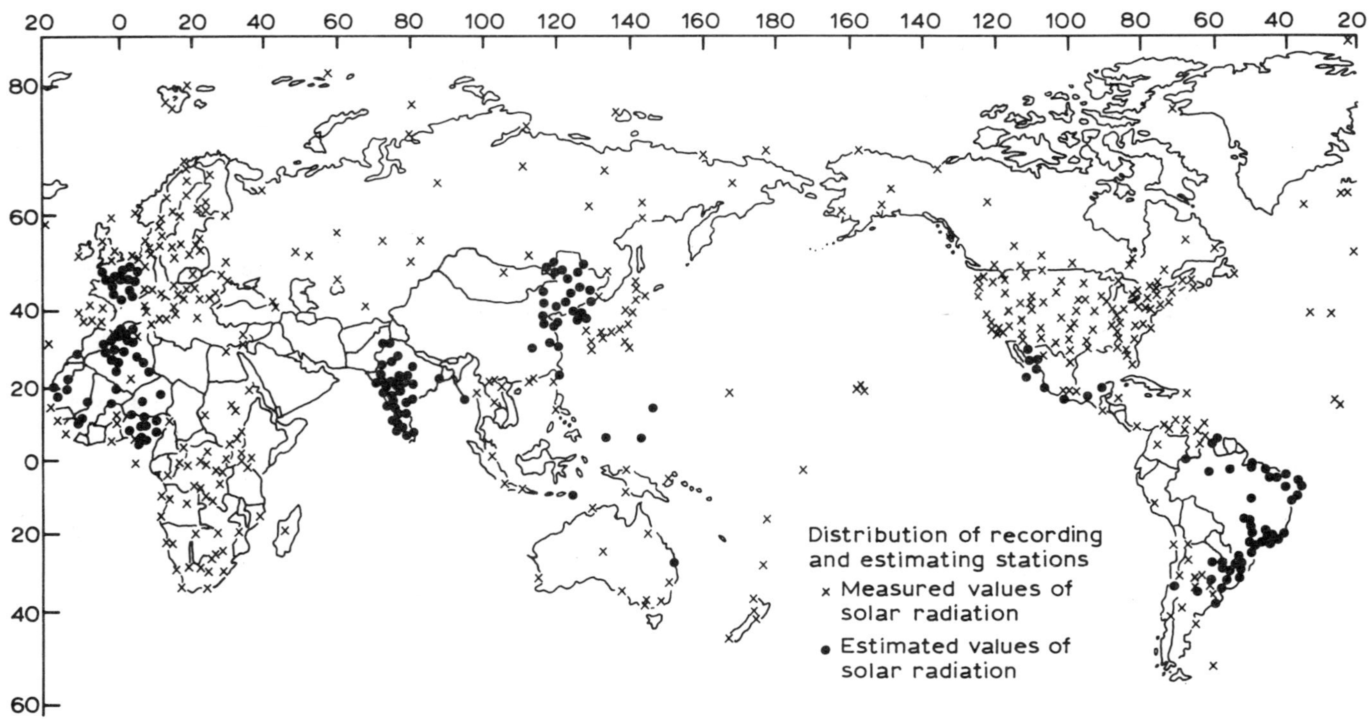

Fig. VI.3. Global distribution of stations where solar radiation data are collected (the crosses indicate pyranometer stations and the dots represent sunshine recorder stations) (source: Thekaekara, 1976).

TABLE VI.1

Solar energy received on horizontal surface, daily average for various cities (sources: Bahm, 1980; Meinel & Meinel, 1976)

Location	*Latitude*	*Jan.*	*Feb.*	*Mar.*	*Apr.*	*May*	*June*	*July*	*Aug.*	*Sept.*	*Oct.*	*Nov.*	*Dec.*
Brussels, Belgium	51	663[a]	1 278	2 438	4 096	4 806	5 220	4 806	4 190	2 971	1 870	900	556
		206	398	759	1 276	1 497	1 626	1 497	1 305	925	528	280	173
Calcutta, India	22	6 297	7 303	8 298	9 245	9 280	9 671	9 659	9 470	7 635	7 386	6 593	5 930
		1 961	2 275	2 584	2 879	2 890	3 012	3 008	2 949	2 378	2 300	2 053	1 847
Dakar, Senegal	15	5 445	6 368	7 493	7 422	7 327	6 865	6 061	5 398	5 492	5 325	5 350	5 563
		1 696	1 983	2 334	2 311	2 282	2 139	1 888	1 681	1 711	1 655	1 666	1 732
Stockholm, Sweden	59	343	923	2 379	3 646	5 528	6 120	5 919	4 640	2 876	1 326	379	213
		107	279	741	1 135	1 722	2 138	1 843	1 445	986	413	118	66
Yanguami, Congo	01	4 841	5 327	5 421	5 279	6 185	4 699	4 167	4 273	4 829	4 853	5 007	4 415
		1 507	1 659	1 688	1 644	1 616	1 464	1 298	1 331	1 504	1 512	1 559	1 375
Tokyo, Japan	36	2 249	2 734	3 243	3 693	4 060	3 587	3 977	4 001	3 007	2 391	2 190	2 000
		700	851	1 010	1 150	1 264	1 117	1 239	1 246	936	745	682	623
Ottawa, Canada	45	1 734	2 697	4 177	4 816	6 100	6 743	6 422	5 779	4 174	2 665	1 477	1 284
		540	840	1 301	1 500	1 900	2 100	2 000	1 800	1 300	830	460	400
La Chorrera, Panama	09	4 816	5 779	6 422	6 100	4 816	5 041	4 495	4 459	5 458	4 816	4 495	5 137
		1 500	1 800	2 000	1 900	1 500	1 570	1 400	1 400	1 700	1 500	1 400	1 600

[a] Upper values in Wh/m^2 per day; lower values in Btu/ft^2 per day.

concentration of these stations appears in western Europe and Japan. The UK, Switzerland, and FRG each have more than 25 solar radiation recording stations. Czechoslovakia has three. In many parts of the world, such as France, West Africa, eastern South America, and India, only sunshine recordings are available (Thekaekara, 1976).

Figure VI.4 is a geographical distribution of the annual solar radiation received at the ground on a global scale. It is useful only for general observations because solar radiation availability can vary significantly between locations that are a short distance from one another.

Table VI.1 lists solar radiation data for several cities around the world. Further data can be obtained from a report published in 1965 at the University of Wisconsin, entitled *World Distribution of Solar Radiation* (Löf *et al.*, 1965). A shorter version appeared the following year in the magazine *Solar Energy* (Löf *et al.*, 1966).

The practical minimum insolation value for the use of solar collectors for heat generation is 2·2–2·8 kWh/m^2 per day. In southern Europe, these values are attained 80–90% of the time during the year, but only 30–35% of the time in northern areas (Pellizzi, 1980).

Figure VI.5 provides a general indication of the hours of annual sunshine in various parts of the world, yet another index of solar availability.

Solar Geometry

Solar energy is not available continuously because of the day/night cycle and cloud cover. Its intensity varies according to season, geographical location, and position of the collector.

The rotation of the earth on its axis every 24 h causes the day/night cycle. Figure VI.6 shows the solar radiation available throughout the day. It peaks when the sun is highest in the sky and decreases proportionally before and after that time. For the location shown, the January curve is proportional to the August curve, it is just much lower.

The seasons are a result of the earth's tilt and rotation around the sun. Figure VI.7 depicts the earth's orbit around the sun. Because the earth tilts, the sun's rays strike the Northern Hemisphere more directly in June and the Southern Hemisphere more directly in December. The earth is actually further from the sun in June than in December. But, because the sun's rays strike the Northern Hemisphere more nearly perpendicular in June, it is warmer in the Northern Hemisphere in June than in December.

Figure VI.8 illustrates the monthly variation of solar radiation available on a horizontal surface at various north latitudes throughout the year. The radiation available at the equator (0°) is high throughout the year but peaks in the spring and fall when the sun is directly above the equator (known as the equinox, Fig. VI.7). Moving north from the equator, the solar radiation striking a horizontal surface decreases during the winter and peaks in the summer (Fig. VI.8).

The ratio between the maximum and minimum insolation throughout the year is around 2·4:1 for the Mediterranean European regions, and as high as 14 or 15:1 in

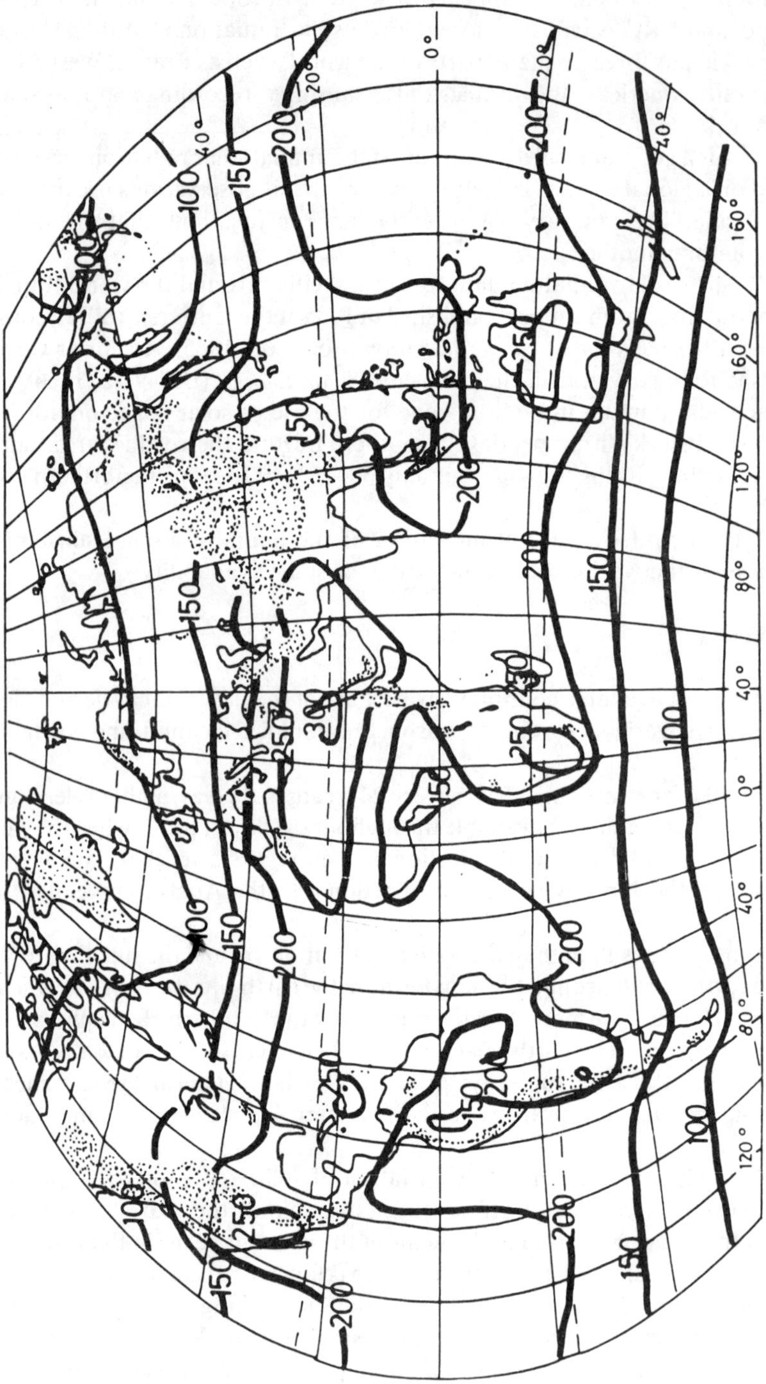

Fig. VI.4. Annual mean global irradiance on a horizontal plane at the surface of the earth (W/m^2 averaged over 24 h) (source: D. W. Robinson & Mollan, 1982).

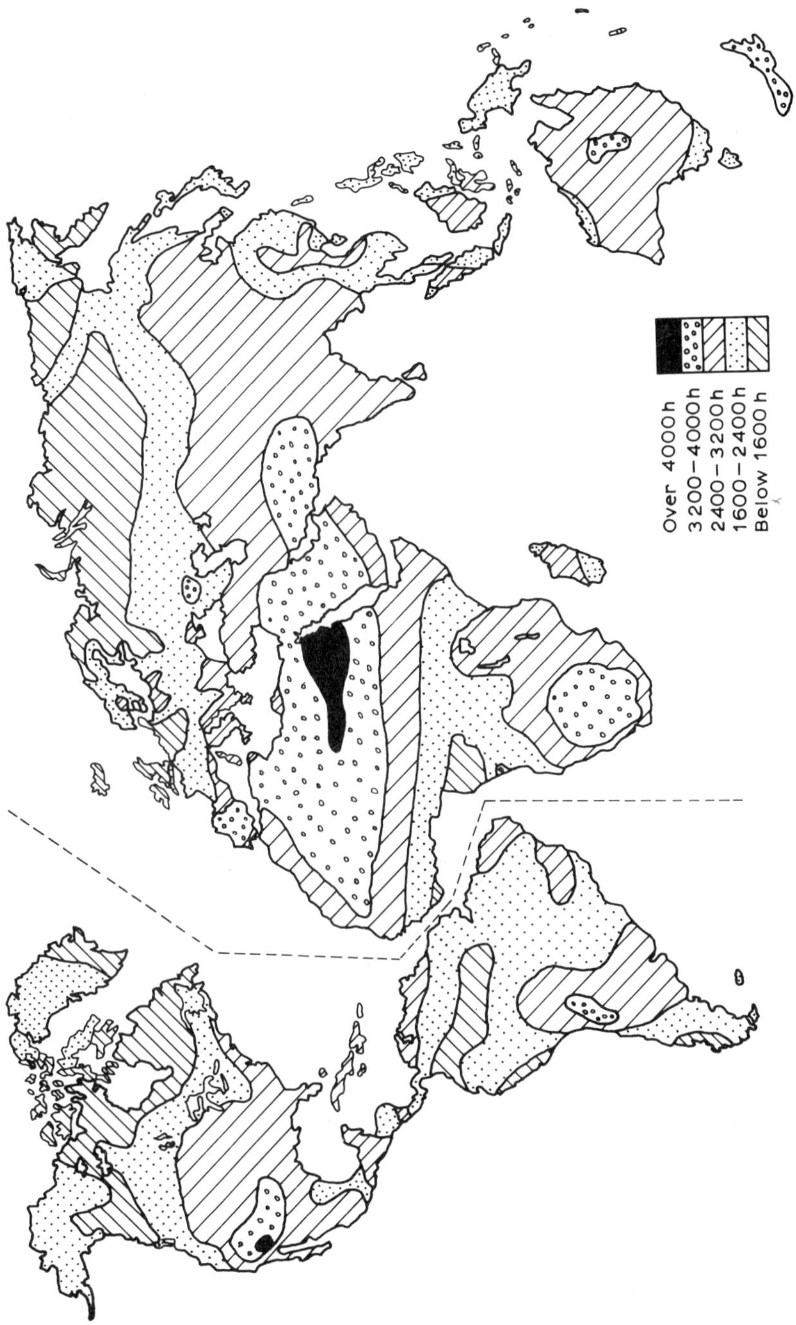

Fig. VI.5. Average annual duration of sunshine (source: Loftness, 1984).

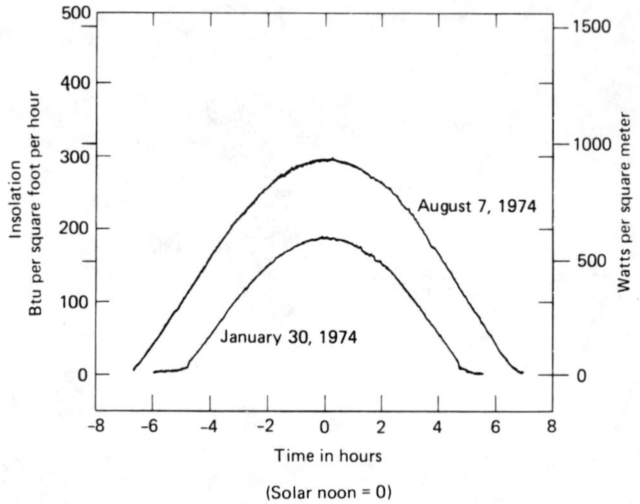

Fig. VI.6. Clear-day insolation measurements on a horizontal surface at Livermore, California (latitude $37\frac{1}{2}°$), at two different times of the year (source: Bush & Richards, 1980).

northern Europe. In summer, a 1 m^2 solar collector placed horizontally at the 32° parallel would yield the same heat as 1·1 m^2 collector at the 56° parallel.

During the winter, the same energy would require a 2·3 m^2 collector placed horizontally at the 32° parallel and a 16·4 m^2 collector at the 56° parallel. Of course, in the Northern Hemisphere, a horizontal configuration during the winter months is the worst possible orientation, but the example does illustrate the dramatic differences in insolation available at different latitudes and months during the year.

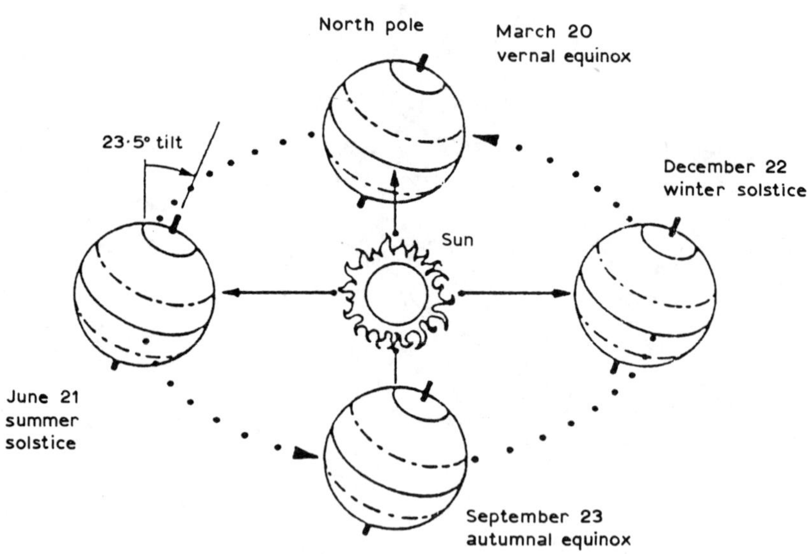

Fig. VI.7. Diagram of the earth's orbit around the sun (source: AIA Research Corp., 1976).

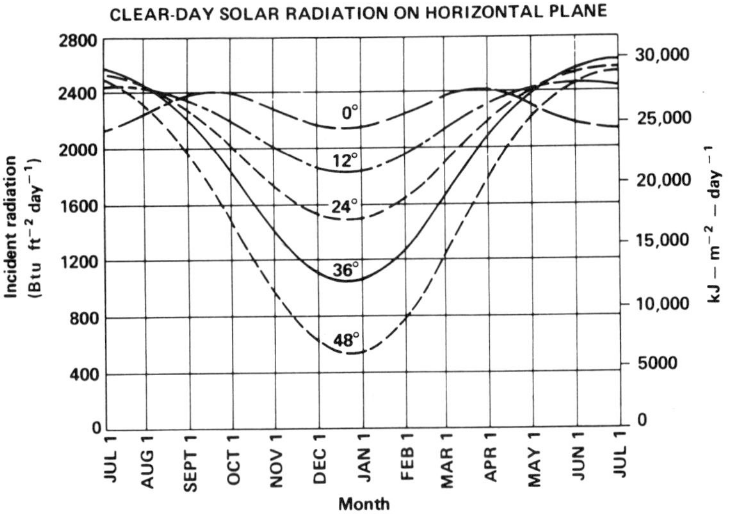

Fig. VI.8. Clear-day solar radiation on a horizontal plane at the earth's surface for various northern latitudes (source: Bahm, 1980).

To avoid oversized collectors and high costs, system uses should be consistent with local insolation (Pellizzi, 1980).

The solar collector's position also has considerable effect on the solar radiation, or insolation, that can be captured. Two primary factors to consider are: (1) collector orientation, i.e. the direction it faces, and (2) the angle from the ground at which the collector is tilted. Figure VI.9 shows the reasoning for these two factors. The two

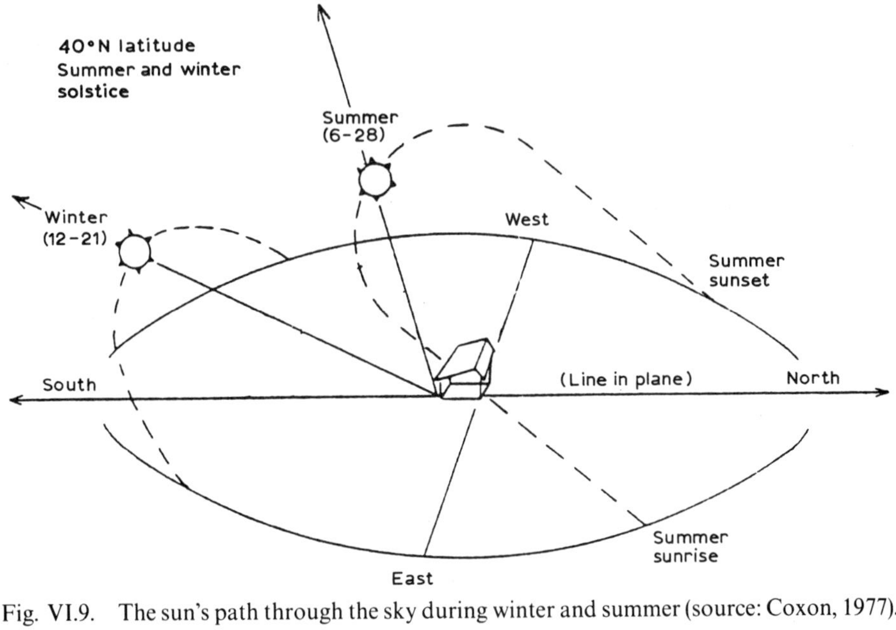

Fig. VI.9. The sun's path through the sky during winter and summer (source: Coxon, 1977).

dotted lines are the paths that the sun travels through the sky during the winter and summer. The collector should face south in the Northern Hemisphere to capture the most solar radiation throughout the day, although a deviation of as much as 15° from due south results in only minor losses in performance.

The greatest amount of solar energy is captured when the sun's rays are perpendicular to the collector surface. Because the sun is high in the sky in the summer (Fig. VI.9), a collector placed almost horizontally would capture more solar radiation at that time. In winter, however, the sun is low in the sky and a collector tilted almost vertically would capture more insolation. Ideally, the collector's tilt angle should be varied from time to time; a north–south tracking collector can accomplish this. Most flat plate solar collectors are fixed in one position, however, to reduce construction costs.

Figure VI.10 shows the monthly variation for a collector at 40° N latitude that is fixed in various positions ranging from horizontal to vertical. Many researchers in the United States believe that the optimum tilt angle (particularly for water heating) is equivalent to the latitude plus 15–20°. For 40° N latitude, the optimum tilt angle is from 55 to 60°. The energy penalty for 10–15° variation is minimal.

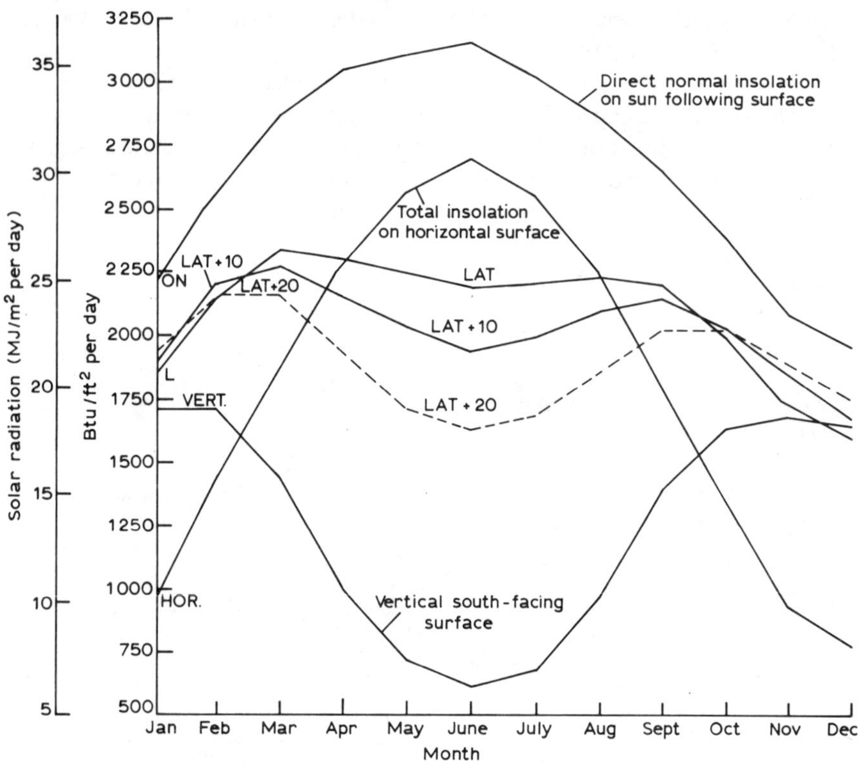

Fig. VI.10. Clear-day solar radiation for sun-facing surfaces at various tilt angles from horizontal to vertical (40° N latitude) (source: Spillman, 1978).

COLLECTION AND STORAGE

Flat Plate Collectors

The simplest and most widely used method of obtaining solar energy is the flat plate collector, a sheet of blackened material positioned so the sun's rays shine on it. Only black absorbs all wavelengths of visible light; other colors reflect certain wavelengths.

Typical collectors are rectangular in shape (to obtain uniform flow) with the black plate sealed by one or two layers of transparent material on the side exposed to the sun and a frame on the edges. Insulation is placed underneath the black plate. A back is also needed unless the collector is built into another structure. There are air spaces between the black material, the transparent side, and the back side. As the sun's rays heat the black surface, the temperature of the black plate increases. A medium, such as water or air, is forced to flow against the black plate so heat is transferred to the water or air.

The flat plate collector designs shown in Fig. VI.11 can be modified depending on use and location. Heat loss, and the heat transfer rate to the water or air, are important considerations. As the difference in temperature between the black plate and the outside air increases, heat lost through conduction, convection, and radiation also increases, reducing collector efficiency. For warmer climates, the examples in Fig. VI.11 may be more elaborate than necessary.

To reduce construction costs, the collector can be built in a fixed position at an optimum tilt for the season of greatest use. Collectors are often set in a fixed vertical or sloped position; this reduces efficiency, but the amount of solar radiation lost depends on the location and season.

In a cold climate, at 40° north of the equator, a properly oriented and designed collector will collect from 20 to 60% of the solar radiation striking its transparent cover (depending on the extent of its use and the time of year). The efficiency might be cut in half for a collector in the same location but on a horizontal surface. In sunny weather, depending on design, the maximum practical operating temperatures for flat plate collectors are 50–90°C above the ambient temperature. Higher temperatures are obtained only at very low efficiencies.

Focusing or Concentrating Collectors

Focusing collectors primarily use direct radiation, unlike flat plate collectors, which can use both direct and diffuse radiation. A sophisticated focusing collector, however, can attain up to 3500°C (Androsky, 1973).

A focusing collector concentrates solar rays by refraction (such as with a lens) or by reflection (Fig. VI.12). The collector area can be any type of reflecting material. Silvered glass mirrors are excellent reflectors, but metal, plastic, or glass-coated aluminum are also effective because they enable the absorber to reach 300°C or higher (Daniels, 1964). The absorber should be as black as possible, capable of

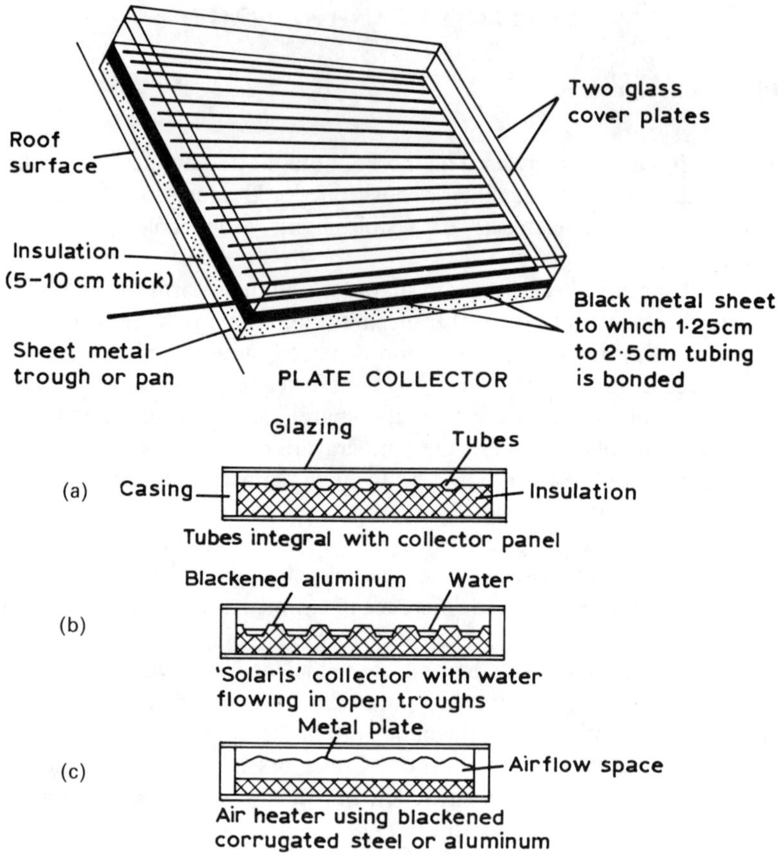

Fig. VI.11. Overall view and various designs of flat plate collectors. The black material is heated by the sun, while fluid is sent over the surface to pick up heat and carry it to where it is needed. On the bottom are three cross-sectional views of various designs: (a) water is carried through the tubes to gain heat from the black surface; (b) water flows in open troughs; and (c) air flows beneath the metal plate to raise the air temperature (source: Merrill & Gage, 1974).

withstanding high temperatures, and insulated (except where the radiation enters) to prevent heat loss.

The most common focusing collectors are cylindrically parabolic in shape with their focus on a line. Energy focused on a line can be effectively absorbed on a pipe containing the heat transfer fluid. During the day, the position of focusing collectors must be changed continually to follow the sun's path. This involves fairly complicated tracking equipment, making the cost prohibitive for many applications. It is possible, however, to track the sun by hand operation, provided the collector is shifted 15° every hour (Daniels, 1964). Some focusing collectors are fixed in position and require no tracking, but need to be shifted with the seasons. Presently, these are inefficient and require more research.

Another problem with focusing collectors is the precision needed in construction.

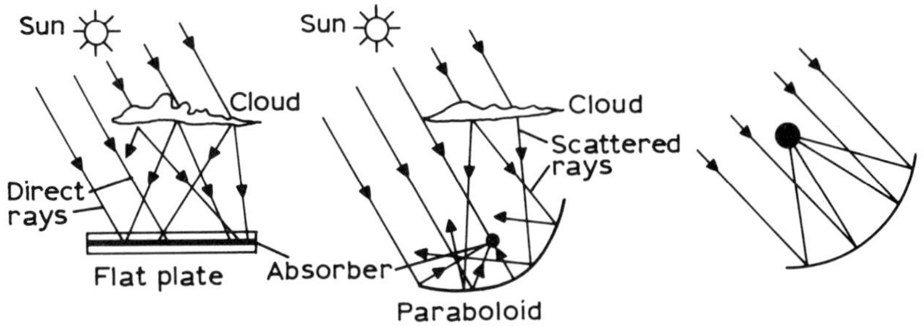

Fig. VI.12. Use of direct radiation by flat plate and concentrating collectors. On the right is a cross-sectional view of a focusing collector. The sun's rays strike the collector area and are redirected to a focal point where the absorber is positioned. The other two figures show the difference in how beam (direct) and diffuse (scattered) radiation react with flat plate and focusing collectors. The flat plate's absorber accepts both direct and scattered rays, while the absorber on the concentrator can only utilize direct rays (sources: Daniels, 1964; Meinel & Meinel, 1976).

Moreover, wind, rain, dust, and other inclement conditions must be properly guarded against. Because these factors increase costs, high-temperature focusing collectors are not economical for most applications today.

A simpler design, the parabolic trough concentrating collector, attains lower temperatures and looks like a long cylinder divided in half lengthwise. The inside surface acts as the reflecting area for the absorber tube running lengthwise through the center (Fig. VI.13). A chain drive rotates the focusing collector to follow the sun's path.

Promising research on the cylindrical, lower temperature, focusing collectors is underway in France. During the past few years, they have tested a new farm collector that produces 80°C water. Researchers feel that this simpler design shows more promise than flat plate collectors, high-temperature concentrating collectors, or photovoltaics (Lucas, J., 1988, pers. comm.).

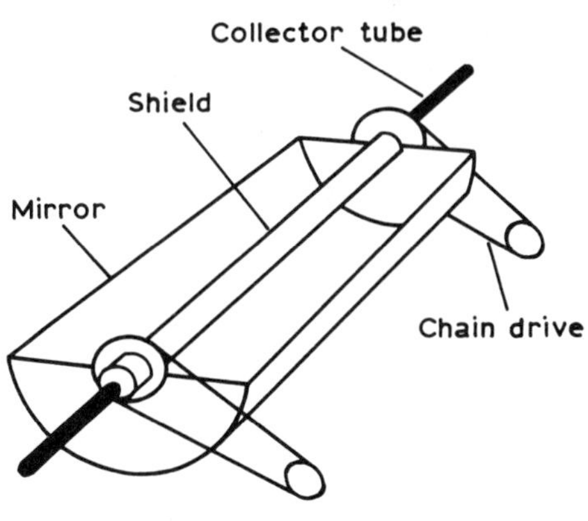

Fig. VI.13. A parabolic trough concentrating collector. Solar radiation strikes the curved surface and is reflected onto the collector tube in a more concentrated state (source: Myers & Mackson, 1981).

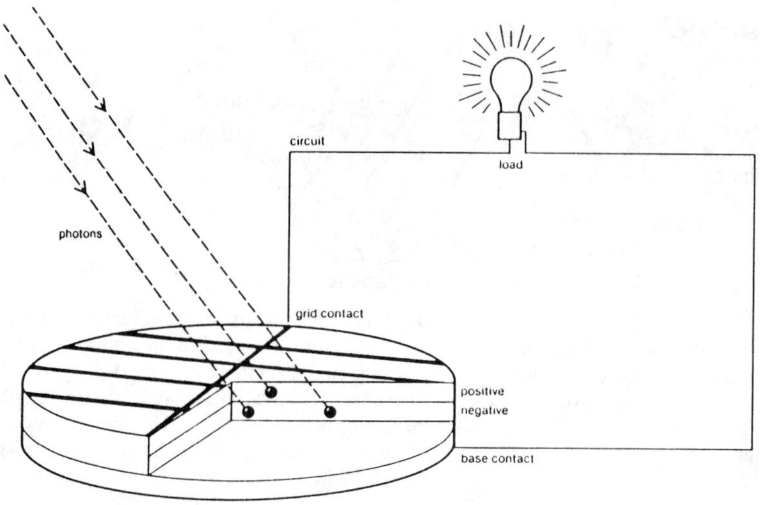

Fig. VI.14. Solar cell cross-section (source: I. Chambouleyron, 1987, pers. comm.).

Photovoltaic Cells†

The photovoltaic solar cell is an electronic device which converts sunlight directly into electricity. The photovoltaic effect was discovered in 1839 by Henri Becquerel. It was used later by Alexander G. Bell in the first photophone. However, photovoltaic technology as we know it today was developed in the USA in 1954 when the first silicon solar cell was produced.

Basically, a solar cell is a large area semiconductor diode. Silicon is used almost exclusively today as the semiconductor in commercially available solar cells. Certain impurities deliberately introduced into the silicon generate excess negative or positive charges which can carry electric current in the silicon. For example, phosphorus atoms give up electrons to the silicon to form excess negative charges (n-type silicon), and boron atoms soak up electrons from the silicon, leaving holes (missing electrons), which behave like excess positive charges (p-type silicon). A p–n junction can be formed close to one surface by allowing boron to diffuse into the surface of an n-type single crystal wafer at high temperatures. The p–n junction formed between the n-type silicon wafer and the p-type surface layer provides the electric field which produces the diode characteristics as well as the photovoltaic effect. Light is absorbed in the silicon generating both excess holes and electrons (one hole–electron pair for each photon absorbed). If this occurs near the p–n junction, the electric field separates the holes from the electrons, causing the holes to build up in the p-type material and the electrons to build up in the n-type material. If wires are connected to the n- and p-type regions, these excess charges generated by light and separated by the junction will flow through an external circuit and provide power to an external load.

The operation of a silicon solar cell is shown schematically in Fig. VI.14. The

† Chambouleyron, I., 1987, pers. comm.

maximum power than can be delivered to an external load is typically 11–15% of the total solar energy incident on the cell (more than 15% in laboratory models). To obtain higher voltages, cells are connected in series; higher currents are obtained by connecting cells in parallel. Such connections are made when cells are packaged in modules. The modules then become the building blocks for arrays designed to meet the specific needs of customers (Fig. VI.15).

The single-crystal silicon cell is the basis for today's technology. Figure VI.16 shows that a module efficiency of 11% was achieved in 1986 at a cost of $550/m^2$ which results in an electrical energy cost of $0·75/kWh. Module efficiencies of 15% and costs of $100/m^2$ will produce electricity at $0·06/kWh, a price that is cost-competitive with many other energy supply options. Research continues on other types of cells with higher efficiency, lower costs, and life expectancies of 20–30 years.

The unique advantages of photovoltaic cells have prompted interest in them, not only in industrialized nations, but also in developing countries. Currently, their main uses are for isolated sites requiring small quantities of electricity. They produce no waste products, require little maintenance, and can apparently operate indefinitely

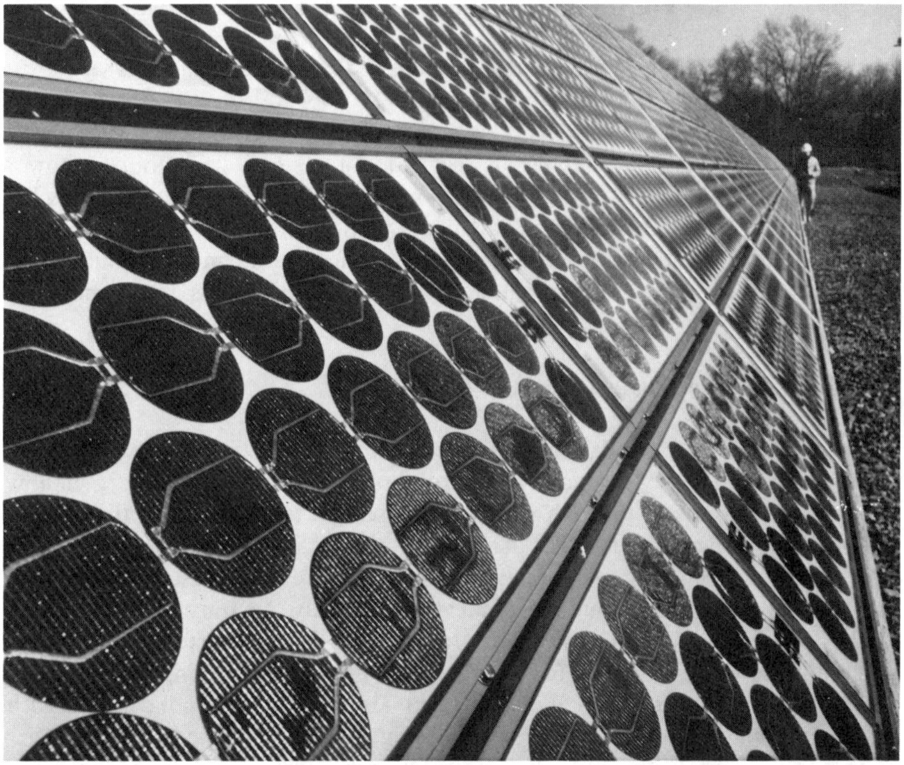

Fig. VI.15. More than 3200 photovoltaic modules assembled in an array to generate 100 kW. Each module contains 35 photovoltaic cells which generate electricity directly from solar radiation. While solar cells are convenient, they are too expensive for most agricultural applications (source: DOE 198 010 005).

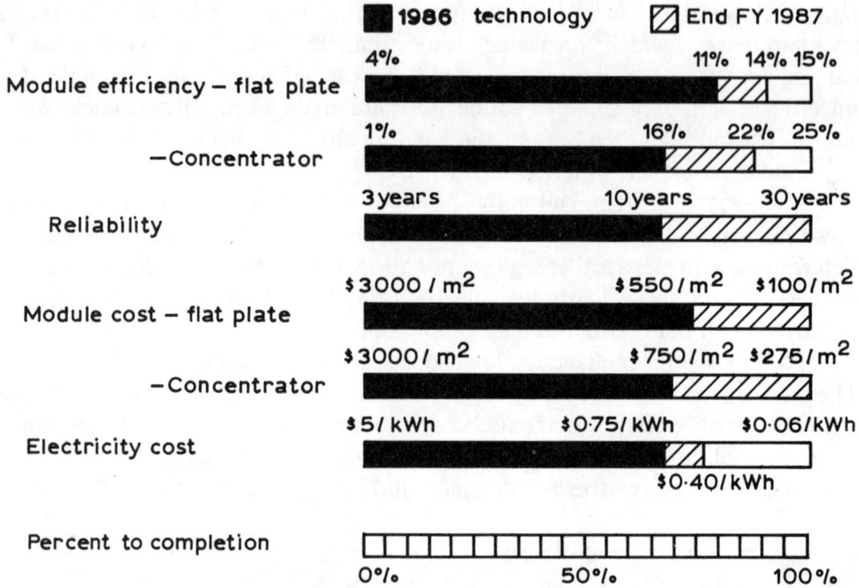

Fig. VI.16. Photovoltaic technology status in 1986 (source: Annan, 1986).

without degradation. New solar cells seem ideal for small, decentralized power sources. They could, for example, provide power for operating offshore lighting for navigation, telecommunications equipment, refrigerators in rural health dispensaries, small electrical appliances in remote hotels and motels, educational television receivers, and audiovisual equipment in rural schools and community centers.

Solar Ponds—Collection and Storage

The principle of solar ponds is simple, but not until the 1950s did a few observers begin to note and relate the underlying natural phenomena. In an ordinary shallow pond, solar heat passes through water and raises the temperature of the bottom layers, thereby lowering the water density and causing it to be replaced by heavier cold water from above. As a result, convection currents are established which disperse the heat throughout the pond. Consequently, no portion of the pond reaches a significantly higher temperature, and heat loss is substantial across the water–air interface (Tabor, 1966*a*). A solar pond is made by forming density layers of salt water in a shallow basin about 1 m deep. The bottom of the pond is lined with a black material to absorb the sun's energy (Fig. VI.17). Thus, the bottom, heavier layer of salt water is heated until the gradient in the pond becomes strong enough to discourage convection currents. Because the heat cannot be dissipated through the pond without these currents, the bottom layer can reach considerably higher temperatures.

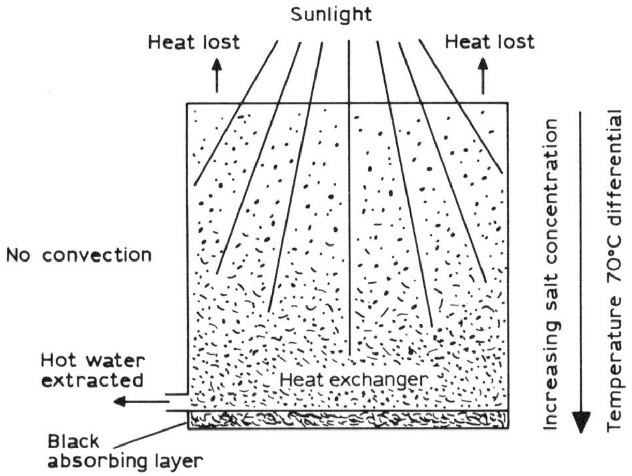

Fig. VI.17. Obtaining heat from a solar pond using a heat exchanger. As sunlight strikes the solar pond it is absorbed by the black material lining the bottom. The bottom layer of salt water is thereby heated. The salt gradient throughout the pond limits convection currents from forming. The temperature differential could reach as high as 70°C in a well-designed and stable pond (source: Tabor, 1966*a*).

Another useful property of salt water gradients is that any layer can be drawn off without disturbing the others. To obtain energy from the pond, the heated bottom layer can be pumped out, sent through a heat exchanger, and pumped back to its original position in the pond. It takes about 2 years for a pond to reach its full value, but substantial energy can be obtained several months after construction. In the winter, only about a fourth as much heat can be withdrawn as in the summer.

Although the first solar pond was built in 1960, only a handful of people have conducted research on solar ponds, including groups in Israel, Chile, Australia, the USA, and the USSR. The Israeli experiments reported an efficiency rating of 25% during 1 year.

Solar ponds show promise, but technical problems remain that are not yet resolved. The salt gradient still exhibits a degree of fragility. Methods of controlling the influences of wind, storms, dust, rain, and evaporation must be found. Furthermore, even though solar ponds are based on simple principles, they must be built by technicians. Nevertheless, the potential of solar ponds is great, especially in developing areas.

The heat extracted from a solar pond can be used, for example, to operate a low-temperature turbine. Calculations indicate that with a 25% efficient 1-km^2 solar pond, a turbine of only 8% efficiency could operate a plant of about 5000-kW capacity during the summer. Generating capacity during the winter would be 20–25% of summer capacity. These calculations assume a 32° N latitude. There would not be such a sharp reduction in winter capacity in areas closer to the equator. Solar ponds might also be used for desalination and any other applications requiring large amounts of inexpensive, low-temperature heat.

Solar ponds differ from other methods of collecting solar energy because they capture solar radiation and store it without external devices.

Heat Storage

A major difficulty with all solar energy systems is that solar radiation is intermittent and variable, particularly in humid climates. Because many applications require a fairly constant energy source, it is necessary to either store heat collected in favorable weather or rely on a conventional source, such as a fossil fuel, for a back-up system.

The storage of solar heat is a problem. Because solar radiation is a dilute energy source, the more economically feasible medium for capturing this energy (water) can only attain temperatures of 100°C or less. Heat storage systems are likewise limited to this temperature range and, consequently, require large volumes. The storage volume is generally filled with water, rocks, or calcium decahydrate (a phase-change material), which retain heat a day or two if well insulated. As solar-heated water or air from the collector passes through the storage area, it raises the temperature of the storage material.

Table VI.2 compares specific heat, heat of fusion, density, and heat storage capacity for rocks, water, and calcium decahydrate. As illustrated, 19 kg of rocks or 4 kg of water can store as much heat as 1 kg of the phase-change material.

Water has the highest heat capacity of any ordinary material. One liter of water requires 1 kcal of heat to raise its temperature by 1°C. Thus, 1 liter of water, as it cools, returns 1 kcal for each degree of cooling. Rocks store about a third as much heat as water for equal volumes (Daniels, 1964).

In most areas, water is inexpensive, but its use as a storage medium necessitates large, expensive tanks to hold the amount of water required. Often, a circulating pump is also needed. For air collectors, a rock bed can be made by simply filling a container with gravel, crushed rock, or brick, preferably of uniform size. Enough air space must remain to prevent excessive resistance to air flow.

All thermal storage systems should be insulated to reduce heat losses. Large water tanks and rock beds can be placed underground because heat conductivity through

TABLE VI.2
Heat capacity of storage materials (source: Spillman, 1978)

	Rocks	*Water*	*Phase-change material*
Specific heat (Wh/kg per °C)	0·2	1·2	0·6 (average)
Heat of fusion (Wh/kg)			64 (average)
Density (kg/m^3)	1 440	990	1 660
Storage of 146·4 kW			
(17°C change in storage temperature)			
Weight (kg)	40 140	7 460	1 970
Weight (relative)	19	4	1
Volume (m^3)	2·63	0·76	0·16
Volume (relative)	16	5	1

the earth is low, although moist soil tends to cause more heat loss than dry soil. For small-scale storage systems, the heat lost to the surrounding soil from underground containers is too extensive. Small heat storages should be located in the space to be heated.

Heat storage media include a number of other phase-change materials, reversible chemical reactions, and metals. Currently, these materials are relatively costly and are seldom used except in more elaborate systems.

AGRICULTURAL APPLICATIONS

Water Desalination

Population increase and economic development is creating a worldwide imbalance between the supply and demand of quality water. The task of providing adequate supplies of fresh water may indeed become the most serious problem facing the world by the end of this century. We must find sources of fresh water, and two of the most likely sources are the oceans and the vast reserves of brackish inland water (Moustafa & Brusewitz, 1978).

Basically, solar stills consist of a blackened tray with a clear glass or plastic cover. Insulation is sometimes needed underneath the tray. As the shallow layer of impure water in the tray is heated by solar radiation, it evaporates and condenses on the cooler transparent cover, eventually running down into collection channels on the lower edges (Fig. VI.18).

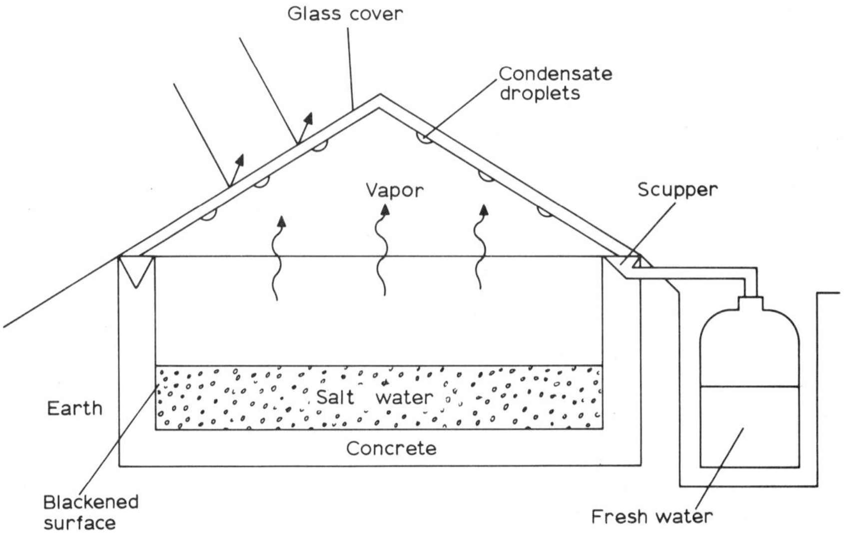

Fig. VI.18. Solar distillation. In this solar still, salt water is placed inside a concrete container lined with black material. As the impure water is heated it vaporizes, leaving the impurities behind as residue. The vapor then condenses on the colder cover and runs down into the collecting troughs (source: Androsky, 1973).

TABLE VI.3
Large basin-type solar stills (source: Howe, 1980)

Country	*Location*	*Date built*	*Basin size* (m^2)	*Productivity* (m^3/day)
Australia	Muresk I[a]	Dec. 1963	372	0·83
	Muresk II	Nov. 1966	372	0·83
	Cooper Pedy	Nov. 1966	3 160	6·35
	Caiguna	Dec. 1966	372	0·78
	Hamelin Pool	Dec. 1966	557	1·21[b]
	Griffith	Aug. 1967	413	0·91
Cape Verde Islands	Santa Maria da Sal	1965	743	2·12[b]
Chile	Las Salinas[a]	1872	4 459	14·76[b]
	Quillagua	1968	100	0·40
	Quillagua	1969	103	0·40
Greece	Symi[a]	July 1964	2 687	7·57[b]
	Aegina	Oct. 1965	1 490	4·24[b]
	Salamis	Oct. 1965	388	1·10[b]
	Patmos	July 1967	8 640	26·11
	Kimolos	1968	2 508	7·57
	Nisiros	1969	2 043	6·06
India	Bhavnagar	Oct. 1965	377	0·83
Mexico	Baja California			
	Nativadad Island	Mar. 1969	95	0·38[b]
	Baja California Sur			
	Puerta Chale	1974	300	1·0
	Punta Eugenia	1974	470	1·5
	Sonora			
	Punta Chucca	1974	470	1·5
Spain	Las Marinas	Mar. 1966	869	2·57
Tunisia	Chakmou	1969	439	0·53
	Mahdia	1968	1 300	4·16
United States	Daytona Beach, Florida[a]			
	Original Deep Basin	Jan. 1959	228	0·53[b]
	Second Deep Basin	Apr. 1961	246	0·57[b]
	Large Inflated Plastic	Jan. 1959	216	0·38[b]
	CWS Inflated Plastic	July 1963	149	0·61[b]
USSR	Balcharden, Turkmena	Apr. 1969	559	1·62[b]
West Indies	Petit St Vincent	Feb. 1967	1 709	4·92
	Haiti	June 1969	223	0·76

[a] No longer operating.
[b] Estimated productivity.

The first large solar distillation plant was built in Chile in 1872 (Tabor, 1966*a*). The total water basin area was about 5000 m^2 and the peak drinkable water output was almost 15 m^3/day (Howe, 1980). It was in operation for about 30 years.

Following drought conditions in many regions of the world in the 1950s, much research was conducted on the desalination of water to provide fresh water supplies. Research on solar desalination reached a peak in the mid-1960s but decreased as large desalination plants powered by fossil fuels became more cost-effective (Howe, 1980).

Table VI.3 lists large solar stills that have been constructed worldwide. The majority were built between 1959 and 1969. Most are still in operation. The large solar stills shown in Table VI.3 average a productivity of about 3·51 m^3/day (Howe, 1980). The overall efficiency of a basin-type solar desalination plant is 30% or lower (Moustafa & Brusewitz, 1978).

In sunny areas, at 30–40° latitude, an annual yield of about 1000 liters/m^2 (1 m^3/m^2 of collector) can be expected (Tabor, 1966*a*; Howe, 1980). A typical lifetime for a still constructed of concrete, glass, and other durable materials is 20 years or more.

If a community or an industry requires small quantities of water (less than 300 000 liters/day), the solar still may be more economical than conventional desalting plants. This is particularly true: (1) for small communities where drinkable water is unobtainable except at a very high cost; (2) for certain industrial and commercial applications where materials must be processed in a region for which all available

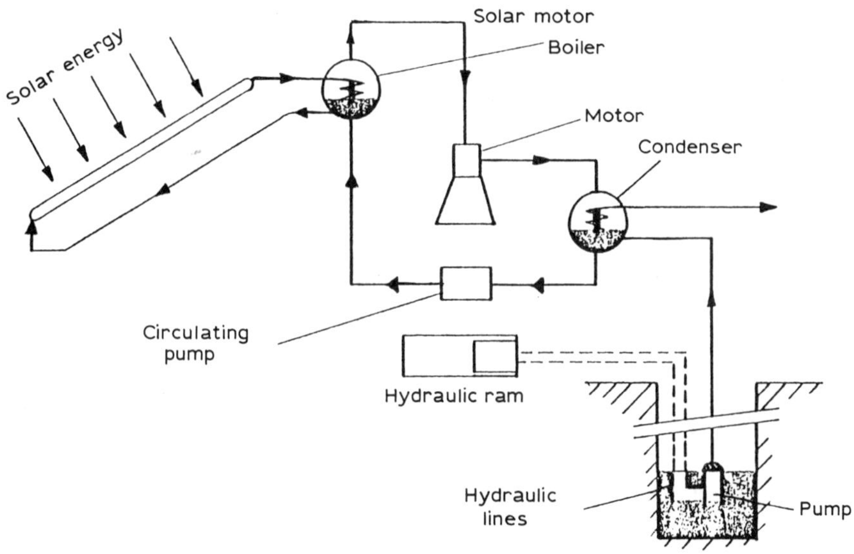

Fig. VI.19. Solar water pump. Heated water, obtained from the solar collector, is circulated through a heat exchanger. The caloric content of the water is transferred to a fluid which circulates through the engine. The expansion of the fluid, resulting from its temperature gain, provides the mechanical work to power the engine. The engine, in turn, drives a hydraulic ram which lifts the water (source: Girardier & Massen, 1980).

water is brackish; or (3) for livestock watering in areas where grazing is possible if water is supplied (UNEP, 1980).

Water Pumping

Solar-powered water pumping could be performed with either solar engines or by photovoltaic systems. In the case of solar engines, collectors convert solar radiation to heat which drives an engine to produce mechanical power (Fig. VI.19).

In Thailand, a low-lift, solar, water pump was developed at the Asian Institute of Technology (Loftness, 1984). The device used an automatic valve mechanism consisting of a set of permanent magnets and springs for low-lift solar water pumping (Fig. VI.20). The pump is based on the thermodynamic conversion method using an intermittent Rankine cycle with freon-113 as a working fluid. Solar radiation collected on a simple flat plate solar collector (size $1{\cdot}4\,m^2$) is used to vaporize the liquid freon. The vapor pushes a rubber diaphragm which, in turn, pumps the water. The exhaust vapor is condensed and collected in a container. The condensate is charged into the flat plate solar collector by gravity for use during the next day's operation. Test results showed that with constant input vapor pressure the discharge of the pump decreased as the head increased. The discharge of the pump was 4·3 liters/min at a 3-m head while it reduced to 2·8 liters/min at a 6-m head with a freon pressure of $2{\cdot}4\,mg/cm^2$ ($240\,kN/m^2$) inside the collector. The head increased as the input vapor pressure increased. Figure VI.21 illustrates yet another solar pumping principle, a solar–thermal system driving a turbine which, in turn, drives a water pump.

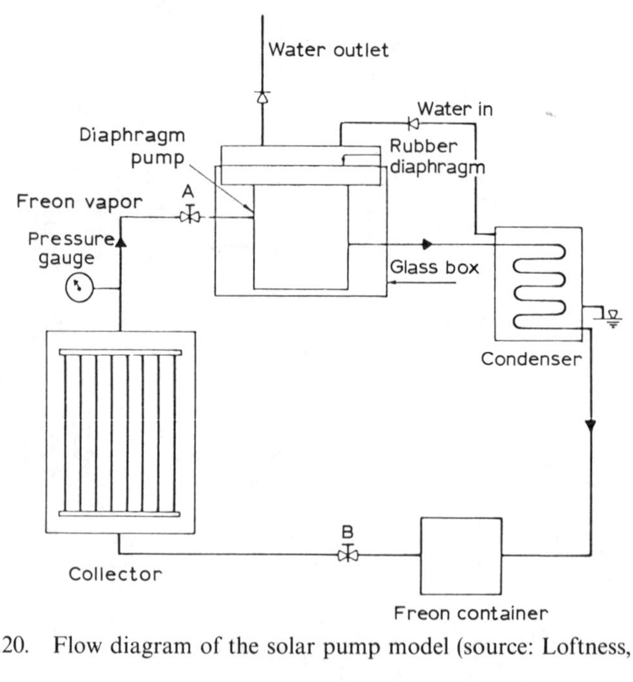

Fig. VI.20. Flow diagram of the solar pump model (source: Loftness, 1984).

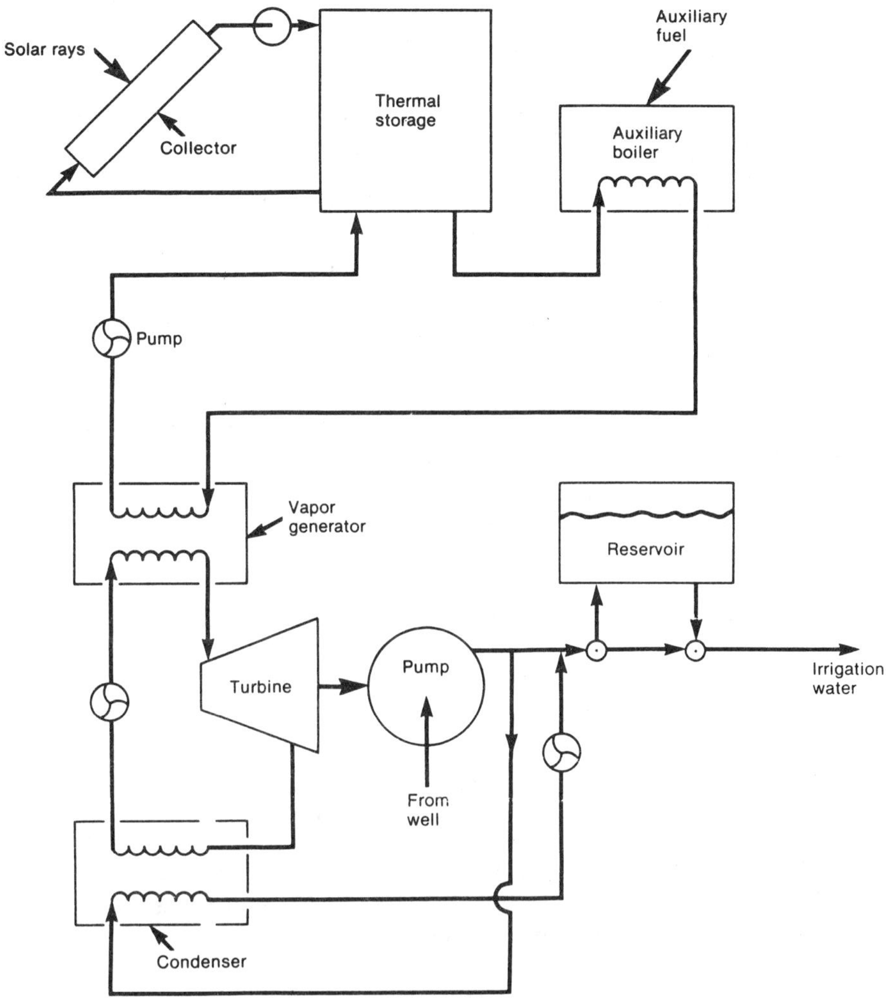

Fig. VI.21. Schematic of solar-powered irrigation pumping system (source: Enochian, 1982).

In the United States, four major solar irrigation pumps are in operation. Photovoltaic cells provide up to 25 kW of electrical energy to drive irrigation pumps at a University of Nebraska farm. Parabolic trough-type solar collectors and Rankine cycle turbine engines comprise the two other systems. The two plants, a 50-kW installation located on a farm in Arizona and a 25-kW unit in New Mexico, are directly coupled to pumps with electric motor back-ups (Larson & Sands, 1979).

A 150-kW plant operated by the University of Arizona was completed in 1979. The solar installation consists of more than 2100 m^2 of single axis tracking, parabolic trough collectors, a 190-m^3 thermal energy storage tank, and a Rankine cycle turbine generator set (Larson, 1981).

The collector system was operated about 95% of the hours, having sufficient insolation in the spring and summer of 1980. The turbine generator was operated 4–5 h/day during the peak period, generating 700–800 kWh. Peak production was 1180 kWh on a sunny June day. The best thermal-to-electrical energy conversion efficiency attained was about 20% (Larson, 1981).

Capital costs of solar water pumping installations can be extremely high. A 1-kW pump installed at Wajir, Kenya, in 1978 cost $90 000 (and $213/day to run) for an output of 5 m^3 of water/h with a 30-m head. This results in a capital investment of $330 000/kW of mechanical work. Serious breakdown and maintenance problems have been reported (G. Goodman *et al.*, 1979).

A more recent report (McGowan, 1987) shows that significant cost reductions in photovoltaic pump systems have occurred during the past few years. A life-cycle cost analysis was used to calculate a benefit/cost ratio for photovoltaic pumps, windmills, and diesel pumps. For the assumptions made, the photovoltaic system compared favorably with the wind and diesel systems (Fig. VI.22).

There are no theoretical obstacles to the development of better working fluids, more robust engines, or more cost-effective solar collectors to increase efficiencies and decrease costs (G. Goodman *et al.*, 1979). Progress in developed countries is rapid in all these areas. With rising fuel prices, worldwide solar pumping systems may approach conventional system costs in the future (Fischer & Talbert, 1980).

Agricultural Crop Drying

Generations of farmers have effectively used the sun and wind to dry crops by spreading them out in thin layers on the ground. Solar radiation heats the surrounding air and changes the equilibrium vapor pressure, causing moisture to diffuse from the plant (Meinel & Meinel, 1976). A moderate wind accelerates the process by forcing the heated air through the crop and carrying the moisture-laden air away. Modern technology still uses heat and air flow for crop drying but provides higher temperatures and better permeation.

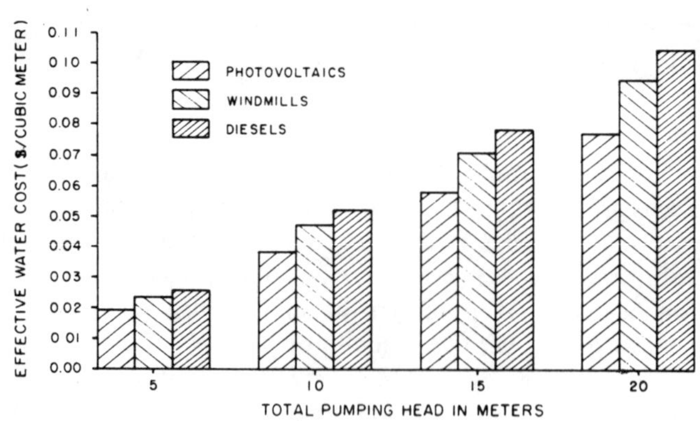

Fig. VI.22. Financial comparison of water pumps: photovoltaics, windmills, and diesels (source: McGowan, 1985).

Two types of crops need drying. Grains usually have a moisture content of 20–30% at harvest, which must be reduced to 12% before storage. Leafy crops and fruit have a moisture content of more than 50% at harvest (Meinel & Meinel, 1976). During the extended drying process, these crops are exposed to dirt, contamination, rain, overheating, and fungal and bacterial growth (Meinel & Meinel, 1976).

Grain Drying

The most energy-efficient solar drying occurs in the field (if weather conditions are favorable) during the late stages of crop maturity. To avoid serious field losses from unfavorable weather, farmers harvest high-moisture-content grains and then dry them to safe moisture levels before storage. Wet harvested grains held in trucks or bins may mold within a few hours in warm weather (H. H. Converse *et al.*, 1981).

How do solar grain drying requirements correspond with the characteristics of solar energy? They coincide quite well, but some problems do exist:

—Grain drying is easily adapted to solar energy because large volumes of low-temperature air are used. Air entering a solar collector is usually at ambient temperature and is heated only a few degrees for in-storage drying systems. In contrast, the air entering collectors for shelter heating is at the minimum comfort level for the occupants, which is normally several degrees above ambient temperatures during the cold part of the year. Since the heat losses from the collector are proportional to the difference between ambient and average collector temperature, heat losses from the collector used in grain drying would obviously be low. Thus, relative low-cost collectors are effective for in-storage grain drying (Hartsock, 1978).

—In-storage grain drying proceeds for several days and is subject to variations in the drying potential of the ambient air. Solar energy, available only during daylight hours, adds to the variation of drying potential between day and night. Fortunately, low-temperature drying systems can tolerate such variable levels of heat input. The grain stores excess energy in the form of over-dried grain. The over-dried grain acts as a desiccant and removes excess moisture from high-humidity night air so drying can proceed. So the grain, in effect, provides its own storage for solar energy.

Several approaches to storing solar heat other than in grain are being studied. The need for external thermal storage for low-temperature bin drying systems has not been fully established, although it is probably needed for higher speed, higher temperature drying systems (Hartsock, 1978).

—The capital intensiveness of solar energy presents a problem in relation to grain drying. Corn requires more drying energy than other grains in the midwestern United States. Corn is dried in the fall during a relatively short period of 6–8 weeks. The annual utilization of solar equipment for corn drying alone is too low for cost-effectiveness. For example, different analyses show that at present prices the fuel savings achieved by the solar heat collected would justify an investment in collectors of less than $\$1/ft^2$. Such a low collector cost is possible only with very simple plastic collectors, and these would need to be stored and protected when not in use to prolong their life. The estimated life of the collector, along with its initial cost and the volume of grain dried, are the three factors contributing most to the cost-effectiveness of solar drying systems.

—The ubiquitous nature of solar energy permits collection and use at the area or areas of greatest need and convenience. There is little economy of scale in the collection and use of solar energy (Hartsock, 1978).

Economic and systems analyses are being emphasized in current solar energy research efforts. Multiple use of collectors is desirable to lower unit cost. Multiple use of the same equipment increases annual use and the amount of energy collected per dollar of investment. Competition also exists for the energy available from competing or conflicting multiple uses, and collectors may need to be portable (Fig. VI.23). However, opportunities to use the same solar equipment to do more than one job should not be overlooked (Peterson, 1983).

In warmer, more humid regions like the southern United States, the problem of spoilage and toxins has prompted most farmers to opt for conventional high-temperature drying to avoid these risks. Using a combination of high- and low-

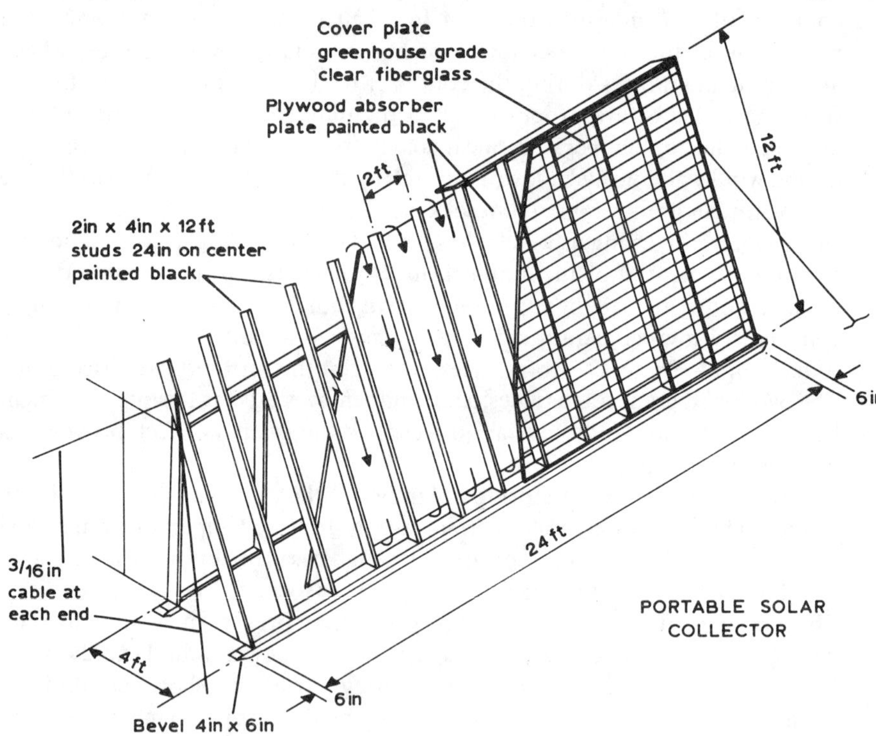

Fig. VI.23. Basic portable solar collector as designed for grain and crop drying in Illinois (source: Peterson, 1983).

temperature drying in these regions, however, can minimize spoilage risks and allow solar energy use (ASAE, 1981).

Farm buildings and grain drying bins have large areas available for the placement of flat plate solar collectors. A simple design was built by the University of Illinois in 1974. It consisted of installing a secondary unpainted metal wall around two-thirds of a grain drying bin circumference (Fig. VI.24). Air was drawn through a 7·5-cm space between the secondary metal wall and the bin wall, and then blown into the bin. Another example of a solar collector incorporated into a farm building is shown in Fig. VI.25.

On clear days the temperature increase averaged 2·2°C and on partly cloudy days it averaged 1·7°C. Once dried, the corn had an average moisture content of 2 percentage points lower than corn in a similar bin without the solar collectors. The bin was later painted black, resulting in a 2·5-fold temperature increase and an efficiency increase of 30% from 12% for the unpainted collector (Silvestrini, 1978).

The next step in increasing solar collector performance is to provide a clear cover over the absorbing surface. For the previous example, it would be more appropriate to paint the bin wall black and construct a clear fiberglass secondary wall over it in place of the sheet metal wall. This type of solar collector provides heat for low-temperature corn drying that normally is provided by a 10-kW electric resistance heater (Silvestrini, 1978).

Solar collectors can be incorporated into the roofs, walls, and attics of farm buildings, such as machine sheds and livestock shelters. These can then be used for multiple purposes, e.g. heat for swine buildings and grain drying. Multiple uses of solar collectors can help justify their cost.

Freestanding collectors can also be used. Figure VI.26 shows several possible arrangements for a solar drying facility using 1·2 m × 2·4 m solar collector sections (Midwest Plan Service, 1980).

Another type of solar collector used for grain drying is shown in Fig. VI.27. The

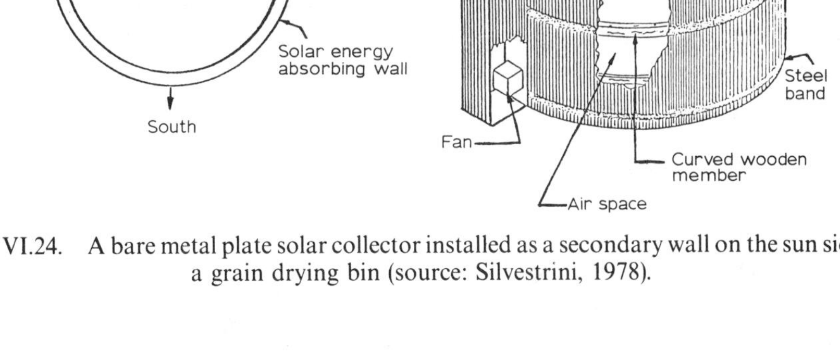

Fig. VI.24. A bare metal plate solar collector installed as a secondary wall on the sun side of a grain drying bin (source: Silvestrini, 1978).

Fig. VI.25. Flat plate solar collector incorporated into the roof of a metal building. Air is drawn into the building between the glazing and the absorber plate for use in grain drying or space heating (source: DOE 007 003 001).

collector, a long, black polyethylene tube, is stretched out in the sun, and a fan forces air through it. As the black material absorbs solar radiation, the warm air circulates through the grain bin. Covering the black polyethylene layer with clear plastic reduces heat loss. Calculations show that in good weather at 45° N latitude, energy equivalent to 109 kWh/day can be obtained from a collector tube 45 m long and 1 m in diameter (Meinel & Meinel, 1976).

Solar grain drying can be cost-effective under certain conditions. Table VI.4 provides comparative costs for a solar drying system and an LP-gas drying system in Illinois. Solar drying costs were 57% of LP-gas costs in 1981 and 37% in 1982 (Peterson, 1983). Solar grain drying can be accomplished in many ways, and much research is underway in this area (Shove, 1977; Hartsock, 1978; Midwest Plan Service, 1980; USDA, 1983, 1983*a*, 1984*a*, 1985).

Rice Drying (Exell *et al.*, 1979)

The traditional methods used by farmers in southeast Asia to dry paddy in the sun after the harvest cannot be used for the second crop in the wet season due to the

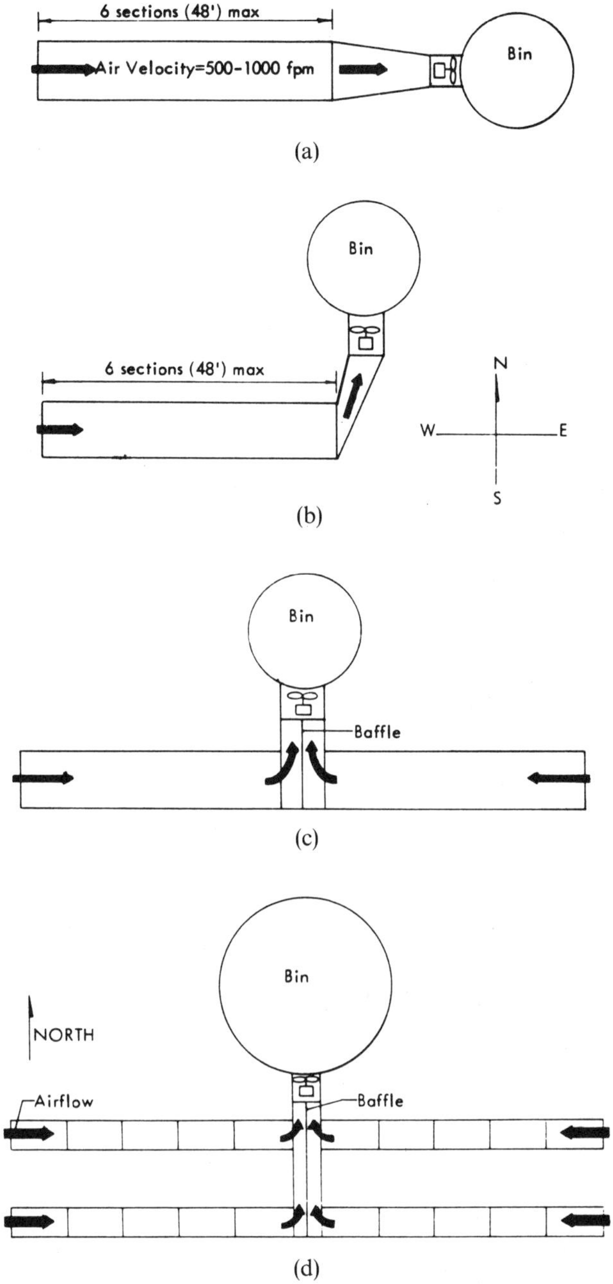

Fig. VI.26. Various solar collector arrangements for a solar drying facility (source: Midwest Plan Service, 1980). (a) In-line: fan on east or west side of bin; only one inlet needed (low air flow). (b) Offset: fan on sun side of bin: obstruction to one side of bin, thus only one inlet needed (low air flow). (c) Split duct: collector more than six sections long; multiple inlets needed to keep air velocity below 1000 ft/min (high air flow). (d) 20-collector layout.

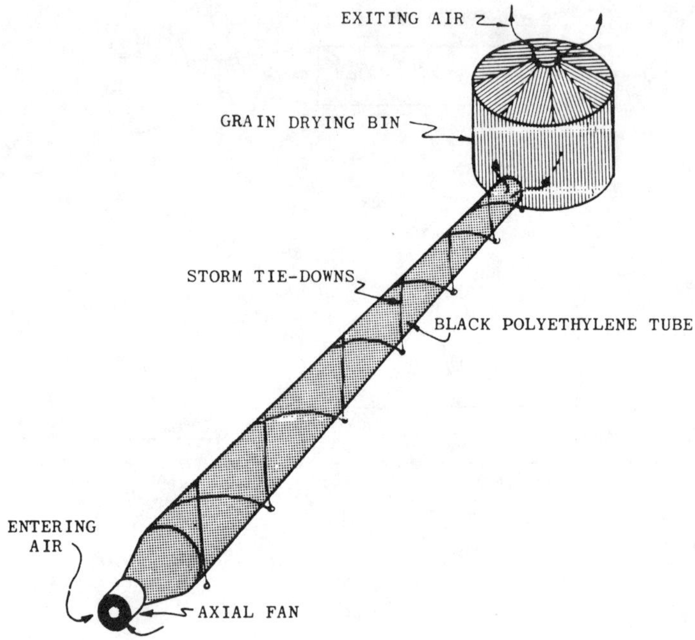

Grain drying	
Latitude	45°
Energy income	$2{\cdot}6 \times 10^4$ kJ/m² per day
Energy income (noon hour)	$2{\cdot}4 \times 10^3$ kJ/m² per hour
Percentage of possible sunshine	80%
Utilization factor	100%
Collector efficiency	0·25
Collector yield	$5{\cdot}2 \times 10^3$ kJ/m² per day
Collector yield (noon hour)	$0{\cdot}6 \times 10^3$ kJ/m² per hour
Mass of air to be heated	$4{\cdot}3 \times 10^3$ kg/h
Heat required for 6°C rise (1·04 J/g°C)	$2{\cdot}7 \times 10^4$ kJ/h
Collector area required	45 m²
Tube length (1·0 m diameter)	45 m
Equivalent energy from collector	$1{\cdot}09 \times 10^2$ kWh/day
Value of heat at \$0·05/kWh	\$3·25/day
Value of heat in 60-day season	\$195·00/season
Value of collector	\$4·34/m² per year

Fig. VI.27. Grain dryer design utilizing a black plastic tube as a solar collector. Air blown through the tube by a fan gains heat from the collector. This heated air is then forced through the grain bin. External power is needed to operate the fan (source: Meinel & Meinel, 1976).

uncertainty in the weather (Fig. VI.28). Cut paddy left in the fields often spoils during alternate rewetting and overdrying, which cracks the kernels and reduces the milling quality of the rice. Ordinarily, the farmer must sell the second crop quickly at a low price before it spoils. The farmer would benefit by either drying the paddy immediately after harvesting so that a better price can be obtained for it, or storing it safely for use at a later time.

TABLE VI.4
Solar drying versus LP-gas drying on an Illinois farm (source: Peterson, 1983)

	1981		*1982*	
	Solar drying	*LP-gas drying*	*Solar drying*	*LP-gas drying*
Bushels dried (bu)[a]	22 700	35 300	21 500	42 700
Initial corn moisture (%)	21·7	21·0	21·0	21·5
Final corn moisture (%)	15·2	15·0	15·7	15·5
Electrical consumption (kWh)	23 200	8 740	14 450	11 170
LP-gas consumption (gallons)	—	3 180	—	4 196
Comparative drying costs (¢/bu—point of moisture removed)	0·61	1·07	0·53	1·43

[a] 1 m^3 = 28·2 bu.
1981 energy cost on Illinois farm: LP-gas = \$0·61/gallon; electricity = \$0·0387/kWh.
1982 energy cost on Illinois farm: LP-gas = \$0·76/gallon; electricity = \$0·0415/kWh.

Fig. VI.28. Rice being sun-dried on mats in Indonesia. Solar dryers and fossil-fuel dryers can increase the drying rate and reduce losses (source: FAO Photo 4609/35).

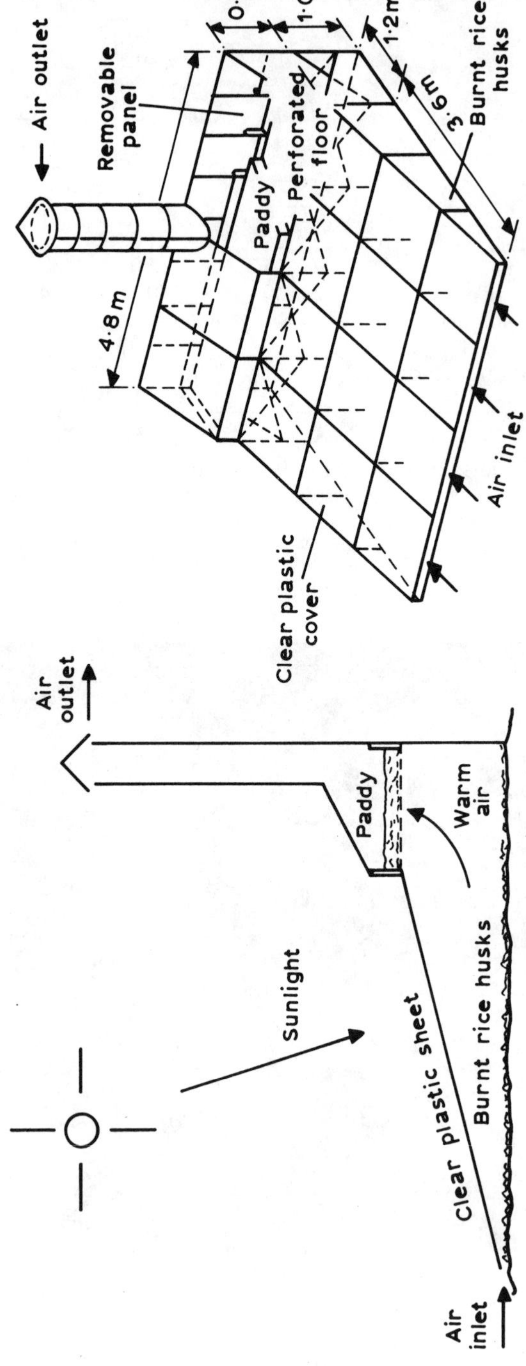

Fig. VI.29. A low-cost solar rice dryer (source: Exell *et al.*, 1979).

Figure VI.29 shows a rice dryer built by the Asian Institute of Technology in 1979. It was designed to be constructed by the farmer at a low cost with local materials.

As shown, sunlight passes through the clear plastic sheet and warms the air inside aided by a layer of burnt rice husks covering the ground below to absorb the radiation. The warm air passes up through the bed of paddy and dries it. The chimney provides a tall column of warm air to increase the flow of the heat through the bed by natural convection. The air inlet at the dryer bottom should face the prevailing wind direction to further increase the air flow.

The energy required to dry the paddy and the intensity of the solar radiation determine the solar air heater size. On a poor day during the wet season, the insolation is typically 15 MJ/m^2. To dry 0·5 t of paddy from an initial moisture content of 22% wet basis to a final moisture content of 14% wet basis for safe storage, about 120 MJ of energy are needed to evaporate the water. Therefore, if drying in poor weather is allowed to require 2 days, and if the solar air heater is 25% efficient, an area of 16 m^2 is required.

The demonstration unit for this rice dryer costs $120. By using less expensive materials and modifying the design, the dryer cost could be cut in half.

Forage Drying

In the United States alone, more than 118 million tonnes of hay are produced on slightly more than 109 million hectares. Most of this forage is field-dried or sun-cured before storage and feeding. Valuable protein, however, is lost during this process. Artificially dried hay is higher in quality and dry matter content, and can be produced with less dependency on the weather than conventional field-cured hay (R. Hansen & Grenard, 1981).

Very little forage is artificially dried. In Italy, for example, only 2% of the total hay production is artificially dried (Pellizzi, 1978). However, solar drying of forage is of significant interest because the drying season consists of four to five harvests during the summer, and tremendous potential exists for improving feed quality.

In 1960, G. P. Shipway studied the use of solar heat for barn hay-drying in the United Kingdom. He concluded the following:

- —Solar energy is sufficient in the United Kingdom for barn-drying hay during the summer.
- —A simple, unprotected solar collector can be constructed consisting of flat and corrugated galvanized steel sheets fixed together. The sheets can be arranged to produce air channels between them and an integral main duct. The solar collector can also form the barn roof.
- —Solar heat for barn hay-drying is technically, but not economically, sound unless the solar collector can be used for more than 5 weeks/year.

Currently, two main research projects are underway on solar hay-drying in the United Kingdom. At the Scottish Institute of Agricultural Engineering, a solar barn uses the entire roof surface and south wall as a solar collector. The Nottingham University project uses a polyethylene semi-circular tunnel (1·6 m × 22 m) as a solar collector duct. The solar-dried grain or hay reached a lower moisture content than the control hay, although both were over-dried. The possibilities of using the tunnel for a heated storage structure or lambing house are being evaluated.

A simple hay-drying solar system, tested in Czechoslovakia, is shown in Fig. VI.30. It has a corrugated metallic sheet absorber with a black non-selective paint; air is the heat medium (Pyke, 1970).

A more complex solar forage-drying system is shown in Fig. VI.31. It was developed and built in Italy using flat plate collectors and counter-current water/air heat exchangers. Initially, the system used a temperature rise of 12–15°C, which resulted in an efficiency of around 25% at an air temperature of 35°C. After the first year of operation, with six loads of hay, the system was modified to attain a temperature rise of 40°C (Pellizzi, 1978).

In the United States, research is underway on solar drying large round bales of hay (Bledsoe & Henry, 1981; Morrison & Shove, 1981). Figure VI.32 shows a solar hay-drying barn built in Illinois. This building was a 14·9 m × 44·2 m rigid frame structure with more than 886·2 m^2 of covered flat plate solar collector built into the north and south slope of the roof and the south wall.

The solar collector was formed by attaching black painted plywood absorber sheets to the underside of the rigid roof frames and to the interior side of the south

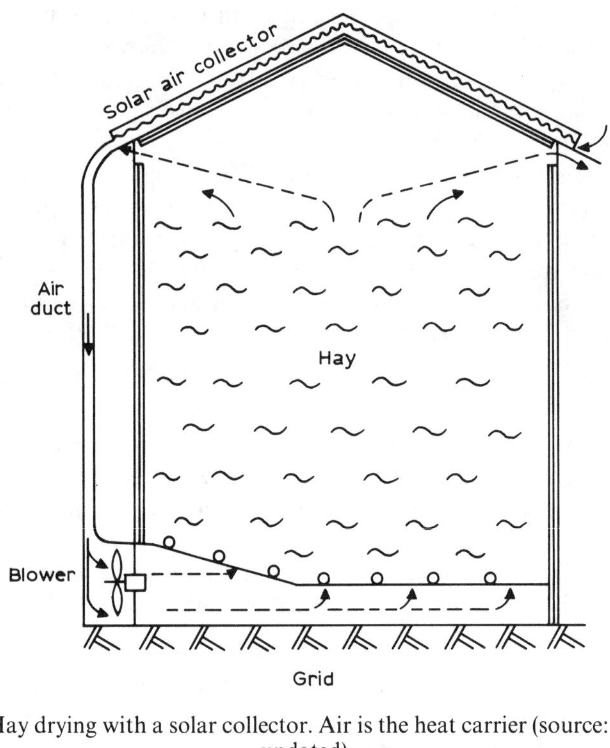

Fig. VI.30. Hay drying with a solar collector. Air is the heat carrier (source: Velebil & Has, undated).

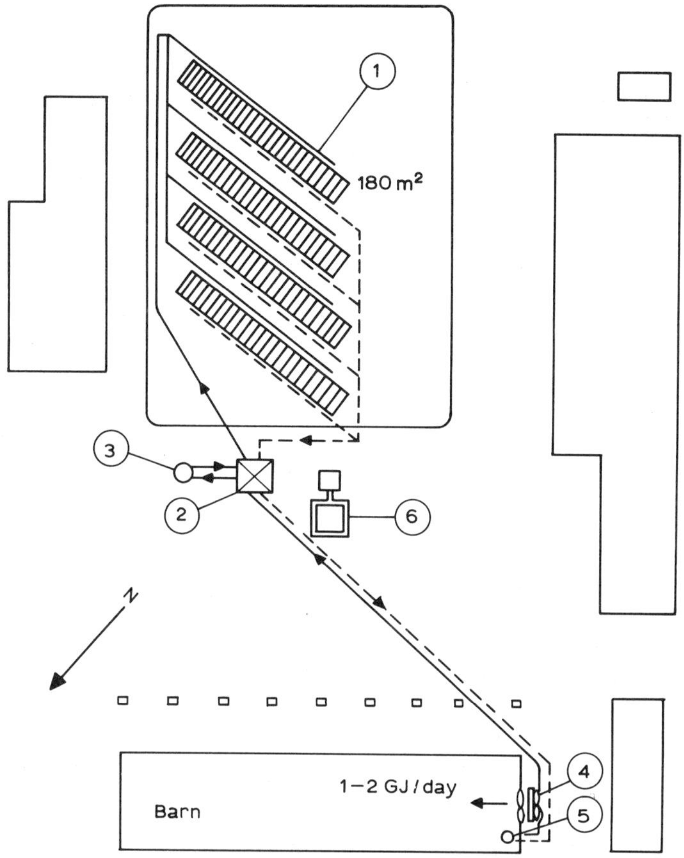

Fig. VI.31. A forage dryer using a group of flat plate collectors (180 m^2) using water as the heat carrier: (1) collectors; (2) control platform; (3) storage tank; (4) fan; (5) water tank for milking machine; and (6) well (source: Pellizzi, 1978).

wall. Tedlar-coated, clear corrugated fiberglass, used as the roofing material and as the south wall exterior covering, served as the solar collector cover material.

Drying fans, located in the south wall of the building, pulled air over the roof and down the south wall through the air space created between the fiberglass and plywood. This solar-heated air then was blown through the air ducts under the floor of the building and into the wet bales stored inside (Morrison & Shove, 1981).

The building was designed to eventually dry sixty 1134-kg bales using fifteen 3·7-kW fans, each fan drying four bales. The air duct consisted of a 0·91 m × 25·4 cm tall steel ring. The wet bales were dried with air forced through the ring and into the bales (Morrison & Shove, 1981).

Compared to unheated air drying, solar drying required 20% less fan energy to remove equal amounts of moisture. Optimum air-flow rates were 0·47–0·57 m^3/s per bale (Morrison & Shove, 1981).

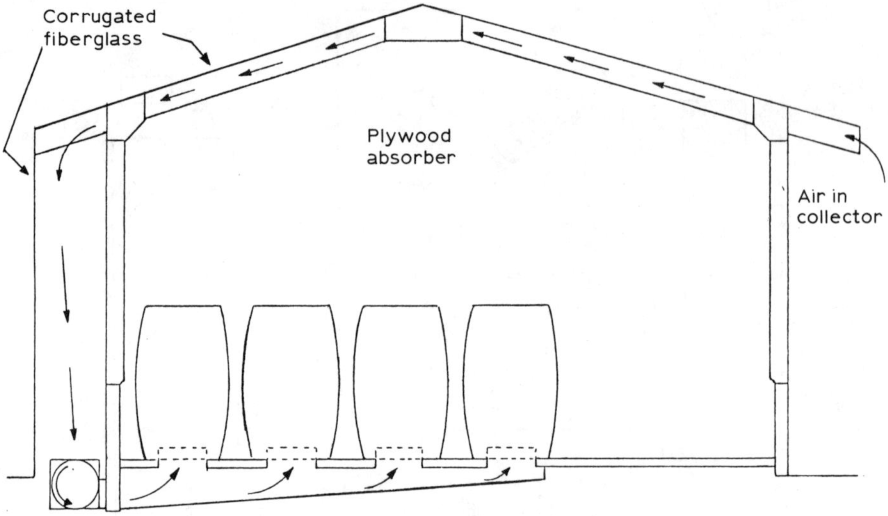

Fig. VI.32. Solar hay-drying barn (source: Morrison & Shove, 1981).

Solar Crop Drying Project

The Solar Crop Drying Project (SCDP) of the African Energy Programme (AEP) is an example of efforts to improve traditional methods of sun-drying crops and commodities in Africa by the development and introduction of appropriate solar drying techniques (CSC, 1985). The following countries participated in the SCDP project: Gambia, Ghana, Kenya, Mauritius, Tanzania, Uganda, and Zimbabwe. An educational workshop was conducted to accomplish the following objectives:

—Provide basic information and instruction on solar crop drying techniques encompassing a range of disciplines.
—Promote the three-stage concept of initial project evaluation, experimental investigation, and project extension.
—Strengthen links and improve liaison activities between researchers through case studies, discussion periods, and design projects.

A 337-page book contains the lecture notes from the workshop (CSC, 1985). It is intended to provide (1) an introduction to basic theories of drying, (2) an outline of the principal operating features that can be used to classify the many designs of solar dryers developed to date, and (3) a detailed account of the construction and operation of the three most popular types of solar dryers. Information is included on the theory and design of flat plate solar collectors, which are an integral part of many dryers. This publication is an excellent supplement to the information contained in this chapter.

Solar Cookers

In India, about 150 million tonnes of firewood, 32 million tonnes of dried dung, and 36 million tonnes of agricultural waste are consumed annually for cooking purposes. This 238 million tonnes of noncommercial fuel constitutes about 45% of the total energy supply in the country. It is absolutely necessary to make the most efficient use of animal manure and agricultural wastes as fertilizer and to stop the destruction of the forests. Inexpensive solar ovens and cookers could supply an alternative heat source for domestic cooking (Garg, 1976).

Several small focusing collectors have been designed for using the sun's energy to cook food. A typical solar cooker consists of highly polished surfaces which reflect sunlight onto a small area. The reflectors concentrate solar radiation and intensify it to the energy level of an open fire (Löf, 1961). A mechanism has also been designed for aiming the reflector at the sun, and a support developed for holding the cooking vessel at the focal point. Solar cookers must be adjusted every 15–30 min to follow the sun (Hughart, 1979). Two designs are illustrated in Fig. VI.33.

Another cooking device is the solar oven—an insulated box with a transparent window on the side exposed to the sun. Collectors are arranged to reflect sunlight through the window and produce, by intensification, the energy equivalent of a fuel-fired oven (Löf, 1961).

Solar ovens require more time to prepare a meal because they do not concentrate the sun as greatly as the parabolic reflectors. However, this means they can be left unattended for a longer time, permitting meal preparation even on days of variable cloudiness and occasional haze (Hughart, 1979).

Solar cookers generally have a collector area of about 1 m^2 or less, a cooking area

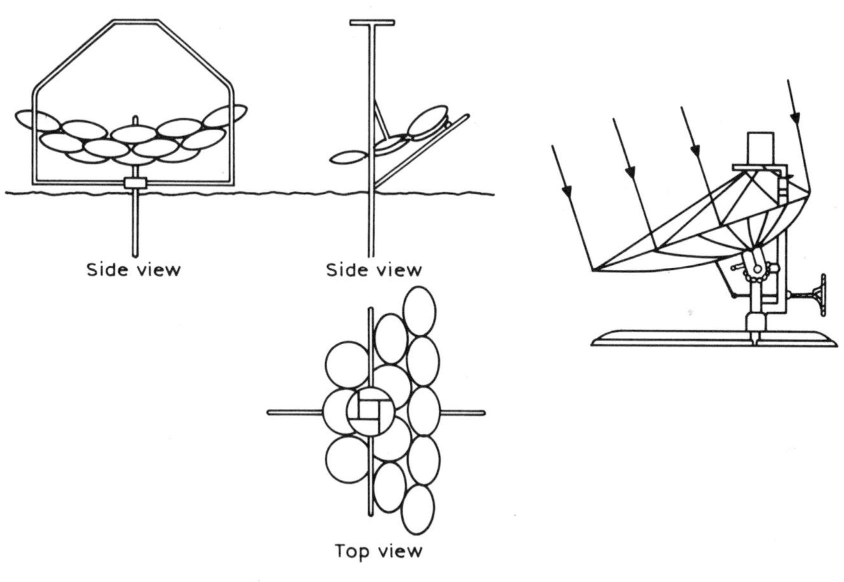

Fig. VI.33. Solar cookers. Shown are three views of a cooker (two side views and one top view) which uses 12 mirrors to collect sunlight and reflect it back onto a small area. The mirrors are arranged parabolically so they all focus on a single spot. The design on the right uses a single parabolic reflector with a focal length of 50 cm. This collector can cook a complete meal in 2 h (sources: Tabor, 1966; Sakr, 1973).

TABLE VI.5
Comparison of solar cooker designs (source: Georgia Institute of Technology, 1978)

	Developer									
	Aboc/Hussein	*Andrasst*	*Prata*	*Chai*	*Tabor*	*Stam*	*Duffie/Löf/Beck*	*Löf/Festor*	*Farber/Ingley*	*Swet*
Collector type	Oven internal flat plates	Oven external flat plates	Combination parabolic cylinder	Focusing paraboloid	Focusing paraboloid	Focusing paraboloid cylinder	Focusing paraboloid	Focusing parabolic umbrella	Hot plate parabolic cylinder	Hot plate parabolic cylinder
Reflector material	Polished aluminum sheets	Anodized aluminum sheets (coated aluminum foil)	Nickel-plated brass sheet (nickel-plated aluminum)	Anodized aluminum sheets	Copper-backed silvered-glaze plane mirrors	Aluminum foil on plastic	Aluminized 'Mylar' plastic film on polyethylene shell	Aluminized 'Mylar' laminated to fabric umbrella frame	Polished aluminum metal sheets	Aluminized 'Mylar' on lightweight formed plastic
Effective solar collection area (m^2)	0·36	0·56	0·74	0·67	0·79	n.r.	1·71	1·02	n.r.	n.r.
Effective cooking power (kW)	n.r.	0·15–0·2	0·15–0·25	0·13	0·5	n.r.	0·4–0·5 @ 0·28 ave. @ 0·32 max.	0·25–0·4	0·27–0·4	0·6
Food-cooking performance	n.r.	Good	Good	Good	Good	n.r.	Good	Good	Good	n.r.
Approximate weight (kg)	n.r.	20	12–18	n.r.	20	n.r.	10	3	n.r.	n.r.
Thermal storage	Yes	Yes	No	No	No	Yes	No	No	Yes	Yes
Total cooking capacity (kg)	n.r.	~2	~4 (2 vessels)	~2	n.r.	n.r.	~4	~2	n.r.	n.r.
Portability	Good	Good	Fair	Fair	Fair	None	Good	Excellent	None	None
Positioning during cooking	Occasional (30–60 min)	Occasional (30–60 min)	Moderate (25 min)	Frequent (15–30 min)	None	None	Frequent (15–30 min)	Frequent (15–30 min)	None	None
Suitability for baking and roasting	Good	Good	Good	Poor	Poor	Fair	Poor	Poor	Good	Good
Suitability for stewing and frying	Fair	Fair	Fair	Fair	Good	Good	Good	Good	Good	Good
Durability	Very good	Very good	Good	Good	Good	n.r.	Good	Fair	Good	Good
Use of native materials	Fair	Fair/good	Fair	Good	Good	Poor	Fair	Fair	Good	Good
Full-scale cookers—constructed and treated	Yes	Yes	Yes	Yes	Yes	No	Hundreds		No	No
Field testing	No	Some	No	Some	Yes	No	Extensive		No	No
Commercial scale	No	No	No	No	No	No	No	Hundreds	No	No
Approximate cost or price	n.r.	n.r.	~$35 (factory)	$14–18	$7–8	n.r.	$16 (factory)	$30 (factory)	n.r.	$100

n.r., not recorded.

20–30 cm in diameter, and an effective cooking power of 0·25–0·50 kW; they can cook enough food for an average family.

Table VI.5 compares various features of 10 different solar cookers for use in sub-Saharan Africa.

Table VI.6 shows the cooking time required in a solar oven for different foods. Most foods (rice, potatoes, chicken, etc.) require 45–60 min to cook. Boiling liquids would be more efficient in a solar cooker than a solar oven (Garg, 1976).

Solar cookers and ovens have never been widely adopted for a variety of technical, economic, and cultural reasons. The key problem is that cooking must

TABLE VI.6
Cooking time in a solar oven in Jodhpur, India (source: Garg, 1976)

Type of food	*Cooking time (min)*
1. Cooking	
(a) Rice (1·0 kg in water)	45
(b) Potatoes (1·0 kg in water)	50
(c) Other vegetables (1·0 kg in water)	60
(d) Tea (4 cups)	25
2. Roasting	
(a) Potatoes (1·0 kg)	60
(b) Chicken (1·5 kg)	60
(c) Fish (1·0 kg)	20
3. Baking	
(a) Cake (1·0 kg)	50
4. Boiling	
(a) Tea (4 cups)	25
(b) Milk (1·0 liters)	45
(c) Water (1·0 liters)	45

take place during midday and outdoors. This inconvenience, and the high initial cost, \$10–\$100, has outweighed the fuel savings in attempts to popularize solar cookers (Hughart, 1979).

Even so, the potential of solar cookers may be realized if researchers can use readily available materials, simplify their construction, and improve their durability. For example, in India, a 'sun basket' utilizing local materials and village technologies has been developed to harness solar energy for household use. Essentially, the sun basket is a bamboo tokra. The inside is papier mâché lined with aluminum foil. The papier mâché (made of wheat flour, methi, and shredded waste paper) and the bamboo are formed over a parabolic plaster of Paris mold. After drying, the mold is removed, the inside of the basket is smoothed, and the aluminum foil is pasted on the basket to form the parabolic mirror (Von Oppen, 1976).

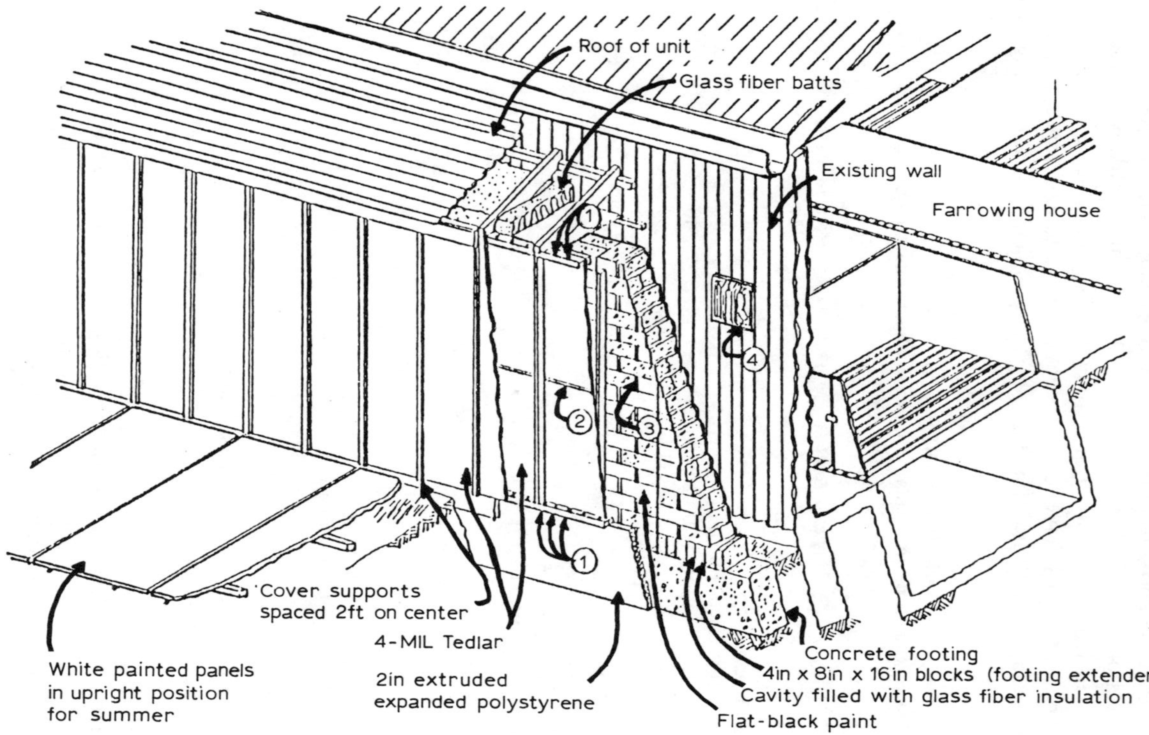

Fig. VI.34. Construction details of a solar collector attached to the side of a swine farrowing house. Ventilating air moving through the system enters at point 1; moves between the covers to point 2; travels through the vertical gaps between blocks (point 3); and to the fan (point 4), where it is moved into the environmental space. The fan would be located on the opposite building wall with an exhaust ventilating system (source: Spillman *et al.*, 1979).

Space Heating

In colder climates, confinement livestock production requires a large amount of low-temperature energy to maintain comfortable environments, especially in buildings that house young animals, which produce little metabolic heat but require a relatively high room temperature. In any animal confinement situation, some fresh air must be admitted to maintain acceptable levels of moisture, dust, and odor, and to remove any toxic gases. Often, the energy needed to heat this incoming air comprises 70% or more of the total heating demand (Spillman *et al.*, 1979).

Considerable research has been performed in the United States on the use of solar energy for livestock-house heating and poultry production (ASAE, 1981). One of the most successful projects was conducted at Kansas State University. The main component of the system is a massive, vertical, concrete wall which faces south and is painted black, through which the ventilating air moves. It is a combined solar collector storage system. There are two transparent covers over the concrete wall (Fig. VI.34).

Ventilating air enters the collector by passing between the two transparent covers before it goes through the concrete block wall. The wall is constructed without mortar in the vertical joints which are kept 3–5 mm wide. The massive concrete wall provides heat storage capacity and heat transfer surfaces. This dampens temperature extremes that can stress young animals and provides solar-heated air throughout the day. The system can be retrofitted or used with new construction, and does not require complicated controls or significant additional supervision by the operator (Spillman *et al.*, 1979; Robbins & Spillman, 1981).

The performance of the solar system for heating swine farrowing buildings was investigated from 1976 to 1980. The 400 mm-thick concrete wall provides a sensible heat capacity of 654 kJ/°C of temperature change for each square meter of collector surface. At high levels of solar insolation, the absorbing surface may reach 60°C. As the air passes through the narrow slots between blocks, much of the heat is deposited in the concrete wall. This reduces the tendency for the building to overheat during the day. Typically, 65% of the collected energy is delivered to the building during the night when demand is highest (Robbins & Spillman, 1981).

In areas where the Kansas State University solar system has been tested, savings of 40–80 liters of propane/m^2 of collector per year can be realized. At an installed cost of \$70–\$100/m^2, the solar system is competitive with propane and electricity. For example, with a fuel saving of 60 liters of propane/year and a system lifetime of 10 years, the solar system is competitive with propane at \$0·40/kg and electricity at \$0·03/kWh and above (Spillman *et al.*, 1979; Robbins & Spillman, 1981).

Another example of space heating is shown in Fig. VI.35. A pile of rocks under the collector provides heat storage. The south-sloped roof is fiberglass. When the sun warms the interior, a differential thermostat turns on a 'charging' fan which draws air down through the rocks and returns it to the space above the rocks. A second, smaller fan draws air from the rock pile to provide a low-volume warm airflow to the building, a pig nursery. The cost of this system was about \$60/m^2 of collector (Peterson, 1983). Figure VI.36 documents the performance of the lean-to collector

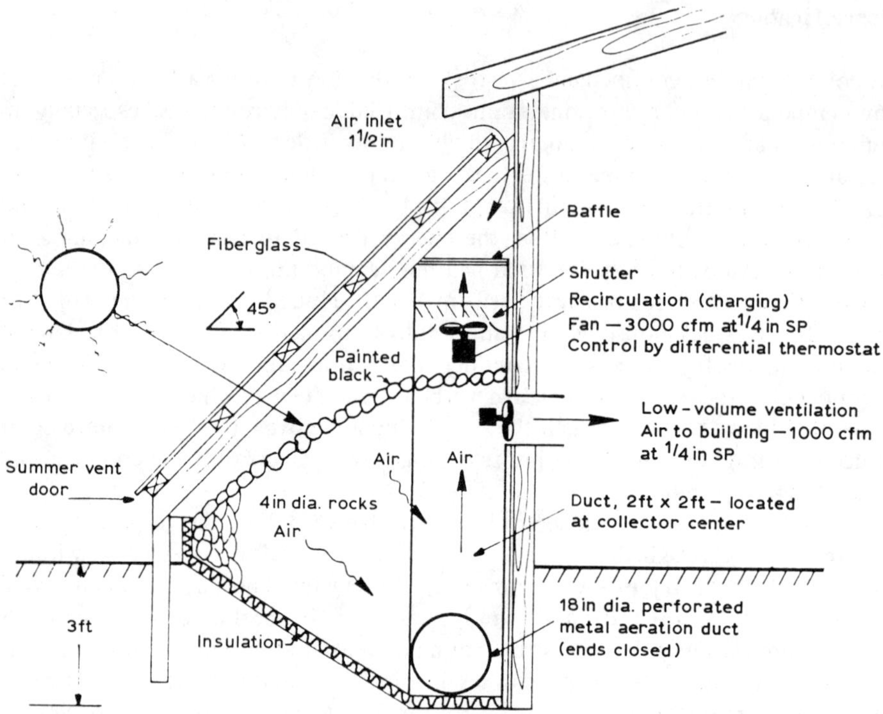

Fig. VI.35. Cross-section of a lean-to solar collector with rock heat storage for reheating ventilation air (source: Peterson, 1983). SP, static pressure; cfm, cubic feet per minute ($1\ ft^3/min = 0{\cdot}000\,47\ m^3/s$).

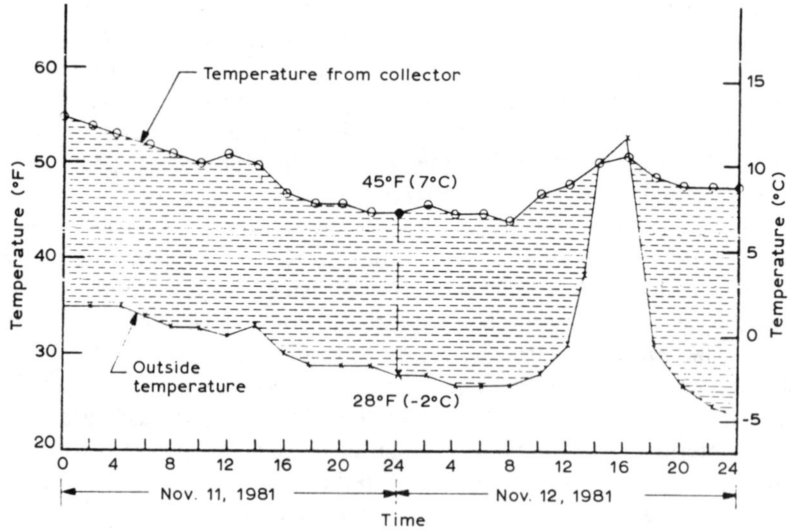

Fig. VI.36. Performance of the lean-to solar collector with rock heat storage during a 2-day period (source: Peterson, 1983).

during a 2-day period. The warm air from the collector ranged from 7 to 12°C when outside air temperatures were generally 0 to −5°C. Under the conditions of this experiment, the fuel savings resulting from use of the solar system provided a payback period of about 6 years (Peterson, 1983).

In Europe, space heating is used mainly for shelters for pigs and chickens. Although this sector is numerically one of the least important types of energy consumption, it is particularly interesting for two reasons. First, consumption in this sector is fairly evenly distributed in all European Economic Community regions. Second, conventional technologies used in these plants are quite standardized, so solar energy systems found to be valid in one region may apply in the others (DiVecchia *et al.*, 1981).

Water Heating

Water heating is one of the few commercially successful solar applications. It is commonly used for residential and commercial purposes. There are more than 2 million units in Japan alone (Kimura, 1976), and about 70 000 have been installed in Australia (UNEP, 1980). Like other solar energy devices, the units vary in efficiency, capacity, and costs, but most use flat plate collectors to capture solar heat.

One of the simplest designs widely used in rural Japan holds 180 liters and attains temperatures of 55°C in summer and 27–35°C in winter (Daniels, 1964). The heater consists of a wooden tray lined with black polyvinyl plastic and covered by a hinged glass plate (Fig. VI.37). It is 0·9 m × 2 m and holds a depth of 10 cm of water.

Another heater consists of black plastic shaped like a pillow. It is filled with water, placed in the sun, and emptied when hot water is needed. A standard size is 90 cm ×

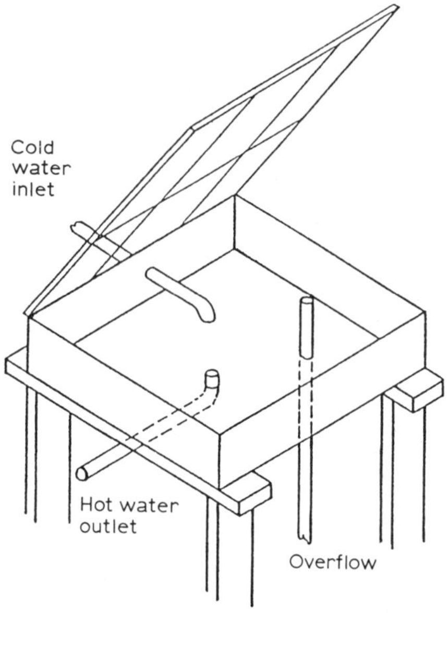

Fig. VI.37. Solar water heating tray made of wood lined with black plastic. Standardized sizes are 0·9 m × 2 m × 10 cm deep. When water is placed inside the box, the glass cover is lowered and solar radiation becomes trapped inside, causing the water to heat to about 50°C in 3 or 4 h in the summer. When hot water is required, it is simply drawn out through the outlet tube (source: Daniels, 1964).

180 cm × 12 cm and holds 190 liters of water. This heater will last about 2 years (Daniels, 1964).

Figure VI.38 shows a more complex solar water heating system than the system in Fig. VI.37. The type of storage tanks, insulation, materials, and design depends on the need and cost. The layer of water must be shallow, preferably 5–10 cm. In sunny, warm weather the water temperature rises about 50°C in 3–4 h. Efficiency ratings vary depending on design, but are good even for simple constructions. Efficiencies of 50–70% are not unusual under good climatic conditions.

The primary use of warm water in agriculture is on dairy farms. It is needed for washing the milking machine and cooling tank, washing the udders, and heating the drinking water in the winter. The water temperature required varies from 50 to 70°C for machine washing, 40°C for udder washing, and just above freezing for drinking water heating. These needs vary depending on the barn type, and on whether parlor, pipeline, bucket, or mobile installations are used for milking (Silvestrini, 1978).

Unlike most other farm operations, milking occurs at least twice a day throughout the year, and the daily demand for energy in terms of hot water remains reasonably constant. Compared to the current price of conventional fuels, however, solar water heating on the farm generally has a payback period in excess of 10 years (ADAS, 1980).

Greenhouses or Glasshouses

This application is of the greatest importance for present energy consumption in the European region. Greenhouses are concentrated in a few areas (central and southern Italy, and The Netherlands) with thermal loads higher in the north. Table VI.7 shows the amount of land under glass in several countries. Italy is second only

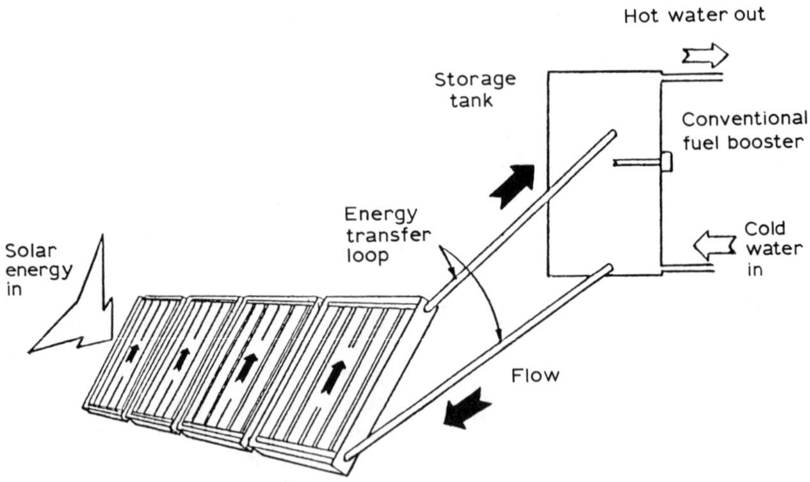

Fig. VI.38. Solar water heater using flat plate collectors to capture the sun's energy. Water is sent through the collector to pick up heat and is then transferred to a storage tank (source: Myers & Mackson, 1981).

TABLE VI.7
Area under glass (in greenhouses) in several countries (source: Silvestrini, 1978)

Country	*Land (ha)*		
	Total	*Food*	*Non-food*
Holland	7 700	5 600	2 100
Italy	5 100	3 700	1 400
Japan	?	10 000	?
USA	2 100	200	1 900
FRG	2 350	850	1 500

to Holland in glasshouse utilization in Europe. It has 5100 ha under glass, of which 3700 ha are for food production. Throughout the world, some 40 countries have greenhouses. Japan has the largest number with more than 10 000 ha in food crops alone.

Two main trends can be identified in cold climates: conservation measures and passive solar techniques. For large greenhouses, structural redesign is complicated by shading problems, so solar interventions should be limited to increasing the heat storage ability and reducing the heating demand. For small greenhouses, designs can be altered to maximize solar energy collection (Badger & Poole, 1979; DiVecchia *et al.*, 1981; NRAES, 1982).

Researchers in England at the University of Cambridge indicate that new greenhouse designs may have significant advantages in comparison to traditional greenhouses. They say that a lean-to greenhouse solar collector oriented east–west, with a sloping insulated back wall lined with reflective material, would have an estimated total transmissivity of approximately 60% more than the traditional greenhouse, and a heat loss of approximately 70% less.

Other solar greenhouse measures include the use of selective films (to transmit photosynthetically useful wavelengths), night insulation, heat storage, and active solar collectors (Silvestrini, 1978).

Night insulation is probably the single most cost-effective measure that can be implemented, because the nighttime heating needs of conventional greenhouses are significant, typically 5–10 times as great as for single family homes (Silvestrini, 1978). In most regions of the United States, enough excess solar energy does not exist during the day, even if captured and stored with perfect efficiency, to meet these heavy requirements. Also, up to a significantly high level, conservation costs less per unit of energy saved than a unit of energy from a solar collection system (Albright *et al.*, 1981). Types of night insulation include movable flaps and panels, roller curtains, inflatable curtains, beadwall, and liquid foams.

A study in New York showed that single-layer night curtains generally save between 35 and 45% of heating needs during the time they are in use. Since part of the heating needs occur during the day, overall savings typically range from 25 to 35% (Albright *et al.*, 1981).

A night curtain was developed in the United States at Cornell University consisting of four layers of material on a roller that is pulled down between tracks at night. Polyethylene tubing on the sides is inflated to keep the four layers of material separated and to seal the edges tightly into the tracks. The payoff period for this curtain is less than 3 years based on current fuel prices, an estimated installation cost of $24·22/m^2$, and an overall saving of 55% (Albright *et al.*, 1981).

FUTURE POTENTIAL OF SOLAR ENERGY

Solar energy is a vast energy source, but it is diffuse, intermittent, and subject to interruption from cloud cover. Through photosynthesis, solar energy produces plants that are vital for animal and human food.

Collection, storage, and utilization of solar heat often requires expensive structures and complex controls. Solar heat is easy to capture, but the design of efficient and affordable systems for widespread usage of solar heat in agriculture continues to be an unachieved goal. The challenge for those who seek to use solar heat in agricultural operations, such as drying and cooking, is to devise simple, inexpensive, and reliable systems. Continued research and development is expected to improve solar technologies and lead to more cost-effective solar heating applications in agriculture.

Solar energy may be utilized for purposes other than heating, e.g. photovoltaic cells that convert sunlight directly to electricity. The cost of photovoltaic systems is decreasing and this is expected to continue. In some situations, photovoltaic-generated electricity is cost-competitive with other primary energy sources.

Wind, biomass, and hydroelectric systems also depend on the sun as their primary energy source. These systems are addressed in other chapters of this book.

CHAPTER VII

Wind Energy†

INTRODUCTION

Wind energy can be used to power mechanical devices such as water pumps, to provide heat or thermal energy, and to generate electricity. Figure VII.1 illustrates a typical fan-type windmill providing water for local livestock.

† By William T. Rose, P.E., Rose Engineering, HC 72, Box 257, East Point, Kentucky 41216, USA.

Financial support for preparing this chapter was provided by the Robert S. Conly Endowment to encourage wind energy research and development.

Fig. VII.1. Fan-type windmill pumping water for livestock in northwest Kenya (source: FAO Photo 11029/I/6).

Wind energy is actually a form of solar energy. Variations in incident solar radiation in relation to time and location cause temperature differences in the atmosphere. These temperature differences lead to variations in atmospheric pressure, resulting in air movement from high to low pressure regions. Air movements which dominate large areas and are relatively constant in direction are called prevailing or planetary winds. Local topography, such as hills or bodies of water, can also cause local winds. The best wind power applications take advantage of both prevailing and local winds.

FUNDAMENTALS OF WIND POWER

Moving air molecules have mass and speed, and, therefore, kinetic energy. This energy can be extracted by the blades or rotor of a wind turbine. Wind power is related to wind speed by the following formula:

$$P = \tfrac{1}{2}\rho A V^3$$

where P is the total instantaneous power available in the wind (W), ρ is the air density 0·05 kg/m^3 at 1 atm and 20°C), A is the cross-sectional area (m^2), and V is the wind speed (m/s). Wind power depends on the air density, the cross-sectional area perpendicular to the wind, and the cubic power of the wind speed. Cubic dependence means that even small differences in wind speed can significantly impact wind power. For example, an increase in speed of only 10% increases wind power by more than 33%. Doubling the wind speed increases its power by a factor of 8. This high sensitivity to speed is an extremely important aspect of wind power.

Variations in air density, ρ, caused by altitude and temperature differences also influence wind power. Air density decreases with increasing temperature or altitude. Thus, in winds with identical speeds, a wind turbine produces more power when located at sea level, or during colder seasons, than on a mountain, or during warmer seasons.

Finally, because wind power depends directly on the cross-sectional area of the rotor, doubling the blade diameter of a rotor with a horizontal axis can produce a fourfold power increase.

WIND ENERGY CONVERSION SYSTEMS

Complete wind power systems, including a wind turbine, tower, and related equipment, such as batteries, pumps, etc., are called wind energy conversion systems (WECS). Generally, the term 'WECS' refers only to large-scale (more than 100 kW) wind–electric systems. Small wind energy conversion systems (SWECS) generally refer to wind–electric systems smaller than 100 kW, and to mechanical power-producing wind systems. (Existing mechanical or thermal wind applications are all small-scale.) However, some use the term 'SWECS' to refer specifically to wind–electric systems, and refer to water pumping systems as SWECS windmills. Finally, wind turbines designed to generate electricity are known as wind turbine generators (WTGs).

Wind Turbines

Classifications

Wind turbines are classified as either horizontal or vertical axis machines (Figs VII.2 and VII.3).

(a) (b) (c)

Fig. VII.2. Horizontal axis wind turbines (source: Surbrook & Friedle, 1978): (a) multi-bladed water pumping windmill; (b) upwind three-bladed rotor; (c) downwind four-bladed rotor.

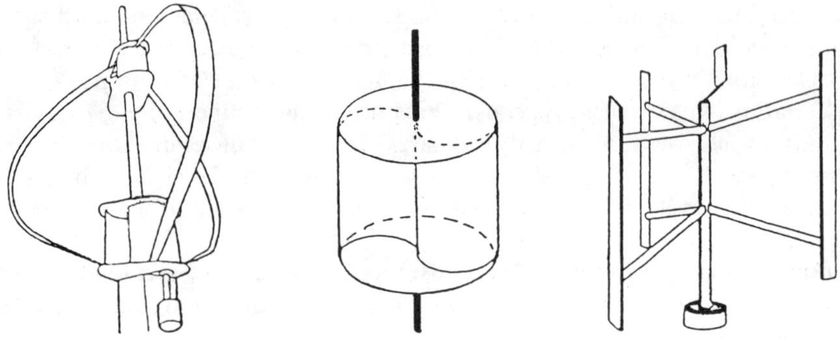

(a) (b) (c)

Fig. VII.3. Vertical axis wind turbines (source: Rose *et al.*, 1980): (a) Darrieus; (b) Savonius; (c) cycloturbine.

On horizontal axis wind turbines, the rotor shaft is located in the horizontal plane; the blades sweep an area in the vertical plane. Such wind turbines pivot freely on a tower to maintain proper orientation into the wind. Upwind or downwind designs are available. A wind turbine with a tail is an upwind design ((b) in Fig. VII.2); a downwind design has a rotor located downwind of the body, which has no tail ((c) in Fig. VII.2).

Vertical axis wind turbines have rotors that rotate around a vertical axis and accept wind from any horizontal direction. Rotors for vertical axis wind turbines are not as efficient per unit of swept area as those on horizontal axis wind turbines. As a result, the vast majority of either mechanical or electrical wind turbines are horizontal axis machines.

Aerodynamic Considerations

Important aerodynamic considerations relating to wind turbine rotors are rotor solidity, lift versus drag, and the tip-speed ratio. Rotor solidity is the ratio of the blade surface area to the total swept area. Rotor solidity influences starting torque. Applications requiring high starting torque, such as water pumping, have rotors with high solidity. The water pumping windmill depicted in (a) in Fig. VII.2 has high solidity, whereas the three-bladed rotor intended for electricity generation in (b) in Fig. VII.2 has relatively low solidity.

Lift and drag are functions of the blade surface interacting with the wind. Lift is a force perpendicular to the wind, whereas drag is a force in the same direction as the wind. The wind actually 'pushes' against the blade or cup of drag devices causing the rotor to rotate about its axis. The Savonius wind turbine depicted in (b) in Fig. VII.3 is a drag device. Lift devices use thin airfoils shaped much like an airplane wing (Figs VII.4, VII.5, and VII.6). The wind does not push against these blades, but flows over and around to create a pressure difference resulting from increased air velocity. The pressure difference causes the perpendicular lift force, which, in turn, 'pulls' the rotor about its axis. Most wind turbines in operation today are lift devices.

Fig. VII.4. Airfoil similar to an airplane wing used to create lift, which results in torque that turns a Darrieus wind turbine (photo by B. A. Stout).

The tip-speed ratio is the ratio of the speed of the blade tip to the free wind speed. If the wind blows at 10 m/s and the outer tip of the blade moves at 40 m/s around its circular path, the tip-speed ratio is 4. Wind turbine rotors can usually extract no more than 40% of the power in the wind. The theoretical upper limit, also known as the Betz efficiency, is 59%. Figure VII.7 illustrates the efficiency of various wind rotors as a function of tip-speed ratios.

Wind Turbine Performance Characteristics

Performance characteristics for wind turbines are usually described by a set of tables or curves relating wind turbine power output to wind speed. In addition, wind turbine manufacturers generally specify the rated power and the cut-in, rated, and cut-out wind speeds for their machines. The cut-in wind speed is the speed at which the wind turbine begins producing power, ranging from 2·0 to 4·0 m/s for water

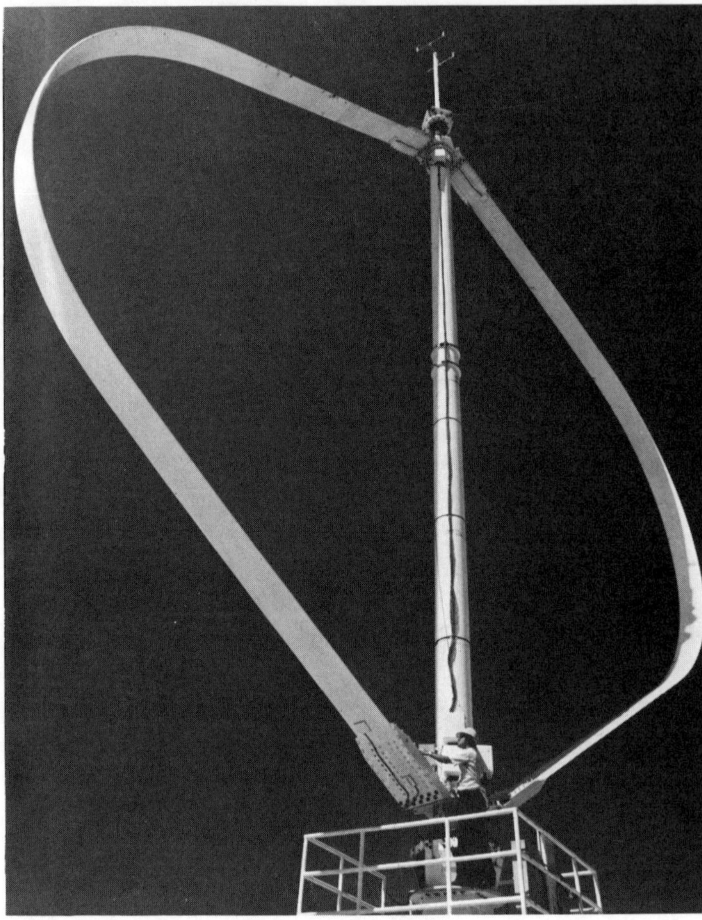

Fig. VII.5. This Darrieus rotor vertical axis wind turbine generator uses airfoil-type blades 17 m in diameter to generate 60 kW of electricity with a 13·4 m/s wind speed (source: DOE Photo 349/009/006).

pumping wind turbines, and from 2·0 to 6·5 m/s for WTGs. The wind turbine produces its rated power capacity at the rated wind speed, ranging from 7·0 to 9·0 m/s for water pumping wind turbines and from 9·0 to 16·0 m/s for WTGs. Finally, the wind turbine turns off at the cut-out wind speed to protect itself from damage caused by high winds. The rated power, along with the cut-in, rated, and cut-out wind speeds are sometimes called the wind turbine machine characteristics.

Water Pumping Wind Turbines

In addition to the machine characteristics, performance data for water pumping wind turbines, or windmills, are usually described by tables or curves listing the water pumping rate at the rated wind speed for different rotor diameters at various total pumping heads (Fig. VII.8). The total head is the sum of the static head, the

Fig. VII.6. A 200-kW airfoil downwind, horizontal axis wind turbine generator located in the United States at Clayton, New Mexico. A fan-type windmill is shown in the foreground (source: DOE Photo 349/002/003).

height the water must be lifted from the bottom of the well to the discharge pipe over the possibly elevated storage tank, and the friction head caused by friction loss in the pipes.

Wind Turbine Generators

Performance characteristics for WTGs are usually described by WTG power curves, such as in Fig. VII.9. Although Fig. VII.9 illustrates two typical WTG power output curves, the exact shape of each curve varies among different WTGs.

WTG performance, in particular, depends on all of the machine characteristics; thus, single characteristic comparisons should be avoided. At sites with low to medium yearly average wind speeds (3·0–5·0 m/s), WTGs with low cut-in and rated wind speeds produce significantly more energy than those with high cut-in and rated

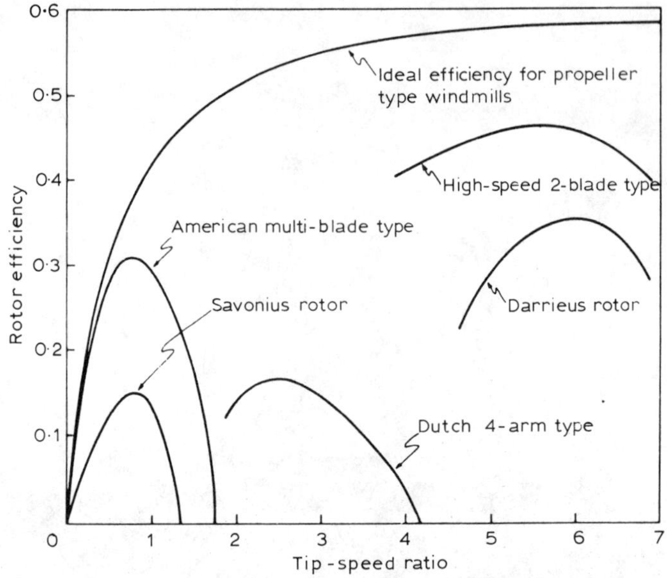

Fig. VII.7. Rotor efficiency versus tip-speed ratio (source: Garling *et al.*, 1980).

wind speeds, even though the two WTGs have identically rated powers. Unless the proposed site has a significantly high yearly average wind speed (greater than 6·0 m/s), WTGs with low cut-in and rated wind speeds usually provide superior performance.

Wind System Applications

Water Pumping Applications

Perhaps the most popular mechanical application for wind power is water pumping, which has been successfully applied for hundreds of years. Many types of systems

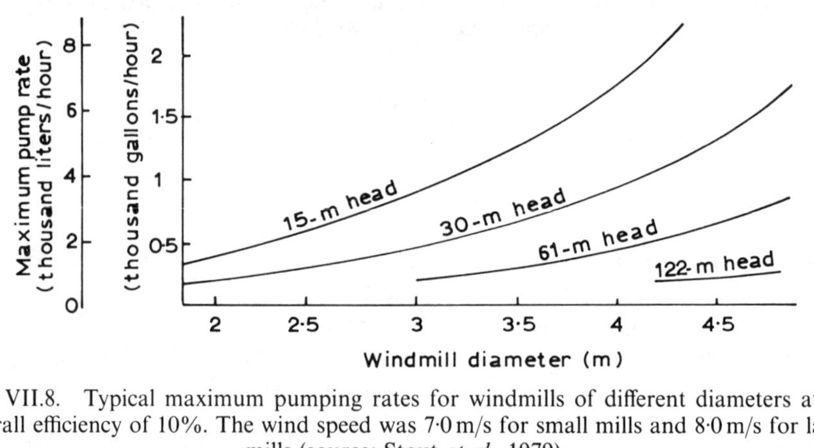

Fig. VII.8. Typical maximum pumping rates for windmills of different diameters at an overall efficiency of 10%. The wind speed was 7·0 m/s for small mills and 8·0 m/s for large mills (source: Stout *et al.*, 1979).

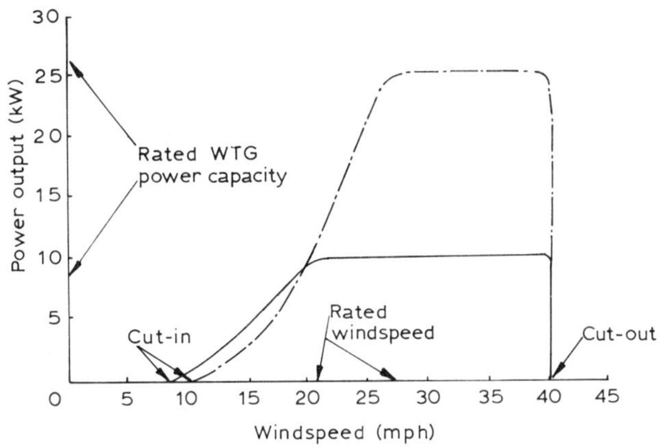

Fig. VII.9. Power curves for typical WTGs. The two curves show two WTGs with different sized generators. Notice that the WTG with the larger generator has a higher cut-in wind speed and is less efficient at lower wind speeds. The rated power capacity is achieved at the rated wind speeds (source: R. N. Clark *et al.*, 1983).

have been designed for this purpose, but the farm, or multi-bladed windmill, and the sailwing are the most common today. A multi-bladed water pumping windmill has a rotor, tower, pump, and water storage tank. A mechanical apparatus near the rotor converts the rotary motion into a reciprocating motion to drive a piston pump. A typical piston pump is illustrated in Fig. VII.10. Perhaps the most popular wind power application today, in terms of the total number of installations, is the sailwing design.

In addition to the piston pump, several other water pumps have been wind-powered. The spiral-screw water pump ((a) in Fig. VII.11) was used in Holland to drain large land areas. The bubble pump ((b) in Fig. VII.11) lifts water as compressed air bubbles expand and rise to the surface. The central water lifting pipe must be sized so that the bubbles are fully expanded before they reach the surface. Although a bubble pump can lift water to approximately twice the height that the pipe extends below the water level, it is inefficient and much useful power is wasted.

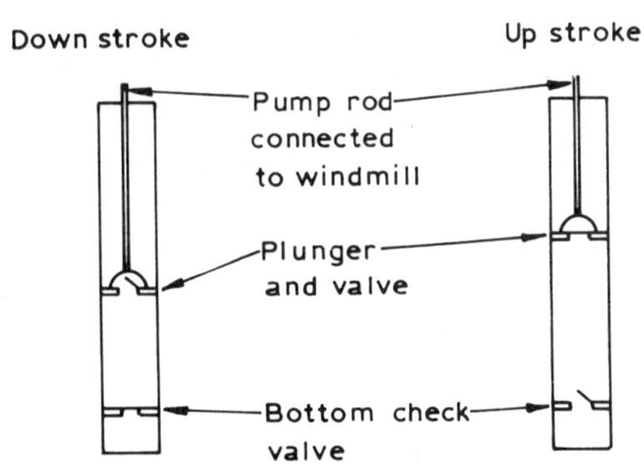

Fig. VII.10. A piston pump for a water pumping windmill (source: V. D. Hunt, 1981).

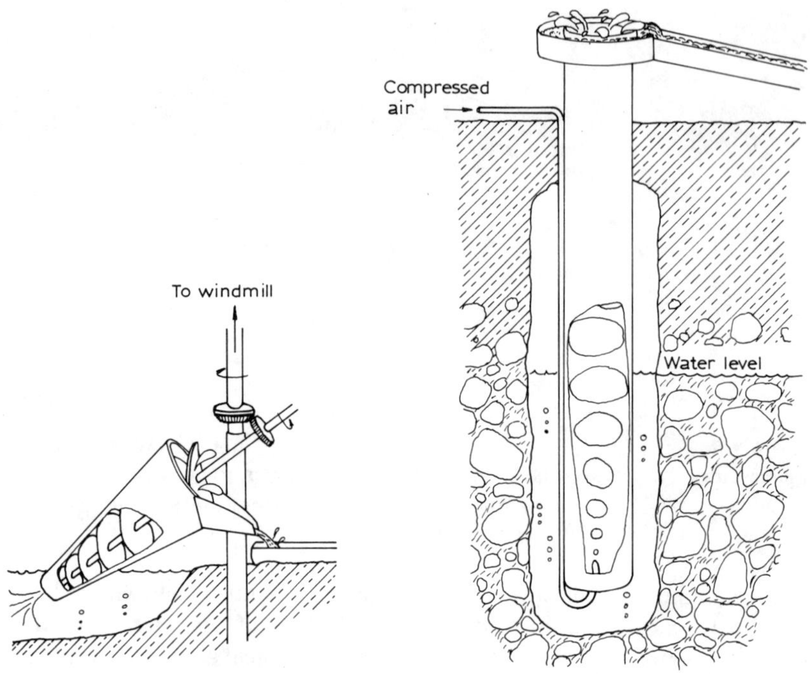

(a) (b)

Fig. VII.11. Unusual types of wind-powered water pumps (source: J. Park, 1981; reprinted by permission of Cheshire Books): (a) spiral screw; (b) bubble pump.

Several experimental water pumping applications use wind turbines to assist diesel- or electric-powered, deep-well irrigation pumps (Garling *et al.*, 1980*a*; Tennyson, 1983; R. N. Clark, 1984, 1985*a*). The wind system assists or provides partial power to the pump during periods of strong winds; the conventional diesel or electric motor provides full pump power during windless periods. The wind system can be mechanically coupled directly to the pump, or an electric generator and motor can be used.

Figure VII.12 illustrates a wind-assisted diesel irrigation pump operated in the United States at Bushland, Texas. Figure VII.13 shows how the power output of the diesel engine decreases as the wind speed increases, while maintaining constant pump power. A corresponding decrease in diesel fuel consumption is shown in Fig. VII.14.

In windy areas, wind-assist devices can provide as much as 35–40% of the total pumping energy requirements of a deep-well irrigation pump (Tennyson, 1983). Oil pumps have also been powered by wind-assist generators (Tennyson, 1983).

Other Mechanical Applications

Several wind turbine manufacturers offer models with air compressors instead of electrical generators. Energy is stored as compressed air in pressurized air tanks. The

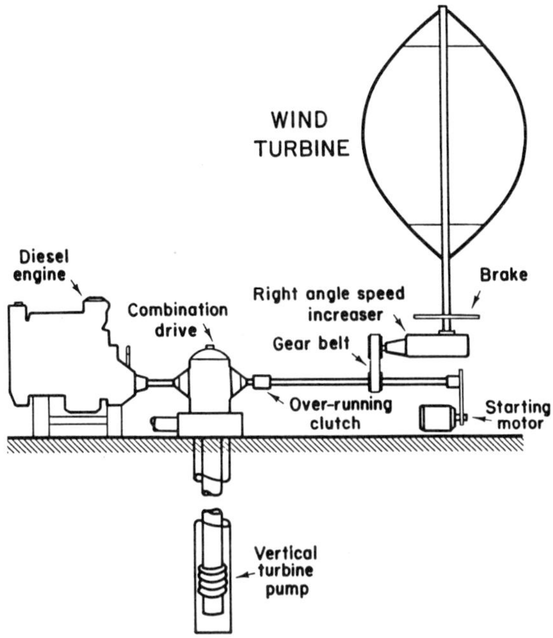

Fig. VII.12. Schematic of mechanical drive wind–diesel irrigation pumping system (source: R. N. Clark, 1984).

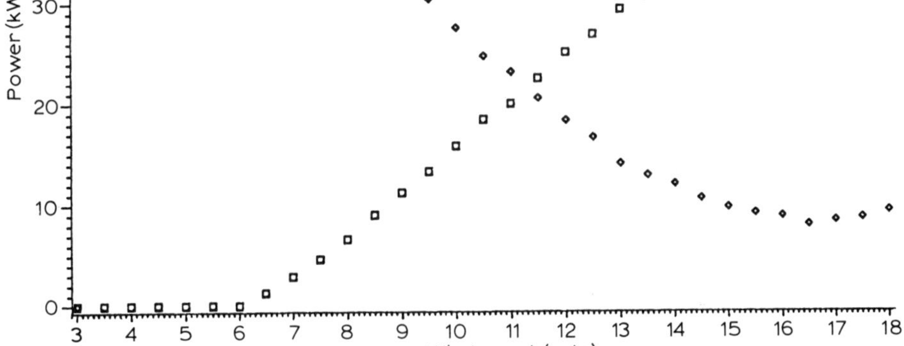

Fig. VII.13. Power output of a diesel engine and wind turbine, and power consumed by the pump as a function of wind speed (source: R. N. Clark, 1984). ◇, Diesel; □, turbine; △, pump.

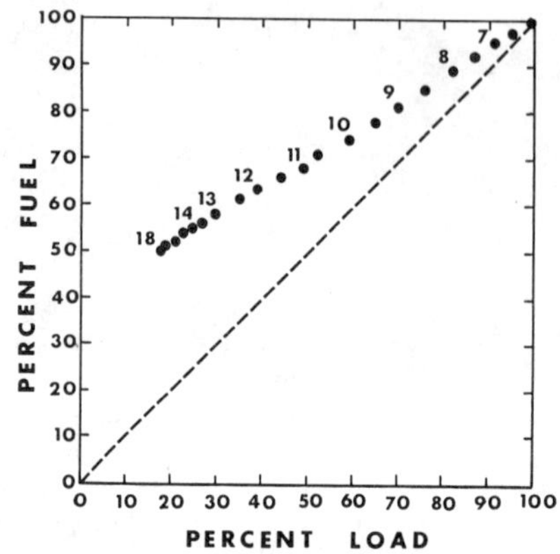

Fig. VII.14. The reduction in fuel consumption when the diesel engine is operated at reduced loads. Numbers indicate the wind speed (m/s) at which the reduction was determined (source: R. N. Clark, 1984).

technology of compressed air-powered tools is adequately developed. Small wind-powered air compressors have also been used to agitate livestock water tanks or troughs to keep water from freezing.

Thermal Applications

Wind power can be used to generate heat or to operate cooling devices. Wind-powered electrical resistance heaters can provide hot water or space heating, and wind-driven heat pumps can provide heating or cooling (Soderholm, 1983). Water can also be heated using wind-powered water churns (Gunkel *et al.*, 1981). Excess wind-generated electricity derived from isolated SWECS can be used to heat water.

Space heating may be a practical wind power application because the highest wind speeds often occur during the coldest months. The heat loss due to wind infiltration can be calculated and then offset by a wind-generated heating system (Soderholm, 1983; G. Park *et al.*, 1984). Wind power has also been used for crop drying and for refrigerating agricultural products (Garling *et al.*, 1980; Gilley, 1980; Soderholm, 1983).

Wind-driven heating devices can be less expensive than other wind systems because fewer components are required. Wind-generated electricity does not need to be modified or conditioned to power a resistance heater, and a water churn can be directly coupled to the WTG drive train. However, heat is a low-grade form of energy; the mechanical power available at a wind turbine rotor shaft can be put to more thermodynamically efficient use if it is producing electricity or directing mechanical power.

Wind–Electric Systems

There are two broad classifications of wind–electric systems: isolated and utility interconnected.

Isolated wind systems. Isolated wind–electric systems stand alone as completely independent power sources. Onsite battery banks store electrical energy for use

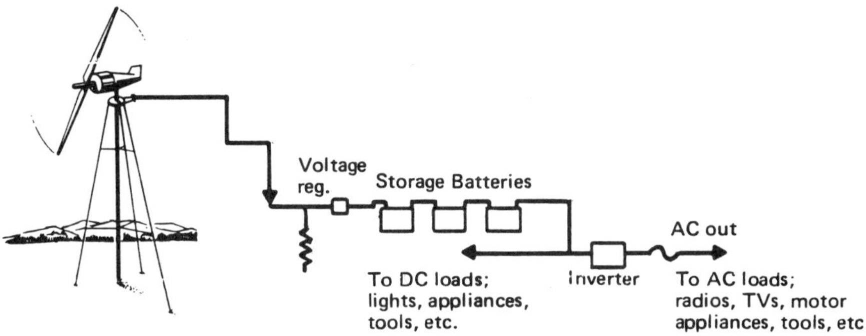

Fig. VII.15. An isolated small wind energy conversion system (source: Eldridge, 1980).

during low wind periods. Figure VII.15 depicts a schematic diagram of an isolated wind–electric system.

Most early wind–electric systems in the United States were isolated designs. Figure VII.16 presents a block diagram of four types of isolated wind–electric systems. In Fig. VII.16(c), for example, the power produced by the rotor is fed into an alternator that generates AC, which is then rectified to DC. A regulator always controls the battery charging DC voltage; energy stored in the batteries is accessed as needed. If the battery storage system is full, the electrical power is diverted to some other useful application.

Isolated systems usually have battery banks for energy storage. Stationary service batteries designed to be repeatedly deep cycled (discharged and charged) should be used. Stationary service batteries have a life expectancy of 10 years. Automobile batteries can be used, but are not designed for deep cycling and will last only 2–3 years with constant use. Stationary service batteries contain glass fiber separators and thicker plates than automobile batteries.

Batteries are rated in ampere-hours (A-h), the ability to deliver a given number of amperes for a specific number of hours. A 160-A-h battery will not deliver 160 A in 1 h; however, stationary service batteries have a maximum discharge rate usually based on 8 h. Therefore, a 160-A-h battery will provide a maximum of 20 A for 8 h. A bank of automobile batteries with a 160 A-h rating will deliver 8 A for 20 h.

To size a battery storage system, suppose the electrical demand is about 300 kWh of electricity per month or about 10 kWh/day. The battery capacity needed to store a day's supply of energy in a 12-V battery bank can be determined using the following formula:

$$A = W/V \qquad A\text{-}h = W\text{-}h/V$$

In this example, 10 000 W-h/12 V = 833 A-h.

If the batteries were rated at 120 V, only 83 A-h would be needed. If voltage is doubled for the same A-h rating, the energy storage capacity doubles, but, generally, so does the cost. For example, a 24-V battery bank can be assembled by connecting two 12-V batteries in series.

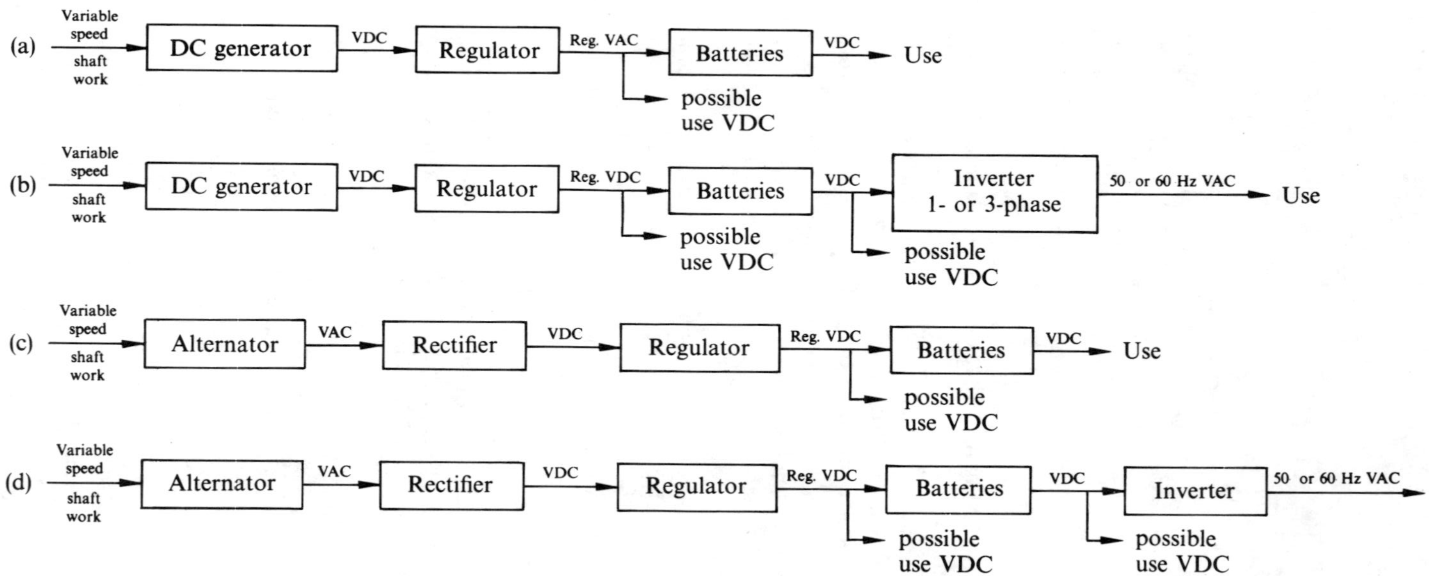

Fig. VII.16. Options for isolated small wind energy conversion systems (source: Rose *et al.*, 1980). Reg. = regulated; VDC = volts direct current; VAC = volts alternating current.

Unless the voltage is known, the A-h rating of a battery is insufficient to determine energy storage capacity. The optimum capacity of a battery bank depends on the nature of the electrical power demand, the frequency of windless periods, battery costs, and the cost of backup generating systems.

New batteries are around 80% efficient, i.e. the battery output energy is only 80% of the battery input energy. Reconditioned batteries are usually less than 70% efficient. Batteries are much less efficient in low temperatures and can freeze if undercharged.

Inverters convert DC electricity to AC electricity. Standalone inverters are used in isolated wind systems to produce an AC output that is not interconnected with any other electrical system. There are two kinds of standalone inverters: static and rotary. A rotary inverter is a combination DC motor and alternator. Regulated direct current powers a DC motor, which drives an alternator to produce AC power, subject to frequency and voltage variations ranging from 10 to 20%. Although they produce a high quality wave form (a sine wave), rotary inverters are only 60% efficient under full loads and even less efficient under small loads. The 'no-load' energy requirement is high to ensure that AC is available on demand; the DC motor must operate even when no AC load is on the alternator. Rotary inverters are relatively inexpensive.

A static inverter, made of solid-state components, is more efficient than a rotary inverter and may require a substantially less 'no-load' current when inactive. Sine wave inverters, the most effective type of static inverters, can be very expensive. Square wave inverters are less expensive, but may not be suitable for wave-sensitive electronic devices such as televisions or stereos.

Utility interconnected wind systems. Utility interconnected wind systems correspond with the utility grid, producing electricity identical in frequency and voltage to utility supplied electricity, i.e. they are synchronized with utility lines. Utility interconnected wind systems do not have onsite energy storage systems, do not constitute independent power systems, and are inoperable when the utility grid is not energized. During periods of little or no wind, the utility company supplies sufficient power to meet any onsite electrical demand that cannot be met by the wind system. Excess electricity, obtained when the wind power output is greater than the electrical demand, is dispersed back through the utility grid. The utility company may purchase this excess wind-generated electricity at special 'buy-back' rates, usually lower than the normal retail rate. Figure VII.17 depicts a block diagram of several possible configurations of utility interconnected wind systems.

Basically, two kinds of utility interconnected wind systems are in existence: (1) systems with a synchronous inverter ((a) and (b) in Fig. VII.17), and (2) systems with an induction generator ((c) in Fig. VII.17). Wind systems with synchronous inverters are sometimes called variable-speed constant-frequency (VSCF) systems. The WTG rotor speed changes with wind speed, but the synchronous inverter produces an AC electrical output of constant frequency, identical to utility current. Wind systems, with WTGs that contain induction generators, are sometimes called constant-speed constant-frequency (CSCF) systems. Because of the nature of the induction

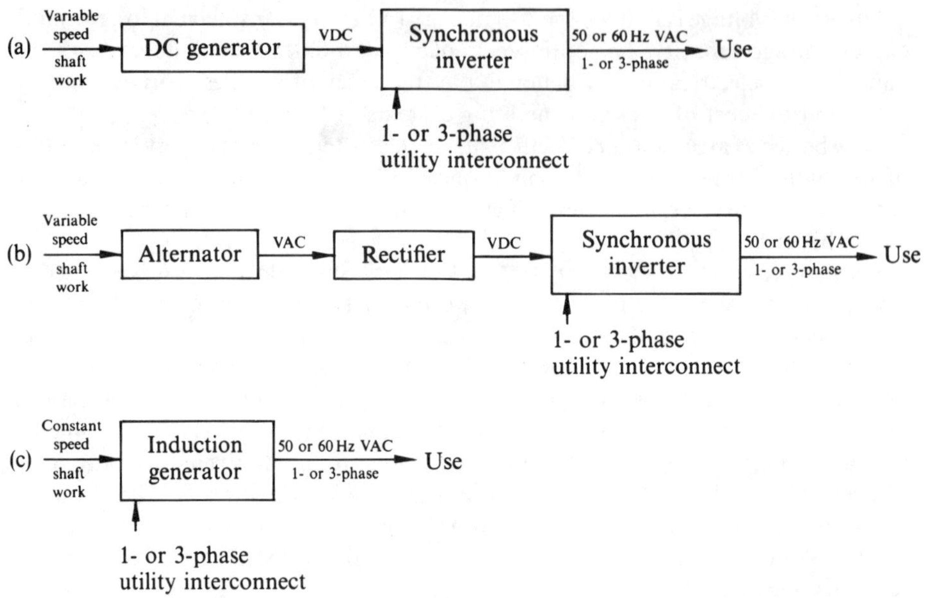

Fig. VII.17. Options for utility interconnected small wind energy conversion systems (source: Rose *et al.*, 1980). VDC = volts direct current; VAC = volts alternating current.

generator, the rotor speed can change by a maximum of 5–10%, regardless of wind speed. Rotors with constant or near constant rotational speeds cannot maintain optimum tip-speed ratios and, therefore, suffer aerodynamic efficiency losses in varying wind speeds. Thus, CSCF systems or WTGs with induction generators generally have larger rotor diameters compared to VSCF systems with similar power ratings.

The solid-state synchronous inverter converts variable voltage DC electricity into AC electricity at the same frequency as the utility line and requires 50 or 60 Hz voltage to operate. The solid-state electronics divide the DC electricity into the proper AC frequency and, under certain conditions, can produce harmonic frequencies higher than utility frequency, and thus may require additional filtering.

Wind farms. Large arrays of utility interconnected wind systems are called wind farms, the bulk of which are located in Denmark and in the United States in California. For example, by the end of 1987, more than 15 000 utility interconnected wind turbines with a combined capacity of more than 1300 MW were installed in California wind farms. The average installed cost of these wind turbines, ranging in size from 65 to 300 kW, dropped from $3100/kW in 1980 to about $1600/kW in 1986. Some of the newest wind turbines can be installed for as little as $900/kW. As a result, the cost of wind-generated electricity dropped from $0·30/kWh in 1975 to around $0·08/kWh in 1986 (ASE, 1987).

Favorable tax credits were responsible for much of this growth, but, as a result, wind technology has become more reliable. Furthermore, performance data for a

Fig. VII.18. US Windpower's Model 56-100 generating power for the Pacific Gas and Electric Company (source: US Windpower Inc.; photographer, Ed Linton).

large variety of wind systems is now available (CEC, various issues; USDOE, various issues). Tax credits for wind systems in the United States were eliminated at the end of 1985 and the rate of new installations has decreased as a result. However, worldwide interest in wind farms is increasing, particularly in Europe, India, and China. Figures VII.18 and VII.19 show examples of wind farms producing electricity for commercial use in California.

WIND CHARACTERISTICS

Some of the most important wind characteristics pertaining to wind power applications include: average wind speeds; yearly, seasonal, and daily wind speed variations; local terrain effects; height effects on wind speed; siting considerations; and using statistical models to describe wind behavior.

Average Wind Speeds

Because available wind power depends on cubic wind speed, the yearly average wind speed is probably the single most important wind characteristic pertaining to wind power applications. Wind power potential at a particular site is often based solely on

Fig. VII.19. Hundreds of SWECS in operation near the Altamont pass of California (source: US Windpower Inc.; photographer, Ed Linton).

the value of the yearly average wind speed. Estimates for the minimum yearly average wind speed needed to support a cost-effective wind power application range from 4·0 to 6·0 m/s but depend on the wind system costs, as well as the cost and availability of conventional power supplies. An accepted rule of thumb for one area may be applicable to other areas.

Yearly average wind speeds are sometimes estimated and compiled into maps for large regions. Such maps generally focus on prevailing wind speeds and are unable to completely account for winds produced by local terrain. Although the chances for a successful wind power application are better in areas with high prevailing winds, local terrain-induced winds are so important that even two nearby areas can have significantly different yearly average wind speeds. To obtain an accurate prediction of the wind power potential at a particular site, measuring wind speeds at that site for an extended period is usually necessary. The average wind speed for a particular year can differ significantly from the long-term average. Figure VII.20 compares the

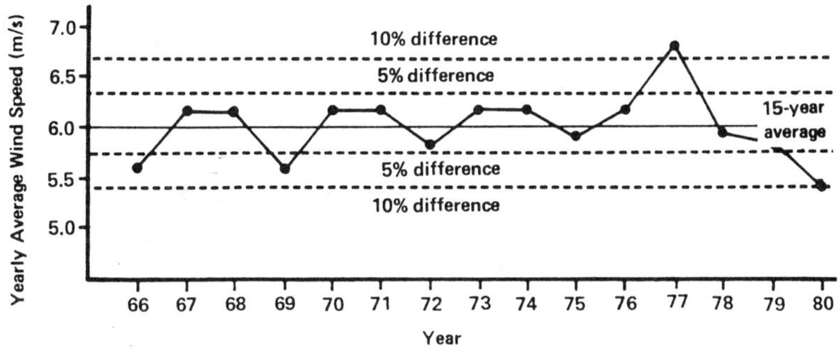

Fig. VII.20. Yearly average wind speeds for Muskegon County Airport, 1966–80 (corrected to 21·3 m using a height correction exponent of 0·16) (source: NOAA, 1980).

15-year average wind speed with the yearly average figure for each year from 1966 to 1980, at the Muskegon County Airport in Muskegon, Michigan, in the United States. During almost 1 of every 3 years, the yearly average wind speed differed from the long-term average by 5% or more (a 15% wind power change).

Seasonal wind speed variations are also important. Colder seasons may have wind speeds that are more than 50% faster than warmer seasons. Because of their cubic dependence on wind speed, wind turbines may produce up to 3·5 times as much power or energy in seasons with higher wind speeds compared to seasons with lower wind speeds. Knowing the average wind speed of the least windy month is particularly important for siting isolated wind power systems. Figure VII.21 illustrates typical seasonal wind behavior for the northern United States.

Finally, wind behavior also displays daily or diurnal speed variations. Figure VII.22 illustrates diurnal wind behavior for Muskegon, Michigan, in 1972. Generally, wind speeds are lowest in the early morning and highest in the mid-afternoon. However, diurnal wind behavior is highly variable and may be unlike Fig. VII.22 for any particular day.

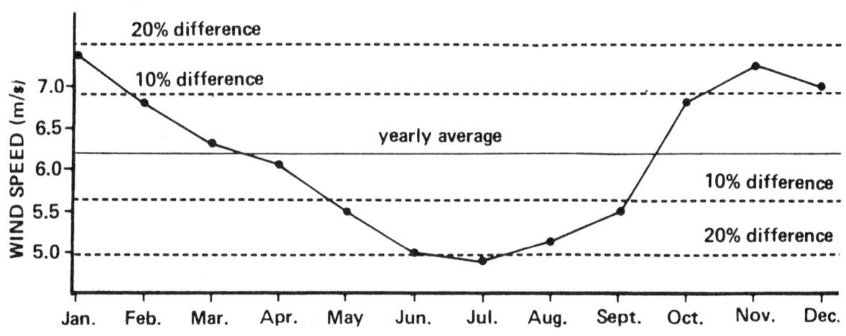

Fig. VII.21. Long-term monthly average wind speeds for Muskegon Coast Guard Station, 1965–75 (corrected to 21·3 m using a height correction exponent of 0·16) (source: NOAA, 1980).

Local Terrain Effects

Local terrain features and land topography strongly affect the suitability of a wind power application at a particular site. Wind displays a characteristic behavior as it blows over smooth open areas; over areas with uniform roughness; over cliffs, hills, and ridges; over mountain passes, saddles, and buttes; and around trees and buildings. A thorough description of the many possible terrain effects is beyond the scope of this book (Heister & Pennell, 1981; Wegley *et al.*, 1978). The several selected examples that follow, however, should illustrate the most important types of terrain-induced wind behavior.

The features of the local terrain impact wind turbulence, wind shear, and wind speed. Turbulent winds swirl or rapidly change direction. Winds that blow smoothly in uniform parallel lines (non-turbulent or laminar flow) are best suited for wind power applications. Wind shear occurs when wind speeds increase rapidly during very short height increases. If the wind blows substantially faster at the top of the wind turbine rotors than at the bottom, unwanted stress is placed on both the rotor blades and the shaft. Areas with turbulent winds or strong wind shear are not suited for wind power applications. Figure VII.23 depicts regions of increased wind speed and those exhibiting turbulence and wind shear as the wind flows over a cliff.

In Fig. VII.23(a), the undisturbed wind approaches the cliff. A wind speed profile indicates relative wind speeds, which increase as the profile curve moves away from the vertical reference line. Turbulent areas are indicated by dotted lines in Fig. VII.23(b) and (c). Strong wind shear occurs in (c). To take advantage of this land formation, the wind system should be placed in the area of enhanced wind speed in (c). The redeveloped wind speed profile continues on in (d).

The actual location of each profile depends on the surface roughness throughout

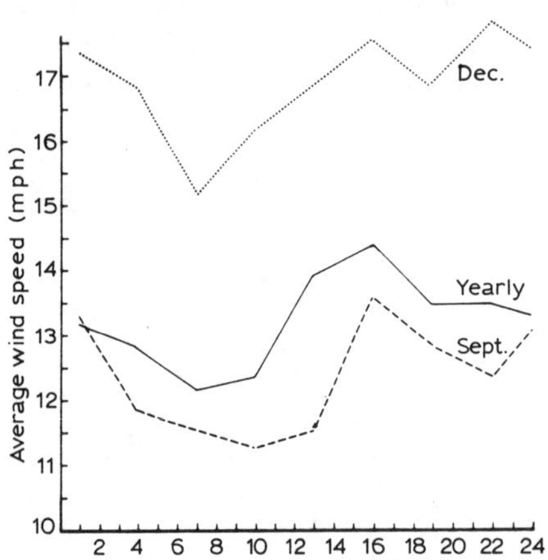

Fig. VII.22. Daily wind patterns for Muskegon, Michigan, 1972 (corrected to 21·3 m using a height correction exponent of 0·16) (source: NOAA, 1980).

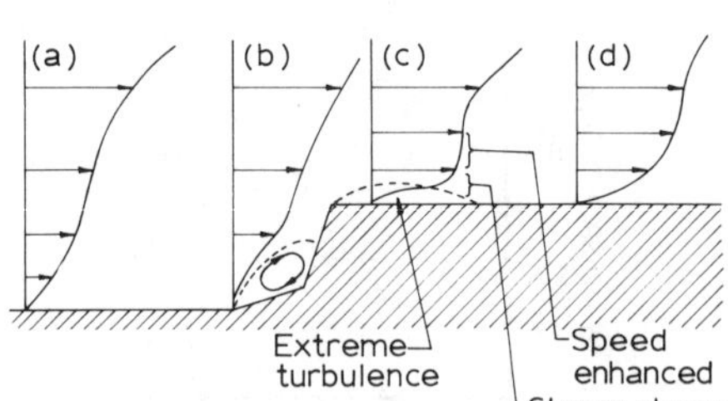

Fig. VII.23. Vertical profiles of air flowing over a cliff (source: Wegley *et al.*, 1978).

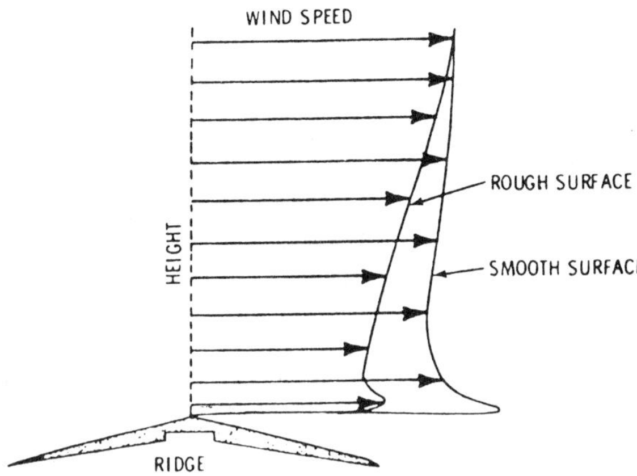

Fig. VII.24. Effects of surface roughness on wind flow over a low sharp-crested ridge (source: Wegley *et al.*, 1978).

the terrain and the height and shape of the cliff. The importance of surface roughness is illustrated in Fig. VII.24. Wind speed increases substantially over the smooth surface.

The effect of the ridge shape is illustrated in Fig. VII.25. The ridge shapes are ranked according to the amount of acceleration they produce. The triangular-shaped ridge causes the greatest acceleration, followed closely by the rounded ridge.

Wind speed increases when wind flows between two hills. Figure VII.26 presents an overhead view of this phenomenon. The narrowing windstream lines indicate increased wind speed.

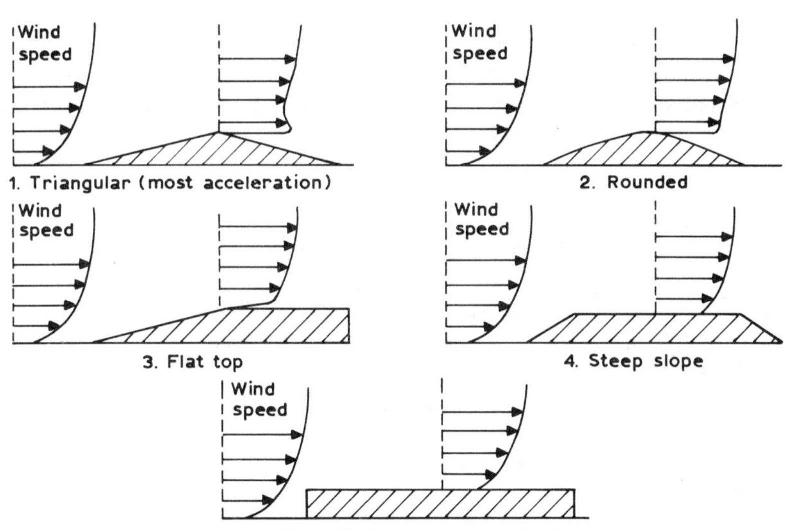

Fig. VII.25. Wind speed acceleration for a variety of ridge shapes (source: Wegley *et al.*, 1978).

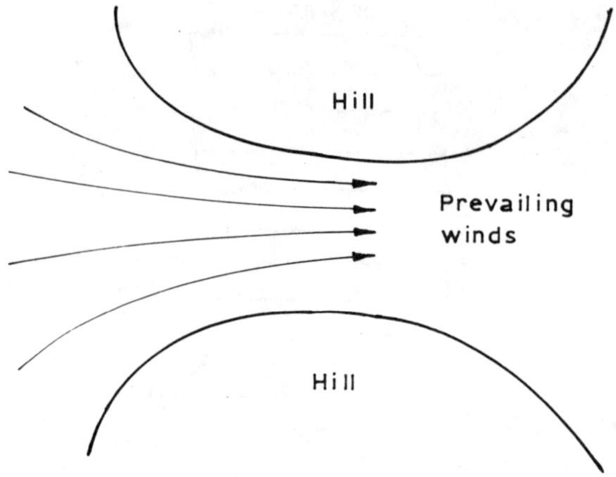

Fig. VII.26. Wind flow between two hills (source: Wegley *et al.*, 1978).

To further illustrate the effect of site-induced wind behavior, consider the wind flow lines around a block building (Fig. VII.27). The region of undisturbed air flow around this building is illustrated in Fig. VII.28. The region of disturbed air flow can be as high as twice the building height and continue on to 20 times the building height. The exact shape of the region of undisturbed air flow depends on the shape, height, and width of the building. Wind flow patterns around trees are similar (Fig. VII.29).

Wind Speed and Height

Wind speed varies with height above ground. Wind speeds high above the ground are similar throughout large regions. Near the ground, however, wind speed decreases because of surface or ground friction. The extent to which ground friction

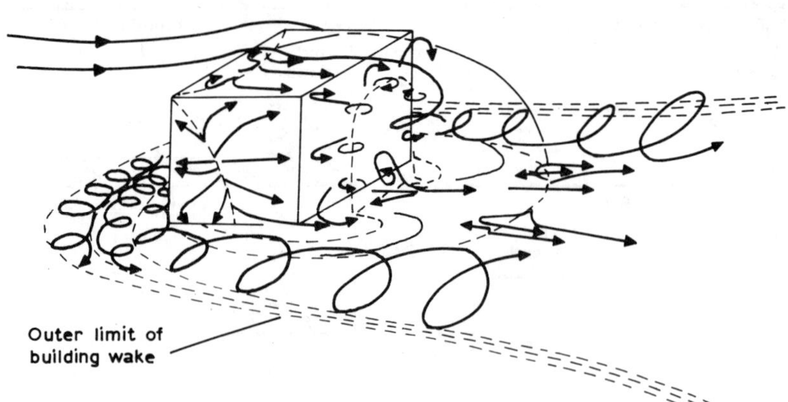

Fig. VII.27. Air flow around a block building (source: Wegley *et al.*, 1978).

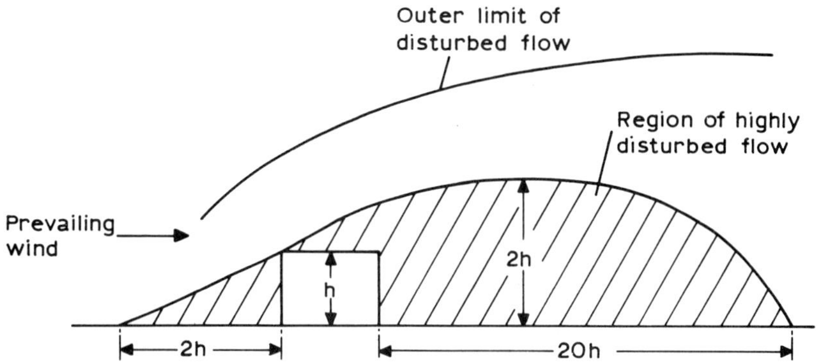

Fig. VII.28. Region of disturbed air flow over a building (source: Wegley *et al.*, 1978).

affects wind speed profile depends on surface roughness. Figure VII.30 illustrates wind speed profile curves for three different surfaces.

Wind speed profile curves can be determined for various surfaces with the following formula:

$$V_2 = V_1(h_2/h_1)^{\alpha}$$

where V_1 is the known average wind speed at a reference height, V_2 is the unknown average wind speed at some new height, h_1 is the reference height, h_2 is the new height, and α is the height correction exponent. The height correction exponent depends on surface roughness. Table VII.1 lists the height correction exponents for a variety of terrains. The values for the height correction exponents used to construct the profile curves in Fig. VII.30 are, from left to right, $\alpha = 0{\cdot}40$, $0{\cdot}28$, and $0{\cdot}16$, respectively.

As an example, suppose the average wind speed was 5·0 m/s at a height of 10 m, in a flat, open field with low grass ($\alpha = 0{\cdot}16$). The value for the average wind speed at 30 m could be estimated as

$$V_2 = 5{\cdot}0(30/10)^{0{\cdot}16}\ \text{m/s} = 6{\cdot}0\ \text{m/s}$$

Because available wind power depends on the cubic wind speed, locating a wind turbine at 30 m rather than 10 m could potentially increase power production by

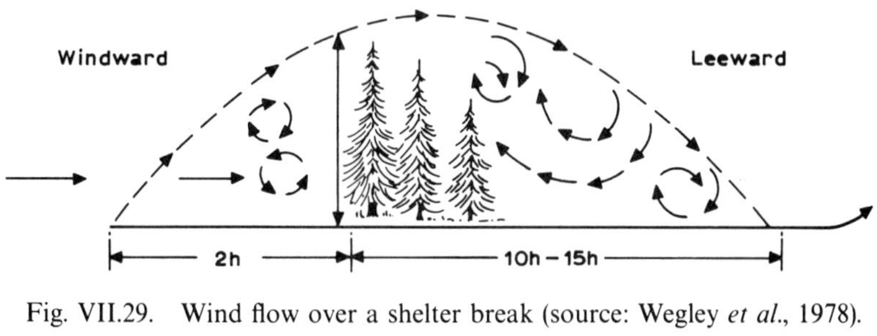

Fig. VII.29. Wind flow over a shelter break (source: Wegley *et al.*, 1978).

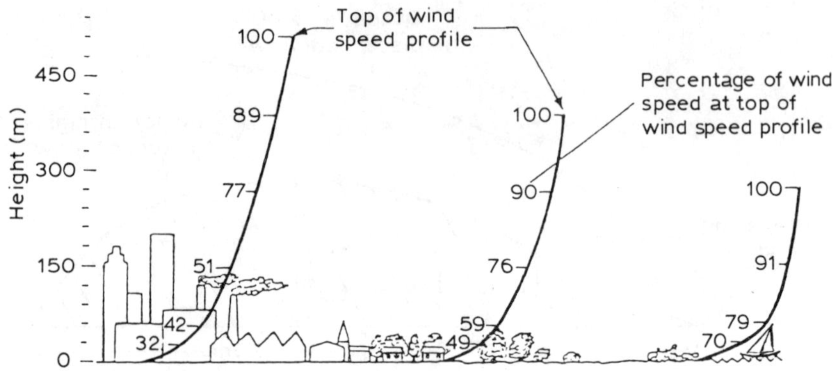

Fig. VII.30. Effects of surface roughness on wind speed profiles (source: Wegley *et al.*, 1978).

73%. Today, WTGs are successfully installed on towers as high as up to 50 m (ASE, 1988).

Because wind speed does, in fact, change with height, wind speed data should always be associated with some reference height. When estimating wind turbine performance, wind speed data should be corrected to the proposed wind turbine hub height.

Statistical Models

To accurately predict the total yearly energy a wind turbine will produce at a given site, the wind speed distribution, or how often the wind blows at a particular wind speed, must be determined. For example, the average wind speed may be 6·0 m/s, but what fraction of the time does it blow at 4·0 or 8·0 m/s?

Statistical models are often used when estimating wind speed distribution. The Weibull distribution model is preferred because it is capable of describing a wide variety of actual wind regimes. Perhaps the most popular statistical model, however, is the Rayleigh distribution model. The Rayleigh distribution model, a special case of the Weibull distribution model, depends on the average wind speed only. That is, if the site has a yearly average wind speed of 6·0 m/s, the Rayleigh distribution

TABLE VII.1
Height correction exponents for various terrains (source: Wegley *et al.*, 1978)

Roughness characteristics	*Height correction exponent, α*
Smooth surface, ocean, sand	0·14
Low grass or fallow ground	0·16
High grass or low row crops	0·18
Tall row crops or low woods	0·21
High woods with many trees	0·28
Suburbs, city	0·40

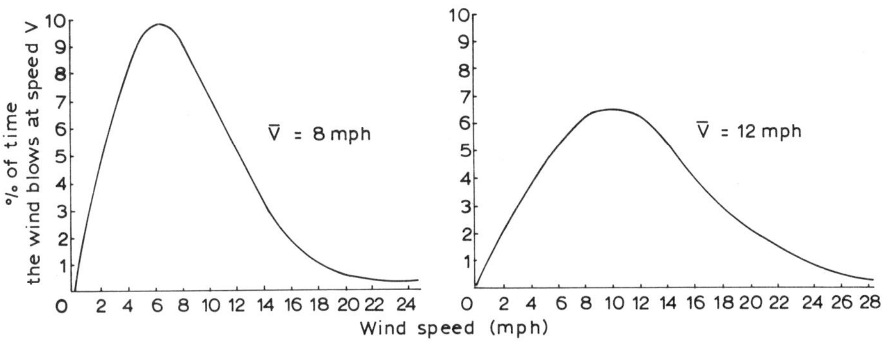

Fig. VII.31. Rayleigh distribution curves (source: Justus *et al.*, 1977).

function can estimate the frequency of occurrence of 4·0 m/s, 8·0 m/s, or other speeds. The formula for the Rayleigh distribution function, in probability density form, is as follows:

$$\rho(V) = (V/a^2)\exp(-V^2/2a^2)$$

where $\rho(V)$ is the probability of the wind blowing at any particular wind speed, V is the wind speed, exp is the exponential function, $a = \sqrt{2/\pi}\,\bar{V}$, and $\bar{V}$ is the average wind speed.

Figure VII.31 illustrates the shape of the Rayleigh distribution curve for two different yearly average wind speeds. The second curve is 'flatter' than the first. If the wind blows part of the time at higher wind speeds, it must necessarily blow less frequently at lower wind speeds.

FEASIBILITY STUDIES

To determine if a proposed wind power application at a particular site is feasible (practical and cost-effective), five site-specific tasks are usually required: (1) determine the site power and energy (or water) requirements, their magnitude, and when they occur; (2) assess the site wind behavior; (3) investigate the costs and availability of various wind systems; (4) estimate the proposed wind system performance based on site wind data; and (5) perform an economic analysis based on the expected wind system performance and costs, as well as the costs of other available energy or power sources. Depending on the proposed application and its projected costs, the feasibility study may range from a few quick calculations to an exhaustive study requiring substantial research.

Site Power and Energy Requirements

What are the site power and energy demands, and how well do they correspond with wind power availability? Can energy-consuming tasks be scheduled for windy periods? Can storage systems be used with sufficient capacity for extended windless periods? Will backup power systems be required? Wind speed and available wind power are highly variable. For example, a 5-kW wind–electric system may produce power at full capacity for 1 week and produce virtually nothing the next. The vast majority of total yearly available wind energy is contained in relatively infrequently occurring high speed winds. The most successful wind power applications capture and use wind energy when it is available, either with an adequate storage system or with sufficient onsite power demand.

Site Wind Assessment

Wind behavior between two nearby locations can change dramatically; therefore, performing some type of onsite wind data assessment is vital. Devices which measure wind speed are called anemometers. Three basic kinds of anemometers are available:

1. *Hand-held anemometers.* These instruments indicate instantaneous wind speed but do not record wind speed through time. Although potentially illustrative, hand-held anemometers have limited use for wind power data assessment.
2. *Wind-run odometers.* Wind-run odometers are cupped rotors connected to an odometer which records the km or mph of wind as it passes the instrument. For example, if 480 km of wind blew by the instrument in a 24-h period, the daily average wind speed would be 20 km/h (5·6 m/s). Wind-run odometers only record average wind speeds and are read daily or weekly, depending on the instrument. Although more frequent readings yield additional information about the actual wind speed distribution, incorporation of a statistical distribution function, such as the Rayleigh model, is usually necessary when predicting wind system energy performance.
3. *Recording anemometers.* Recording anemometers record wind speed through time either in digital form, using microprocessors, or using strip charts or tapes. Recording anemometers record the actual wind speed distribution, so relying entirely on statistical models to estimate the wind system energy performance at the site is not necessary.

For low-cost wind data assessment programs, wind-run odometers are the appropriate instruments to purchase because they are significantly less expensive than recording anemometers. Recording anemometers yield potentially more useful information (actual wind speed distribution, as well as average wind speed) and are desirable for more sophisticated site surveys.

If a weather station is located within 20 km of the proposed wind power site, and the terrain is similar, weather station data may be applicable to the site. If the monthly site data and weather station data are either extremely similar or different by some constant amount, obtaining an accurate estimate for the wind behavior on the proposed site may be possible with only several months of onsite measurement. Otherwise, onsite wind data measurement should be continued for at least a full year, and longer if possible.

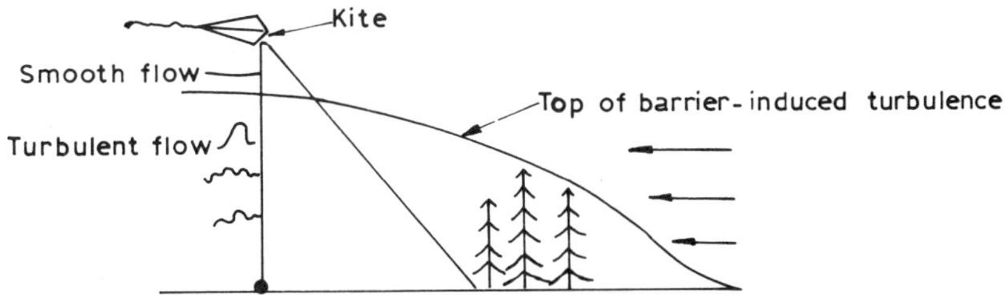

Fig. VII.32. Simple method of detecting turbulence (source: Wegley *et al.*, 1978).

When evaluating a proposed wind power site, determining where local terrain might enhance wind speed and where turbulence and wind shear might occur is necessary. Generally, the following should be considered:

—Trees, buildings, and rough terrain reduce wind energy, so wind systems are most effective when placed on smooth open ground on top of towers at least 20 m high.
—Wind energy is enhanced at a site that is on top of a gradual slope open to the prevailing wind.
—Large bodies of water enhance wind speed. The windy side of a large lake often contains excellent wind power sites.
—If trees are unavoidable, the tower height should be increased so the blades are completely in the clear—at least 10 m above the tree tops.

Areas with turbulent winds can be located by a kite equipped with streamers, as illustrated in Fig. VII.32.

Commercially Available Wind Systems

The total installed wind system costs, including towers, controls, energy or water storage systems, installation, shipping, etc., as well as estimated annual maintenance costs, must be determined. Today, a large number of commercially available wind systems are manufactured all over the world. Many have detailed performance histories, including maintenance problems and costs, resulting from a variety of testing programs (CEC, various issues; USDOE, various issues). Also, many plans are available showing how to build a wind system. Wind system manufacturers are currently restructuring because of the sudden drop in wind system demand caused by the expiration of United States tax credits in 1985. Several resources can provide more information about up-to-date, commercially available wind systems (ASE, 1987; CEC, various issues; for additional wind energy resources, see *Sources of Further Information* in Appendix 1).

Wind System Power and Energy Output Calculations

To perform an economic analysis, the yearly amount of useful energy the wind system can produce at the proposed site must be estimated. For a wind–electric system, the yearly energy output in kWh must be determined; for water pumping windmills, the total amount of water pumped should be estimated. Ideally, wind turbine manufacturers should be able to accurately predict the energy their machines can produce in a given wind regime. However, until calculation methods are standardized, manufacturers' estimates will not always be reliable (Asmussen, 1980; Rose *et al.*, 1980; G. Park *et al.*, 1984). In the absence of reliable manufacturers' estimates, a number of calculation techniques have been devised to predict wind system performance.

Water Pumping Wind Systems

A relatively simple, yet fairly accurate, technique for estimating the energy or hydraulic output for water pumping wind systems is as follows:

$$P = 0{\cdot}1AV^3$$

where P is the net power output in watts, 0·1 is the estimated overall system efficiency, A is the swept area of rotor (m^2), and V is the wind speed (m/s).

Energy equals power multiplied by time; therefore, daily energy output is

$$E = 24 \times 0{\cdot}1AV^3$$

where E is the net energy output (Wh/day) and V is the daily average wind speed.

To convert E to hydraulic energy output, E should be divided by 2·725 (Lysen).

Example

Given:

daily average wind speed $= 5{\cdot}0$ m/s

rotor swept area, $A = 3{\cdot}1\ m^2$ (rotor diameter, $d = 2$ m and $A = \pi d^2/4$)

then

$E = 24 \times 0{\cdot}1 \times 3{\cdot}1 \times (5)^3 = 930$ Wh/day.

Converting to hydraulic energy,

$E = 930/2{\cdot}725 = 340\ m^3/m$ per day.

Such a wind system would pump $34\ m^3$/day at a 10-m head, $3{\cdot}4\ m^3$/day at a 100-m head, etc.

For any isolated wind power system, the number and frequency of occurrence of windless or low-wind periods must be investigated in order to properly size adequate storage systems.

Wind–Electric Systems

A technique for calculating the energy output of wind–electric systems similar to the technique just described is as follows (Lysen):

$$E = 24 \times 0{\cdot}2AV^3$$

where E is the energy output (Wh/day), 0·2 is the estimated system efficiency, A is the swept area of rotor (m^2), and V is the daily average wind speed.

The 0·2 estimated efficiency is the amount of wind energy converted to electricity and does not take into account additional losses due to batteries or inverters. Although the preceding technique is useful for estimating daily energy output, it may underestimate yearly energy output by as much as a factor of 2 if V is assumed to be the yearly average wind speed. Correction factors, or 'cube factors', have been developed which can be multiplied by the preceding formula when V is the yearly average wind speed. The cube factors are based on the Rayleigh distribution function and the machine characteristics of a given WTG. Wolff and Meyer (1978) provide more information on this process.

Wind speed distribution functions provide a more accurate and sophisticated approach for calculating the annual energy output for wind–electric systems using calculus techniques. This method 'multiplies' the wind speed frequency distribution curve by the power output curve. In other words, the number of hours per year that each wind speed is expected to occur is multiplied by the wind turbine generator power output at the wind speed. The wind speed distribution can be the actual measured distribution or, if unavailable, the distribution based on the Weibull or Rayleigh distribution functions. Although a number of techniques are available for performing these calculations, ranging from hand-tabulated approaches to sophisticated computer programs, a thorough description is beyond the scope of this book. Rose *et al.* (1981) provides a more detailed explanation of these approaches.

Economics

The costs of wind energy systems must be compared with the total costs of other forms of available energy. A lifecycle approach should be followed, i.e. a comparison of the costs and savings of each candidate throughout the life expectancy of the systems. Lifecycle cost methods account for the time value of money, or the reality that the value of money decreases through time because of inflation or the cost of capital. Lifecycle cost analyses also incorporate certain costs that are expected to escalate, such as the price of utility supplied electricity. All future costs and savings of two competing energy systems are evaluated based on their present worth or value in today's dollars. Stout (1984) provides further discussion and examples of lifecycle cost analyses.

The relative economic merit of competing wind systems can be determined by calculating the ratio of the total installed cost (IC) of each wind system with its respective annual energy outputs (AEO), or IC/AEO. The system with the lowest IC/AEO ratio has the most economic merit. However, both the IC and the AEO must be site-specific calculations.

The feasibility of wind systems is extremely site-dependent. Wind system economic analyses must be based on the wind behavior at a selected site, the annual energy output of a particular wind system, and the costs of a specific competing energy source. Wind behavior, in particular, and the cost and availability of competing energy sources can differ dramatically even between two nearby sites. An accurate feasibility study must consider all the unique characteristics of each wind system application.

CHAPTER VIII

Hydropower

INTRODUCTION

Hydropower is derived from the natural movement or flow of water masses. This power is commonly harnessed by taking advantage of water falling from one level to another, thus exploiting the effect of gravity. A turbine converts the energy of falling water to mechanical energy. Microhydropower turbines come in many shapes and sizes—from water wheels to pumps used as turbines (in which water is forced through the pump in the opposite direction), and squirrel-cage turbines, known as crossflow turbines. Once the turbine converts water energy into mechanical energy, it, in turn, can be used to perform work or can be converted to another energy form, such as electrical energy, or hydroelectric energy. The energy-producing potential at any given hydropower site depends on the energy of the water which depends on the distance the water falls (the head—in hydraulics, each energy type is referred to as 'head'—examples include potential head due to water height, pressure head, velocity head, and friction head) and on the amount of water flow (McKinney *et al.*, 1983).

The amount of mechanical or hydroelectric energy produced at a hydropower site also relies on the efficiency at which the turbine or turbine-generator unit can convert the water energy to other energy forms. Modern microhydroelectric systems have efficiencies ranging from 40 to 75%. In other words, 40–75% of the potential energy in the water is actually converted into useful energy (McKinney *et al.*, 1983). Water is an important energy source; it provides about 23% of the world's electrical power. Hydropower produces more than 1·5 billion megawatt-hours (MWh) annually (J. J. Fritz, 1984).

FUNDAMENTALS OF HYDROPOWER

A hydropower plant uses energy from falling water to produce shaft power. A suitable quantity of water is set up to fall from a higher to a lower level. A water wheel or a turbine placed at the lower level captures the energy in the water, converting it to mechanical shaft power. The turbine rotates, providing torque to drive other equipment such as mills, grinders, or lathes, or to operate an electric generator.

Calculating Power

The power available from a hydropower plant is provided as follows:

$$\text{Power (kW)} = \rho g Q H \eta / 1000$$

where ρ is the density of water (kg/m^3), g is the acceleration of gravity (9·8 m/s^2), Q is the portion of total flow actually passing through the turbine (m^3/s), H is the height from which water falls, or head (m), and η is the plant efficiency (decimal).

Estimating Discharge

Power potential is estimated by measuring the water discharge at the proposed site and the available head. Several methods can be used to measure these quantities (Alward *et al.*, 1979). The float method provides a simple, fairly reliable way to determine the discharge. With this method the cross-sectional area of the stream and the velocity of the stream are measured. The cross-sectional area is determined by measuring the width of the stream and the water depth. Because most natural streams have irregular widths, and the water depth also varies from point to point, measuring the width and depth in several areas along the stream and using the average values is recommended. The cross-sectional area (A) is

$$A \text{ (m}^2) = wd$$

where w is the average width of stream (m) and d is the average depth of water (m).

To determine the stream velocity, a straight section of water about 10 m long should be selected. The stream in this section should be free from any obstruction. By floating a piece of wood in midstream, the time required to travel the 10-m stretch can be determined. Again, calculating the average after recording a few observation times is recommended.

To calculate the stream velocity, divide the distance the float covers by the time required (expressed in meters per second). The stream velocity, however, does not represent the velocity of water at the sides and bottom of the stream. The *actual velocity* is measured by the float method multiplied by a correction factor of 0·6–0·8, depending on the roughness of the stream bed. The flow rate of the stream (Q) is then estimated as

$$Q\ (\mathrm{m^3/s}) = \text{cross-sectional area} \times \text{velocity}$$

Estimating the Head

Several methods are available for measuring the head. A topographical area map can be used to obtain the elevation difference between the proposed intake and tailwater level. This method is useful for high-head systems. Altimeters can also be used to obtain the effective head at the proposed site. Figure VIII.1 demonstrates an easy measurement method using a carpenter's level, rod, and measuring tape.

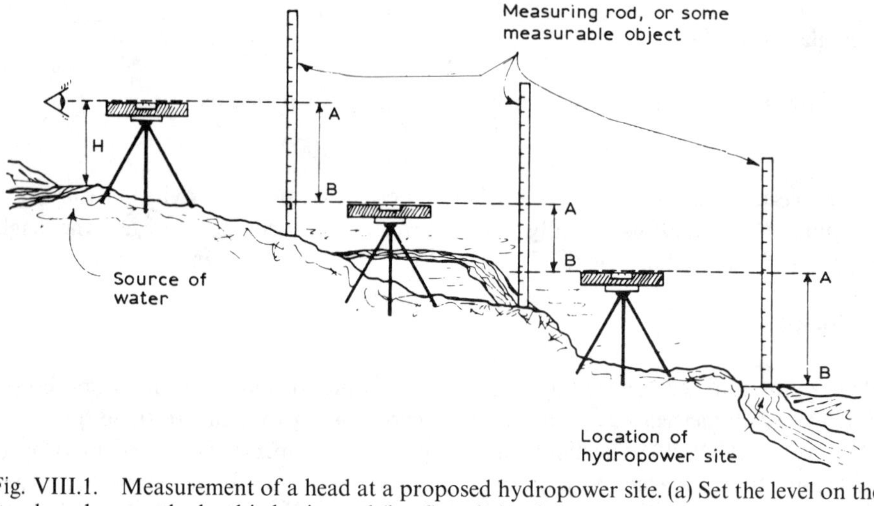

Fig. VIII.1. Measurement of a head at a proposed hydropower site. (a) Set the level on the stand; make sure the level is horizontal (level) and that its upper edge is either at the same elevation as the water source, or a known vertical distance above the water surface. (b) Sight along the upper edge of the level to a spot on a nearby object (tree, rock, or building) that is further down the hill and can be reached for measuring. (c) Note this precise spot on the object and mark it (point 'A' in the diagram). (d) Move the level and stand down the hill slope and erect it again so the upper edge of the level is below point A on the first object, as in the drawing. Mark this point 'B' and measure and record the vertical distance A to B. Now sight along the upper edge of the level in the opposite direction to another object further downhill. (e) Repeat this procedure until you end up at the same elevation as the proposed power plant site. (f) If more than one setup was required, add all the vertical distances A–B. If your first setup was above the water surface, subtract the vertical distance between the water surface and the upper edge of the level from the sum of the vertical distances. You now have the total head. (Source: Alward *et al.*, 1979.)

In planning a hydroelectric site, hydrologic, geological, and environmental factors all must be considered. The purpose of any hydrologic study is to estimate the power and energy potential at the proposed site. In addition to measuring the discharge and head, the peak flow of the stream and the flow duration curve must be estimated. The peak flow or expected maximum water level is needed to size a spillway, to position machinery above the highest expected water level, and to design diversion structures or canals.

The flow duration curve gives the statistical distribution of monthly discharge in the stream. It is needed to estimate the site's reliability for producing a given amount of power. A typical flow duration curve is shown in Fig. VIII.2. These data can be

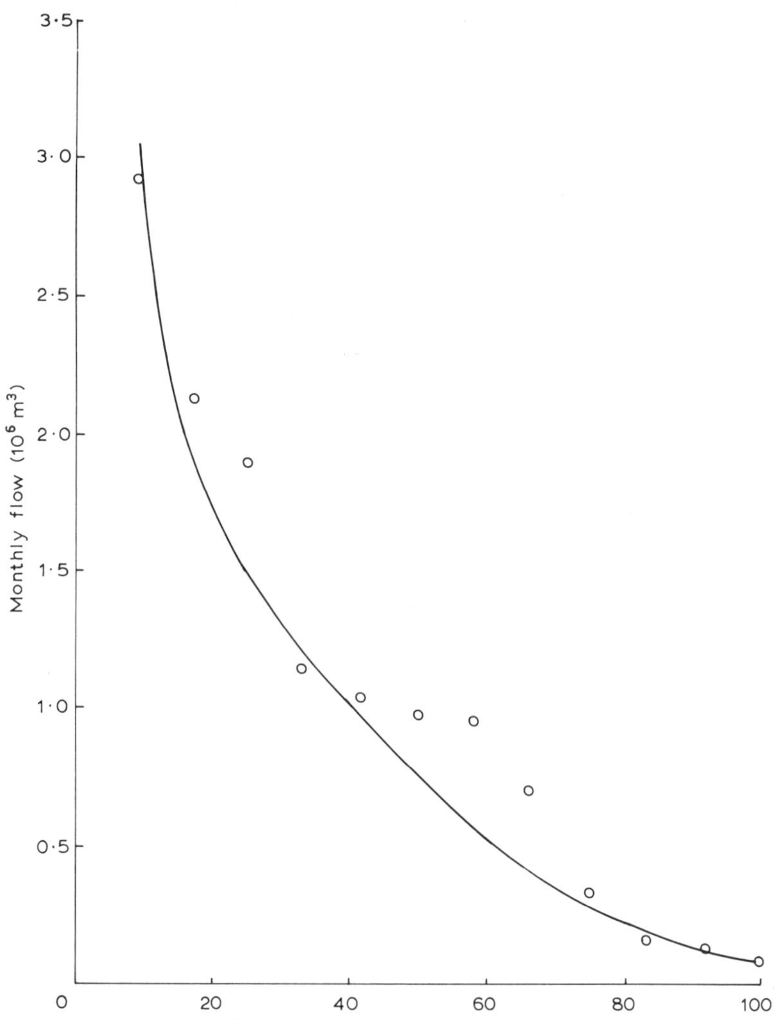

Fig. VIII.2. Flow duration curve for monthly flow volumes (source: Crawford, 1981).

used to estimate the discharge available for power production for any desired percentage of time. If the discharge rate and head are known, the power output can be calculated.

Major geological factors to be considered in siting hydroelectric projects are stream sediment loads, condition of the valley sidewalls and the foundation at dam sites, and earthquake conditions. The assessment should identify any adverse environmental impact in the planning stage when it can be mitigated by a suitable design.

Tables and charts are available showing the relationship between the various

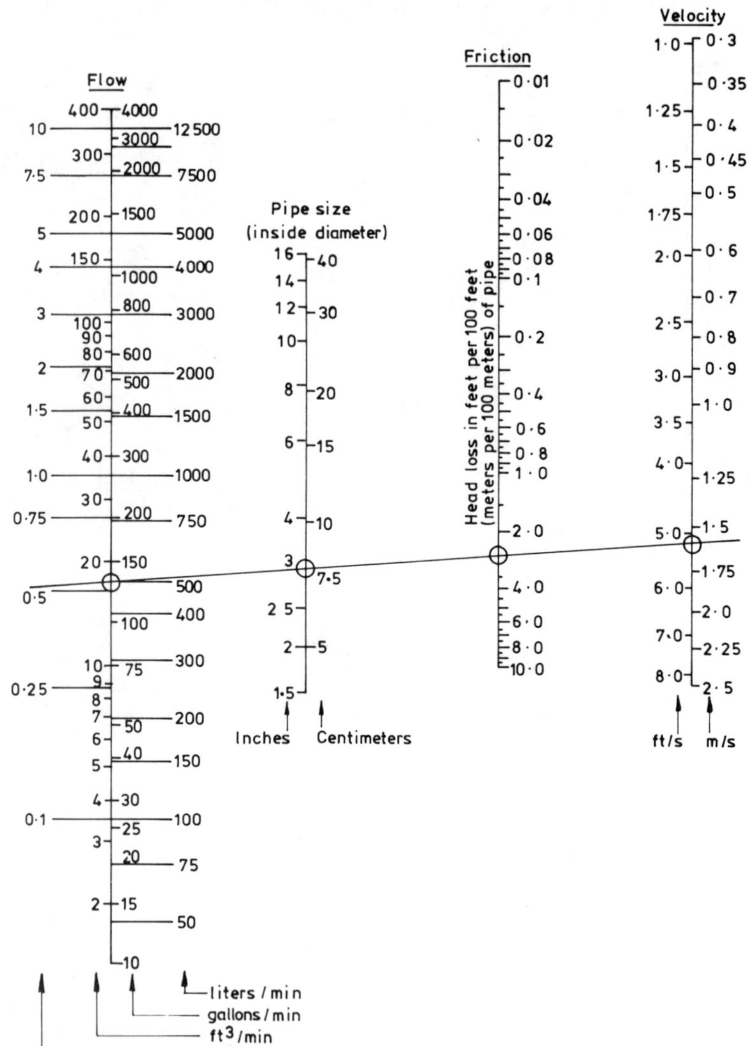

Fig. VIII.3. Hydroelectric system sizing nomograph relating flow, pipe size, frictional losses, and velocity (source: Alward *et al.*, 1979).

parameters of a hydroelectric system. Selection of penstock pipe for a given amount of flow can be made using the nomograph shown in Fig. VIII.3. Another nomograph (Fig. VIII.4) provides an estimate of the electric power in terms of the effective head and the discharge available.

The lines drawn on the nomograph illustrate a site with an effective head of 122 m and a discharge of 8·2 liters/s. If selecting a PVC pipe of 7·5 cm diameter as penstock, the head loss due to pipe friction is estimated at 0·84 m for each 30-m length of pipe. From the nomograph relating the available power in terms of effective head and flow, the power potential is 5·0 kW (Fig. VIII.4).

Since ancient times, harnessing water power has been a primary concern of

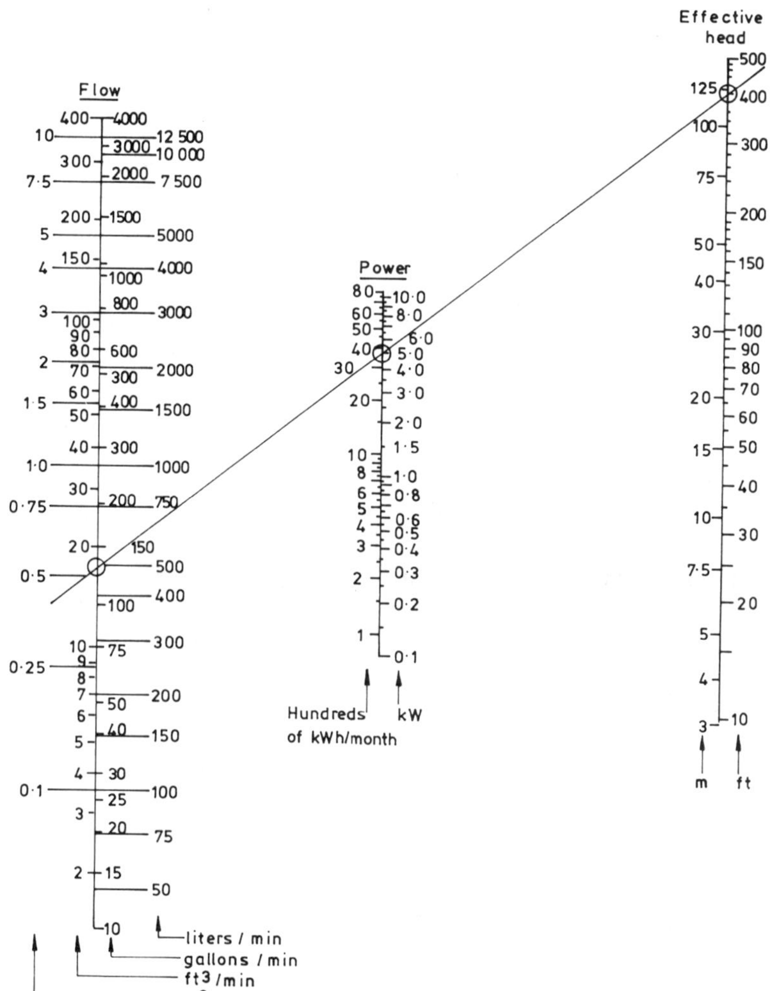

Fig. VIII.4. Hydroelectric system sizing nomograph relating flow, power available, and effective head. For larger hydroelectric power requirements, for any given head, power availability is directly proportional to water flow (source: Alward *et al.*, 1979).

human life. Water wheels were used for grinding purposes in the kingdom of the Pharaohs (2000 BC). By about AD 1000, water-powered mills had become quite common. Water wheels provide high torque with slow rotational speed. They operate with large variations in water flow and require minimal maintenance and repair.

A water-pumping network in Versailles was among the greatest engineering feats of the 17th century. Fourteen water wheels operated 221 pumps and raised water, step-by-step, through 162 m. The plant produced about 75 kW. Water wheels contributed significantly toward producing motive power and alleviating the use of human and animal power.

Electric generators require rotational speeds from 720 to 3600 rpm. Generators operating at higher speeds are smaller and cost less than those operating at lower speeds. Thus, the modern turbine is favored for electric generation. Electric generators for the conversion of mechanical power to electrical power developed from a crude machine in 1832. Around 1880, the availability of more efficient generators and incandescent filament lamps led to the rapid expansion of electric lighting. Most generators were driven by steam engines.

The world's first hydroelectric power station was built in England in 1880 to supply lighting for a private home. In 1881, the world's first hydroelectric station for public lighting of a town was also built in England. From 1880 to 1894, several other such installations were built in England, France, and the United States. In 1882, the first hydroelectric plant (125-kW capacity) was installed in the United States in Wisconsin.

By 1925, hydropower supplied nearly 40% of the world's electric power. The availability of abundant and cheap fossil fuel in the late 1930s favored large-scale centralized power stations. Thermal power plants with higher output were rapidly built and networks of transmission and distribution lines supplied electricity to cities and towns. With widespread use of fossil fuel, the share of hydropower in supplying the world's electric power dropped to 23% in 1980.

With the depletion of fossil fuels and rising fuel prices, a worldwide awareness is developing of the need to look for renewable energy sources. Hydropower, a viable alternative, offers several advantages. It uses a clean, continually renewable energy source and the technology is well developed and proven. Units ranging in capacity from a few kilowatts to several thousand kilowatts are commercially available.

Hydropower installations have a long life expectancy and can be integrated into multipurpose water resource development projects. Small-capacity hydroelectric plants can be constructed in remote areas in developing countries with a relatively simple and reliable technology. They make maximum use of local materials and labor, and, therefore, are more appropriate for developing countries.

HYDROPOWER POTENTIAL

The potential for developing hydropower in most regions of the world is vast (Tables VIII.1 and VIII.2, and Fig. VIII.5). Only 15–16% of the world's hydropower

TABLE VIII.1
World hydroelectric power generation, 1977–87 (billion net kilowatt-hours; source: USDOE, 1988)

Area and country	*1977*	*1978*	*1979*	*1980*	*1981*	*1982*	*1983*	*1984*	*1985*	*1986*	*1987*
North, Central, and South America											
Argentina	6	8	11	15	15	17	18	20	20	21	21
Brazil	94	103	115	127	130	140	150	165	177	181	185
Canada	220	234	243	251	263	255	263	283	301	308	313
Colombia	11	12	13	15	18	18	19	19	19	20	20
Mexico	19	16	18	17	25	23	21	24	26	28	28
United States	224	284	283	279	264	312	335	324	284	294	253
Venezuela	12	12	13	15	15	16	18	20	21	21	22
Other	20	23	25	28	28	34	37	39	42	44	46
Total	606	692	721	747	758	815	861	894	890	917	887
Western Europe											
Austria	25	25	28	29	31	31	30	29	31	32	32
Finland	12	10	11	10	13	13	13	13	12	12	13
France	76	69	67	69	73	71	71	67	64	64	72
Italy	53	47	48	49	45	44	44	45	44	44	42
Norway	72	80	88	83	92	92	105	105	102	96	103
Portugal	10	11	12	8	5	7	8	10	11	8	9
Spain	40	41	47	31	23	28	29	33	33	27	28
Sweden	53	57	60	59	60	55	64	67	70	60	70
Switzerland	36	33	32	34	36	37	36	31	32	34	35
West Germany	17	18	18	21	20	19	19	18	17	18	20
Yugoslavia	24	25	26	28	25	23	22	25	23	27	27
Other	19	20	24	23	27	29	25	30	28	29	35
Total	437	436	461	444	450	449	466	473	467	451	486
Eastern Europe and USSR											
Romania	9	11	11	13	13	12	10	11	12	12	12
USSR	146	168	170	182	185	173	179	201	212	214	215
Other	13	12	14	15	14	13	15	15	16	15	15
Total	168	191	195	210	212	198	204	227	240	241	242
Middle East and Africa											
Egypt	9	9	9	10	10	10	10	10	11	11	11
Zambia	9	8	9	9	10	10	10	10	10	10	10
Other	36	39	46	50	43	42	41	37	37	37	37
Total	54	56	64	69	63	62	61	57	58	58	58
Far East and Oceania											
Australia	14	15	16	17	15	14	13	13	15	15	14
China	47	44	50	58	65	74	86	86	92	99	100
India	38	47	45	46	49	48	50	53	51	53	54
Japan	76	74	84	91	90	83	87	73	87	86	86
Korea, North	17	19	20	22	23	25	26	27	28	29	30
New Zealand	14	16	15	16	19	18	19	20	20	22	22
Other	24	26	32	31	33	35	43	49	50	56	58
Total	230	241	262	281	294	297	321	321	343	360	364
World total	1 495	1 616	1 702	1 750	1 776	1 820	1 913	1 971	1 998	2 026	2 038

TABLE VIII.2
People's Republic of China water turbine generator sets, 12–500 kW (source: Kiely, 1981)

Item number	Turbine				Generator				
	Type	Model	Head (m)	Flow (m^3/s)	Capacity of generator (kW)	Speed (rpm)		Voltage (V)	
						50 Hz	60 Hz	50 Hz	60 Hz
1	Axial-flow	ZD760-LM-60	2–6	0·80–1·80	18, 30, 40, 55, 75	1 000	1 200	400	460
2	reaction	ZD760-LM-80	2–6	1·40–3·50	30, 40, 55	1 000	1 200	400	460
					75, 100, 125	750	900	400	460
3		GD560-WZ-60	8–14	0·90–2·20	40, 55, 75, 100, 125, 160, 200	750	720 (900)	400	460
4	Mixed-flow	HL210-WG-20	6–15	0·12–0·19	12, 18	1 000	1 200	400	460
5	reaction	HL310-WG-30	6–18	0·32–0·53	18, 30, 40, 55	1 000	1 200	400	460
6		HL260-WJ-35	10–25	0·50–0·80	55, 75	750	900	400	460
					100, 125	1 000	1 200	400	460
7		HL260-WJ-42	8–25	0·65–1·15	75, 100, 125, 160	750	900	400	460
					200	1 000	1 200	400	460
8		HL110-WJ-42	30–80	0·36–0·59	100, 125	750	900	400	460
					160, 200, 250, 320	1 000	1 200	400	460
9		HL240-WJ-50	15–40	1·20–2·00	125, 160, 200	600	600	400	460
					250, 320 (400)	750	720	400 (6 300)	460 (6 300)
					500	100	900	6 300	6 300
10	Mixed-flow reaction	HL220-WJ-50	25–59	1·50–1·80	320 (400)	750	720	400 (6 300)	460 (6 300)
					500	1 000	900	6 300	6 300
11		HL110-WJ-50	35–80	0·56–0·85	200, 250, 320	750	900	400	460
					400 (500)	1 000	1 200	400 (6 300)	460 (6 300)
12		HL110-WJ-60	30–70	0·70–1·10	160, 200	600	600	400	460
					250, 320 (400)	750	720	400 (6 300)	460 (6 300)
13		HL240-WJ-71	23–32	2·90	500	500	514	6 300	6 300
14	Impulse	CJ-W-55/1 × 58	100–260	0·12–0·19	75, 100, 125, 160	750	720	400	460
					200, 250, 320, 400	1 000	900	400	460
15		CJ-W-65/1 × 72	100–260	0·18–0·29	125, 160	600	600	400	460
					200, 250, 320 (400)	750	720	400 (6 300)	460 (6 300)
					500	1 000	900	6 300	6 300
16		CJ-W-92/1 × 11	138–145	0·50	500	500	514	6 300	6 300

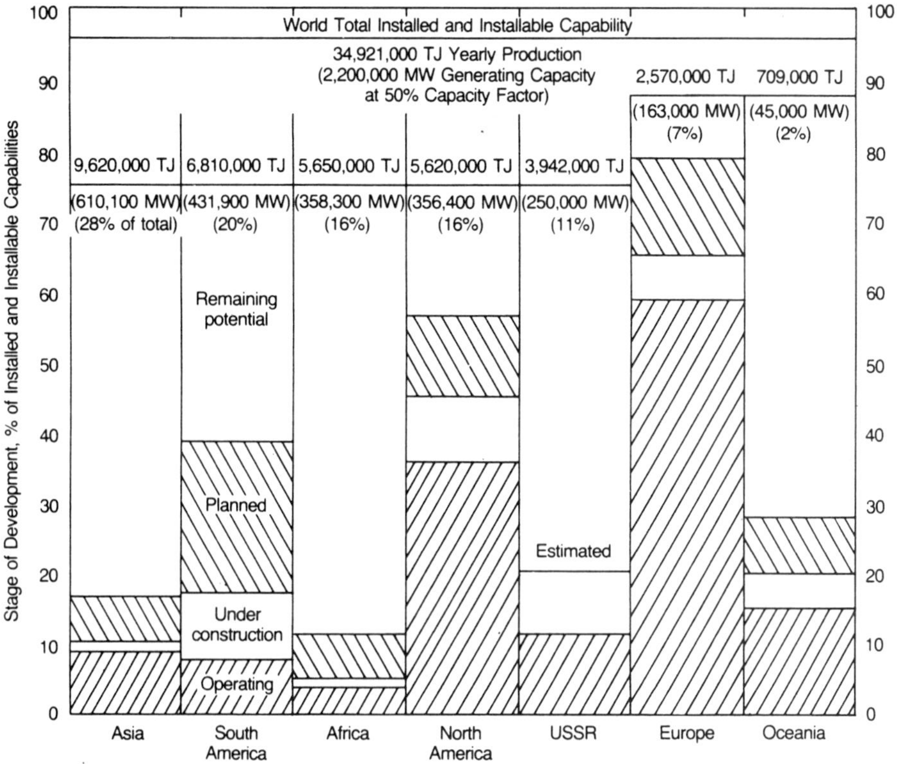

Fig. VIII.5. World's installed and installable hydropower capacity (source: J. J. Fritz, 1984).

potential has been developed thus far. Several countries are now planning to increase their hydroelectricity production. In the United States, 49 000 existing dams have been identified for possible hydroelectric power production (US Army Corps of Engineers, 1977).

The World Bank has forecast the role of hydroelectricity in the power development programs of 100 developing countries from 1982 to 1995, and provided estimates of the cost to identify and prepare hydroelectric projects for financing consideration by sources such as development finance institutions.

Tables VIII.3, VIII.4, and VIII.5 show the hydropower potential, hydropower development status, and electricity supply forecast for each country identified, as well as aggregates for each of the six World Bank geographical regions and combined regions. The database used for the background study summarized by the paper was prepared in early 1982, so the electricity supply forecasts are often somewhat overstated, at least for the short term, and may not adequately reflect the continuing economic stress in many countries. Nevertheless, the results provide a guide to the probable future role of hydropower (World Bank, 1984*d*).

The survey indicates that hydropower will continue to play a significant role in the power programs of developing countries, accounting for 43% of electricity

TABLE VIII.3
Hydro development status by region, 1982 (source: World Bank, 1984*d*)

Annual technical potential	*West Africa*	*East Africa*	*South Asia*	*East Asia*	*EMENA*[c]	*LAC*[d]	*100-Country total*
TWh[e]							
Installed	14	45[a]	60	78	76	221	494
Under construction	9	37	47	72	48	165	378
Under investigation	48	85	90	790	47	563	1 623
Remaining	301	826	526	1 752	186	1 716	5 307
Total	372	993	723	2 692	357	2 665	7 802
Percentage of 100-country total	5	13	9	34	5	34	100
1995 demand[a]	85	63[a]	381	1 019	540	916	3 004
1995 hydro production	38	43[a]	163	263	140	630	1 277
Percentage in 1995							
Hydro share	45	69	43	26	26	69	43
Potential developed	10	4[b]	23	10	39	24	16
GWh[f]							
Installed	3	9	14	26	19	51	122
Under construction	2	8	12	17	16	38	93
Under investigation	10	17	21	155	15	126	344
Remaining	70	182	124	356	50	335	1 117
Total	85	216	171	554	100	550	1 676

[a] The demand and hydro usage refer to regional and 100-country use. In East Africa, some hydro is dedicated for export to South Africa. Therefore, the total 1982 installed potential for East Africa is larger than the 1995 hydro usage to meet East African demand because of the substantial export.
[b] Five % usage if exports to South Africa are included.
[c] Europe, Middle East, and North Africa.
[d] Latin America and the Caribbean.
[e] TWh = terawatt-hour.
[f] GWh = gigawatt-hour.

TABLE VIII.4
Countries with significant hydro potential (source: World Bank, 1984*d*)

	Annual technical hydro potential	
	(TWh)	*(%)*
China	1 923	24·6
Brazil	934	12·0
Zaïre	530	6·8
Colombia	521	6·7
India	396	5·1
Peru	390	5·0
Subtotal	4 694	60·2
Remaining 94 countries	3 109	39·8
Total	7 803	100·0

TABLE VIII.5

Installed capacity by region, in gigawatts (source: World Bank, 1984*d*)

	West Africa	*East Africa*	*South Asia*	*East Asia*	*EMENA*[a]	*LAC*[b]	*100-Country total*	%
1980								
Thermal	3·0	2·0	23·1	67·7	32·6	31·3	159·7	57·5
Geothermal	—	—	—	0·4	—	0·2	0·6	0·2
Nuclear	—	—	1·0	0·6	—	0·4	2·0	0·7
Hydro	2·8	7·5	14·2	25·3	18·4	47·4	115·6	41·6
Total	5·8	9·5	38·3	94·0	51·0	79·3	277·9	100·0
1995								
Thermal	12·0	4·4	51·0	134·4	84·8	60·8	347·4	47·7
Geothermal	—	0·1	—	4·0	negligible	1·3	5·4	0·7
Nuclear	—	—	4·9	12·3	6·5	12·4	36·1	5·0
Hydro	9·7	18·0	43·8	79·4	45·3	142·9	339·1	46·6
Total	21·7	22·5	99·7	230·1	136·6	217·4	728·0	100·0
Added 1980–95								
Thermal	9·0	2·4	27·9	66·7	52·2	29·5	187·7	41·7
Geothermal	—	0·1	—	3·6	negligible	1·1	4·8	1·0
Nuclear	—	—	3·9	11·7	6·5	12·0	34·1	7·6
Hydro	6·9	10·5	29·6	54·1	26·9	95·5	223·5	49·7
Total	15·9	13·0	61·4	136·1	85·6	138·1	450·1	100·0
(%)	(3·5)	(2·9)	(13·7)	(30·2)	(19·0)	(30·7)	(100·0)	

[a] Europe, Middle East, and North Africa.
[b] Latin America and the Caribbean.

production by 1995. Preparation and engineering expenditures of about $10 billion are needed for the projects required to support this growth. By 1995, only about 16% of the technical hydropower potential in developing countries will have been developed. The study concludes, however, that an intensified hydropower program would add only 3% to the capacity otherwise planned.

The main obstacles to hydropower development are economic constraints and lack of power markets rather than lack of knowledge about resources and prospective projects. This contradicts the prevailing viewpoint that aggressive development of hydropower resources could be the key to energy problems in developing countries. Nevertheless, the study identifies specific actions many countries can take to accelerate hydropower development (World Bank, 1984*d*).

After the 1973 energy crisis, Sweden launched a program to develop small (100–1500-kW) hydropower stations along with plans for establishing 1300 mini-hydropower stations. In Latin America, decentralized hydropower development is underway to supply electricity to rural areas.

In 1980, Bolivia operated 183 hydropower stations from 20 to 500 kW. In Peru, 185 small hydropower plants provided public service in 1976, and plans are proceeding to use the country's 58 000 MW hydropower potential. Ecuador has nearly 1000 hydropower sites, each capable of generating 100 kW.

China operated more than 87 000 hydropower stations in 1980 with the rainy provinces south of the Yangtze accounting for most of the total—approximately 69 600. Kwantung alone has almost 20% of all the small hydropower plants operating throughout China. The typical installed station capacity is very small. Provincial figures result in the weighted average of about 45 kW per hydropower station.

Prefecture and county figures, collected for 10 500 stations during 1972–77, indicate the highly skewed size distribution and significant local variations: most stations are very small (less than 25 kW per station) and higher averages are rather meaningless because of the few medium-sized plants in the totals. Extreme turbogenerator sizes range from miniature devices (0·4–12 kW capacity) produced by the Tientsin Electro-Driving Research Institute, to tap spring and creek waters, to units in excess of 1000 kW.

The total generating capacity of small- and medium-sized hydropower plants rose sharply from 500 MW in 1969 to almost 3000 MW in 1975. As expected, the load factor of small stations is considerably lower than the average for large installations: available figures translate into the mean time of about 2700 h/kW per year (a load factor of 31%) in Kwantung, and 2000 h/kW per year in Chekiang and Honan (a load factor of merely 22·8%), with some county averages as low as 1250 h. Nevertheless, the small stations accounted for one-third of total hydropower generation—or at least 10 billion kWh—in 1975, contributing considerably to the rudimentary electrification of the Chinese countryside.

Besides providing power for small local industries, as well as irrigation and drainage, food and fodder processing, and timber sawing, small hydropower reservoirs regulate water supplies, help prevent floods, and breed fish and other aquatic commodities. This multipurpose nature of small hydropower projects seems to be evidence of their advantage and assurance of their continuous development (Smil & Knowland, 1980).

Several other Asian countries are developing potential mini- and micro-hydropower for rural areas. The Philippines, Pakistan, Thailand, Indonesia, Nepal, and Sri Lanka already have some similar schemes in operation.

Interest in hydropower is increasing in African countries as well. In Morocco and Zaïre, small hydropower plants are in the planning stages. In the future, they will probably be a primary source of electricity in many rural areas of the world.

CLASSIFICATION OF HYDROPOWER PLANTS

Hydropower plants are generally classified as either micro, mini, small, or large. No distinctly defined range is identified for these terms, but the following values are commonly used:

micro = less than 100 kW
mini = 100 to 1000 kW
small = less than 30 MW
large = greater than 30 MW

Until recently, only large hydroelectric schemes were considered feasible by planners in developing countries. However, Third World countries are discovering more applications for small hydropower stations today. These systems have special interest features which attract developing countries. Whereas large projects often rely on imported equipment and expertise, small projects generally utilize local materials and skills. Simple turbine and generator designs can be fabricated in existing industries and workshops.

Because of the extensive network of transmission and distribution lines required, supplying electricity to rural areas is costly. Most of the electricity demand in rural areas is domestic, so the economic base to pay for electricity from the grid may not be available. Small, decentralized power systems built with local input can supply electricity at lower costs.

Small hydropower projects also permit direct mechanical drives for agro-processing, small industries, and workshop equipment. Direct mechanical drives can be installed in the powerhouse and operate directly from the turbine, thus avoiding costly electric motors.

These plants can be developed for use with irrigation systems without heavy civil construction works. Furthermore, they can be geared to the capital, technical, and managerial resources of the local people in developing countries.

REASONS TO CONSIDER SMALL HYDROPOWER STATIONS

Small hydropower stations have several characteristics that make them attractive (J. J. Fritz, 1984):

—Hydropower is a non-consumptive electrical energy generator, utilizing a renewable resource that is continually available through the hydrologic cycle activated by the sun.

—Hydropower is essentially non-polluting and releases no heat. Adverse environmental impacts are negligible and, for small installations, may be non-existent.

—The design and construction time for hydropower plants is 1–2 years. Licensing requirements are minimal, equipment is readily available, and construction procedures are well known.

—Some type of water control is required for hydropower plants, up to and including full regulation of watershed discharge. Water control is an important element in the multipurpose utilization of water resources and can reduce potential flood damage. When storage facilities are involved, floodwaters are retained and more efficiently directed to agricultural production, river regulation, improved navigation, fish and wildlife protection, recreation, municipal use, and improved wastewater control.

—Within the hydrologic limitations of the site, hydropower is a reliable resource. The relative simplicity of hydraulic machinery makes energy instantly available as needed. Since no heat is involved, hydropower equipment is durable and malfunctions are rare.

—Using relatively simple technology, hydropower can be a catalyst in mobilizing productive resources and creating enhanced economic opportunities for local residents in remote areas.

—Hydropower plants are characterized by reliability and flexibility of operations, including fast startup and shutdown in response to rapid changes in demand. Thus, hydropower plants become a valuable part of any large electrical system, increasing overall economy, efficiency, and reliability.

—Hydropower has an excellent peak-power capability. The alternative for handling peak loads in a large electrical system may be the utilization or costly expansion of old and relatively inefficient thermal units.

—The technology for hydropower is well developed and thoroughly tested, with turbine efficiencies running as high as 90%. While the equipment must be adapted to the specific site for greatest efficiency, its performance will generally comply with the manufacturer's claims.

—Hydropower facilities have a long life expectancy. As a rule, dams and control works perform for a century or more with minimal maintenance.

—Few operating personnel are required for hydropower stations. Some small-scale installations are operated entirely by remote control. Freedom from fuel dependence together with the longevity of equipment make hydroelectric power installations resistant to inflation.

—Small hydropower plants can fully utilize local materials and labor. They usually provide more local employment for civil works construction projects than thermal facilities.

—Hydropower resources remain untapped, especially in developing countries, where less than 7% of the actual potential has been developed. In some countries, the figure is less than 2%.

—With more realistic methodologies for economic evaluation, the economic feasibility of hydropower plants is improving compared to other energy sources that use finite fuels. Advantages such as full recognition of the value of non-consumptive water use, freedom from fuel dependence, and minimal environmental impact have made small hydropower stations increasingly desirable.

—The potential for hydropower in industrialized countries can be developed to expand hydropower capacity at existing powerhouses and dams. The possibility of retrofits and additional turbines and generators makes upgrading present installations attractive.

PLANNING A HYDROPOWER PROJECT

Several types of studies should precede the actual construction of a hydropower project:

—A reconnaissance study (also called appraisal or pre-feasibility study) to determine if the project under investigation merits a full feasibility analysis (Table VIII.6).

TABLE VIII.6

Elements of a reconnaissance study in a developing country (source: J. J. Fritz, 1984)

I. Site-development rationale
  - Economic profile of region
  - Energy demand

II. Site description
  - Community description
  - Physical site characteristics

III. Power and energy potential
  - Hydraulic resource
  - Power determination
  - Annual energy production and profile

IV. Physical configuration
  - Turbogenerating equipment
  - Electrical equipment
  - Distribution system
  - Civil works

V. Project costs
  - Capital costs
  - Annualized operation and maintenance costs
  - Annual investment cost

VI. Project benefits
  - Revenue streams
  - Indirect benefits

VII. Economic feasibility
  - Cost/benefit analysis or rate-of-return

VIII. Financial feasibility
  - Funding mechanism
  - Covenants of loan agreement
  - Cash flow analysis

IX. Environmental impact
  - During construction
  - During operation

X. Social analysis
  - Management and operation
  - Local participation opportunities
  - General community impact
  - Institutional arrangements

—A feasibility study to detail the proposed project and assess its investment potential (Table VIII.7).

—Definitive or final site studies performed after making the decision to proceed with the project, including permit applications, licensing negotiations, financing arrangements, marketing agreements, and engineering designs and specifications (J. J. Fritz, 1984).

TABLE VIII.7

Elements of a feasibility study in a developing country (source: J. J. Fritz, 1984)

I. Introduction
  - Minihydro in national energy strategy
  - Rationale for development of site
  - Sociocultural profile of community
  - Economic/energy profile of community

II. Site description
  - Physical features
  - Community energy-use profile
  - Existing diesel generation

III. Power and energy production
  - Description of the hydrological resource
  - Power capacity and annual energy production
  - Alternative plant schemes

IV. Physical configuration
  - General arrangement of power systems
  - Distribution systems
  - Powerhouse details
  - Details of civil works
  - Electromechanical equipment specifications
  - Safety equipment
  - Excavation plan

V. Project costs
  - Engineering and design costs
  - Capital cost breakdown
  - Operation and maintenance costs

VI. Project benefits
  - Tariff structure
  - Direct revenue streams
  - Increased community productivity
  - Total community benefit

VII. Economic feasibility
  - Cost/benefit analysis, or
  - Lifecycle cost analysis of alternative

VIII. Financial feasibility
  - Loan or grant covenants
  - Schedule of payments
  - Financial arrangements

IX. Institutional analysis
  - System operation and maintenance
  - Field or service arrangements
  - Tariff collection and administration

X. Environmental impact
  - Description of flora and fauna
  - Impact during construction
  - Impact during operation

XI. Schedules
  - Construction
  - Cash flow analysis

XII. Summary recommendations

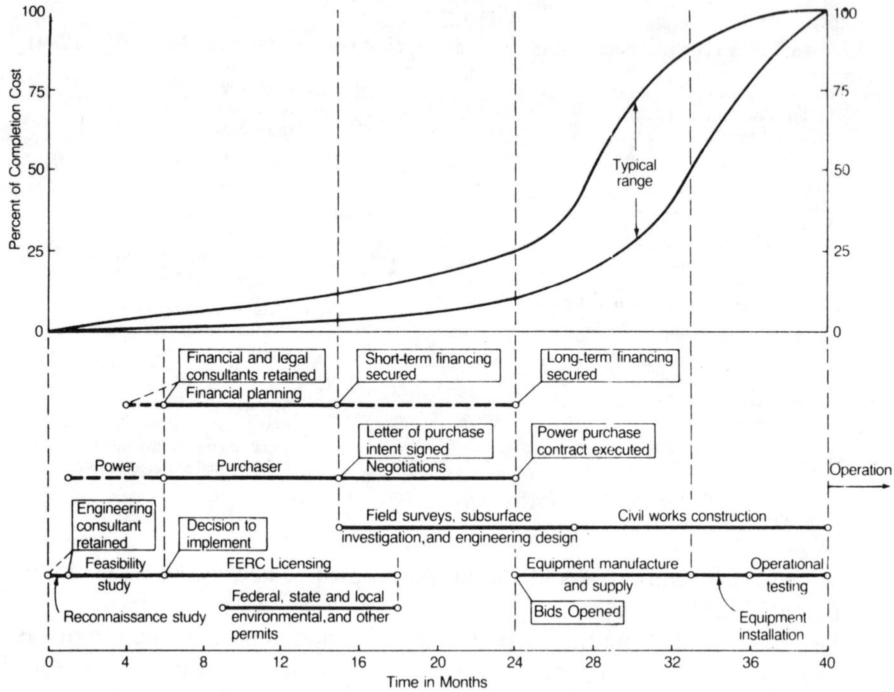

Fig. VIII.6. Typical project implementation schedule and expenditure patterns (source: J. J. Fritz, 1984).

Figure VIII.6 illustrates a typical project implementation schedule and indicates the relative amount of time devoted to each of the activities relating to the studies described.

THE HYDROELECTRIC POWER SYSTEM

A hydroelectric power system broadly consists of: (1) water supply; (2) turbogenerator; and (3) electricity distribution. The water supply system depends on the method by which the head is developed. In large hydroelectric plants, a dam is usually constructed across the river and water is conveyed to the turbine through a penstock (Fig. VIII.7). Water can be stored upstream from the dam site and released when required. This is of special benefit, particularly for a multipurpose irrigation-and-power system. Such large systems, however, require a detailed engineering study of the dam structure, often adding to the overall cost.

Several hydraulic devices are shown in Fig. VIII.8. A water wheel already in use for centuries is a low-speed, high-torque machine. Modern turbines are high-speed, low-torque machines, designed especially for electric power generation.

Dam construction, exclusively for electric power generation, is generally cost-prohibitive. Figures VIII.9–VIII.12 illustrate methods of using existing dams,

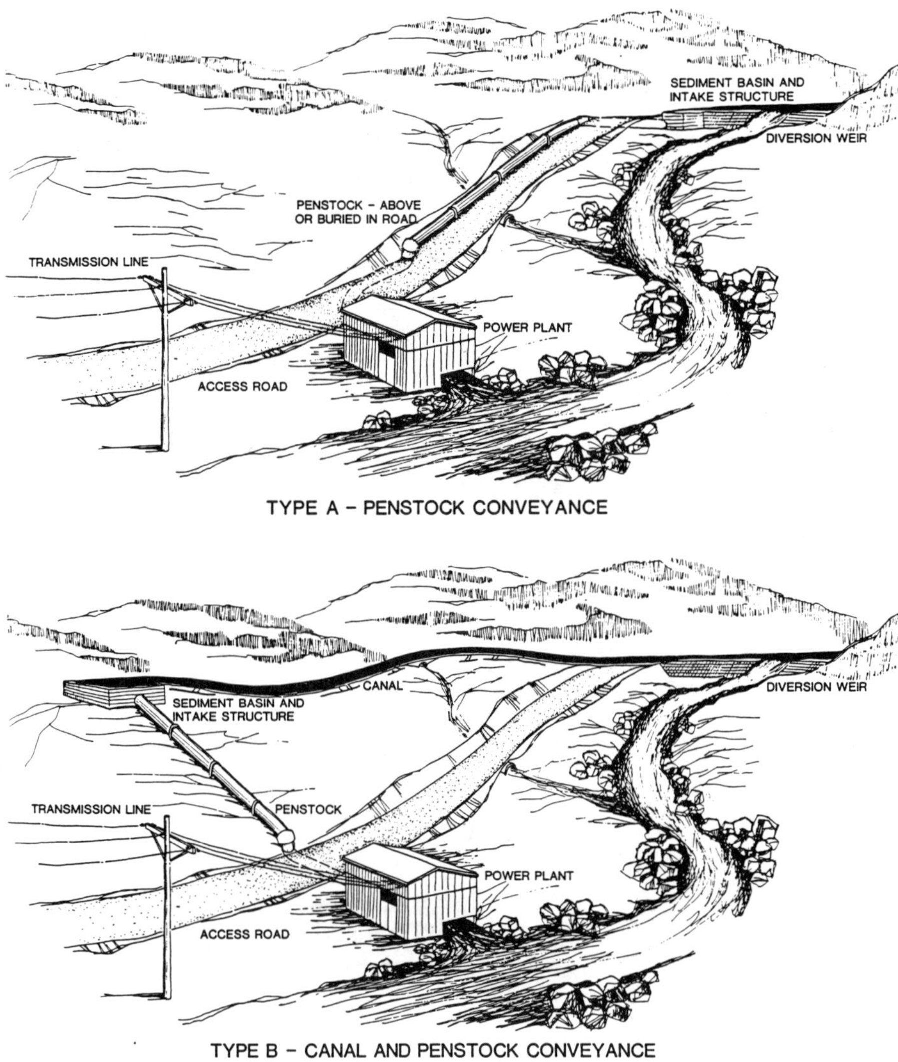

Fig. VIII.7. Typical new sites for hydro projects (source: World Bank, 1984*e*).

canals, or wastewater discharge lines for electric power generation (McKinney *et al.*, 1983).

Several types of dams can be constructed: arched, buttressed, gravity-filled, rock-filled, earthen, and timber cribs (Fig. VIII.13). Each design requires special foundation and sidewall considerations. The most common structure for small-capacity, low-head installations is the earthen or rock-filled dam (Fig. VIII.14). A spillway is needed to carry the flood water. Usually, this is provided on one abutment and the powerhouse near the opposite abutment. In some designs, however, spillways and turbogenerators are housed in the same structure.

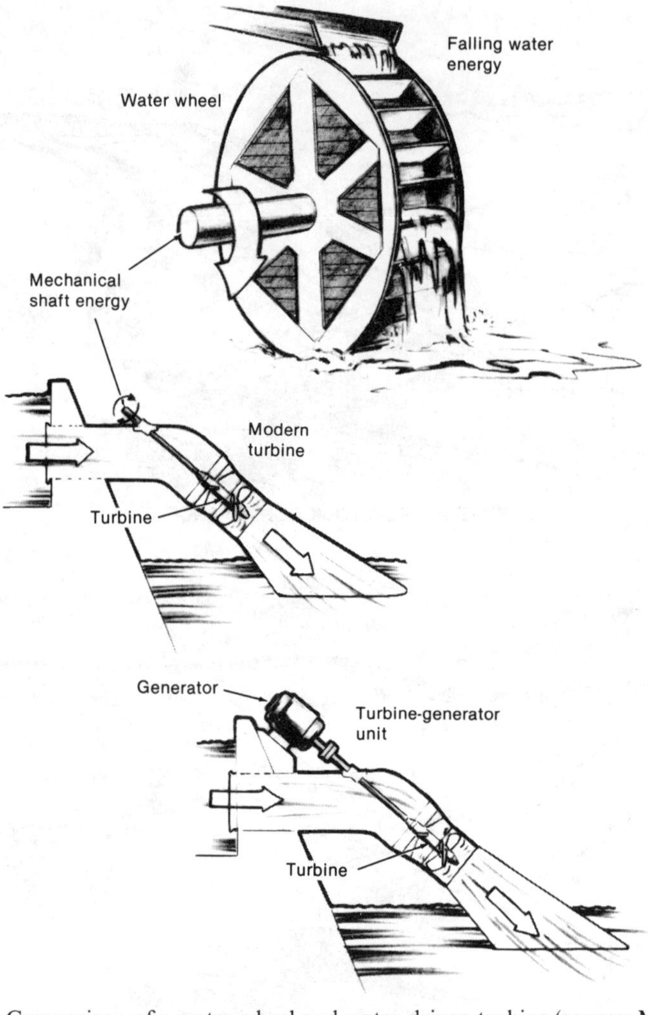

Fig. VIII.8. Comparison of a water wheel and water-driven turbine (source: McKinney *et al.*, 1983).

Gates are also needed to isolate the turbine for inspection and maintenance. Trash racks installed upstream protect the turbine unit from falling debris. To gain maximum effective head, penstocks are designed for low friction loss. In this type of low-head facility, flood water flowing over the spillway may increase tailwater levels downstream, reducing the effective head on the turbine. Simple low-head installations ranging from 2 to 20 m are illustrated in Fig. VIII.15.

Water Supply

A microhydropower system can be developed from either a natural source or a man-made structure. Natural sources include a stream without a dam, a waterfall, a

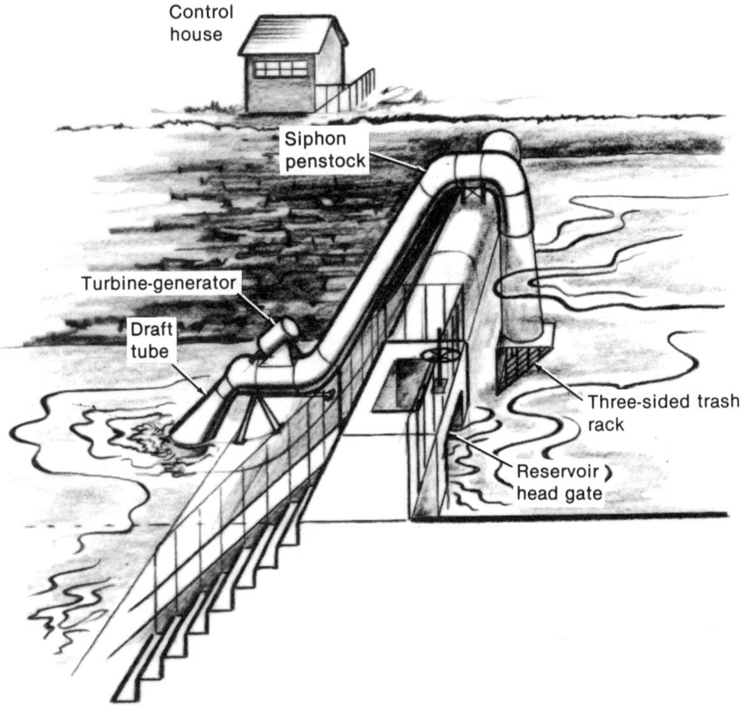

Fig. VIII.9. Siphon penstock at an existing dam (source: McKinney *et al.*, 1983).

Fig. VIII.10. Siphon penstock installation at Wasdell Falls in Ontario, Canada. The rate is $5{\cdot}7\ m^3/s$, the head is $3{\cdot}6$ m, and the generator capacity is 145 kW (source: EG&G Idaho, Inc.).

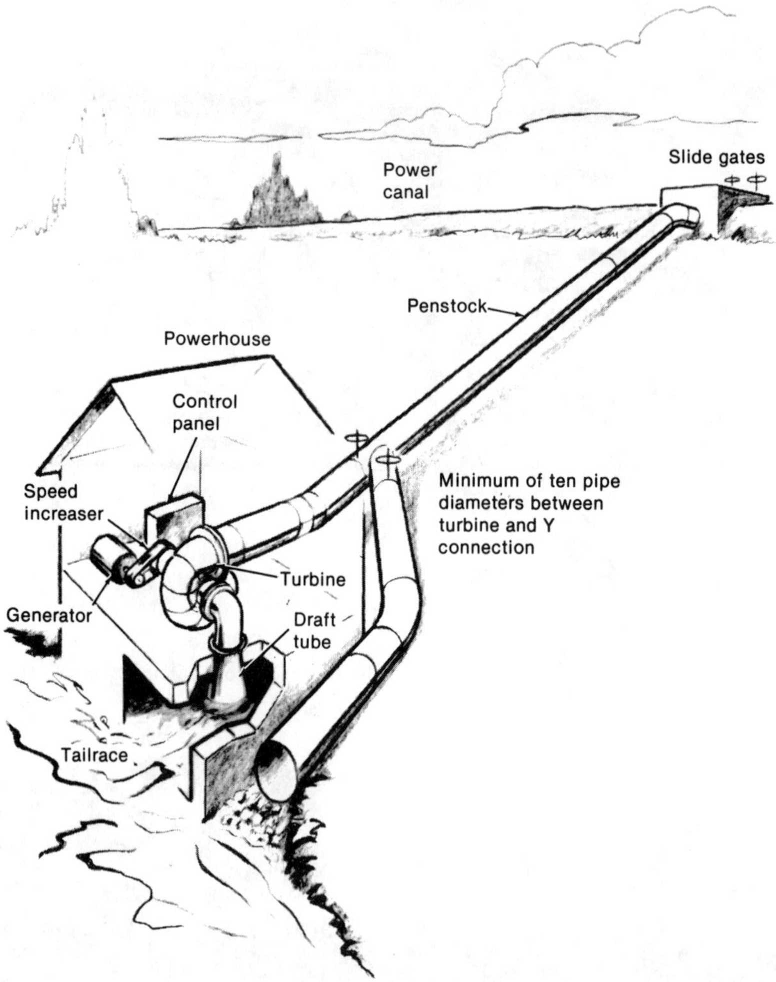

Fig. VIII.11. Electric power generation installation at a canal (source: McKinney *et al.*, 1983).

spring branch, or even a natural lake. A man-made source includes any structure used to increase the head or provide a source of water other than a natural one. Examples include dams, canal drops, and industrial or domestic wastewater discharge (McKinney *et al.*, 1983).

A microhydropower system developed on a natural stream is referred to as 'run-of-the-stream'. This approach eliminates the need for costly dam construction. Figures VIII.16 and VIII.17 show two such systems with appropriate nomenclature. In a run-of-the-stream development, an intake structure diverts water from a stream to a penstock. The structure consists of the following (McKinney *et al.*, 1983):

—*Diversion works.* The diversion works divert water from the stream into the intake system.

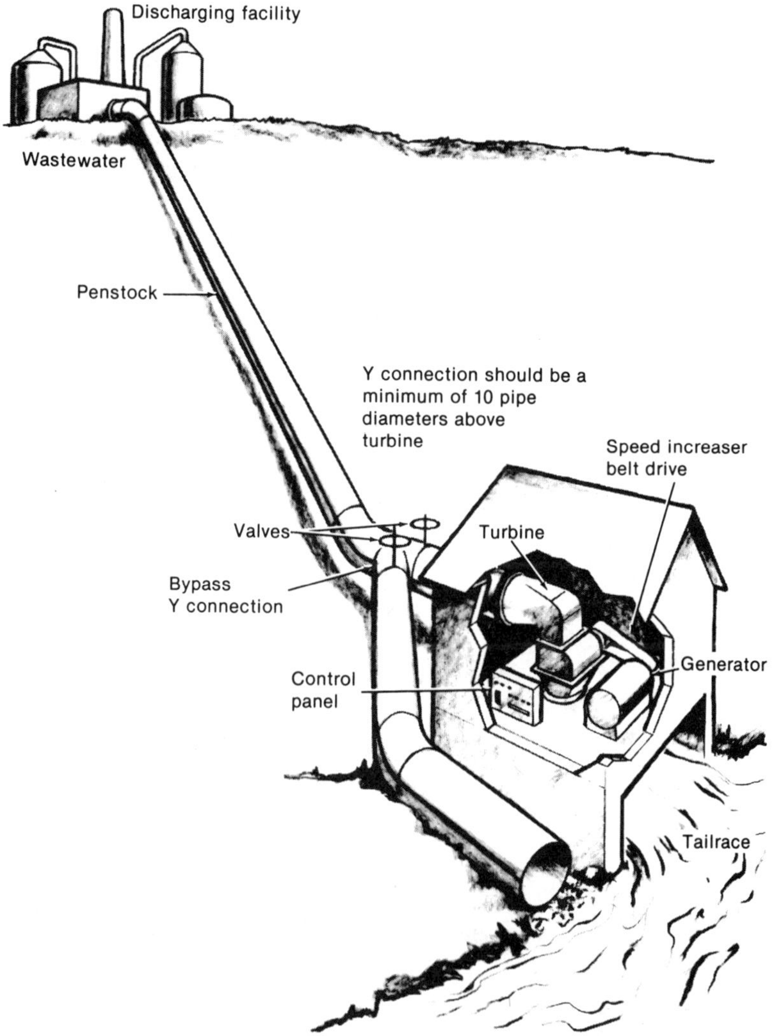

Fig. VIII.12. Electric power generation installation using wastewater discharge (source: McKinney *et al.*, 1983).

—*Settling basin.* The settling basin allows suspended material to sink slowly to the bottom of the stream bed before any water enters the power canal.

—*Power canal.* The power canal carries water from the diversion works and settling basin to the forebay. A canal is useful if water can be carried at approximately the same elevation to a point from which the penstock can be made as steep, straight, and short as possible.

—*Forebay.* A forebay is a settling basin in which suspended material can sink to the bottom before the water enters the penstock. A forebay is required in all run-of-the-stream developments.

—*Penstock intake structure.* This structure provides the transition from the forebay to the penstock and the framework for the trash racks and intake gates.

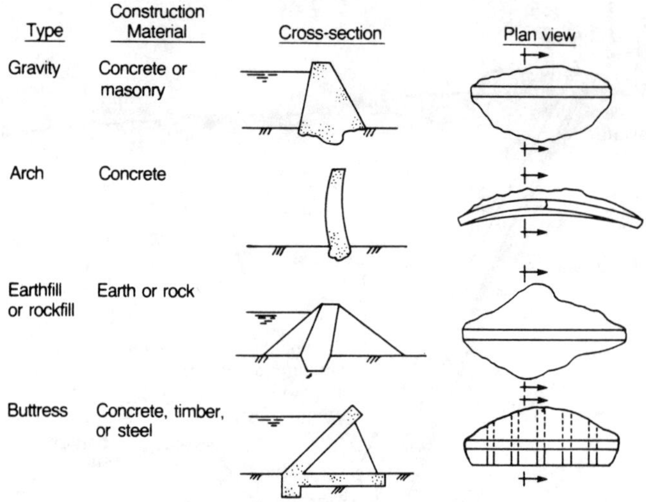

Fig. VIII.13. Basic types of dams (source: J. J. Fritz, 1984).

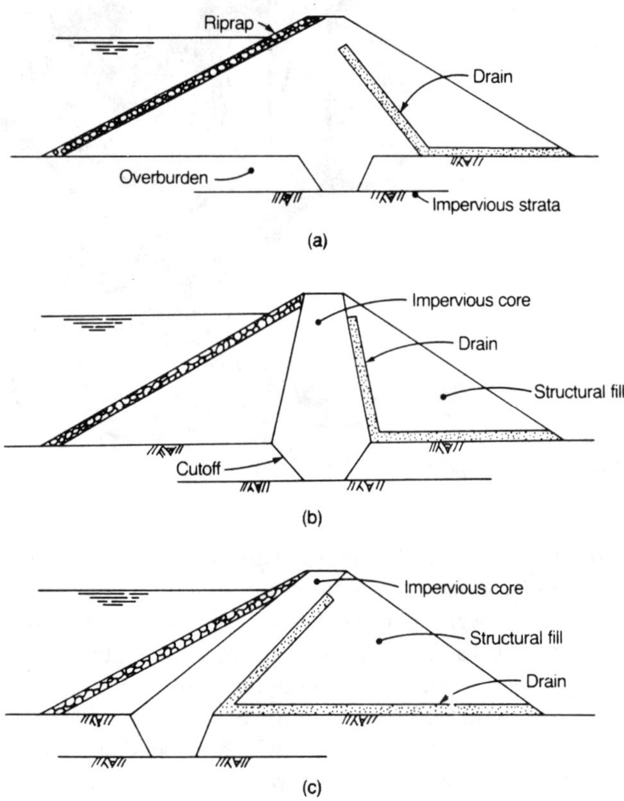

Fig. VIII.14. Typical earthen dam sections. (a) Homogeneous earthen; (b) zoned earthen with central core; (c) zoned earthen with upstream core (source: J. J. Fritz, 1984).

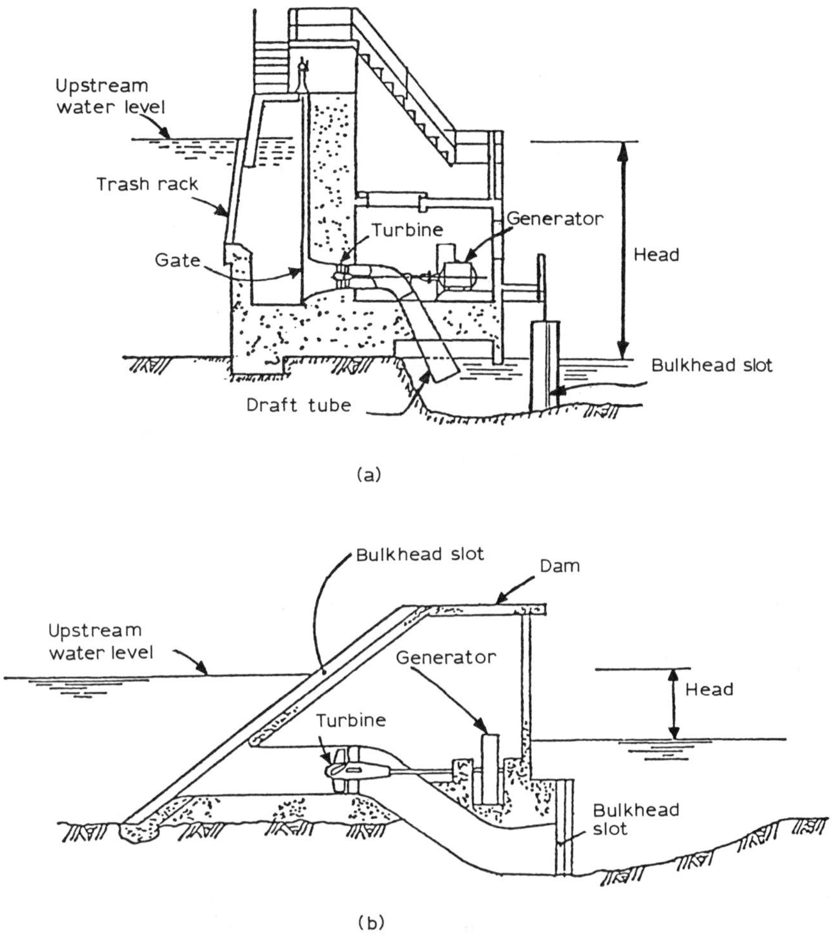

Fig. VIII.15. Low-head dams and turbines. (a) Horizontal propeller unit installation; (b) tube turbine installation (source: Cassidy, 1980).

After passing through the turbine, the water is returned to the original stream, or can be used for irrigation purposes. The spillway canal returns the excess water from the forebay to the river. The size of the powerhouse depends on the size and number of machines. Some useful information on design and construction of public works is available in an NRECA publication (Dunkerly *et al.*, 1981).

Turbines and Generators

Today's water turbines have developed as a result of evolutionary processes from ancient water wheels. Originally used for the direct drive of milling machinery, the application of turbines for electricity production is relatively recent. Various types of turbines are shown in Figs VIII.18–VIII.25.

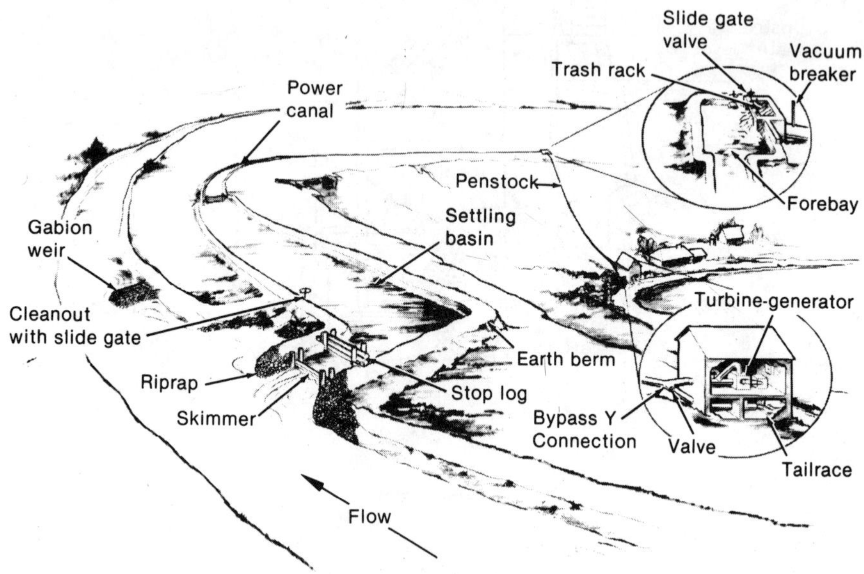

Fig. VIII.16. Run-of-the-stream development (source: McKinney *et al.*, 1983).

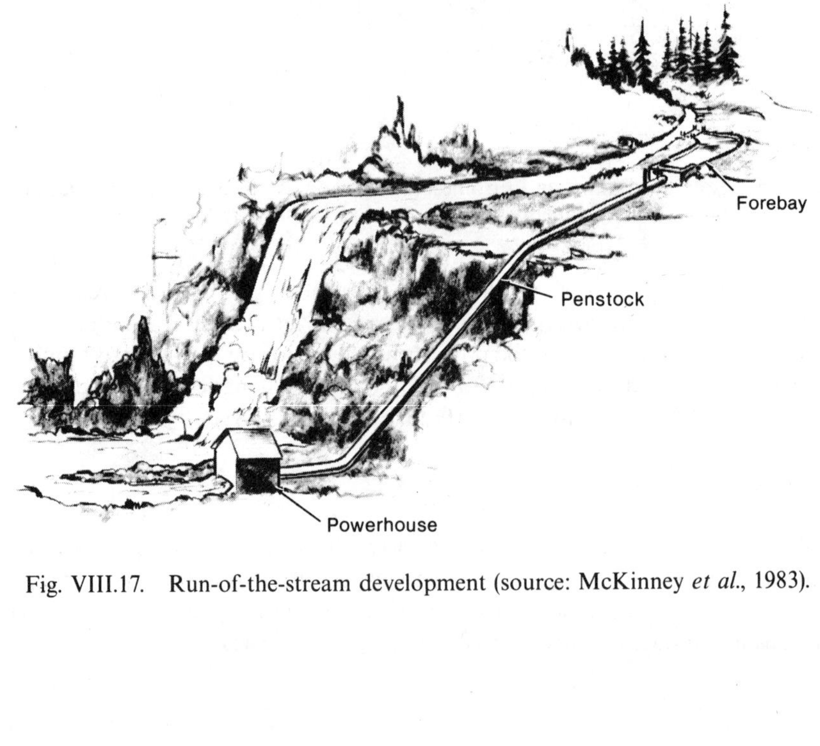

Fig. VIII.17. Run-of-the-stream development (source: McKinney *et al.*, 1983).

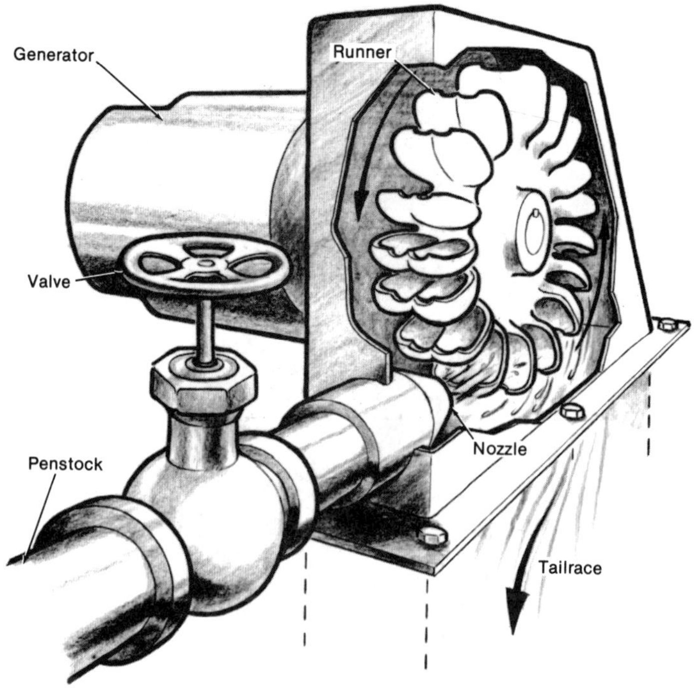

Fig. VIII.18. Impulse turbine (Pelton wheel) (source: McKinney *et al.*, 1983).

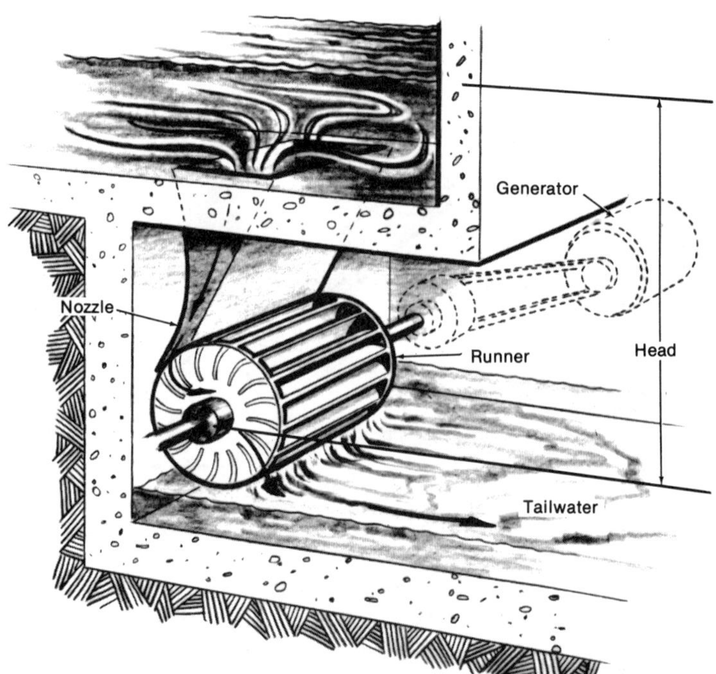

Fig. VIII.19. Crossflow turbine (source: McKinney *et al.*, 1983).

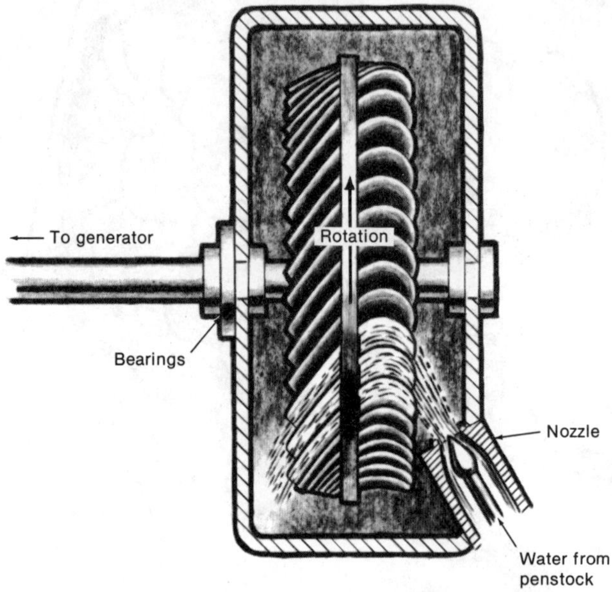

Fig. VIII.20. Turgo impulse turbine (source: McKinney *et al.*, 1983).

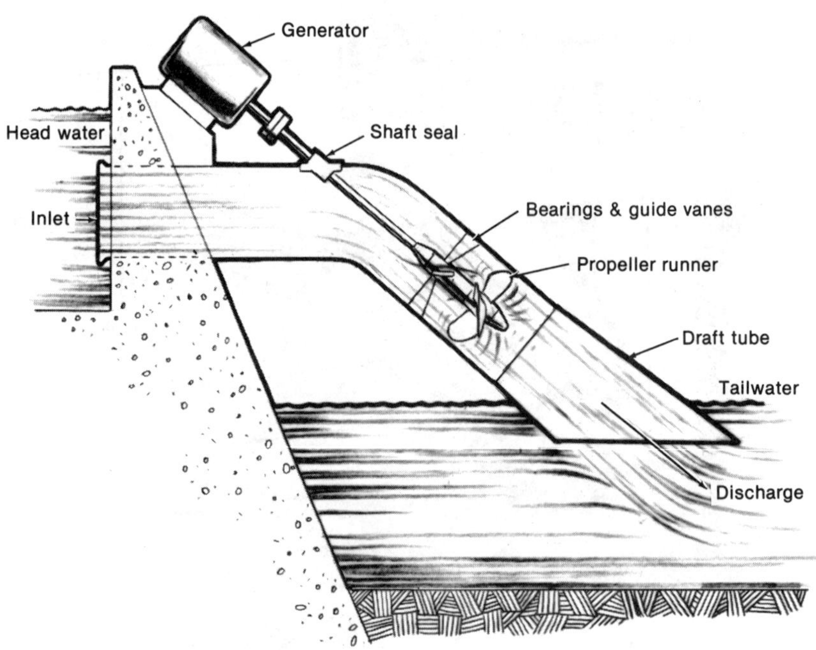

Fig. VIII.21. Reaction turbine (source: McKinney *et al.*, 1983).

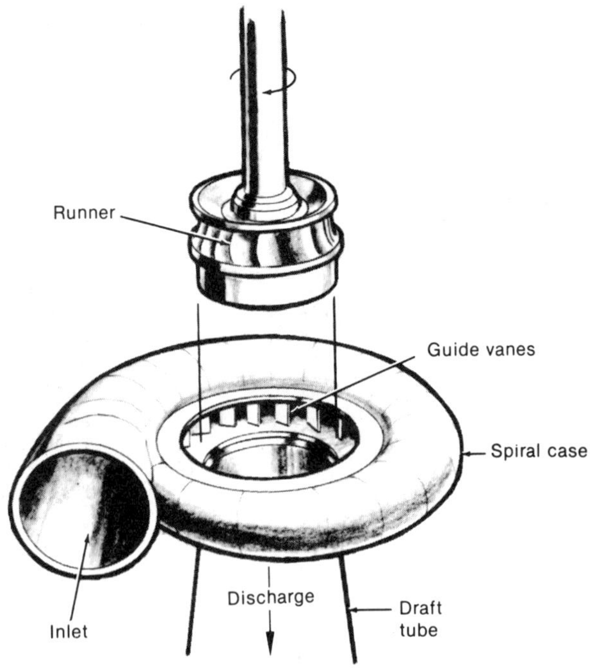

Fig. VIII.22. Francis reaction turbine (source: McKinney *et al.*, 1983).

Fig. VIII.23. Vertical Francis pump serving as a turbine for an installation in Idaho in the United States. The flow rate is 175 liters/s, the head is 43 m, and the generator capacity is 50 kW (source: EG&G Idaho, Inc.).

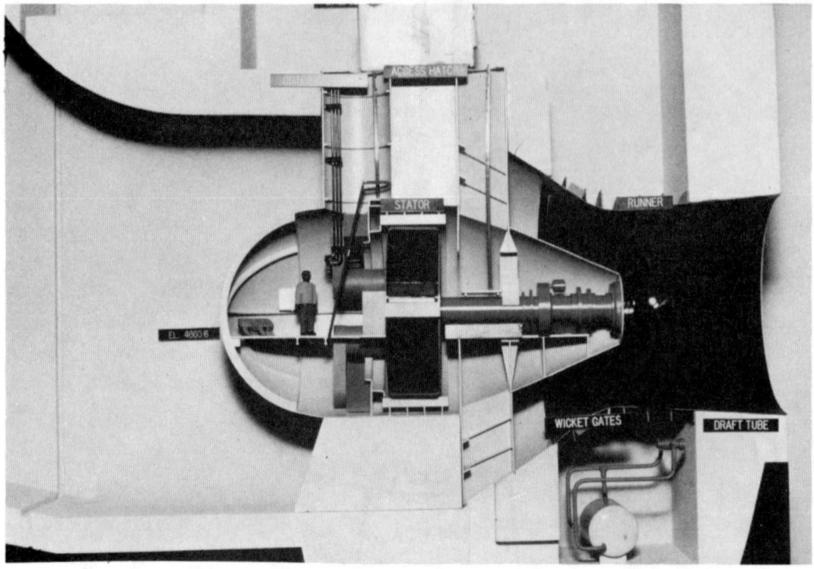

Fig. VIII.24. A bulb turbine designed to produce 7·2 MWe (source: US DOE Photo 163 005 001).

Fig. VIII.25. A low-head hydroelectric power plant utilizing a bulb turbine (source: US DOE Photo 163 004 001).

Two basic turbine types are in existence: impulse turbines and reaction turbines. In an impulse turbine, the velocity of a water jet drives the turbine runner. The force resulting from deflection of the water jet off the curved buckets on the runner creates torque and power.

Modern impulse turbines consist of Pelton, Turgo, and crossflow types. The Pelton turbine is used mostly on high-head installations and is as high as 90% efficient. The crossflow impulse turbine operates satisfactorily at heads ranging from 1 to 200 m and has an efficiency of about 85%. The part-load capacity of the crossflow turbine results in efficiency ratings comparable to the other types.

In a reaction turbine, the runner is completely submerged in water and both the pressure and velocity decrease from inlet to outlet. Reaction turbines are either radial (mixed-flow Francis turbines) or axial-flow propeller turbines. Several variations of the axial-flow turbines are available, such as Kaplan, tube, and bulb turbines.

The Kaplan turbine has adjustable blades and wicket gates. This turbine is efficient throughout a wide range of head and discharge capacities. The tube turbine can be fixed-propeller, semi-Kaplan, or fully adjustable. The externally mounted generator is driven by the turbine shaft. The bulb turbine is a recent development, in which the turbine and the generator are housed within the flow.

Reaction turbines are about 80% efficient. The propeller turbines are used for low-head and high discharge, whereas the Francis turbine is most efficient in the intermediate range of heads. Interest is increasing in reversing the water flow and using pumps as turbines. While such units have low efficiencies, they have the advantage of simplicity and economy.

Impulse turbines are used for higher head sites (more than 30-m heads) and reaction turbines are more appropriate for lower head sites. Various turbines are described in detail by Alward *et al.* (1979), UNEP (1980), J. J. Fritz (1984), and McKinney *et al.* (1983).

Water turbine selection is dictated by the flow and head characteristics of the site. Turbines suitable for various combinations of head and discharge are shown in Fig. VIII.26. The chart can be used for preliminary turbine selection. Figure VIII.27 indicates various commercially available turbines and the conditions for which each is suitable. The final selection should be made by considering turbine characteristics. The no-load turbine speed can be higher than the design speed by a factor as high as 2·5. Thus, a device for regulating speed is necessary. The significance of speed regulation depends on the machine and installation, the type of load, and whether the plant is connected to an electrical grid.

Speed Regulation

Speed regulation is normally achieved by controlling the flow. In a Francis turbine, speed is controlled by opening and closing the guide vanes which, in turn, control the flow of water according to the magnitude of the load. Impulse turbines are controlled by deflecting the jet or by using an auxiliary jet which bypasses the flow from the main jet.

Several governor designs are available: mechanical, mechanical–hydraulic, and

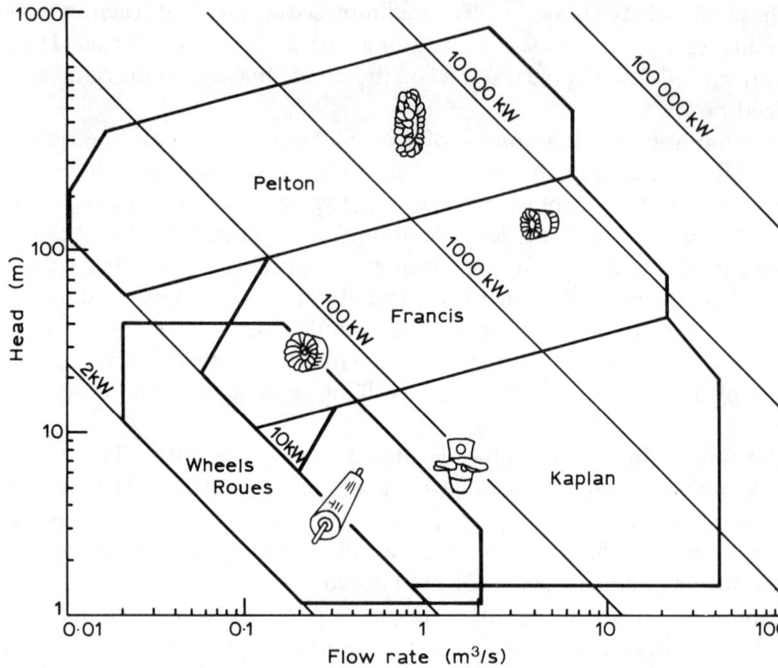

Fig. VIII.26. Theoretical useful range of water wheels and turbines in rural areas and on farms (source: Pellizzi, 1980).

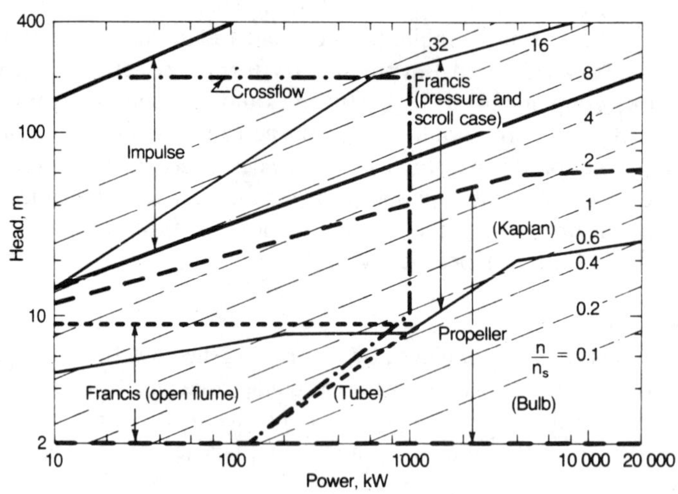

Fig. VIII.27. Summary chart of commercially available turbines (source: J. J. Fritz, 1984).

electrohydraulic. In small hydroelectric plants, the cost of a governing system becomes disproportionately high and may not be justified. New methods are being developed to regulate the power, rather than the water flow. In these new designs, the load on the generator is maintained constant by the automatic switching of a dummy load. When the consumer load is reduced, the dummy load is switched on, and vice versa; thus, the load on the generator remains constant.

Generators

The generator converts the mechanical energy of the turbine into electrical energy. Depending on speed, the generator may be coupled directly to the turbine shaft, or assembled through a gear or a pulley-belt system. Generator selection is governed primarily by the type of load it supplies. In isolated installations needing electricity only for lighting, a direct current generator can be used. In most applications, however, a three-phase synchronous generator is used.

These machines are commercially available with various power ratings. They produce voltage at the standard power frequency; the voltage rating may be selected according to system needs. Very small systems usually need a generator that produces power at the consumer voltage level, eliminating the need for a transformer. Synchronous generators can be operated either in isolation or in connection with an existing grid system.

Most designs use a static excitation device to convert alternating current to direct voltage using a solid-state device. This eliminates the need for a separate, expensive exciter for the generator field system. Generator voltage remains constant using a regulator to control the machine excitation and efficiency ranges from 85 to 90%.

Interest in induction generators in small hydroplants is extensive. A standard, low-cost induction motor can be operated as an induction generator if an external source provides the excitation voltage. Induction generators offer the advantage of simple construction, absence of a direct current excitation system, and low cost. However, because an existing power system supplies the magnetizing force, induction generators cannot be used in an isolated system. Furthermore, their efficiency values are generally lower than synchronous generators.

Generators should be equipped with the necessary control and protective devices. Plants connected to a grid system require automatic or manual synchronizing equipment. The exciter–regulator system maintains the generator terminal voltage within a narrow band. Generators are provided with circuit breakers which automatically disconnect the unit from the system under abnormal conditions. Protection is provided against overloads, short circuits, and ground faults. Hydroelectric plants can be designed to operate automatically by remote control in isolated areas.

Electric Distribution System

In a hydroelectric station, electric power is usually produced at an alternating voltage. If the service area is large and power is supplied to consumers located several kilometers from the powerhouse, a high voltage distribution system is needed. The power produced in the generator is transformed to a higher voltage by using a step-up transformer located close to the generator. High voltage lines supply power to the areas where consumers are located.

Another transformer, installed at the load end of the line, transforms power into a lower voltage for consumer use. Every country has its own standard consumer

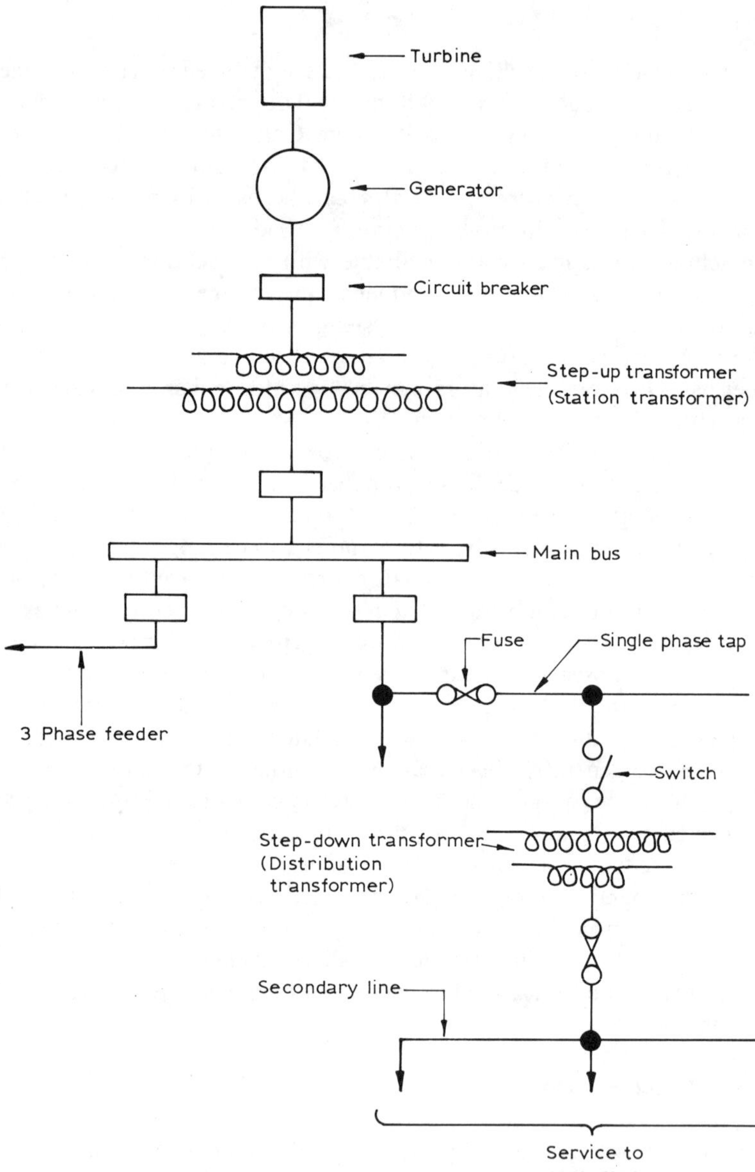

Fig. VIII.28. General electric distribution scheme.

voltage. For domestic consumption, the voltage is usually either 220 or 110 V. Industrial electric motors may require higher voltages. Small motors usually operate on the same voltage as the domestic supply.

The cost of a distribution system depends on the power transmitted and the extent of the service area. In a small hydroelectric system, these considerations may constitute a substantial portion of the total cost. In a well-designed distribution system, capital and operating costs, as well as energy losses, are considered. Proper selection of voltage levels, conductor sizes, and system configurations is essential.

A general distribution scheme is illustrated in Fig. VIII.28. The station transformer steps up the generator voltage to a level suitable for the distribution lines. The distribution transformer steps down the voltage to supply power to consumers. A higher distribution voltage requires a smaller conductor size to distribute the same quantity of power. Conductor sizes are selected to maintain voltage regulation within a reasonable limit when supplying maximum power demand. Selection of the optimum conductor size for a given voltage level, conductor length, desired voltage regulation, and future extensions can be made from charts and graphs available in electrical engineering handbooks.

The distribution system may be located overhead, using bare conductors, or underground, using insulated cables. The underground system is expensive and, therefore, used only where necessary. Overhead lines are attached to steel, concrete, or wooden poles. In areas where timber is available, wooden poles are more economical than concrete structures. Wooden pole attachments are also simple and do not require special foundations. Span lengths of 100–150 m are used in most rural areas.

In a small (10-kW) system, the electrical distribution scheme should be simple and inexpensive. Transformers are not needed because consumers are supplied directly from the generator panel. The load center of small systems should be within 1 km of the powerhouse, otherwise the cost for lines may be too expensive.

Distribution lines, transformers, and consumer premises are equipped with protective devices. Consumers may choose to install energy meters at their facilities and pay for electricity on a monthly basis. To help defray the cost of installing the meters, consumers may be charged a flat rate, depending on the connected load.

COST OF A SMALL HYDROELECTRIC PLANT

Hydropower is one of several alternative renewable energy sources under consideration to replace fossil-fuel-fired electricity generation systems. The comparison between small hydropower systems and petroleum-fired generation systems or alternative sources of energy should be made within the context of small-scale dispersed electricity production. System configuration and operating conditions are site-specific.

Site conditions may include resource availability—such as hydrology, insolation, geology, land and biomass availability, and fossil fuel availability—equipment and construction material availability, financing, market demand, and local social structure. These conditions vary extensively from country to country and often within countries (J. J. Fritz, 1984).

A hydroelectric plant incurs capital costs and operating costs. Total plant expenditures include costs for planning, designing, constructing, and installing the plants, as well as purchasing the distribution system. The operating cost of the plant includes funds for plant operation, repairs, and maintenance.

The capital cost of a hydroelectric system is highly site-specific. It depends on the work involved in diverting and carrying water to the turbine. In some schemes, a dam accounts for most of the civil engineering cost. The materials used and the technology applied influence the cost significantly. The cost of the turbogenerating equipment depends on the hydraulic characteristics of the site.

Low-head sites are more expensive because they require large machinery. The cost of low-head plants may be double the cost of systems with higher heads. Site access can influence the cost as well, particularly if heavy equipment and material transportation are required; building roads to the site could be cost-prohibitive.

For isolated stations, the cost of transmitting and distributing electrical energy depends on the area served and the powerhouse distance from the load center. If the

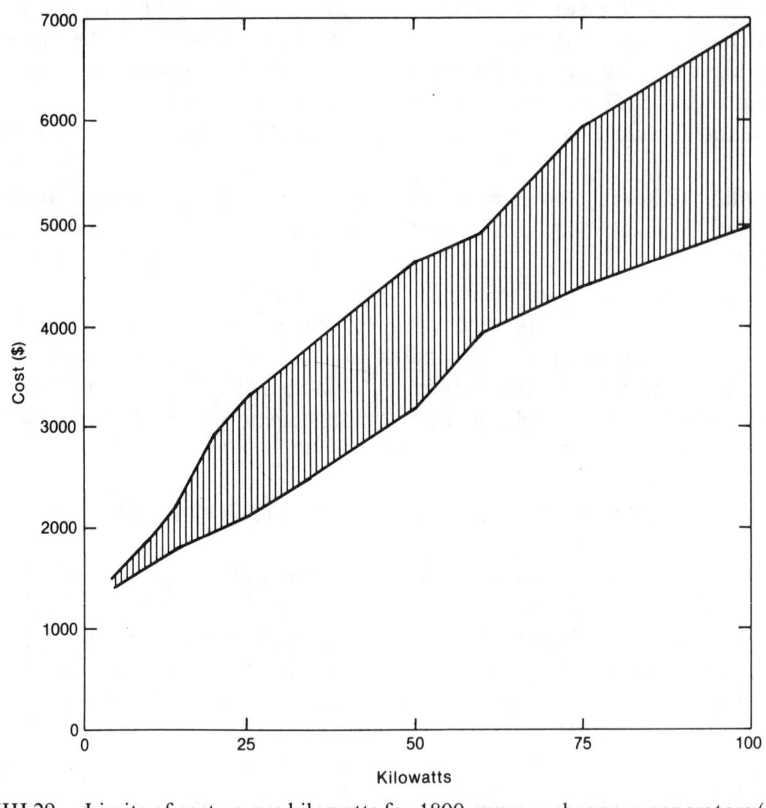

Fig. VIII.29. Limits of cost versus kilowatts for 1800-rpm synchronous generators (source: McKinney *et al.*, 1983).

distance is far, the cost will necessarily increase because high voltage transmission lines requiring step-up and step-down transformers are needed. When the distance between the plant and the load center is close, a low voltage distribution network is used to supply power to the consumer.

For small hydropower schemes, the ranges of the component costs as a percentage of the total expenditure are as follows: civil works, from 25 to 50%; turbogenerating plant, from 30 to 40%; transmission and distribution, from 10 to 30%; and engineering and contingencies, from 20 to 30%. These values vary depending on site conditions. Figure VIII.30 illustrates the cost of induction

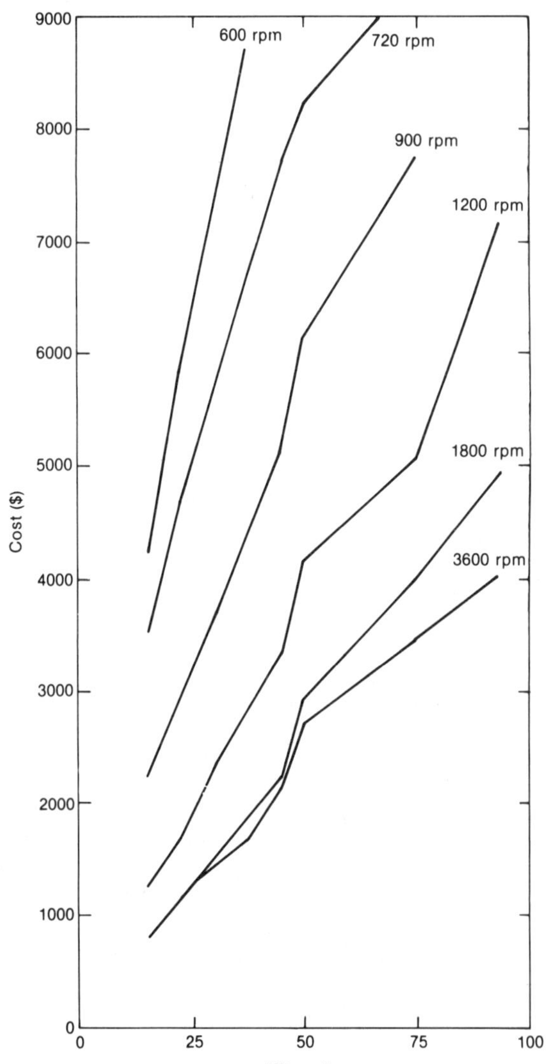

Fig. VIII.30. Cost versus kilowatts for induction generators (source: McKinney *et al.*, 1983).

TABLE VIII.8
Installation costs of small hydroplants in developing countries (source: P. J. Clark, 1981)

Country	*Capacity (kW)*	*Head (m)*	*Cost ($/kW)*[a]	*Equipment as % of cost*	*Comments*
Thailand	800	40	2 850	15	Concrete construction for civil works
Ecuador	400	40	2 200	36	Lined canal; imported equipment
Ecuador	400	19	2 700	30	Lined canal; direct intake; imported equipment
Nepal[b]	200	58	900	36	Partially lined canal; loose-packed rock dam
Nepal	120	14	900	21	Masonry-lined headrace
Indonesia	120	15	1 300	58	No headrace; local turbine
Thailand	100	79	950	21	Concrete construction for civil works
Philippines[b]	100	27	750	39	Local construction materials and turbine
Indonesia	90	19	1 050	54	Concrete-lined headrace
Nepal	80	16	1 650	23	Local materials and turbines
Nepal	80	34	1 050	31	Local materials and turbines
Nepal	25	22	650	28	Existing irrigation works
Indonesia[b]	15	22	600	30	No headrace or governor, used alternator
Thailand	15	10	2 350	41	Earth dam and headrace
Thailand	10	6	1 500	27	Lined canal; local turbine
Pakistan[b]	10	—[c]	270	61	Local materials and equipment
Nepal	9·5	9	600	46	Local materials and equipment, except penstock
Pakistan[b]	7·5	—[c]	320	60	Local materials and equipment

[a] Not including transmission and distribution costs.
[b] Projects implemented and managed locally.
[c] Precise heads unknown, but generally between 5 and 10 m.

generators. Cost data collected from actual experience in developing countries are provided in Table VIII.8. Note that the cost per kilowatt ranges from a few hundred dollars when local materials and equipment are used to $3000 or more in more sophisticated installations. Many site-specific factors cause actual costs to deviate considerably from the baseline costs shown.

Technology and materials obviously control overall cost. Higher cost values usually reflect conventional design and imported equipment. On the other hand, nonconventional approaches implemented by some developing countries result in low installation costs. Civil works are completed using local materials and techniques familiar to villagers. Simply designed water turbines are fabricated locally using labor-intensive methods.

The conventional approach emphasizes permanent structures based on sound engineering design and more sophisticated machinery. Such hydropower schemes are more efficient and reliable than nonconventional designs. The main difference between the two approaches is the trade-off between cost and performance.

The cost of turbogenerating equipment is influenced considerably by the operating head of the turbine. For a given power, as the head increases, the amount of necessary flow needed is reduced. This, in turn, reduces turbine size, permitting higher operating speeds. Both factors help reduce the total cost of the system.

According to a study by Stone and Webster Engineering Corporation (1981), the cost of a package unit decreases from $500–$1300/kW to $200–$700/kW when the operating head increases from 4 to 15 m. The specific cost per kilowatt for a given operating head generally decreases as power output increases.

For example, an impulse turbine package unit may cost \$1500/kW for a 5-kW unit and only \$300/kW for a 50-kW unit. Because the cost of fabricating turbines and generators does not increase in proportion to their power rating, installing large capacity machines to achieve maximum economy is recommended, provided sufficient water is available.

Substantial cost savings for such turbogenerator equipment may be realized by eliminating the adjustable flow control devices, such as wicket gates, guide vanes, propeller runner blades, needle valves, and speed control governors.

The cost of energy per kilowatt-hour is obtained from the annual cost of operating the plant and the amount of electrical energy produced during the same period. Plant operating costs are calculated by including annual expenses of capital investment, operation, and maintenance. The annual cost of capital investment depends on the interest rate and the useful life of the plant. Small hydropower plants have a life expectancy of 25–30 years. In estimating the operation and maintenance cost, all expenses to maintain the waterways, civil works, machinery, and accessories are included. The salary and wages of the operating staff are also incorporated into the total cost of operating and maintaining the plant.

The amount of energy produced by the plant depends on the power drawn from the generator and the time for which the power is supplied. The energy produced is expressed in kilowatt-hours. In a normal situation, a power plant rarely operates at full capacity for the entire year. The load fluctuates during the day and at various times throughout the year, depending on the type of consumer. To account for load variation, load factor must be determined. Load factor is the ratio between the actual energy generated and the energy which the plant can generate if operating at full capacity. The load factor for an isolated small hydropower plant may range from 0·2 to 0·6.

The amount of energy produced and the cost per kilowatt-hour, therefore, strongly depend on the load factor. Running the plant at a high load factor is recommended. This could be achieved by introducing small-scale industries, e.g. agro-based processing units, workshops, water pumps, etc. These units can operate during the day, and the plant will supply the lighting load during the night. The improved load factor reduces the cost of generation.

Small hydropower plants often compete with diesel–electric installations. Figure VIII.31 summarizes the results of an analysis performed by the Intermediate Technology Development Group in England, comparing a 40-kW micro-hydropower system to a 40-kW diesel generator installed in Nepal. The total operating cost of the hydropower system is essentially constant regardless of the percentage capacity utilized. The operating cost for diesel consists of both a fixed and variable cost component (primarily fuel cost). Thus, the total cost for a diesel generator increases linearly with work output. In this example, the break-even cost occurred at 32 000 kWh or about 10% of the hydropower capacity.

This example is noteworthy because in many less developed countries the demand for electricity may be high during certain periods, such as for lighting at night, and

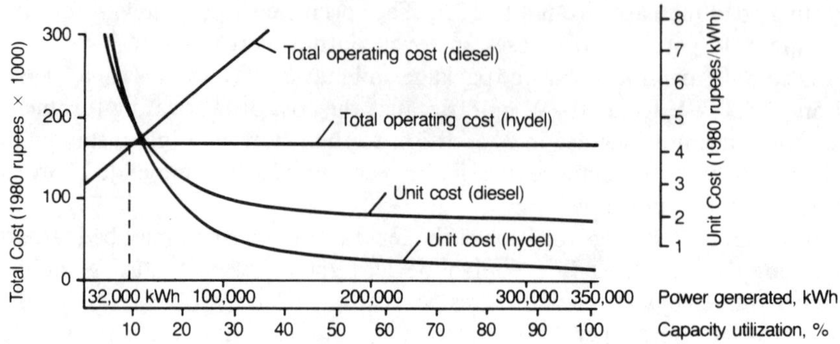

Fig. VIII.31. Total annual cost and unit cost versus power generated and capacity utilization for 40-kW diesel and hydroelectric installations (source: J. J. Fritz, 1984).

very low the remainder of the day, thus creating a high peak demand but a low capacity factor. Under such circumstances, the diesel generator system remains economically attractive. This is an example of site-specific conditions which could reverse the general perception that hydropower is more attractive than diesel.

Another example of the relative costs of diesel and hydropower is provided in Table VIII.9. Under these conditions, diesel-generated electricity costs three times more than hydropower electricity. While the overall cost of producing electricity from a hydropower plant may be less than from a diesel plant, the initial capital cost of such plants is relatively high. Figure VIII.32 illustrates the additional capital cost of hydropower plants relative to oil-fired plants.

The Mbulu project in Tanzania consists of three 500-kW turbine generators operating under a 420-m head. A distribution system supplying power to two communities, Mbulu and Babati, is also planned. Planners anticipate that 8·0 GWh

TABLE VIII.9

Comparison of hydropower and diesel generated electricity costs (source: J. J. Fritz, 1984)

Cost type	*Hydropower*	*Diesel*
1. Investment cost ($/kW)	2 000·00[a]	700·00[b]
2. Useful life of equipment (years)	35	15
3. Capital recovery factor for an interest rate of 10% per annum	0·103 7	0·131 45
4. Power generated per year at a load factor of 45% (kWh)	3 942	3 942
5. Annual fixed costs		
Debt service ($)	207·40	92·00
Annual cost for maintenance at 1 and 2·5% of capital investment ($)	20·00	17·50
Annual cost for operation at 0·5% of investment ($)	10·00	3·50
Annual charges for administration and overhead at 0·5% of investment ($)	10·00	3·50
Total annual fixed cost ($)	247·40	116·50
6. Annual fixed cost ($/kWh)	0·062 7	0·030
7. Variable cost		
Cost of fuel in remote localities ($/kWh)	—	0·18[c]
8. Total cost of generation at 45% load factor ($/kWh)	0·062 7	0·21

[a] Assumption for mid-1981 price level.
[b] Assumption on the basis of average investment cost of diesel power stations up to 1000 kW capacity.
[c] Based on the price of diesel fuel at $0·55/liter delivered at site.

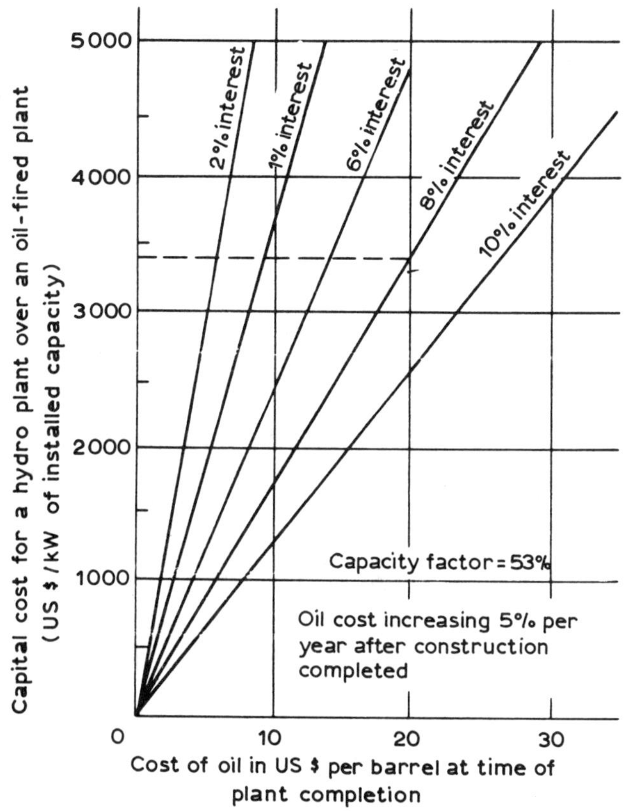

Fig. VIII.32. Additional capital cost of a hydroplant over an oil-fired plant as a function of oil cost (source: J. J. Fritz, 1984).

TABLE VIII.10
Capital cost of Mbulu project[a] (source: J. J. Fritz, 1984)

Item	*Depreciation period (years)*	*Capital investment[a]*	*Capital recovery factor, 10% annual interest*	*Annual cost[a]*
Generation				
Civil works	35	2 250·00		
Plant and equipment	35	750·00		
Subtotal		3 000·00		
Interest during construction		210·00		
Subtotal generation		3 210·00	0·103 7	330
Transmission, transformation, and distribution	25	1 107·00		
Interest during construction		77·50		
Subtotal		1 184·50	0·110 2	130
Grand total		4 394·50		

[a] Thousand US dollars.

TABLE VIII.11
Annual fixed costs of the Mbulu project[a] (source: J. J. Fritz, 1984)

Cost item	Generation works	Transmission and distribution
Maintenance, 1% for generation and 2% for transmission and distribution	30 000	22 000
Operation, 0·5%	15 000	5 500
Overhead, 0·5%	15 000	5 500
Capital investment	330 000	130 000
Total	390 000	163 000

[a] US dollars.

of energy at a 60% load factor will be available. Construction is estimated to take 3 years with capital costs expended at 10, 60, and 100% cumulatively.

Tables VIII.10 and VIII.11 show capital cost and annual fixed costs for the Mbulu project. The cost of generation was computed as follows:

$$\frac{\text{Annual fixed costs}}{\text{Annual energy production}} \; \frac{\$556\,000}{8{\cdot}0\,\text{GWh}} = \$0{\cdot}070/\text{kWh}$$

If a 10% transmission loss occurs, leaving 7·2 GWh, then the cost to the consumer would be $0·077/kWh (J. J. Fritz, 1984).

Unfortunately, hydropower systems are not portable and may be far removed from the end-user of electricity. In such cases, the cost of a distribution system may be a significant part of the basic electricity cost.

SMALL HYDROPOWER STATIONS VERSUS OTHER ALTERNATIVE SOURCES

Table VIII.12 presents estimated capital costs of several electricity producing systems. Figure VIII.33 provides a method of estimating generating costs for various system sizes operating at different load factors. One of the most promising options for meeting rural electrical energy needs is mini- and microhydropower stations, particularly in the tropical highland areas of Latin America, Asia, and Africa. In these regions, economic development is often limited by the availability of electrical power and the dependence on diesel fuel (J. J. Fritz, 1984).

Although minihydropower units are usually more expensive initially than diesel-powered generators, the lifecycle economic picture is generally more attractive. Estimates of total installed costs (1978) for micro- and minihydropower units in Asia ranged from $400/kW to $2000/kW. In Papua New Guinea, for example, the capital cost of a 5-kW microhydropower plant was $438/kW compared to $580/kW for the same size diesel generator. The economics of minihydropower units can be enhanced if additional benefits accrue such as irrigation, water supply, or flood control (J. J. Fritz, 1984).

TABLE VIII.12
Comparative system costs per kilowatt of generating capacity (source: J. J. Fritz, 1984)

Power source	*Cost ($)*
Hydro	1 296
Geothermal	1 564
Thermal	863
Photovoltaics	10 000–20 000
Wind	
Less than 1 kW	3 000–6 000
5–20 kW	1 000–2 000
Biomass	
Thermal	1 000
Gasifier/engine–generator	700–1 500
Ethanol	1 550

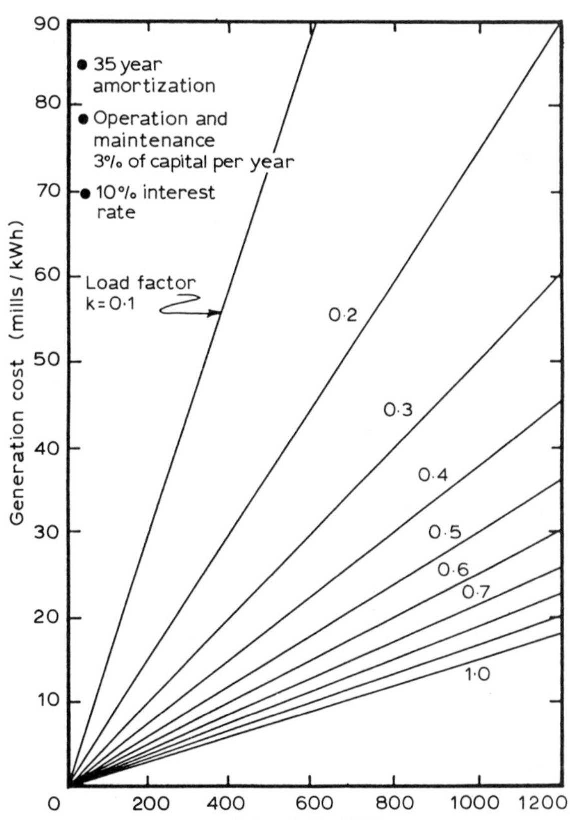

Fig. VIII.33. A chart for estimating generating costs for hydropower plants of various sizes operated at various load factors (source: Alward *et al.*, 1979). (1 mill = US$1/1000.)

Operating costs depend primarily on the capacity factor, which may be as low as 15–20%. Methods must be sought to utilize electricity on a 24-h basis.

EXAMPLES OF HYDROPOWER END USES

A hydroelectric power scheme can be planned and designed as a multipurpose project, in which electricity generation is simply one component. Multipurpose water projects can be developed, particularly with low-head sites, in which the water can also be used for irrigation. Dams constructed for flood control, fish breeding, or recreational purposes can also be used to operate turbogenerator plants, thereby significantly improving the cost-effectiveness of such a project.

In selecting a system size, both power and energy needs must be estimated. Power is usually referred to as 'demand' and represents the instantaneous power required by all the electrical load devices connected to the system at a given time (in watts or kilowatts). The second parameter, electrical energy, often called 'production', involves not only the demand, but the length of time each device is connected to the system. Electrical energy production is given in watt-hours or kilowatt-hours. The physical size of the turbine and generator is determined primarily by the power demand, whereas the flow of water and head requirement is determined by the energy demand (J. J. Fritz, 1984).

In China, which has many small hydroelectric plants, electricity generation is a part of water resource development schemes. Most hydraulic structures are designed for flood control. The reservoirs supply water for irrigation purposes and are also used as ponds for fish breeding. Electricity production through these schemes is a lower priority.

The nature of the energy output of a small hydropower plant is particularly suited to a variety of applications, such as small and medium-sized industry, agricultural processing, agricultural production, and commercial and residential usage. The power and energy requirements of various operations in rural areas of India are presented in Table VIII.13. Table VIII.14 gives another example of the load demand for small-scale enterprises in Latin America. A small amount of power can be very useful for agricultural activities in these rural areas. For example, a 5-kW motor can pump domestic water to a community of 200 households. Similarly, lift irrigation can be provided for small land areas by using 10–15-kW motors.

Another example of multipurpose water turbine usage is illustrated in Fig. VIII.34. The multipurpose power unit developed in Nepal utilizes a 'bucket-type' turbine to drive a milling device or an optional power take-off (Nakarmi & Backmann, 1982).

With judicious energy use, plant load factors should improve greatly, thus reducing electrical generating costs. For example, agro-processing and other industrial activities can be performed during the day, and lift irrigation during the night, after area lighting needs are met.

For several applications, the shaft power of the turbine can be used for grinding, milling, and similar tasks. Machines can be installed in the powerhouse and

TABLE VIII.13
Technical characteristics of agricultural processing machinery in India (source: General Electric Company, 1963)

Process or operation	*Equipment*	*Motor rating (hp)*	*Output (lb/h)*	*Power consumption (kWh/100 lb)*
Oil pressing and extraction	Hydraulic press			
	Large	22	1 500	1·3
	Medium	16	1 000	1·5
	Small	8	130	4·5
	Screw expeller			
	Large	20	2 000	1·0
	Medium	15	1 000	1·5
	Small	5	100	5·0
Rice hulling	Disk sheller (German)	25	2 400	0·8–1·0
	Japanese rubber roller sheller	3	300	2·0–2·5
Cereal milling	Large	20	800	1·5–2·0
	Small	5	180	2·0–3·0
Cotton ginning	Large	20		
	Small	5	78	2·75–3·2
Fodder chopping	Rotary cutter	2–10		0·25–1·0
Sugar cane processing	Crusher refining pan	3–15		0·50–3·0

operated from the shaft using a suitable pulley-belt coupling system. When compared with electrically driven machines, shaft-driven plants are inexpensive. Often, the cost of an electric motor may be comparable to the processing unit cost. Some schemes in Colombia, Nepal, and Pakistan are hybrid, producing both electrical and mechanical power. In Nepal, water turbines are used for milling and rice hulling in remote areas. Some systems also have wood-working equipment, such as table saws, connected through mechanical drivelines to water turbines.

TABLE VIII.14
Demand load for small-scale enterprises in Latin America (source: J. J. Fritz, 1984)

Activity	*Required capacity (kW)*	*Activity*	*Required capacity (kW)*
Carpentry	3–15	Brick factory	1–5
Bakery	2–5	Restaurant	1–2
Crafts	1–2	Vegetable canning	5–20
Small sawmills	15–30	Dairy products	2–10
Grinders	10–20	Milk (cooling and pre-evaporation)	5–20
Flour mill	3–20	Electrical and mechanical workshops (repairs)	5–15
Looms	0·5–6	Gas pumps	2–100
Coffee grinder (commercial)	5–30	Silos	3–5
Quarry mill	6–30		
Ice making	6–60		
Irrigation pump	2–100		

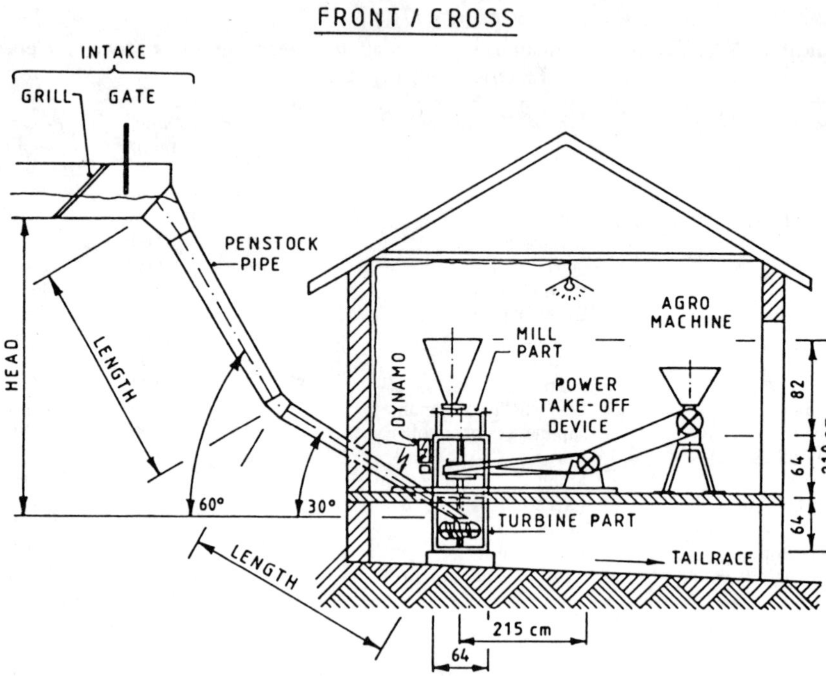

Head (*m*)	*Power* (*hp*) *at given flow rate*		
	80 liters/s	*100 liters/s*	*120 liters/s*
3·0	2·4	3·0	3·6
3·5	2·8	3·6	4·3
4·0	3·2	4·0	4·9
4·5	3·6	4·5	5·4

Note: Minimum water quantity required is 60 liters/s.
Oil expelling needs a minimum of 3 hp.

Fig. VIII.34. Multipurpose water turbine developed in Nepal (source: Nakarmi & Backmann, 1982). Drive shaft speed, 250–300 rpm (the speed remains constant ±20%, with or without load; overspeeding will not occur).

In addition to operating agro-based industries, the electricity produced can be used for other purposes. Using a rectifier, alternating current can be changed into direct current and storage batteries can be charged. A battery-charging facility is especially valuable for people who are operating buses and trucks in isolated rural areas. Storage batteries can also provide electricity for lighting low-power, low-voltage bulbs in individual dwellings or shops.

Special energy applications include the usage of mechanical heat generators and electric arc welders. A mechanical heat generator, which provides hot air at 100°C, is used for crop drying and several other processes (Metzler, 1982). The Kettering

Research Laboratory has developed an experimental 3-kW electric arc reactor that produces sufficient nitrogen fertilizer for 5–20 ha.

OBSERVATIONS ON THE STATUS OF HYDROPOWER

The World Bank, after a major study in 100 developing countries, made these general observations about hydropower (World Bank, 1984*d*):

—Capital costs for current hydropower projects range from $1000/kW for large, high-head projects in Colombia to $6000/kW for a small, low-head project in the Upper Volta. The average range is $1500–$2000/kW. This implies a power cost of about $0·05/kWh at the generating station for an average project.

—The recent downturn in oil prices and resource constraints may slow the rate of hydropower installation in certain situations, particularly in countries where hydropower projects are marginal. In such countries, hydropower projections probably represent an upper limit of the hydropower development that will ultimately be implemented.

—In most large countries, the decision to shift completely away from oil has already been made, so the overall hydropower installation totals will not change significantly in such countries as China, Brazil, India, Colombia, and Thailand. In these countries, the ratio of hydropower development is dictated by load growth and resource availability, not by oil prices.

—Smaller countries face the continuing dilemma of deciding when to shift from oil to hydropower because small system sizes limit coal-thermal as a viable alternative. These countries must meet both the oil expenses for thermal and the investment requirements for hydropower during the hydropower plant construction period which usually lasts several years. However, any delay in initiating hydropower projects may only aggravate the future situation.

—In areas where hydropower can displace oil, each kilowatt of hydropower can be substituted for the electricity generated from 1 t of oil annually, assuming that the hydropower station operates an average of 4000 h annually.

—Future oil price uncertainties make any economic comparison difficult, but 1 t of oil savings might be worth $200–$500 annually (1981 prices), depending on the type—residual or gas oil, and the location—coastal or inland.

—The increasing concern about environmental considerations for hydropower projects, including resettlement issues, requires close attention during the earliest investigating stages. Factors such as resettlement could preclude development of some sites that were otherwise very attractive (e.g. Lokoja in Nigeria and Pa Mong in Laos and Thailand).

—Further technological advances in hydropower are still possible, particularly for low-head projects. In circumstances such as Gambia and Bangladesh, where large flows are available at very low-head conditions, economical hydropower installations may be possible if inexpensive generating equipment can be devised.

SUMMARY

Hydropower is an important energy source for developing countries. This chapter reviews the fundamentals of hydropower and focuses on micro, mini, and small hydropower stations, i.e. less than 30 MW. Essential steps are presented for planning a hydropower project. Components of the entire hydroelectric system are discussed, and the cost of electricity from hydroelectric systems is compared with the cost from other energy sources. Examples of hydropower applications in agro-industries are presented.

CHAPTER IX

Alternatives

INTRODUCTION

The energy situation varies for each developing country, but all countries have a commercial and a noncommercial energy component. The world price of oil, the major indicator of commercial energy prices, varied dramatically during the 1970s and 1980s. Oil prices are not determined simply by physical supply and demand, but are affected significantly by political and/or military decisions that are usually impossible to forecast. Therefore, continued gyrations in world oil prices seem likely as physical supply constraints are coupled with other unpredictable and intangible factors.

Commercial energy prices in developing countries are normally set by government policy, and thus may not reflect current world market prices. For example, in India, the price of ordinary petrol was 6·12 rupees (Rs)/liter in 1984, rose to 7·43 Rs/liter in 1986 and to 8·50 Rs/liter in 1988 even though world oil prices tumbled. Diesel fuel prices remained constant at 3·22 Rs/liter from 1984 to 1986 and increased to 3·50 Rs/liter in 1988.

An important consideration to remember is that most rural people in developing countries depend primarily on noncommercial energy sources and are therefore quite insulated from fluctuations in world oil prices. Because this book addresses the total energy needs of agriculture in the rural sectors of developing countries, both commercial and noncommercial energy sources are discussed. However, the correlation between supply and demand price relationships for one form of energy sometimes seems quite unrelated to other forms.

Nevertheless, agriculture needs energy in the right form at the right time to maximize production. Chapters II and III provided numerous examples of the crucial nature of energy for agriculture. Additional energy is required to transport, store, and process food, and to prepare it for human consumption.

Agriculture is but one sector in national economies and cannot be examined in isolation. Economic objectives for all countries usually include the following (Starr, 1981):

—Food in the form of a balanced diet for all.
—Physical comfort including shelter, heat, sanitation, etc.
—Physical security, including police and military protection.
—Economic growth.
—Amenities, including improved health services, education, and recreational and leisure opportunities.

ENERGY OPTIONS

An overriding concern of the nations of the world is that the agricultural sector must be provided with the small but crucial supplies of commercial energy, as well as the noncommercial energy required for food production. What range of options or strategies are available? Can a balance between energy and food supplies be maintained as the world population continues to grow? Should price be the primary determinant of allocation, or should other mechanisms be considered? On a global scale, how can poor nations compete with rich ones for petroleum products? What alternatives to petroleum are feasible? Will renewable alternative energy resources be cost-effective? These and many other complex and difficult questions must be answered for all nations to survive and prosper.

The options for fulfilling the energy needs of agriculture fall into two broad categories:

—Increasing energy supply by developing renewable energy alternatives including energy cropping, integrated energy systems, and priority allocation.

—Decreasing energy demand by designing more efficient energy management systems and reduced-energy farming alternatives.

A FRAMEWORK FOR POLICY ASSESSMENT AND FORMULATION†

The purpose of a framework for analysis is to evaluate the potential of various energy resource and technology options needed to meet national economic and social development goals. Two categories of options are considered: (1) improvement of existing elements in the energy system, and (2) introduction of new technology such as solar cells and biomass conversion.

These options are diverse—ranging from increased end-use efficiency (e.g. better stoves), to centralized technologies (e.g. large-scale electric generation), to centralized technologies (e.g. farm-level gasifiers). They also encompass commercial fuels as well as noncommercial fuels, such as animal dung. These resource and technology options must be analyzed in relation to a number of criteria:

—Total demand and fuel mix.
—Reduction of imported oil.
—National and social goals, such as improved nutrition, housing, and rural development, including electrification and increased food production.
—Conservation of total energy costs.
—Maintenance or improvement of environmental quality, such as reducing soil erosion, air and water pollution, etc.

Computer simulation techniques are used to quantify relationships and extrapolate beyond observed data. Scenarios can be developed to illustrate various combinations of assumptions.

India (ABE, 1985)

A first step towards selecting options and formulating energy policies is to analyze current and future energy supply and demand. An example is the study, *Towards a Perspective on Energy Demand and Supply in India in 2004/05*, published by the Indian government's Advisory Board on Energy (ABE, 1985). This comprehensive study forecasts energy supply and demand in 2004 and 2005. It concludes that 'Indian energy problems are acute in nature and likely to accentuate with the passing of time unless we take the most resolute and sustained steps to tackle them.' Assumptions included:

A minimum level of energy consumption in the household sector.

—2596 kJ/person per day for cooking;
—126 kJ/person per day for space heating;
—126 kJ/person per day for lighting.

† Adapted from Mubayi *et al.* (1981).

Irrigation using a mixture of electric, diesel, and animal power for pumping.
Three population values in the year 2004 and 2005.

—1·003 million;
—1·046 million;
—1·115 million.

Indian energy demand by the year 2004 and 2005 was projected as follows:

—Fuelwood: 300–330 million tonnes (120–130 million tonnes in 1985). This assumes a cooking stove (chulha) of 8% efficiency.
—Dung cake for fuel: 200 million tonnes (the report concludes that 'this cannot be allowed because there would be no dung for the biogas program. Dung cake for fuel should be stabilized at 75–80 million tonnes).
—Agricultural residues for fuel: 100 million tonnes.
—Electricity: 139 000–188 000 MW.
—Coal: 450–550 million tonnes.
—Oil: 94–123 million tonnes.

Proposals to meet the demand include:

Sixty million hectares of marginal land could be used for fuelwood/fodder plantations. (This recommendation will be difficult to implement because of other land needs.)
As many as 16–22 million households (out of 121 million in India) could meet energy needs for cooking through biogas. So far, only a fraction of this potential has been achieved. Research and development needs include methanogens active at temperatures less than 20°C, processes requiring less water, and multi-feedstock policy, including human feces and human and animal urine.
The electrical mix should include:

—10 000 MW nuclear thermal;
—6570 MW lignite;
—77 000–112 000 MW coal;
—1750 MW combined cycle lean gas;
—57 450 MW hydroelectric;
—5000 MW micro/mini hydropower.

The use of solar cells for electrification of remote villages should be promoted vigorously (the estimated potential is 75 MW).
Wind power sites along the coasts where average windspeeds exceed 22 km/h should be exploited.
Coal exploration should be intensified and household use promoted.
Sixty million cubic meters per day of natural gas is projected to be available, but large amounts are needed for ammonia production.
To manage the oil supply deficit, kerosene should be phased out as a lighting fuel, natural gas should be used as a diesel fuel, and methanol should be explored as a transport fuel. Greater attention to fuel efficiency is essential.

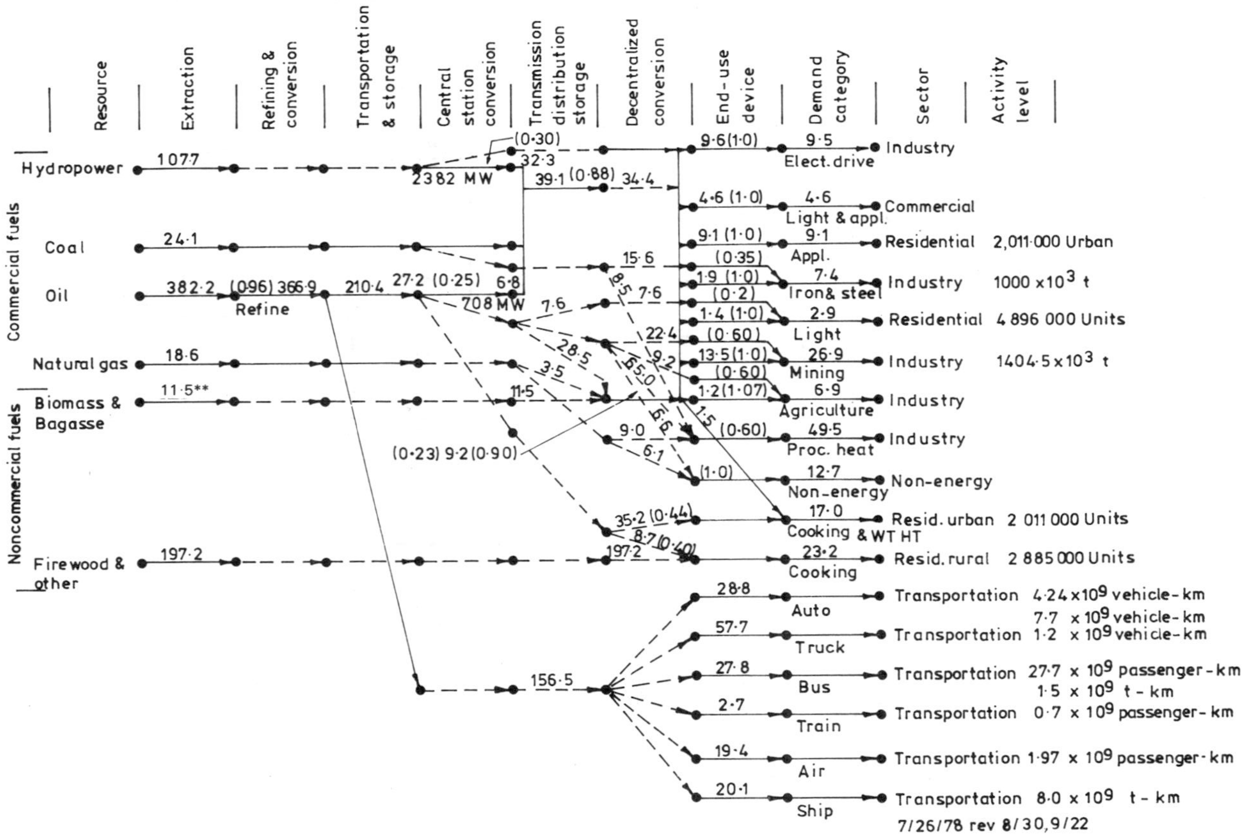

Fig. IX.1. Reference energy system for Peru, 1985. Energy resource consumption: 741×10^{15} J. Notes: 1. Flows in 10^{15} J. 2. Solid line indicated real process. 3. Conversion efficiencies are shown in parentheses. Connected to grid before 1985.

** This energy input is used for electric generation. In addition, approximately 20 GJ are required for process heat in the production of sugar. The source of this process heat is not known, but is believed to be bagasse. This accounts for up to 90% of the energy content of the bagasse produced (source: Mubayi *et al.*, 1981; © 1981 John Wiley & Sons, Inc.).

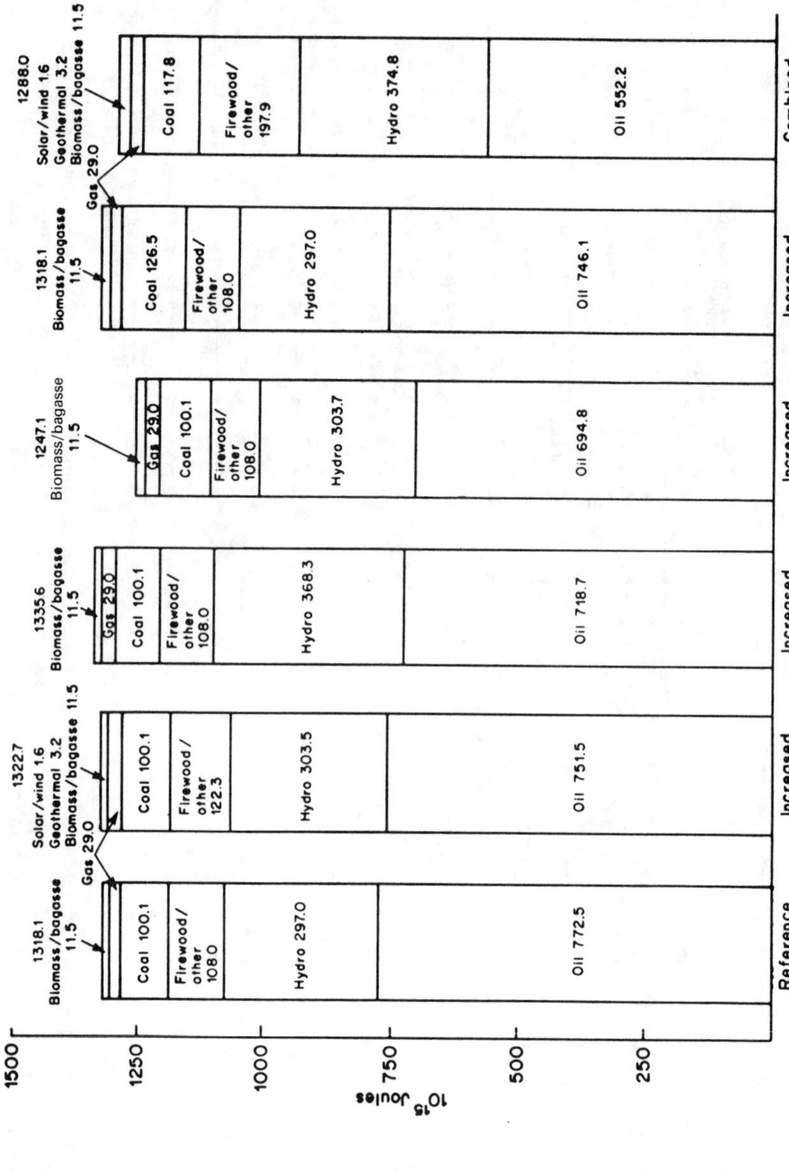

Fig. IX:2. Resource consumption in the year 2000 under alternatives (source: Mubayi *et al.*, 1981).

This look into the future of India is an excellent example of the kind of analysis that should be undertaken in all nations. The results will provide a basis for future energy plans and policy formulation.

Peru (Mubayi *et al.*, 1981)

An analytical approach to energy resource and technology assessment was developed by the Brookhaven National Laboratory in New York. It included a reference energy system (RES), which integrates a set of energy demands, energy conversion technologies, fuel allocations, and energy resources into an overall energy supply–demand balance. The cost of various energy forms, environmental emissions, and other factors associated with all elements in the system are part of the supporting database for the RES.

Figure IX.1 presents the RES for Peru in 1985. Similar analyses were made for 1976 and 2000. The energy resource in the left column represents the raw energy input necessary to satisfy a set of basic energy demands. For example, oil, natural gas, and biomass/bagasse are input resources for agriculture, whereas oil and firewood are resources for cooking.

Five basic energy strategies were analyzed to evaluate their impact on the reference case (Fig. IX.2). These strategies included increased use of renewables, such as solar, wind, and biomass; increased use of hydropower; increased energy efficiency in industry and transportation; and increased use of coal. The combined strategies (right-hand bar) would reduce both overall energy and oil usage for the year 2000 as shown in Fig. IX.2.

INCREASING THE ENERGY SUPPLY

More Energy Is Needed

To adequately feed the expanding world population and to meet other social and economic goals, the amount of energy used per person per hectare in agricultural production, processing, and distribution should be significantly increased. Extension of the current low-energy techniques commonly used—the man or woman with only a hoe, a bullock and plow, and ox-cart transportation to the nearest village—will not produce the surpluses needed to feed rapidly growing urban populations which are often far away from agricultural production areas.

Poor countries need to obtain as much energy as possible, as inexpensively as possible. Previously, policy-makers in most developing countries were able to ignore techniques for the proper management and effective use of noncommercial energy sources because oil was inexpensive and readily available and technology was usually borrowed from industrialized countries which had failed to develop and use noncommercial energy sources. Energy from biomass, solar energy applications in agriculture, small-scale wind energy, and hydropower were discussed in Chapters V, VI, VII, and VIII, respectively. Other energy sources and processes of special importance to agriculture, such as human labor and animal draft power, were discussed in Chapter II. Detailed cost–benefit analyses are needed to determine the feasibility of each process under particular conditions.

In his analysis of energy use in rural India, Revelle (1976) emphasizes that a considerable increase in energy use is necessary to meet future food needs—primarily for irrigation, chemical fertilizers, and additional draft power to cultivate fields.

The climate and water supply permit two crops per year on most of India's arable land, but double-cropping is possible only if facilities for surface and groundwater irrigation are significantly expanded and if an abundance of nitrogen fertilizers are available so fields do not have to be left fallow to accumulate nitrogen. According to estimates by the Indian Irrigation Commission, full irrigation development requires pumping about 46 million net ha-m of water annually from wells, which would utilize at least $4{\cdot}2 \times 10^{14}$ kJ of fuel energy—four times the bullock, diesel, and electrical energy currently used. Applications of nitrogen fertilizer need to be increased to about 100 kg/ha per crop (or 20 million tonnes) for 100 million double-cropped hectares. This increase in applications would result in an energy requirement of $1{\cdot}5 \times 10^{15}$ kJ.

Bullocks cannot provide enough draft power to prepare seed-beds rapidly enough to grow two crops per year. Makhijani and Poole (1975) estimate that an additional $2{\cdot}1 \times 10^{6}$ kJ/ha per crop are required to construct and operate small tractors, or $4{\cdot}2 \times 10^{14}$ kJ/ha per crop are needed if 100 million hectares are to be doubled-cropped.

Cultivating two crops per year would increase farm employment by at least 50% because of the added human energy requirement of $1{\cdot}26 \times 10^{14}$ kJ/year. The total additional energy requirement would be $2{\cdot}43 \times 10^{15}$ kJ more than the energy now used in Indian agriculture. With these added energy inputs, India could increase its foodgrain yield from the present 0·8 t/ha per crop to near the average United States yield of 3·28 t/ha per crop.

Assuming a yield of 2 t/ha per crop and double-cropping on 100 million hectares, foodgrain production could be increased to between 300 million and 400 million tonnes (depending on the amount of farmland devoted to other crops), or between three and four times the present production. The energy input from all sources per tonne of foodgrains would be near $7{\cdot}54 \times 10^{6}$ kJ, significantly less than the current situation. If India attained the same average yields per hectare in foodgrain production as the United States, the energy input would be about $4{\cdot}61 \times 10^{6}$ kJ/t of foodgrains.

Revelle (1976) concludes: 'The men and women of rural India are tied to poverty and misery because they use too little energy and use it inefficiently, and almost all the energy they use is secured by their own physical efforts. A transformation of rural Indian society could be achieved by increasing energy quantity and improving the energy technology.' This conclusion is applicable to virtually all non-oil-producing developing countries; however, among the obstacles are landlord–tenant relations, the lack of political support for national resource allocations to agriculture, and inquality of educational opportunities between city and countryside.

Energy Cropping (FAO, 1981)

The changing energy situation has created incentives in many countries to search for renewable sources of energy. Biomass, including the option of producing biomass as an energy crop, has received significant attention. The idea of energy self-sufficiency in agriculture is appealing to many, and if farmers produced their own fuel crop, it would be a step closer to reality. But use of land and other resources would compete with food production and create hardship if higher food prices resulted.

The situation varies widely from one country to another. In very large countries such as Brazil, where only a fraction of cultivatable land is required to grow enough food, production of alcohol from sugar cane or other crops may raise the real incomes of rural inhabitants by providing increased employment. But in Asian countries with high population densities on cultivatable land, growing biomass fuels for transportation and electricity may actually increase food prices and reduce the real incomes of the poor in spite of the added employment provided by rural industry (Revelle, 1980).

FAO organized an Expert Consultation on energy cropping versus food production in 1980. Experts from countries already engaged in large-scale alcohol production programs, as well as authorities from countries planning such programs and from technical institutes, met with FAO staff members to share experiences and evaluate the benefits and risks of energy cropping. Specific recommendations included the following (FAO, 1981):

—Individual countries planning biomass energy programs as part of their agricultural energy systems should carefully examine the range of factors relevant to their situation, as well as the timing and flexibility of policy options. They should seek assistance in this endeavor from international technical assistance agencies in areas such as monitoring and assessing relevant and appropriate technology; providing a clearer understanding of the complex interrelationships influencing food/energy decisions; and assisting individual country levels.

—Trade and aid mechanisms transmit the influence of food/energy conversion programs throughout the world; therefore, countries should consider the potential effects of their actions on other countries. To analyze these effects, the necessary information must be available to make a meaningful assessment of worldwide implications. International organizations interested in tracking the world supply/demand balance, such as FAO, the World Bank, the World Food Council, and specific commodity-oriented organizations, should assess and monitor possible outcomes.

—Underlying the previous two recommendations is the importance of production response to the changed circumstances created by the increased demand for food/feed commodities for energy conversion. While short-term production responses are reasonably well-known, uncertainties surround the potential long-term reponses. Thus, the Consultation recommended that more long-term analysis be conducted to improve the understanding of future policy impacts related to food/energy conversion programs. The Consultation determined that evaluation of the impacts of existing biomass energy programs and their future development in both developed and developing countries is necessary to guide future actions.

Numerous studies have been conducted on the food versus fuel issue in general and specific analysis for a given country or region. A primary concern is that growing crops for fuel will have an adverse effect on food supplies and prices (L. R. Brown, 1980; Kenney & Erichsen, 1983).

One study was performed in the Dominican Republic (Trehan *et al.*, 1980). The conclusions were as follows:

—About 0·5 million hectares of arid and semi-arid land is available for biomass farming and has the potential of providing wood for 300–1800 MW of electrical generating capacity.
—Biomass farming in dry forest areas should be based on minimizing production costs rather than maximizing yields by intensive agriculture. Local farming methods with high labor input and minimum commercial energy input should be implemented.
—Direct wood combustion to produce steam for driving turbines is the most logical energy conversion method.
—The major challenge in implementing a biomass farming program is acquiring and motivating a large force of workers to plant, tend, and harvest the biomass crops.
—Wood-fired generating plants appear to be economically attractive.
—Wood should also be considered for other uses, including charcoal production.

Another study examined energy farming to produce transport fuels in New Zealand (Harris, 1980). This detailed, three-volume report concluded that energy farming could provide substantial quantities of transport fuels, perhaps as much as 30% of New Zealand's needs by the year 2000. The cost was projected to be similar to imported gasoline costs. Both agricultural and forest crops were considered. Beet-to-ethanol and wood-to-methanol both appear to be viable alternatives for producing large quantities of transport fuels. Small-scale, on-farm plants did not appear to be competitive with larger plants. Byproduct utilization is important and needs serious consideration.

Many other examples of energy cropping could be cited, such as the Brazil case in which billions of liters of liquid fuels have been produced from sugar cane and other biomass feedstocks. Nevertheless, the economic energy picture can change unpredictably and make energy cropping less attractive. For example, lower world oil prices in the mid-1980s tended to discourage energy farming and made it more difficult to justify. A well-conceived, up-to-date feasibility study is necessary before proceeding with energy farming proposals.

Other Solar Options

Solar heating, wind power, and hydropower were discussed in Chapters VI, VII, and VIII, respectively. These options and others have their niche in the energy supply mix. All are highly site-specific. The technical aspects of these technologies have been thoroughly researched and are thoroughly understood. In isolated, rural areas, where energy resources are plentiful, any of these technologies may prove technically and economically viable. Again, a current, scientifically based feasibility study is absolutely essential to determine if a renewable energy system is a wise investment.

Integrated Energy Systems (NRC, 1981*c*)

Use of a mixture of energy technologies in crop and/or livestock production is referred to as an integrated energy system. The primary output or byproducts from one component or element become the input for another element so overall energy system output is optimized. Figure IX.3 illustrates a simple integrated energy system consisting of a home, crop and livestock production, and a compost pile. The composting unit could be replaced by a biogas plant or a fish pond. Several feedback loops are illustrated.

Figure IX.4 depicts a small-scale integrated energy system in Papua, New Guinea which includes a biogas generator, an algae pond, a fish and duck pond, and an irrigated garden. Tilapia yields of up to 1 t were estimated, and about 200 ducks were

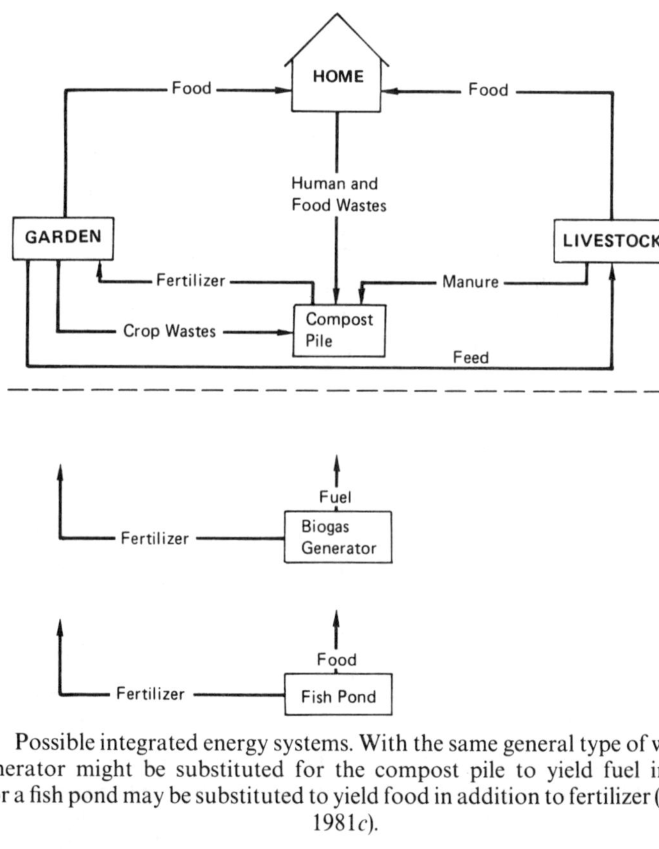

Fig. IX.3. Possible integrated energy systems. With the same general type of waste inputs, a biogas generator might be substituted for the compost pile to yield fuel in addition to fertilizer, or a fish pond may be substituted to yield food in addition to fertilizer (source: NRC, 1981*c*).

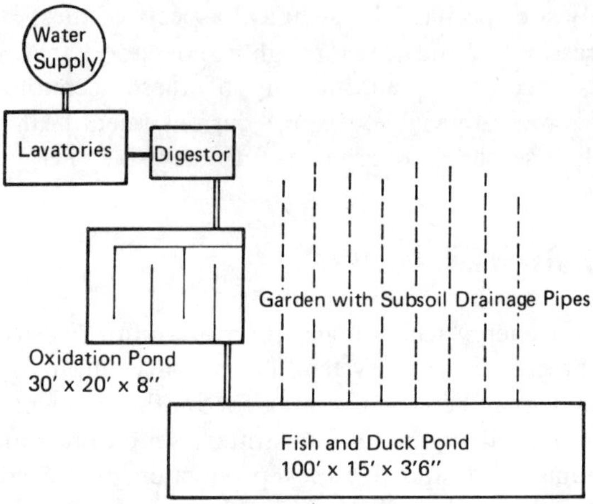

Fig. IX.4. An operating integrated system in Papua, New Guinea (source: NRC, 1981*c*).

raised each year. Returns from this simple, but unconventional waste system, were impressive.

Another example of an integrated energy system includes a greenhouse that uses solar and wind energy and produces fish and vegetables (Fig. IX.5). Wind power pumps water through a solar water heater from the lowest to the highest pond. In the first pond, water is purified in a filter system. In the second pond, daphnids, cyclopids, and algae are cultured. The second pond overflows into the last and largest pond, sweeping these organisms to the tilapia. The water in these ponds helps provide a heat reservoir for the plants growing in the enclosure, and plants are used partially as fish feed. Some of the nutrient-rich water is also used to fertilize the garden outside the enclosure.

The integrated system shown in Fig. IX.6 has been proposed in the Philippines for locations where space is limited. It includes a swine production operation, algae ponds, a biogas generator, fish ponds, and vegetable plots. Manure from the hogs and wastes from the home are used in a biogas generator to produce methane for power and light. The liquid effluent from the biogas unit is used for algae culture, fish farming, and garden irrigation. A portion of the algae produced is used for pig food. A windmill pumps the effluent.

An even more compact design is illustrated in Fig. IX.7. This design is based on a single family unit plus one cow and 50 chickens. The system includes biogas generation, algae production, and rainwater collection. Algae in slurry form is fed to the cow and constitutes its sole source of drinking water to force algae consumption in wet form. Slurry not consumed is dried on sand-beds and fed to the chickens. All human and animal wastes are fed to the biogas unit to produce fuel and the biogas effluent is used for algae culture.

An integrated dairy farm is shown in Fig. IX.8. Manure is used for biogas generation and the effluent is used to raise duckweed. Several species of duckweed

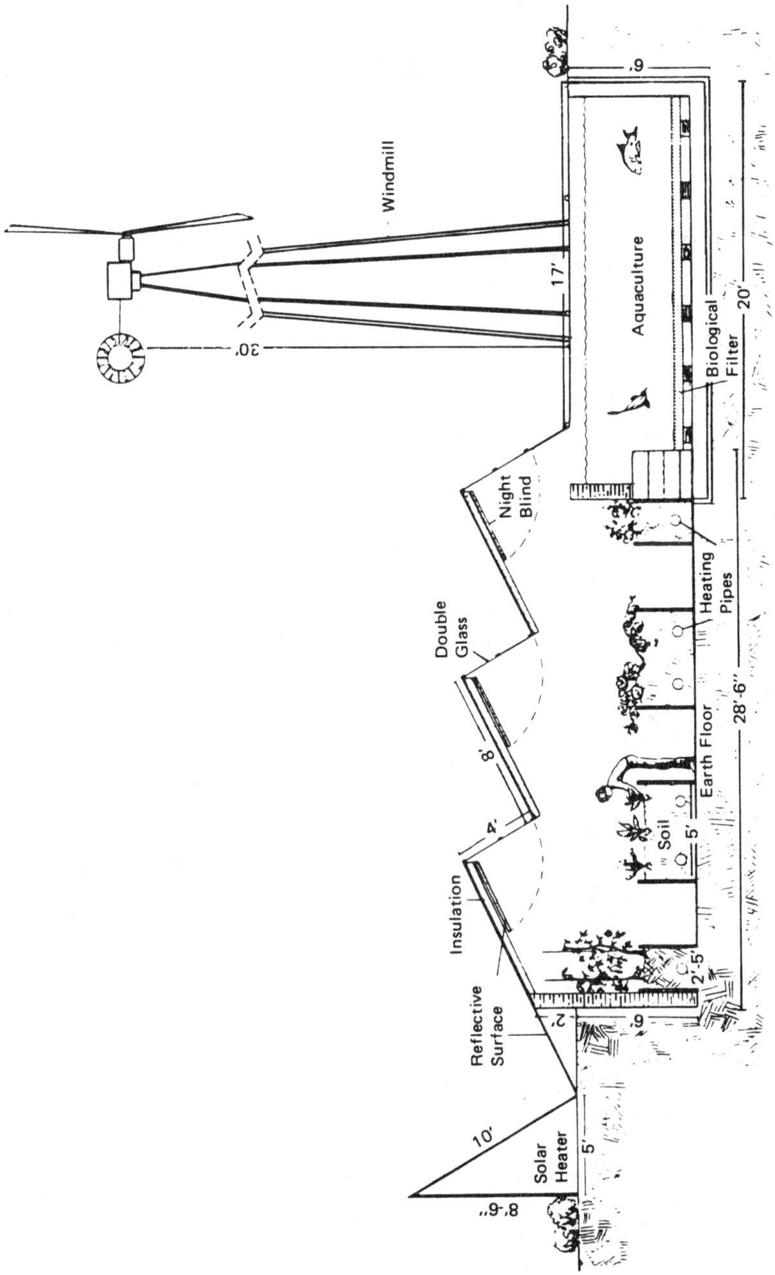

Fig. IX.5. A solar-heated greenhouse–aquaculture complex for cooler climates (source: NRC, 1981*c*).

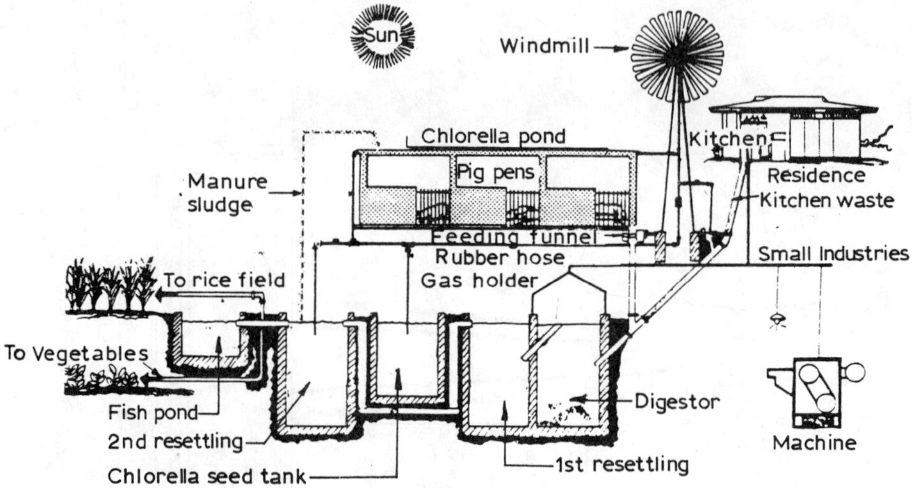

Fig. IX.6. A proposed integrated system including pigs, fish, biogas, *Chlorella*, and vegetables (source: NRC, 1981*c*).

are nutritious as food and feed. Overflow from the duckweed lagoons could be used for fish culture and irrigation.

An ethanol production and anaerobic digestion system is illustrated in Fig. IX.9. Biogas from cattle or swine manure provides heat for the ethanol distillation system. Ethanol is used to power the farm equipment. Maize, partially fertilized by effluent from the biogas unit, is grown as a feedstock for ethanol production, and byproduct stillage is fed to the livestock.

The preceding examples have focused on small-scale systems. A large-scale project in Thailand is illustrated in Fig. IX.10. It includes a rice mill and an agriculture–aquaculture complex that produces about 450 t/day of parboiled and polished rice from purchased paddy. The byproducts and co-products of rice processing are used in the production of swine, poultry, eggs, fish, vegetables, bricks, bran oil, and energy. Rice husks are formed to produce heat for parboiling and drying. Waste heat from flue gases is also used for drying paddy. Chicken coops are located above the pigsties so that wasted feed and droppings can be consumed by the pigs. Some pig manure is used in the biogas plant; the remainder is used as fertilizer.

In China, the Zinbu Brigade of the Liliu Commune is working toward self-sufficiency through an integrated farming process. Figure IX.11 illustrates their system, which includes production of fish, crops, silkworms, and pigs. Much of the energy and fertilizer requirements are met by recovering animal, human, and agricultural wastes. Nearly all of the 89 families have individual biogas units and seven other digesters are operated on a communal basis. Gas from family units is used for cooking and gas lighting, whereas communal digesters are used to produce electricity.

The United States Department of Energy sponsored a series of energy-integrated

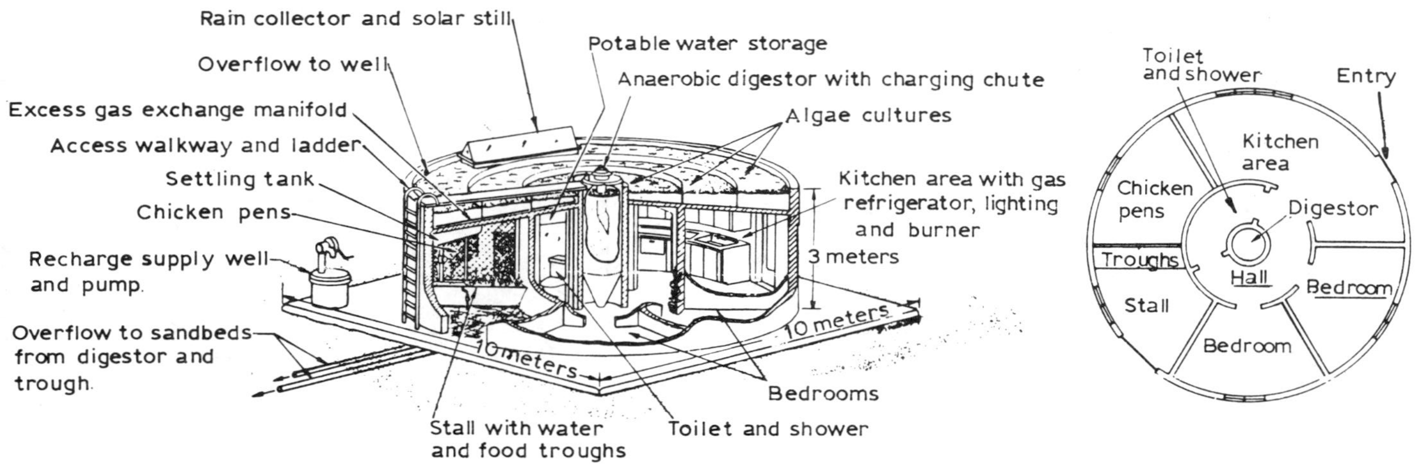

Fig. IX.7 A proposed dwelling unit for a family of four and their livestock with a recycle system for water, nutrients, and energy (source: NRC, 1981*c*).

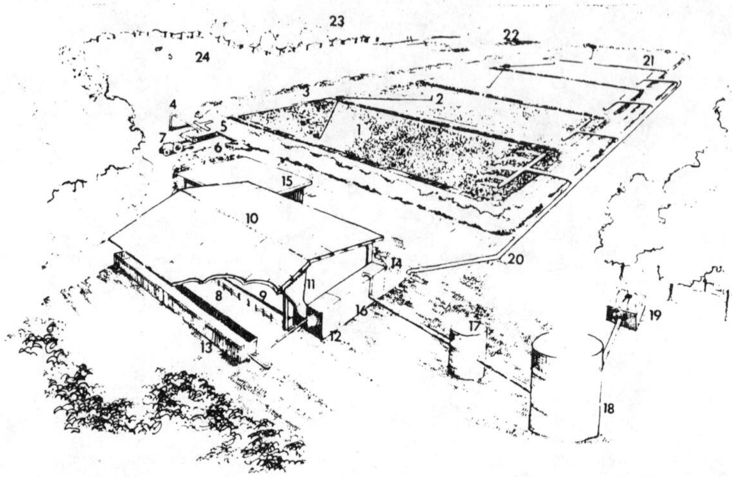

Fig. IX.8. An integrated dairy farm based on the culture of duckweed (source: NRC, 1981*c*). 1, Two acre lagoons; 2, skimmer; 3, harvesting canal; 4, washwater line; 5, conveyor/strainer; 6, washwater collection pit; 7, feed trailer; 8, feeding alley; 9, feeding trough; 10, feedlot covered with solar collector system; 11, duct drawing heated air from roof; 12, heat storage; 13, drying tunnel; 14, waste pit; 15, milking center; 16, fermentation unit; 17, gas scrubber; 18, methane gas storage; 19, generator; 20, waste sludge to lagoons via heat exchange in waste pit; 21, final stage lagoon treatment; 22, aquaculture system; 23, orchard; 24, pasture.

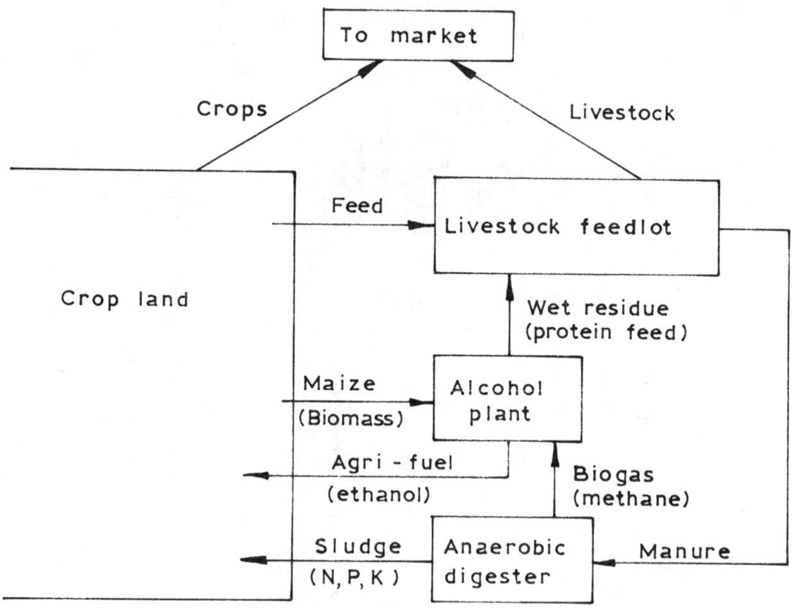

Fig. IX.9. Integrated farm fuel system (source: NRC, 1981*c*).

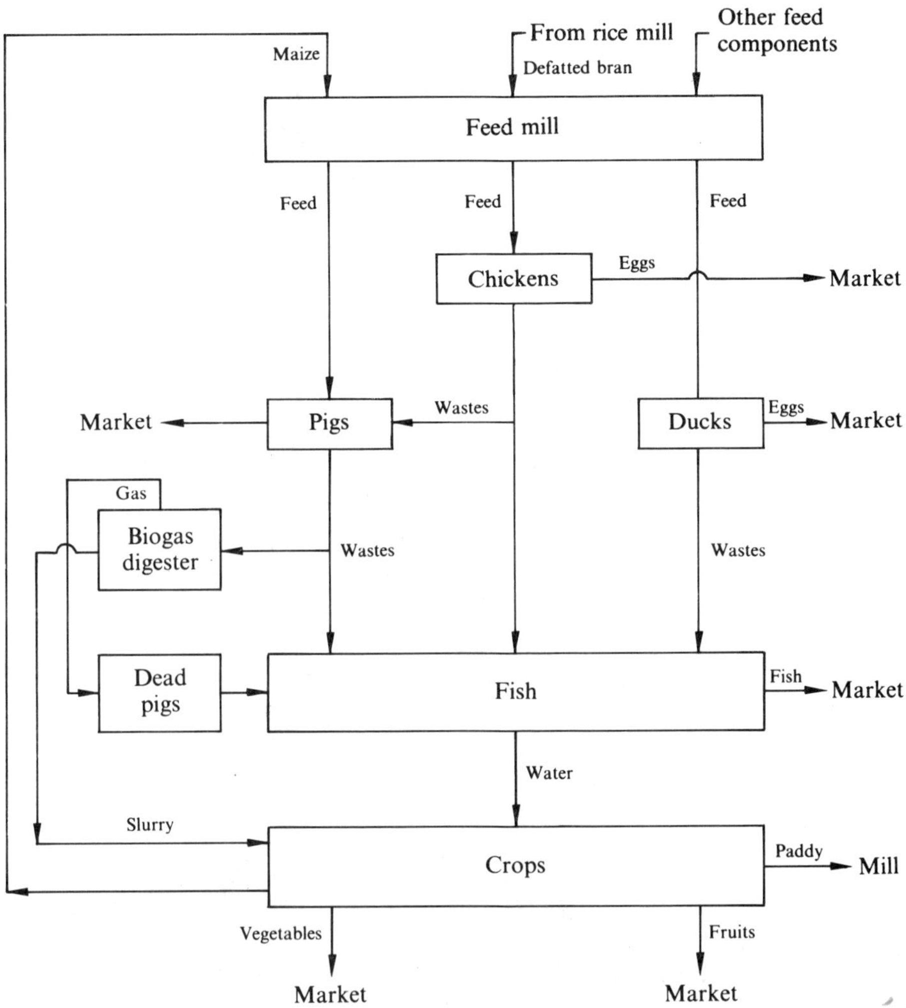

Fig. IX.10. Schematic of M/S Kamol Kij Co., Ltd farm, Thailand, in which livestock, crops, and fish are integrated (source: NRC, 1981*c*).

farm studies described in Table IX.1. These demonstrations involved many of the elements previously mentioned plus solar PV cells to generate electricity to pump irrigation water. Many other examples could be cited, but these illustrate a wide range of small- and large-scale approaches.

There are two major limitations to integrated energy systems. The first is the obvious capital investment required, although some of the systems described are designed specifically to keep costs to a minimum. The second limitation is the inherent complexity of integrated energy systems, which may require a high level of management skill to operate them efficiently. Most energy technologies are relatively complex even in a standalone situation. When two or more energy

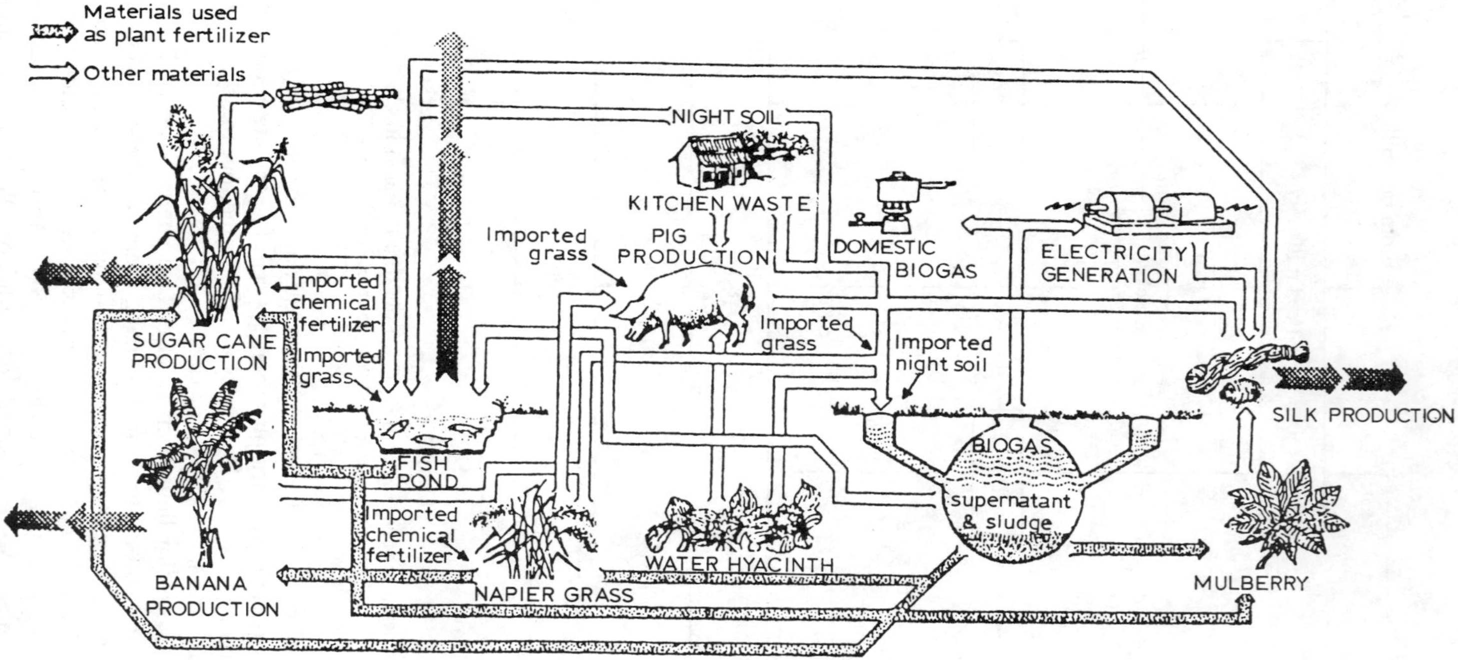

Fig. IX.11. Fish, pigs, sugar cane, bananas, and silk are produced on this integrated farm in rural China (source: NRC, 1981*c*).

TABLE IX.1
Summary of United States Department of Energy—integrated energy farm systems, 1980–84 (source: Hiler & Stout, 1985)

	Puerto Rico	*Georgia*	*Nebraska*	*North Dakota*	*Texas*	*Virginia*	*New York*
Location	Granja Caribe Poultry Farm and University of Puerto Rico	Aubrey Farm and Georgia Institute of Technology	University of Nebraska Field Laboratory Farm	North Dakota State University Agricultural Experiment Station	Del Valle Hog Farm and SUMX Corporation	Foxlease Farm and Archbold Investment Company	Millbrook Farm and Cornell University
Contact	University of Puerto Rico, Rios Piedras, Puerto Rico 00931	Georgia Institute of Technology, Atlanta, Georgia 30332	University of Nebraska, Department of Agricultural Engineering, Lincoln, Nebraska 68508	North Dakota State University, Department of Agricultural Engineering, Fargo, North Dakota 58102	SUMX Corporation PO Box 14864, Austin, Texas 78761	Foxlease Farms and Archbold Investment Co., PO Box 2394, Upperville, Virginia 45357	Cornell University, Department of Agricultural Engineering, Ithaca, New York 14850
Farm characteristics	16-acre farm; 36 000 hens	600 dairy cattle; 250 milk cows	Rotating corn, soybeans, and sweet sorghum	500 acres cropland; 100 milk cows, 980-ft^2 residence	2 990 acres grain	1074 acres cropland; 200 acres pastureland; 200 acres woodland	455 acres total; 300 dairy cattle; 180 milk cows
Energy alternatives demonstrated	Anaerobic digestor; solar	Alcohol still; anaerobic digestor; solar; electricity generation from methane	Anaerobic digestor; solar; alcohol still	Solar; anaerobic digestor; alcohol still	Anaerobic digestor; electricity generation from methane; alcohol still	Electricity generation from methane; solar; anaerobic digestor; alcohol still	Anaerobic digestor; alcohol still; electricity generation from methane
Expected energy savings	NA	16 000 kWh/month; 10 000 gallons of diesel/year; 9 700 gallons gasoline/year; 2 000 gallons propane/year	NA	3 940 gallons fuel oil/year; 5 000 kWh to residence	5 500 gallons diesel/year; 7 000 gallons gasoline/year; 47 kWh day	5 900 gallons diesel/year; 6 400 gallons gasoline/year; 184 000 kWh/year	438 barrels oil/year
Total cost	$1 013 000	$688 500	$1 720 400	$797 400	$818 000	$1 353 900	$1 532 000
Government share	$638 000	$378 700	$860 200	$398 700	$409 000	$677 000	$766 000
Private funds	$375 000	$309 800	$860 200	$398 700	$409 000	$676 900	$766 000

technologies are combined with crop or livestock production and the human element, the overall system may become too complex to manage. Simplicity of design and operation is a prime consideration in planning integrated energy systems.

Priority Allocation of Scarce Fuels

One way to increase the supply of scarce fuels to agriculture and the rural sector is by priority allocation. If determined to be in the national interest, allocation mechanisms other than price can be devised to ensure the fulfillment of agriculture's energy needs.

DECREASING ENERGY DEMAND THROUGH EFFICIENT MANAGEMENT

Many publications have addressed energy conservation. Energy problems in the rural areas of developing countries cannot be solved by energy conservation alone. The primary challenge is to make more energy available for food production and other elements of the food system. Rural inhabitants need more energy for improved housing, transportation, and other aspects of quality living.

The rural sector of developing countries is comprised of individuals and families in three categories:

1. *Subsistence farmers*: people for whom farming is a means of survival; that is, existence in an austere manner.
2. *Rural non-farmers*: inhabitants of the villages and small towns, such as shop keepers, traders, and laborers who are not actively involved in farming.
3. *Commercial farmers*: people who use modern inputs and operate their farms as businesses.

Energy management procedures for each of these groups must be addressed separately.

Subsistence Farmers

This group uses little or no commercial energy. Therefore, energy management for them means efficiently applying their own manual efforts, or, for some, efficiently using animal power. Limited commercial fertilizers or pesticides may be used in this environment. Dung cakes and firewood are the primary fuels for cooking. Efficient energy management for the subsistence farmer denotes survival with few comforts and barely enough food at various times during the year.

Rural Non-farmers

These individuals use some commercial energy for lighting, heating, and cooking. Common energy forms include kerosene and electricity for lighting, and purchased wood for cooking. Perhaps the greatest energy savings in this environment could be achieved through the development and widespread use of improved cooking stoves. Traditional stoves achieve very low efficiencies (8–12%); thus, the potential for stretching energy supplies through improved stove efficiency is significant.

Commercial Farmers

This sector requires modern inputs to enhance production. Energy in the right form at the right time is a vital input. Commercial farming is a business; thus, agricultural energy decisions are made in a professional manner, emphasizing reduced operating costs and increased return on investment. Reduction of risk is a reality that may encourage these farmers to invest more energy to complete a task in good weather and to minimize crop losses.

Energy is a relatively small item in the total agricultural production budget of commercial farmers. Therefore, many commercial farmers are not exceedingly concerned about energy as long as plentiful supplies are available when needed. Nevertheless, using energy as efficiently as possible is good business.

Chapter IV addressed efficient energy management practices in agriculture and presented numerous examples of energy-saving techniques.

POLICY IMPLICATIONS†

The formulation of agricultural and rural energy policies within the overall context of national and international energy policies, is obviously a complex challenge, but one that deserves high priority.

Energy Use in Agricultural Production and the Food System

The agricultural sector must receive the small but crucial supplies of commercial energy needed for production and use of energy-intensive inputs. Until new technologies are developed, these energy-intensive inputs are essential to continue the expansion of agricultural production, and, in particular, to accelerate production in developing countries to keep pace with the demands of growing populations, without increasing imports. A number of countries will depend on the agricultural resource to provide part of the renewable energy resources, through energy cropping, to supplement fossil fuels. Therefore, the agricultural industry should have priority in energy use for such essential purposes as chemical fertilizers and pesticide production, and power for farm machinery and irrigation equipment.

In many countries, price policies will be a major consideration in ensuring that farmers can both obtain and afford the energy-intensive inputs they require. In general, such inputs increase yields so significantly, that their use should continue to be profitable even at higher prices. However, observing farm price relations closely in association with the volatile energy situation is critical. Selective subsidies must be

† Adapted from Hrabovszky (1982).

used, in some cases, to raise food prices, while compensatory measures are implemented in favor of poorer consumers. Policies such as transport equalization schemes for fertilizers and other inputs may require reconsideration in relation to higher transportation costs.

At the same time, agriculture, like all other sectors, must emphasize economical energy and use the gradual transition from nonrenewable to renewable sources of energy. These measures not only reduce the depletion of irreplaceable resources, but also improve the environment.

A major task, particularly for agricultural research and extension services, is developing and promoting farming systems that save fossil fuel. These systems would implement more efficient usage of chemical fertilizers, increased usage of organic fertilizers, crop rotations, catch cropping (especially with legumes capable of fixing atmospheric nitrogen), improved tillage, and other practices to save tractor fuel. In addition, they would provide methods for more efficient use of irrigation water, weed control by improved tillage and mechanical or hand-weeding, integrated methods of pest control that rely as much as possible on natural elements, greater use of draft animals, and improved hand implements and animal draft equipment. Basic agricultural research should be accelerated to enhance photosynthesis efficiency in crop plants, to reduce requirements for and increase crop responsiveness to energy-intensive inputs like fertilizers, to breed pest- and disease-resistant varieties, and to transfer the capacity to fix atmospheric nitrogen in such basic food crops as cereals and roots.

The fisheries sector also offers considerable opportunities for energy savings. Measures to control the number and size of fishing vessels would not only save fuel and reduce operating costs, but would also contribute to the conservation of an important natural resource by limiting access to heavily exploited stocks.

In addition, an obvious, but overlooked area for saving energy is the non-farm areas of the food systems in developed countries. In some instances, relocating the production of certain agricultural commodities because of higher transportation costs, may be necessary.

Fuelwood and Charcoal Production for Rural Households and Other Uses

The evolving energy situation is placing new emphasis on every possible energy resource—especially renewable resources such as wood, which is traditionally one of the most important. In addition to its attributes as a renewable and versatile form of stored energy, wood is socially and environmentally acceptable. When available and locally accessible, its sustainable supply can contribute substantially to energy self-reliance, at least in rural energy systems. Because a large number of people in developing countries depend on wood for energy, the demand for wood is likely to be reinforced in the foreseeable future by the increasing difficulty in shifting to alternative fuels. The maximum contribution of forest biomass to the basic energy requirements of these populations must be accomplished through carefully planned, environmentally compatible integration into rural energy systems.

Production and utilization of wood for energy on an industrial scale also deserves consideration because the cost of energy obtained from wood is continuously improving, while the price of fossil fuels increases and supplies decrease. Either through direct conversion or through alternative fuel production, the contribution of wood to new energy systems could be substantial either in decentralized energy systems or when important land and capital resources are available. The use of wood for energy does not have the limitations associated with fossil fuels.

The potential production and use of wood for energy are related to the following major issues:

—Wood-related energy must be examined in the context of national energy policies and energy balances within the forest industry. Wood is advantageous as a low energy requirement material as well as a renewable source of energy. This new energy dimension may require some adjustment in forest policies.

—Most fuelwood and charcoal supplies presently meet the basic energy requirements of rural energy systems and should be closely integrated into these systems and into other rural development activities. Farmers who grow trees for fuel in areas outside the forests need forest services to provide the necessary research and extension support.

—Research and extension are necessary to promote more efficient use of available wood fuel resources. Specific concern should be directed toward areas that depend on wood as a primary source of energy for domestic requirements and areas where severe energy shortages are devastating the poor and deteriorating the environment.

—The available information on producing and using wood for energy needs to be substantially improved to include enhanced information on available resources; potential supply, including wood residues; consumption; and present and future requirement levels. Accurate information should form the basis for planning future developments and identifying and formulating programs and projects.

—A plan for organizing and distributing the supply of wood for energy deserves particular emphasis. Wood should be made available on a physical and economic level which meets the requirements of the users.

—For energy to be a primary wood production objective, the multiple use and management of forest resources must be accelerated. Other production and environmental functions should also be examined for energy-related advantages. For example, the forest biomass, which is suitable for energy use, but is not actively managed, should be investigated.

—The potential for energy plantations requires careful assessment based on possible biomass productivity, availability of adequate land, and comparative costs in relation to competition with other land uses. Perhaps growing trees for fuel could be integrated with other agricultural practices.

—The possibility of using of wood for the industrial production of energy is based mainly on comparative economics in relation to other fuels. Therefore, continual analyses of the advantages of energy from wood, as well as the implications of new technological developments, are critically important.

The renewed importance of wood as a source of energy is increasing the role forestry plays in meeting the basic requirements of human life, especially for billions of people in rural areas. This new energy dimension requires closer integrated resource management in conjunction with other rural development activities within the framework of comprehensive energy policies. Foresters and forestry institutions may need to enhance their professional capabilities in order to meet the new challenges created by the demand for wood as an energy source.

General Considerations on Energy Policy and Planning

National energy policy in developing countries must address the needs of poor rural communities. Officials need to ensure that traditional sources of fuel are not made unavailable because of wholesale conversion to more costly commercial energy that is too expensive to obtain. Resources are necessary for the development of renewable sources of energy such as windmills and small hydroelectric units, which, although insignificant at the national accounting level, can make a substantial impact on the standard of living for the rural poor.

National energy policy-makers in developing countries may tend to view the agricultural and rural sectors primarily as a plentiful source of renewable energy to provide substitute fuels for the predominantly urban, affluent minority, thereby easing the balance of payments. This danger evolves because the agricultural ministry is usually in a relatively powerless position within the government hierarchy and the responsibility for rural development is often dispersed among various departments. One way of avoiding this danger is to establish the institutional machinery to formulate energy policy that involves the many interests concerned, including agriculture, fisheries, forestry, rural development, nutrition, and consumer organizations.

Finally, a number of international policy aspects are involved. The international research effort should focus strongly on the various national policy concerns discussed in this chapter. For example, the increased demand that energy cropping places on fragile soil resources contributes to the urgency of establishing procedures to monitor the condition of the world's soil resources at an international level. The potentially important role of natural fibers and rubber in replacing the more popular petrochemically synthetic products could be expedited by establishing more formal and informal international commodity agreements, including price stabilization measures.

Perhaps the most important international policy issues are those concerning world food security. The world food situation is already dangerously insecure, even without the new competition of energy cropping for scarce land and other agricultural resources. While some of the present fears may be exaggerated, the situation should be carefully monitored, especially in relation to the long-term implications once large-scale programs have been implemented. International agreement is needed, especially in the event of a scarce food supply situation that could occur as a result of widespread bad harvests. Priority must always be given to food exports, and food aid commitments must always be honored to ensure that food security reserves will not be jeopardized.

While these policy recommendations emphasize specific areas, effective solutions to the energy problems of rural areas can be implemented only if policy-makers fully consider the individual implications of each solution on all parts of rural energy systems, including the technical, economic, and social aspects.

RESEARCH NEEDS†

Following are some major research needs related to energy in agriculture. Most of these needs have been thoroughly discussed in this publication. Numerous research projects are underway to determine methods to best meet these needs.

Increasing the Energy Supply

More than half of the world's inhabitants cook and heat with wood, charcoal, animal dung, and other residues. Human labor and animal draft power account for a large share of the energy for producing crops.

Renewable energy in the form of solar heating, wind power, and hydropower, as discussed in Chapters VI, VII, and VIII, respectively, can be used to increase the energy supply. Biomass conversion processes discussed in Chapter V can also increase the energy supply by utilizing biomass feedstocks which are currently wasted. Energy research needs are many and varied. Technological improvements can lower the cost and improve the reliability of renewable energy systems.

Renewable energy systems require an assessment of the quantity of the resource. This assessment includes the quantities of biomass available in various locations and months of the year, solar insolation and how it is distributed, wind speeds, and hydropower potential. With biomass resources, the question of competing uses must be addressed. If biomass is used for energy, it will probably be diverted from other uses. What will be the consequences?

The question of biomass collection, densification, and transportation to a conversion facility also exists. In the most primitive case, biomass may be collected by hand, bundled or packed into a basket, and carried to the home where it is burned to cook food. But as the activity increases in size, collection may occur throughout an extensive area, and densification becomes necessary for efficient transportation. Because biomass production is usually seasonal in nature, storage of large quantities may be necessary. An example is a rice-straw-fired gasifier operating an electric generator year-round. Because rice straw may be produced during 1 or 2 months per year, large quantities must be stored for continuous power plant operation. Every aspect of renewable energy systems requires comprehensive research and development to ensure reliable, long-term operation.

Reducing Energy Needs

Fertilizers and other chemicals represent a tremendous energy input for crop production. Fertilizer-responsive varieties, more timely fertilizer applications, slow

† Adapted from Pimentel (1984).

release fertilizers, and increased legume production can all increase the energy efficiency of crop production, but all of these aspects require ongoing research. Development of biological nitrogen fixation in crops other than legumes could significantly reduce fertilizer energy inputs.

In some areas, pumping and moving irrigation water consumes a large amount of energy. Irrigation timing and application rates affect water use efficiency. New approaches to surface water application can improve energy and water efficiency. New technologies developed through research, such as drip irrigation and low pressure sprinklers, are important energy-savers.

Reducing crop losses is important, not only to maintain the world food supply, but also to ensure that energy used to grow crops is not wasted. Research directed towards reducing crop losses results in increased energy efficiency.

Genetic engineering, or biotechnology, has exciting potential for increasing food productivity and energy efficiency. Research activities targeted towards plants that can use water more efficiently, have better pest resistance, or produce more biomass, can help maximize energy use efficiency. For example, development of perennial grain crops will eliminate the current energy investment in annual plants.

Research studies on low-energy farming systems are underway at several locations. Decreasing energy input, while maintaining or increasing crop yields, is the ultimate objective of low-energy farming systems.

Nevertheless, food production is only one part of the food chain. Research to identify more energy-efficient methods of food transportation, processing, storage, marketing, and cooking can reap large dividends for both the producers and consumers of the world food supply.

TECHNOLOGY ASSESSMENT

Two general aspects of energy-related technology need to be considered—the direct and indirect effects.

Direct costs and benefits involve calculations such as payback periods, returns on investment, lifecycle cost analyses, etc. Generally, engineers and scientists have accurately identified these direct effects of technology and established the procedures for calculating them.

The indirect effects, however, are much more difficult to determine. Technology is developed to serve people, but the social impact is harder to define and quantify. Who gains and who loses? What is the effect on the environment? What are the implications to the user, the community, and the nation?

For example, if animal dung is used in a Gobar gas plant, it is then not available as dung cakes for cooking food. Therefore, is a Gobar gas plant a technological improvement, or not? If wheat or rice straw is used as a feedstock for a gasifier, will enough organic matter be returned to the soil? In the Punjab of India, the answer is probably yes, but what about in other developing countries? A given technology is never used in isolation. In the real world, a single energy source or system would rarely be used alone. Solar energy is always used in multiple ways to produce food. In addition, fossil fuels and electricity may be part of the energy system.

Therefore, gasifiers, anaerobic digesters, solar collectors, and other energy technologies should be part of an integrated system, with energy as just one component. Integrated systems are by nature complex and hard to understand and quantify. Non-numerical factors such as trade-offs and interactions among people and their politics, judgments, values, and opinions all have to be considered. However, people and their continued welfare are the only reasons for developing new technologies that incorporate more productive and efficient energy systems. As a result, the approach to an integrated systems analysis and the outcome may be controversial.

A technology assessment includes calculating the direct effects, establishing relationships between the quantifiable aspects, and clearly identifying the non-quantifiable aspects. Sophisticated computer simulation techniques for integrated energy system models have been developed and can help extrapolate beyond the observed data. Various scenarios can be developed to project the consequences of alternative courses of action. Although the consequences of alternative policies and actions will not be identified easily, questions must be asked and answers sought.

THE INDIAN ENERGY CENTER: A MODEL FOR DEVELOPING COUNTRIES

The Center for Advanced Studies on Energy Management in Agriculture located in Ludhiana, India, in the Punjab, is an example of how India is tackling its agricultural energy problem. It provides a model that may be useful for other developing countries.

With funding from the United Nations Development Program and the Indian Government, this national energy center seeks to ensure adequate energy for agriculture through a series of educational and research activities. One component, a School of Energy Studies in Agriculture, offers a PhD program in Energy Science and Technology for candidates from the engineering, agricultural economics, biochemistry, and microbiology disciplines. Research projects address all aspects of energy efficiency and energy supply for agriculture, including agro-industries. Pilot projects demonstrate new and improved technologies emanating from the research program. Contacts have been established with other Indian institutions, as well as relevant foreign institutions. Dissemination of scientific information and technology on energy application and management in the agricultural sector occurs through publications, seminars, workshops, symposiums, and extension programs.

Examples of recent projects at the Indian Energy Center include:

- —Study of energy flows and their socio-economic implications in two representative villages.
- —Anaerobic digestion of agricultural residues and cattle dung with emphasis on high solids content.
- —Instrumentation of draft animals to monitor their performance.
- —Evaluation of a commercial 45-kW gasifier.
- —Design of a 1-MW air-blown fluidized bed gasifier fueled with paddy straw.

Twenty specialized courses in the PhD program have been developed, including:

—Bioconversion of agricultural residues.
—Benefit/cost analysis.
—Production economics.
—Automatic digital computation.
—Energy management in agricultural production and processing.
—Fermentation technology advancements.

The Indian Energy Center has funds for sending faculty members and researchers abroad to receive short-term education and obtain specialized consultants. The Center operates under the auspices of the Indian Council of Agricultural Research as a part of its nationwide research and education program.

SUMMARY

There are no easy solutions to the energy supply and conservation dilemmas of developing countries. However, effective use of both commercial and non-commercial energy will significantly increase crop yields and improve production efficiency. Other elements of the food system, such as food processing, transportation, marketing, and preparation also require more efficient energy use to adequately feed all the people of the world. In addition, less expensive, renewable energy forms, such as direct solar heating, wind power, hydropower, and biomass fuels, must be tapped to ensure sufficient energy availability for the generations to come.

This publication, which discusses these alternative energy forms and efficiency measures, is designed to help developing countries in their quest to ensure a better quality of life for their citizens through effective energy strategies.

APPENDIX 1

Sources of Further Information

Alternative Energy Institute
West Texas State University,
PO Box 248,
Canyon, TX 79016,
USA

Alternative Sources of Energy (Magazine)
107 S. Central Ave,
Milaca, MN 56353,
USA

Annual Review of Energy
Annual Reviews, Inc.,
4139 El Camino Way,
Palo Alto, CA 94306,
USA
(Annual volumes beginning in 1976)

American Society of Agricultural Engineers
2950 Niles Road,
PO Box 410,
St Joseph, MI 49085–9659,
USA

American Society of Heating, Refrigeration, and Air Conditioning Engineers (ASHRAE)
1791 Tullie Circle, NE,
Atlanta, GA 30329,
USA

American Wind Energy Association
1730 N. Lynn Street, #610,
Arlington, VA 22209,
USA

Beijer Institute
International Institute for Energy and Human Ecology,
Royal Swedish Academy of Sciences,
Fach 1–104 05,
Stockholm, Sweden

Biomass (Journal)
Elsevier Applied Science Publishers Ltd,
Crown House, Linton Road,
Barking, Essex IG11 8JU, UK

Brace Research Institute
Macdonald Campus of McGill University,
PO Box 900,
Ste. Anne de Bellevue,
Quebec, Canada H9X IC0

Danish Wind Energy Association
Lille Sct. Hansgrade 20,
Wilborg, Denmark 8800

Electric Power Research Institute (EPRI)
3412 Hillview Ave,
PO Box 10412,
Palo Alto, CA 94303,
USA

European Cooperative Network on Rural Energy
FAO Regional Office for Europe,
Via delle Terme di Caracella,
00100 Rome, Italy

European Wind Energy Association
PO Box 1, 1755 26 Petten,
The Netherlands

Hangzhou Regional Center for Small Hydropower
Newsletter of the Asia–Pacific Regional Network for Small Hydropower,
PO Box 544,
Hangzhou, China

Intermediate Technology Development Group Ltd
Myson House,
Railway Terrace,
Rugby CV21 3HT, UK

Publications Office
9 King Street,
Covent Garden,
London WC2E 8HW, UK

Kern Wind Energy Association
208 S. Green Street, #5,
Tehachapi, CA 93561,
USA

Microfilming Corporation of America
200 Park Ave,
New York, NY 10166
(*Energy in Agriculture* microfilm collection contains more than 1800 papers and reports)

National Rural Electric Cooperative Association (NRECA)
1800 Massachusetts Ave, NW,
Washington, DC 20036,
USA

Renewable Energy Institute
1001 Connecticut Ave, NW,
Suite 719,
Washington, DC 20036,
USA

Solar Energy (Journal)
The Journal of Solar Energy Science and Technology,
Pergamon Press Ltd,
Headington Hill Hall,
Oxford OX3 0BW, UK

Solar Energy Research Institute
Document Distribution Service,
1617 Cole Blvd,
Golden, CO 80401,
USA

The United Nations University
Toho Seimei Building,
15-1 Shibuya 2-chome Shibuya-Ku,
Tokyo 150, Japan

United Nations Education, Scientific, and Cultural Organization
7 Place de Fontenoy,
75700 Paris, France

United States Department of Agriculture (USDA)
National Agriculture Library,
Reference Section, Room 111,
Beltsville, MD 20705,
USA

United States Department of Energy
1000 Independence Ave,
Washington, DC 20585,
USA

Volunteers for International Technical Assistance (VITA)
1815 N. Lynn Street,
Suite 200,
Arlington, VA 22209,
USA

Wind Power Digest (Journal)
Wind Power Publishing Co.,
PO Box 700,
Bascom, Ohio 44809,
USA

Worldwatch Institute
1776 Massachusetts Ave, NW,
Washington, DC 20036,
USA

World Bank
1818 H. Street, NW,
Washington, DC 20433,
USA

APPENDIX 2

Conversion of Units

The International System of Units (SI Metric System) is used in most of this book. The system consists of seven base units and a series of derived units as follows:

Physical quantity	*Base unit*
Length	meter (m)
Time	second (s)
Mass	kilogram (kg)
Temperature	kelvin (K)
Electric current	ampere (A)
Luminous intensity	candela (cd)
Amount of a substance	mole (mol)

Physical quantity	*Derived unit*
Acceleration	m/s^2
Area	m^2 or ha (hectare)
Density	kg/m^3
Electric energy	kWh
Flow rate	kg/s or m^3/s
Force	newton (N)
Energy or work	joule (J)
Frequency	hertz (Hz)
Head	m
Heat capacity	J/K
Heat content	J/kg or J/m^3 or J/liter
Heat flow rate	W
Insolation	W/m^2
Linear measure	μ (micron) or m or km
Power	watt (W) = J/s

(*continued*)

Physical quantity	*Derived unit*
Pressure	pascal (Pa)
Pumping head	pascal (Pa)
Specific heat	J/kg·K
Thermal conductivity (k-value)	W/m·K
Thermal resistivity	m·K/W
Thermal resistance (R-value)	m^2·K/W
Torque	N·m
Velocity	m/s
Viscosity, kinematic	m^2/s
Volume	m^3 or liters

The energy content of fuels used in this book are as follows:

No. 2 Diesel fuel = 39·0 MJ/liter or 140 000 Btu/gal
Gasoline = 34·6 MJ/liter or 124 000 Btu/gal
LP gas = 26·6 MJ/liter or 96 000 Btu/gal
Propane = 25·5 MJ/liter or 91 600 Btu/gal
Natural gas = 37·3 MJ/m^3 or 1000 Btu/ft^3
Crop residues = 10–15 MJ/kg or 4500–6500 Btu/lb
Wood = 23 MJ/kg or 10 000 Btu/lb
Coal = 30 MJ/kg or 13 000 Btu/lb
Solar energy (40° lat.) = 10 MJ/m^2·day in Dec. or 30 MJ/m^2·day in June
Crude oil, 1 t equivalent = 42·2 GJ
Crude oil, 1 barrel = 42 US gal = 6·6 GJ
Coal, 1 t equivalent = 29·3 GJ
Biogas = 11–22 MJ/m^3 or 300–600 Btu/ft^3
Producer gas = 5·6 MJ/m^3 or 150 Btu/ft^3

Common prefixes, their exponential equivalent, and their number meaning are as follows:

Prefix	*Exponent*	*Number*
hecto (h)	10^2	hundred
kilo (k)	10^3	thousand
mega (M)	10^6	million
giga (G)	10^9	billion
tera (T)	10^{12}	trillion
peta (P)	10^{15}	quadrillion
exa (E)	10^{18}	—

Selected conversion units that may be helpful in interpreting data throughout the book follow:

1 metric ton (tonne) = 1·1 short tons
1 ha = 2·47 acres
746 W = 1 hp
4186 J = 1 kcal
1055 J = 1 Btu
1 kcal = 3·968 Btu
1 kWh = 3413 Btu
1 kg = 2·2 lb
1 lb = 4·45 N
3·785 liters = 1 US gal
28·3 liters = 1 ft^3
$m^3 = 35{\cdot}31\ ft^3$
$m^2 = 10{\cdot}76\ ft^2$
m = 3·28 ft = 39·37 in
3·6 MJ = 1 kWh
101·2 kPa = 1 atmosphere
6·89 kPa = 1 psi
2·54 cm = 1 inch
Specific heat 4·89 J/kg·K = 1 Btu/lb·°F
Temperature change 1·8°F = 1·0°C = 1·0 K
Acceleration of gravity = 9·807 m/s
°C = 5/9 (°F − 32)
1 quad = 10^{15} Btu
1 therm = 10^5 Btu

Bibliography

ABE (1985). *Towards a Perspective on Energy Demand and Supply in India in 2004/05.* Advisory Board on Energy, Government of India.

Abeles, T. P., Friedman, D. F., DeBaere, L. A., & Ellsworth, D. A. (1978). *Bibliography of Anaerobic Digestion.* Oasis 2000, Rice Lake, WI 54868

Abelson, P. H. (1976). Energy from biomass. *Science*, **191**(4233), 1.

ADAS (1980). Solar energy for heating in the farm dairy. Agricultural Development and Advisory Service Report MEO 3401 (DC06), Woodthorne, Wolverhampton, UK.

AIA Research Corp. (1976). Solar dwelling design concepts. US Department of Housing and Urban Development, Washington, DC.

Albright, L. D., Langhans, R. W., White, G. B., & Donohoe, A. J. (1981). Enhancing passive solar heating of commercial greenhouses. In *Proceedings National Energy Symposium*, **II**, 509–12. American Society of Agricultural Engineers, St Joseph, MI.

Alonso, M. (1981). Energy strategies for oil-importing developing countries. In *Energy and the Developing Nations*, ed. P. Auer. Pergamon Press, Oxford, pp. 344–58.

Alward, R., Eisenbart, S., Volkman, J., & Haumberger, H. (1979). *Microhydropower: Reviewing an Old Concept.* The National Center for Appropriate Technology, PO Box 3838, Butte, MT 59701.

AMBIO (1985). Special issue on energy in developing countries. **14**(4–5). Royal Swedish Academy of Sciences, Box 50005, S-10405 Stockholm, Sweden.

Amr, A. T., Golden, J., Ouellette, R., & Cheremisinoff, P. (1981). *Energy Systems in the United States.* Marcel Dekker, New York.

Anderson, E. E. (1983). *Fundamentals of Solar Energy Conversion.* Addison-Wesley, Reading, MA.

Anderson, L. L. & Tillman, D. A. (Eds) (1977). *Fuels from Waste.* Academic Press, New York.

Anderson, R. E. (1979). *Biological Paths to Energy Self-Reliance.* Van Nostrand Reinhold, New York.

Androsky, A. (1973). Uses of the sun in the service of man. Aerospace Corporation, El Segundo, CA, Report No. ATR-74-(9470).

Annan, R. H. (1986). Photovoltaic progress: an incremental process. *Alternative Sources of Energy*, **78**, 31–3, 51.

Anon. (1957). Appropriate technologies for semiarid areas: wind and solar energy for water supply. Conference Report, German Foundation for International Development, Berlin.

Anon. (1980). International workshop on energy survey methodologies for developing countries. In *Proc. BOSTID.* National Academy Press, Washington, DC.

Anon. (1983). Alcohol as an octane enhancer—and profit extender. Proceedings of a conference sponsored by *Alcohol Week* and Renewable Fuels Association, San Francisco, 23–24 March, PO Box 7167, Benjamin Franklin Station, Washington, DC.

Anon. (1984). Defining the relations between energy and agriculture. Final report submitted to USAID. Development Sciences Inc., Washington, DC.

Anon. (1986). Solar salt ponds: a low-tech energy phenomenon. *Compressed Air Magazine*, **91**, 22–30.

Armstrong-Evans, R. (1979). Micro-hydro as an appropriate technology in developing countries. In *Waterpower 79, International Conference on Small Scale Hydro*. US Army Corps of Engineers, Washington, DC.

ASAE (1981). *Agricultural Energy*. Vol. 1, *Solar Energy, Livestock Production*; Vol. 2, *Biomass Energy, Crop Production*; Vol. 3, *Food Processing. Proceedings of the National Energy Symposium*. American Society of Agricultural Engineers, St Joseph, MI.

ASAE (1981*a*). *The Biological Liquid Fuels Alternative*. American Society of Agricultural Engineers, St Joseph, MI.

ASAE (1982). Vegetable oil fuels. In *Proceedings of the International Conference on Plànt and Vegetable Oils as Fuels*. American Society of Agricultural Engineers, St Joseph, MI.

ASE (1987). Windpower. *Alternative Sources of Energy*, **96**, 82.

ASE (1988). Taller trees give higher performance. *Alternative Sources of Energy*, **98**, 47.

Ashworth, J. & Neuendorffer, J. (1980). Matching renewable energy systems to village level needs. Solar Energy Research Institute, Report TR-744-514.

Asmussen, J. Jr (1980). *Wind Power, its Problems and Promises*. Center for Industrial and Institutional Development, University of New Hampshire, Durham, NH 03824.

Astrand, P. & Rodahl, K. (1977). *Textbook of Physiology*. McGraw-Hill, New York.

Atwood, J. & Fischer, L. (1980). Cost of production of fuel ethanol in farm size plants. Report No. 115, Department of Agricultural Economics, University of Nebraska, Lincoln.

Auer, P. (1981). *Energy and the Developing Nations*. Pergamon Policy Studies, Pergamon Press, Oxford.

Avlani, P. K. & Chancellor, W. J. (1977). Energy requirements for wheat production and use in California. *Trans. ASAE*, **20**, 429–37.

Azevedo, J. & Stout, P. R. (1974). *Farm Animal Manures: An Overview of their Role in Agricultural Environment*. Manual 44, Agri. Publication, University of California, Berkeley, CA 94720.

Bacon, D. M., Brear, F., Mancrieff, I. D., & Walker, K. L. (1981). The use of vegetable oils in straight and modified form as diesel engine fuels. In *Proceedings of the Third International Conference on Energy Use Management*, Berlin, 26–30 October. Pergamon Press, Oxford, pp. 1525–33.

Badger, P. C. & Poole, H. A. (1979). Conserving energy in Ohio greenhouses. Extension Bulletin 651, Ohio Cooperative Extension Service, Columbus, OH.

Bahm, R. J. (1980). Terrestrial solar radiation availability. In *Solar Energy Technology Handbook*, Part A: *Engineering Fundamentals*, ed. W. C. Dickinson & P. N. Cheremisinoff. Marcel Dekker, New York.

Bailie, R. C. (1976). Technical and economic assessment of methods for direct conversion of agricultural residues to usable energy. Prepared under contract by ERDA/USDA, 1964/76/FR-4 by West Virginia University, Morgantown, WV. Available from National Technical Information Service, US Department of Commerce, Springfield, VA 22161.

Bailie, R. C. (1980). Economic justification for small scale biomass fueled electrical generating system. In *Proceedings, BIO-ENERGY '80*. The Bioenergy Council, Washington, DC 20006.

Bandel, W. (1981). A review of the possibilities of using alternative fuels in commercial vehicle engines. Presented at the 3rd International Conference on Energy Use Management, Berlin, 26–30 October. Pergamon Press, Oxford.

Bartolelli, M., Bartolelli, U., Giordani, R., Liguori, F., Pellizzi, G., Peri, C., Pieretti, G., & Riva, M. (1982). The agro-forestry-food system from an energy angle. CNR—Progetto finalizzato meccanizzazione agricola—Quaderno n. 24, Rome, Italy.

Batty, J., Hamad, S. N., & Keller, J. (1975). Energy inputs into irrigation. *J. Irrigation and Drainage Division, ASCE*, **101**, 293–307.

Beadle, C. L., Long, S. P., Imbamba, S. K., Hall, D. O., & Olembo, R. J. O. (1985). Photosynthesis in relation to peanut production in terrestrial environments. UNEP. Tycooly Publishing, Oxford.

Beagle, E. C. (1978). Rice husk conversion to energy. FAO Agricultural Development Paper No. 31.

Beckman, A., Klein, S. A., & Duffie, J. A. (1977). *Solar Heating Design*. John Wiley and Sons, New York.

Beijer Institute (1983). *Producer Gas 1982*. The Beijer Institute, Stockholm, Sweden.

Bene, J. G., Beall, H. W., & Marshall, H. B. (1979). *Energy from Biomass for Developing Countries*. International Development Research Center, Ottawa, Ontario.

Bente, P. F. (1984). *Bio-Energy Directory and Handbook—1984*. The Bio-Energy Council, Arlington, VA 22209.

Berdoll, D. & Raley, D. (1985). Del Valle hog farm energy integrated farm system Phase III: demonstration. USDOE Contract No. DE-AC07-84 ID12493.

Bergey, M. L. S. *et al.* (1980). Performance rating document. Prepared by the American Wind Energy Association (AWEA) for USDOE under Cooperative Agreement #DE-FC04-80AL12926. AWEA, 1609 Connecticut Avenue NW, Washington, DC 20009.

Bernardini, O. (1983). Local energy systems in Ethiopia. In *Proceedings of the 1st International Conference on Energia e Agricoltura*, Milan, Italy, 27–29 April, ed. B. Guerra. CESAT, Gruppo dell'Energio, Milan.

Bhatnagar, A. P., Panesar, B. S., Gupta, P. K., & Jain, A. K. (1986). Collection, storage, and pre-conditioning of paddy straw as fuel for a 10-MW power plant in Patiala district. SER 5/86. Center for Advanced Studies of Energy Management in Agriculture, Punjab Agricultural University, Ludhiana, India 141004.

Bledsoe, B. L. & Henry, Z. A. (1981). Solar drying of large round bales of high moisture hay. In *Proceedings National Energy Symposium*, **I**, 66–72. American Society of Agricultural Engineers, St Joseph, MI.

Bodria, L. & Riva, G. (1980). Richerche su un collettore solare a basse costo per la produzione di aria calda. *Energie Alternative, HTE*, **2**, 262–73.

Bonny, S. (1986). *L'Energie et sa Crise de 1974 a 1984 dans l'Agriculture Française*. Institut National de la Recherche Agronomique, 78850 Thiveral–Grignon, France.

Braunstein, H. M., Kancirak, P., Roop, R. D., Sharples, F. E., Tatum, J. S., & Oakes, K. M. (1981). *Biomass Energy Systems and the Environment*. Pergamon Press, Oxford.

Breckenridge, R. P. *et al.* (1987). Energy integrated farm system technical summary report. USDOE/ID-10167.

BRI (1975). A survey of solar agricultural dryers. Technical Report T99, Brace Research Institute, McGill University, Montreal.

Briscoe, J. (1979). Energy use and social structure in a Bangladesh village. *Population and Development Review*, **5**, 615–41.

British Petroleum Company (1981). *BP Statistical Review of World Energy*. Britannic House, Moor Lane, London EC2Y 9BU.

British Petroleum Company (1984). *BP Statistical Review of World Energy*. Britannic House, Moor Lane, London EC2Y 9BU.

British Petroleum Company (1988). *BP Statistical Review of World Energy*. Britannic House, Moor Lane, London EC2Y 9BU.

Brooks, H. (1984). Scientists and policy makers in an international context. *IIASA Options*, Issue No. 1, p. 12.

Brown, B. (1981). *Package Systems for Small-Scale Hydroelectric Installations*. 1460 Livingston Avenue, North Brunswick, New Jersey 08902.

Brown, L. R. (1980). Food or fuel: new competition for the world's cropland. Worldwatch Paper No. 35, Worldwatch Institute, Washington, DC.

Brown, L. R. (1984) *State of the World 1984*. Worldwatch Institute, W. W. Norton & Co., New York.

Brown, L. R. *et al.* (1985). *State of the World 1985*. Worldwatch Institute, Washington, DC, W. W. Norton & Co., New York.

Brown, L. R. *et al.* (1986). *State of the World 1986*. Worldwatch Institute, Washington, DC, W. W. Norton & Co., New York.

Brown, L. R. *et al.* (1988). *State of the World 1988*. Worldwatch Institute, Washington, DC, W. W. Norton & Co., New York.

Brown, N. L. (1980). Renewable energy resources for developing countries. *Ann. Rev. Energy*, **5**, 389–413.

Brown, N. L. (Ed.) (1978). *Renewable Energy Resources and Rural Applications in the Developing World*. AAAS Selected Symposium 6, Westview Press, Boulder, CO.

Brown, N. L. (1977). *Methane Generation from Human, Animal, and Agricultural Wastes*. National Research Council, National Academy of Sciences, Washington, DC.

Brown, R. J. & Yanuck, R. R. (1980). *Life Cycle Costing*. The Fairmont Press, Atlanta, GA.

Brown, S. J. & Batty, J. C. (1976). Energy allocation in the food system: a microscale view. *Trans. ASAE*, **199**, 758–61.

Bruin, S. (1980). Biomass as a source of energy. In *Proceedings of the 13th International TNO Conference*, Rotterdam, 27 and 28 March.

Bruwer, J. J. & Hugo, F. J. C. (1981). World overview of plant oils for fuel: status of research and applications technology. In *Proceedings of the Third International Conference on Energy Use Management*, Berlin, 26–30 October. Pergamon Press, Oxford, pp. 287–95.

Bullard, C., Hannon, B., & Herendeen, R. (1975). Energy flow through the United States economy. Chart published by Center for Advanced Computation, University of Illinois, Urbana, IL.

Bungay, H. R. (1981). *Energy: The Biomass Options*. Wiley–Interscience Publication, John Wiley and Sons, New York.

Bush, G. E. & Richards, L. M. (1980). Solar geometry and time. In *Solar Energy Technology Handbook*, Part A: *Engineering Fundamentals*, ed. W. C. Dickinson & P. N. Cheremisinoff. Marcel Dekker, New York.

Büttner, S. (1981). Biogas—an alternative fuel for automotive application. In *Proceedings of the Third International Conference on Energy Use Management*, Berlin, 26–30 October. Pergamon Press, Oxford, pp. 1553–6.

Butts, C. L., Harner, J. P., Baker, J. L., & Vaughan, D. H. (1981). Combination drying of grain using solar energy. *Proceedings National Energy Symposium*, **I**, 148. American Society of Agricultural Engineers, St Joseph, MI.

Carillon, R. (1981). L'agriculture et l'energie. *Revue d'Economie Industrialle*, **18**, 110–23.

Cassidy, J. J. (1980). *Civil Works Design, Operation, Maintenance for Small Hydropower Facilities, Small Hydroelectric Plants*. National Rural Electric Cooperative Association (NRECA), 1800 Massachusetts Avenue NW, Washington, DC.

CEC. Various issues. California Energy Commission, wind energy performance rating system, annual reports. Publication Office, 1516 Ninth Street, MS-13, Sacramento, CA 95814.

CEC (1980). Energy. Solar energy programme of the Commission of the European Communities. EUR 6959. Batiment Jean Monnet, Luxembourg.

CEEMAT (1972). *The Employment of Draught Animals in Agriculture*. Issued by arrangement with Centre d'Etudes et d'Experimentation du Machinisme Agricole Tropical by the FAO of the United Nations, Rome.

CEQ (1980). The Global 2000 Report to the President: The Technical Report, Vol. 2. Council on Environmental Quality and the Department of State, Washington, DC.

Cervinka, V. & DeForest, R. (1980). Hunger, energy and agricultural development. In *Energy and the Developing Nations*, ed. P. Auer. Pergamon Press, Oxford, pp. 123–5.

Chancellor, W. J. (1978). The role of fuel and electrical energy in increasing production from traditionally based agriculture. *Trans. ASAE*, **21**, 1060–7.

Chancellor, W. J. (1983). Efficient use of energy in rice farming systems. Presented at the 1983 Rice Research Conference, International Rice Research Institute, Los Banos, Laguana, Philippines.

Chandler, W. V. (1985). Energy productivity: key to environmental protection and economic progress. Worldwatch Paper No. 63, Worldwatch Institute, Washington, DC.

Chatterjee, A. K. (1981). State-of-the-art review on the pyrolysis of wood and agricultural biomass. Final Report, Contract No. 53-319-R-0-206. AC Project PO380, USDA Forest Service, Washington, DC.

Chatterji, M. (Ed.) (1981). *Energy and Environment in the Developing Countries.* John Wiley and Sons, New York.

Cheremisinoff, N. P. (1979). *Gasohol for Energy Production.* Ann Arbor Science Publishers, Ann Arbor, MI.

Cheremisinoff, N. P. & Cheremisinoff, P. N. (1981). *Gasohol Sourcebook.* Ann Arbor Science Publishers, Ann Arbor, MI.

Cheremisinoff, N. P., Cheremisinoff, P. N., & Ellerbusch, F. (1980). *Biomass: Applications, Technology and Production.* Marcel Dekker, New York.

Chilcott, R. E. (1985). *Wind-Powered Water Pumping Workbook.* Department of Agricultural Engineering, Lincoln College, University College of Agriculture, Canterbury, New Zealand.

Choucri, N. (1982). *Energy and Development in Latin America.* Lexington Books, D. C. Heath and Co., Lexington, MA.

Christensen, R. L., Colwell, H. T. M., Turner, F., & Wade, J. C. (1982). Evaluating energy management for agriculture within a policy framework: a discussion outline. In *Energy Management in Agriculture*, ed. D. W. Robinson & R. C. Mollan. *Proceedings 1st Summer School in Agriculture*, Dublin, Ireland. The Royal Dublin Society, Dublin, pp. 35–42.

Church, D. C. (1977). *Livestock Feeds and Feeding.* O and B Books, Oregon.

Claar, P. W. II, Buchele, W. F., & Marley, S. J. (1981). Development of a concentric-vortex agricultural residue furnace. In *Proceedings National Energy Symposium*, **I**, 349–56. American Society of Agricultural Engineers, St Joseph, MI.

Clark, P. J. (1981). Cost implications of small hydropower systems. Workshop on small-scale hydropower in Africa. National Rural Electric Cooperative Association (NRECA), 1800 Massachusetts Avenue NW, Washington, DC.

Clark, R. N. (1984). Co-generation using wind and diesel for irrigation pumping. ASAE Paper No. 84-2603, ASAE, St Joseph, MI.

Clark, R. N. (1985). Wind energy systems for pumping water. In *Proceedings of the International Conference on Water and Water Policy in World Food Supplies*, Texas A&M University, 26–30 May, ed. W. R. Jordan. Texas A&M University Press.

Clark, R. N. (1985*a*). Irrigation pumping with mechanical wind systems. In *Proceedings 5th Annual Solar and Biomass Energy Workshop*, USDA/DOE, 23–25 April, Atlanta, GA, pp. 257–60.

Clark, R. N. & Vosper, F. C. (1984). Long-term performance of wind-driven induction generators. ASAE Paper No. 84-4548, American Society of Agricultural Engineers, St Joseph, MI.

Clark, R. N., Hellickson, M. A., & Shelton, D. P. (1983). *Proceedings Solar and Wind System Workshop*, Great Plains Agricultural Council, University of Nebraska, Lincoln.

Claus, H. G. (1979). Einstaz von solar-kollektoren im bereich der landwirtschaft. In *Elektrizitat—KTBL, AEL, HEA—Symposium*, 12–13 September, Koblenz, S.93–109.

Claus, H. G. & Hartmann, W. (1978). Bin drying of cereals with supplemental heating by solar energy. 2. Internat. Sonnenforum, 12/14 Juli, Hamburg, Bd. III, S.87/98.

CNEEMA (1974). L'agriculture et les 'carburants'. Etudes du CNEEMA No. 391. Centre National d'Etudes d'Experimentation de Machinisme Agricole. Parc de Tourvoie, 92160 Antony, France.

CNEEMA (1975). Essai sur l'energie dans l'agriculture ou dans le systéme agro-alimentarie en France. Etudes du CNEEMA No. 404. Centre National d'Etudes d'Experimentation de Machinisme Agricole. Parc de Tourvoie, 92160 Antony, France.

CNEEMA (1975*a*). L'activité agricole et l'energie. Etudes du CNEEMA No. 408. Centre National d'Etudes d'Experimentation de Machinisme Agricole. Parc de Tourvoie, 92160 Antony, France.

Collins, G. S., Griffin, R. C., & Lacewell, R. D. (1982). Impacts of substituting plant oils for diesel fuels. *Proceedings International Conference on Plant and Vegetable Oils as Fuels*, American Society of Agricultural Engineering, St Joseph, MI, pp. 138–48.

Considine, D. M. (Ed.) (1977). *Energy Technology Handbook*. McGraw-Hill, New York.

Converse, H. H., Lai, F. S., Aldis, D. F., & Sauer, D. B. (1981). Application of solar energy in grain drying. *Proceedings National Energy Symposium*, **I**, 131–6. American Society of Agricultural Engineers, St Joseph, MI.

Converse, J. S. & Graves, R. E. (1974). Facts on methane production from animal manure. Fact Sheet A-2636, University of Wisconsin, Madison.

Coxon, D. A. (1977). *Here Comes the Sun*. Jordan College, Cedar Springs, MI.

Crawford, N. H. (1981). Hydrologic estimates for small hydroelectric projects. National Rural Electric Cooperative Association (NRECA), 1800 Massachusetts Avenue NW, Washington, DC.

Cruz, I. E. (1985). Producer-gas technology for rural applications. FAO Agricultural Services Bulletin 61, FAO, Rome.

CSC (1985). Solar dryers. Their role in post-harvest processing. Commonwealth Science Council, Commonwealth Secretariat, Marlborough House, Pall Mall, London SW1Y 5HX, UK, 337 pp.

Cullison, A. E. (1979). *Feeds and Feeding*. Reston Publishing Co., Reston.

Cullumbine, H. (1950). Heat production and energy requirements of tropical people. *J. Appl. Physiol.*, **2**, 640–53.

Cunney, B. (1977). Possibilities for solar energy on the farm. University College, Dublin. Seminar organized by the Solar Energy Society of Ireland and An Foras Taluntais (The Agricultural Institute).

Da Silva, J. *et al.* (1978). Energy balance for ethyl alcohol production. *Science*, **201**, 903–6.

Damrath, J. (1979). Solarversuchsanlage für die gewachsausheizung. In *Kurfzfassung zur Internationalen Tagung Landtechnik*, VDI, 7–9 November, Braunschweig, S.15–16.

Daniels, F. (1964). *Direct Use of the Sun's Energy*. Yale University Press, New Haven.

Davis, C. H. & Blouin, G. M. (1977). Energy consumption in the US chemical fertilizer system from the ground to the ground. In *Agriculture and Energy*, ed. W. Lockertz. Academic Press, New York.

Dekkers, W. A., Lange, J. M., & Dewit, C. T. (1974). Energy production and use in Dutch agriculture. *Neth. J. Agr. Sci.*, **22**, 107–18.

deLucia, R. *et al.* (1981). The economic cost of renewable energy. Final report prepared for USAID, Contract No. AID/SOD/PDC-C0306.

deLucia, R. J. & Jacoby, H. D. (1982). *Energy Planning for Developing Countries: A Study of Bangladesh*. Johns Hopkins University Press, Baltimore.

Denton, F. H. (1985). A revisionist view of the traditional energy sector. *BOSTID Developments*, **5**, 10–13. National Research Council, 2101 Constitution Ave NW, Washington, DC 20418.

Dernedde, W. (1979). Troknung von getreide mit solarenergie. In *Kurfzfassung zur Internationalen Tagung Landtechnik*, VDI, 7–9 November, Braunschweig, S.13–14.

Dernedde, W. (1979*a*). Wirkungsgrade von luftkollektoren unterschiedlicher bauweise zur trocknung landwirtschaftlicher produkte. 9. International CIGR Congress, Michigan State University, East Lansing, MI.

Dernedde, W. & Peters, H. (1978). Wirkungsgrade von solar-kollektoren für trocknungs-anlagen. *Landtechnik*, S.29/30.

Deudney, D. (1981). Rivers of energy: the hydropower potenial. Worldwatch Paper 44, Worldwatch Institute, Washington, DC.

Dickson, W. C. & Cheremisinoff, P. N. (Eds) (1980). *Solar Energy Technology Handbook*, Part A: *Engineering Fundamentals*; Part B: *Applications, Systems Design, and Economics*. Marcel Dekker, New York.

DiVecchia, A. *et al.* (1981). Possibilities for the application of solar energy in the European agricultural community. *Solar Energy*, **16**, 479–89.

DNES (1985). Annual Report. Department of Non-Conventional Energy Sources. Ministry of Science and Technology. Block No. 14, Central Government Offices Complex, Lodhi Rd, New Delhi 110003, India.

Doering, O. C. (1977). An energy based analysis of alternative production methods and cropping systems on the corn belt. NSF/RA-770125, Agricultural Experimental Station, Purdue University, Lafayette, IN.

Doering, O. C., Considine, T. J., & Harling, C. E. (1977). Accounting for tillage equipment and other machinery in agricultural energy analysis. NSF/RA-770128, Purdue University, Lafayette, IN.

Dovring, F. (1988). *Farming for Fuel.* Praeger, New York.

Duff, B. (1978). Augmenting human energy supplies for agricultural development. Paper No. 78-0005, Philippine Society of Agricultural Engineers, Manila.

Duisberg, C. R. (1982). Biofuels: international concerns in a world of uncertainty. AAAS Symposium, Washington, DC.

Duke, J. A. (1987). *Agricultural Energy Potential of Developing Countries.* CRC Press, Boca Raton, FL 33431.

Dunkerly, J. & Ramsey, W. (1982). Energy and the oil-importing developing countries. *Science*, **216**, 590–5.

Dunkerly, J., Ramsey, W., Gordon, L., & Cecelski, E. (1981). *Energy Strategies for Developing Nations.* Resources for the Future, Washington, DC.

Dvoskin, D. (1983). *Energy and Israeli Agriculture.* Heshev—The Inter Kibbutz Unit for Management Service, 61400 Tel-Aviv, Israel.

Earl, D. E. (1974). *Charcoal.* FAO, Rome.

Eckholm, E. (1976). *Losing Ground.* W. W. Norton and Co., New York.

Edelson, E. (1976). New ways to make solar cells trim cost of future sun power for your home. *Popular Science*, **208**, 74–7, 152, 156.

Egneus, H. & Ellegard, A. (1985). Bioenergy '84. *Proceedings of International Conference on Bioenergy*, 15–21 June, Goteborg, Sweden. Vol. I, Bioenergy state of the art; Vol. II, Biomass resources; Vol. III, Biomass conversion; Vol. IV, Bioenergy utilization; Vol. V, Bioenergy in developing countries. Elsevier Applied Science Publishers, London.

El Amani, S. & Bahri, A. (1980). Essai d'estimation de la consommation energetique de l'agriculture Tunisienne. Republique Tunisienne, Ministere de l'Agriculture, Centre de Recherche du Genie Rural.

El Halwagi, N. M. (Ed.) (1984). *Biogas Technology and Diffusion.* Proceedings of an International Conference held at the National Research Center, Cairo, Egypt. Elsevier Applied Science Publishers, London.

Eldridge, F. R. (1980). *Wind Machines.* MITRE Corp., Van Nostrand Reinhold Co., New York.

EMA (1982). *200-Hour Screening Test for Alternate Fuels.* Alternate Fuels Committee, Engine Manufacturers' Association, Chicago, IL.

Engler, C. R. *et al.* (1982). Economic and engineering evaluation of plant oils as a diesel fuel. Final Report, Project No. 80-B-4-4a. Texas Energy and Natural Resources Advisory Council, Agricultural Engineering Department, Texas A&M University, College Station, TX 77843-2117.

Engler, C. R. *et al.* (1983). Engineering evaluation of plant oils as diesel fuel. Final Report, Project No. 80-B-4-4c. Texas Energy and Natural Resources Advisory Council, Agricultural Engineering Department, Texas A&M University, College Station, TX 77843-2117.

Enochian, R. V. (1982). Solar- and wind-powered irrigation systems. Agricultural Economics Report 482, Economic Research Service, US Department of Agriculture.

Ermenc, J. J. (1978). Small hydraulics prime movers for rural areas of developing countries: a look at the past. In *Renewable Resources and Rural Applications in the Developing World*, ed. N. L. Brown. Westview Press, Boulder, CO, pp. 89–114.

Exell, R. H. B. *et al.* (1979). A low cost solar rice dryer for farmers in Southeast Asia. *Agricultural Mechanization in Asia*, Autumn 1979, pp. 75–82.

Exxon Corporation (1980). *World Energy Outlook.* Exxon Corporation, 1251 Avenue of the Americas, New York 10020.

Exxon Corporation (1984). *Middle East Oil and Gas*. Exxon Corporation, 1251 Avenue of the Americas, New York 10020.

Faidley, L. W. (1985). Energy for world agriculture: water implications. In *Proceedings of the International Conference on Water and Water Policy in World Food Supplies*, Texas A&M University, 26–30 May, ed. W. R. Jordan. Texas A&M University Press.

Fang, K. S. (1949). Vegetable oils as diesel fuel for China. MS thesis, Department of Agricultural Engineering, University of Nebraska, Lincoln.

FAO (1950). Calorie requirements. FAO Nutritional Studies No. 5.

FAO (1975). Shifting cultivation and soil conservation in Africa. FAO Soils Bulletin No. 24. FAO, Rome.

FAO (1976). Energy and agriculture. FAO Annual Report, The State of Food and Agriculture, Chapter 3, pp. 82–111.

FAO (1977). China: recycling of organic wastes in agriculture. Soils Bulletin No. 40. FAO, Rome.

FAO (1979). Agriculture towards 2000. Preliminary Report. FAO, Rome.

FAO (1980). Power and heat plants. Study prepared for FAO by EKONO Oy, PO Box 27 SF-00131, Helsinki, Finland.

FAO (1981). Energy cropping versus food production. Agricultural Services Bulletin 46. FAO, Rome.

FAO (1981*a*). *Agriculture: Towards 2000*. FAO, Rome.

FAO (1982). Appropriate use of animal energy in agriculture in Africa and in Asia. Report of Expert Consultation, Rome.

FAO (1982*a*). Agricultural residues. Agricultural Services Bulletin No. 21—World Directory of Institutions. No. 33—Compendium of Technologies. No. 47—Bibliography and Quantitative Survey. FAO, Rome.

FAO (1983). *1982 FAO Production Yearbook*, Vol. 36. FAO, Rome.

FAO (1984). FAO activities in the field of new renewable sources of energy. Prepared for the second session of the Intergovernmental Committee on the development and utilization of new and renewable sources of energy, New York, 23 April–4 May.

FAO (1984*a*). Animal energy in agriculture in Africa and Asia. FAO Animal Production and Health Paper 42. Technical papers presented at FAO Expert Consultation held in Rome, 15–19 November 1982.

FAO (1984*b*). New and renewable energy in agriculture. Bibliography extracted from AGRIS 1975–83.

FAO (1984*c*). *World Food Report*. FAO, Rome.

FAO (1984*d*). *1983 FAO Production Yearbook*, Vol. 37. FAO, Rome.

FAO (1985). *FAO's Activities in the Field of New and Renewable Sources of Energy*. FAO, Rome. W/R 3098.

FAO (1985*a*). *World Food Report*. FAO, Rome.

FAO & WHO (1974). Handbook on human nutritional requirements. FAO Nutritional Studies No. 28.

Fazzolare, R. A. & Smith, C. B. (1981). Beyond the energy crisis. In *Proceedings of the Third International Conference on Energy Use Management*, Berlin. Pergamon Press, Oxford.

FEA (1975). Energy consumption in the food system. A report prepared for the Industrial Technology Office, Federal Energy Administration, by Booz, Allen and Hamilton, Inc.

Fekete, F., Benet, I., & Sebestyen, K. (1981). Energy problems in Hungarian food economy. Presented at 3rd Congress of European Association of Agricultural Economists, 31 August–4 September.

Finnell, J. (1984). Energy and agriculture in Haiti: linkages in the food system. Draft report, Bureau for Science and Technology, Office of Energy, USAID.

Finnell, J. (1984*a*). Energy and agriculture in Senegal: linkages in the food system. Draft report, Bureau for Science and Technology, Office of Energy, USAID.

Fischer, R. D. & Talbert, S. G. (1980). Irrigation pumping. In *Solar Energy Technology*

Handbook, Part B: *Applications, Systems Design, and Economics*, ed. W. C. Dickinson & P. N. Cheremisinoff. Marcel Dekker, New York.

Flavin, C. (1981). Wind power: a turning point. Worldwatch Paper 45, Worldwatch Institute, Washington, DC.

Fluck, R. C. (1985). Energy sequestered in repairs and maintenance of agricultural machinery. *Trans. ASAE*, **28**, 738–44.

Fluck, R. C. (Ed.) (in press). *Energy in Farm Production*. Energy in World Agriculture Series, Vol. 5. Elsevier Science Publishers, Amsterdam.

Fluck, R. C. & Baird, C. D. (1980). *Agricultural Energetics*. AVI, Westport, CT.

Foley, G. & Barnard, G. (1983). Biomass gasification in developing countries. Technical Report No. 1, Earthscan, 10 Percy Street, London W1P 0DR, UK.

Foley, G. & Moss, P. (1983). Improved cooking stoves in developing countries. Earthscan, 10 Percy Street, London W1P 0DR, UK.

Free, J. (1974). Solar cells. *Popular Science*, December, pp. 52–5, 120, 212.

Fritz, J. F. (1985). Recent experience in the application of small hydropower as a principal energy source in rural areas of developing countries. *Proceedings of the International Conference on Water and Water Policy in World Food Supplies*, Texas A&M University, 26–30 May, ed. W. R. Jordan. Texas A&M University Press.

Fritz, J. J. (Ed.) (1984). *Small and Mini Hydropower Systems*. McGraw-Hill, New York.

Fry, J. L. (1973). *Methane Digesters for Fuel Gas and Fertilizer*. The New Alchemy Institute, Woods Hole, MA.

Fuls, J. & Hugo, J. C. (1981). On-farm preparation of sunflower oil esters for fuel. In *Proceedings of the Third International Conference on Energy Use Management*, Berlin, 26–30 October. Pergamon Press, Oxford, pp. 1595–600.

Garg, H. P. (1976). Solar energy research at the Central Arid Zone Research Institute, Jodhpur—Achievements and Future Plans. Jodhpur, India.

Garg, H. P. & Aminullah (1977). *World Bibliography on Biogas and its Utilization*. CAZRI Library Publication Series No. 4. Central Arid Zone Research Institute, Jodhpur, India.

Garling, W. S., Harper, M. R., Merchant-Geuder, L., & Welch, M. (1980). Economic analysis of wind-powered crop drying. Prepared by Science and Education Administration, US Department of Agriculture, Beltsville, MD 20705. Prepared for US Department of Energy. Report No. DOE/SEA-1109-20401/81/2.

Garling, W. S., Harper, M. R., Merchant-Geulder, L., & Welch, M. (1980*a*). Economic analysis of wind-powered refrigeration cooling/water heating systems in food processing. Prepared by Science and Education Administration, US Department of Agriculture, Beltsville, MD 20705. Prepared for US Department of Energy. Report No. DOE/SEA-1109-20401/81/1.

Gavett, E. E., Grinnell, G. E., & Smith, N. L. (1986). Fuel ethanol and agriculture: an economic assessment. Agr. Econ. Rep. No. 562.

General Electric Company (1963). Analysis of demand for electric power in rural communities. USAID Contract AID REPAS-I, Washington, DC.

Georgia Institute of Technology (1978). A state of art survey of solar powered irrigation pumps, solar cookers, and wood burning stoves for use in sub-Saharan Africa.

Gever, J., Kaufmann, R., Skole, D., & Vörösmorty, C. (1986). *Beyond Oil*. Ballinger, Cambridge, MA.

Gifford, R. M. (1984). Energy in different agricultural systems: renewable and nonrenewable sources. In *Energy and Agriculture*, ed. G. Stanhill. Springer, Berlin, pp. 84–112.

Gifford, R. M. (1984*a*). Energy in Australian agriculture: inputs, outputs, and policies. In *Energy and Agriculture*, ed. G. Stanhill. Springer, Berlin, pp. 154–68.

Gifford, R. M. & Millington, R. J. (1975). Energetics of agriculture and food production. CSIRO Bulletin No. 288, Australia.

Gilley, J. R. (1980). On-farm US irrigation pumping plants. Prepared by Science and Education Administration, US Department of Agriculture, Beltsville, MD 20705. Prepared for US Department of Energy. Report No. DOE/SEA-7315-20741/81/1.

Gilley, J. R. & Supalla, R. J. (1985). Energy requirements and management for pumping irrigation water. *International Conference on Water and Water Policy in World Food Supplies*, Texas A&M University, 26 May.

Gilliland, M. W. (1975). Energy analysis and public policy. *Science*, **189**, 1051–6.

Gilliland, M. W. (Ed.) (1978). *Energy Analysis: A New Public Policy Tool.* AAAS Selected Symposium 9, Westview Press, 5500 Central Avenue, Boulder, CO.

Gilmore, E., Barieau, R. E., & Nelson, V. (1977). Feasibility of using wind to pump irrigation water. Final report to Governor's Energy Advisory Council, Austin, TX. Under Contract IAC-76-77(1143).

Gipe, P. (1985). Overview of the US wind industry. Presented at the 12th Energy Technology Conference and Exposition, Washington, DC, 25–27 March.

Gipe, P. (1988). Taller towers give higher performance. *Alternative Sources of Energy*, February, 47.

Girardier, J. P. & Massen, H. (1980). Utilization of solar energy for lifting water. International Solar Energy Conference, Melbourne, Australia.

Giraud, A. L. (1983). Energy in France. *Ann. Rev. Energy*, **8**, 165–91.

Gladwell, J. S. (1980). *Small Hydro: Some Practical Planning and Design Considerations.* Idaho Water Resources Research Institute, University of Idaho, Moscow, ID 83843.

Goe, M. R. (1983). Current status of research on animal traction. *World Animal Review*, No. 45. FAO, Rome, pp. 2–17.

Goe, M. R. & McDowell, R. E. (1980). *Animal Traction: Guidelines for Utilization.* Cornell University, Int'l Mimeo, Ithaca, NY.

Goldemberg, J. (1978). Brazil: energy options and current outlook. *Science*, **200**, 158–64.

Goldemberg, J. (1980). Energy consumption and decentralized solutions. In *Energy and the Developing Nations*, ed. P. Auer. Pergamon Press, Oxford, pp. 98–109.

Goldemberg, J. (1980*a*). Brazil: energy options and current outlook. In *Energy in the Developing World*, ed. V. Smil & W. Knowland. Oxford University Press, Oxford, pp. 225–39.

Goldemberg, J. (1982). Energy issues and policies in Brazil. *Ann. Rev. Energy*, **7**, 139–74.

Goldemberg, J. (1984). Energy problems in Latin America. *Science*, **223**, 1357–62.

Goldemberg, J., Johansson, T. B., Reddy, A. K. N., & Williams, R. H. (1987). *Energy for Development.* World Resources Institute, 1735 New York Avenue NW, Washington, DC 20006.

Goldemberg, J., Johansson, T. B., Reddy, A. K. N., & Williams, R. H. (1987*a*). *Energy for a Sustainable World.* World Resources Institute, 1735 New York Avenue NW, Washington, DC 20006.

Goodman, G. *et al.* (1979). Report of the energy resources working group at the International Conference, Agricultural Production: Research and Development Strategies for the 1980s, Bonn.

Goodman, L. J. & Love, R. N. (Eds) (1981). *Biomass Energy Projects: Planning and Management.* Pergamon Press, New York.

Goodman, L. J., Hawkins, J. N. & Love, R. N. (1981). *Small Hydroelectric Projects for Rural Development.* Pergamon Press, New York.

Goswami, D. Y. (1986). *Alternative Energy in Agriculture*, Vols I and II. CRC Press, Boca Raton, FL 33431.

Graef, M. (1981). Alcohol as an alternative fuel in agriculture. In *Proceedings of the Third International Conference on Energy Use Management*, Berlin, 26–30 October. Pergamon Press, Oxford, pp. 1603–10.

Gray, J. E., Davis, W. K., & Harned, J. W. (1988). *Energy Supply and Use in Developing Countries.* University Press of America, 4720 Boston Way, Lanham, MD 20706.

Green, M. B. (1978). *Eating Oil.* Westview Press, Boulder, CO.

Green, M. K., Griffin, R. C., & Stout, B. A. (1983). Life-cycle economic analysis for ethanol production. ASAE Paper 83-3566, American Society of Agricultural Engineers, St Joseph, MI.

Guerra, B. (Ed.) (1983). Energia e agricoltura. In *Proceedings 1st International Conference CESAT.* Largo Richini, 2/A 20122 Milano, Italy.

Guerra, B. (Ed.) (1986). Energia e agricoltura. In *Proceedings 2nd International Conference CESAT.* Enerconsult SRI, Via Trento 33d 25125 Brescia, Italy.

Gunkel, W. W., Lacey, D. R., Neyeloff, S., & Porter, T. G. (1981). Wind energy for direct water heating. Final report on Contract No. 12-14-3001-12 prepared for SEA, USDA.

Gustavson, M. R. (1979). Limits to wind power utilization. *Science*, **204**, 13–17.

Halacy, D. S. Jr (1980). Solar energy and the biosphere. In *Solar Energy Technology Handbook*, Part A: *Engineering Fundamentals*, ed. W. C. Dickinson & P. N. Cheremisinoff. Marcel Dekker, New York.

Hall, C. W. (1981). *Biomass as an Alternative Fuel.* Government Institutes, Inc., PO Box 1096, Rockville, MD 20850.

Hall, C. W. & Hinman, G. W. (1983). *Dictionary of Energy.* Marcel Dekker, New York.

Hall, D. O. (1979). Issue paper on biomass energy. UN Conference on New and Renewable Sources of Energy. King's College, University of London.

Hall, D. O. & de Groot, P. J. (1985). Biomass for fuel and food—a parallel necessity. Presented at the World Resources Institute Symposium on Biomass Energy Systems, Building Blocks for Sustainable Agriculture, January 1985.

Hall, D. O., Barnard, G. W., & Moss, P. A. (1982). *Biomass for Energy in the Developing Countries.* Pergamon Press, Oxford.

Hamad, A. M., Abdel Dayem, A. M., & El Halwagi, M. M. (1982). Evaluation of the performance of two rural biogas units of Indian and Chinese design. *Energy in Agriculture*, **1**, 235–50.

Hamm, H. W. (1971). *Low-Cost Development of Small Water-Power Sites.* Volunteers in Technical Assistance, 1815 N. Lynn Street, Suite 200, Arlington, VA 22209.

Hansen, A. C., Meiring, P., & Lyne, P. W. L. (1984). Evaluation of ethanol mixed with diesel fuel in agricultural tractors. Final report, University of Natal Pietermaritzburg, Republic of South Africa.

Hansen, R. W. (1975). Methane generation from livestock waste. In *Great Plains Beef Cattle Handbook.* Cooperative Extension Service, Great Plains States.

Hansen, R. W. & Grenard, J. L. (1981). High quality forage production from solar energy drying systems. In *Proceedings National Energy Symposium*, **I**, 78–83. American Society of Agricultural Engineers, St Joseph, MI.

Harris, G. S. (1980). The potential of energy farming for transport fuels in New Zealand. New Zealand Energy Research and Development Committee, Report No. 46 (3 volumes). University of Auckland, Private Bag, Auckland, New Zealand.

Harris, G. S. *et al.* (1979). The potential of energy farming for transport fuels in New Zealand. New Zealand Energy Research and Development Committee, University of Auckland, Private Bag, Auckland, Vols I and II.

Hartsock, J. G. (Ed.) (1978). *Solar Grain Drying Conference Proceedings*, CONF-7805129, Purdue University, 2–3 May. West Lafayette, IN. Cosponsored by the US Department of Energy, Washington, DC.

Hashimoto, A. G., Chen, Y. A., & Prior, R. L. (1978). Methane and protein products from animal feedlot wastes. Presented at the 33rd Annual Meeting of the Soil Conservation Society of America, Denver, CO, 30 July–2 August.

Hawkins, C. S., Fuls, J., & Hugo, F. J. C. (1981). Laboratory procedures for investigating some fuel properties of sunflower oil esters in diesel engines. In *Proceedings of the Third International Conference on Energy Use Management*, Berlin, 26–30 October. Pergamon Press, Oxford, pp. 1611–16.

Hayes, D. (1977). Energy for development: Third World options. Paper No. 15, Worldwatch Institute, Washington, DC.

Heichel, G. H. (1973). Comparative efficiency of energy use in crop production. Bulletin 739, The Connecticut Agricultural Experiment Station, New Haven.

Heichel, G. H. (1976). Agricultural production and energy resources. *American Scientist*, **64**, 64–72.

Heister, T. R. & Pennell, W. T. (1981). *The Siting Handbook for Large Wind Energy Systems.* Battelle Memorial Institute, Pacific Northwest Laboratory, Richland, WA 99352.

Helsel, Z. R. (Ed.) (1987). *Energy in Plant Nutrition and Pest Control.* Energy in World Agriculture Series, Elsevier Science Publishers, Amsterdam.

Herendeen, R. A. & Dovring, F. (1984). Liquid fuel from Illinois sources. Agri. Econ. Research Report No. 192, University of Illinois, Urbana.

Hiler, E. A. & Stout, B. A. (Eds) (1985). Biomass energy: a monograph. Texas A&M University Press, College Station, TX 77843.

Hiler, E. A. *et al.* (1982). Ethanol production in small-to-medium size facilities. Final Report, Project No. 80-B-1-1. Texas Energy and Natural Resources Advisory Council, Agricultural Engineering Dept, Texas A&M University, College Station, TX 77843.

Hirsch, R. L. (1987). Impending United States energy crisis. *Science*, **235**, 1467–73.

Hirst, E. (1974). Food-related energy requirements. *Science*, **184**, 134–8.

Hoffmann, P. (1981). *The Forever Fuel: The Story of Hydrogen.* Westview Press, Boulder, CO.

Hofman, V., Kaufman, K., Helgeson, D., & Dinusson, W. E. (1981). Sunflower for power. North Dakota Cooperative Extension Service, Circular AE-735, Fargo.

Hofman, V., Dinusson, W. E., Zimmerman, D., Helgeson, D. L., & Fanning, C. (1979). Sunflower oil as a fuel alternative. North Dakota Extension Service Bulletin 13 AENG-5.

Hokanson, A. E. & Rowell, R. M. (1977). Methanol from wood waste: a technical and economic study. USDA Forest Service Tech. Report No. FPL12.

Hopfen, J. H. (1969). Farm implements for arid and tropical regions. Agricultural Development Paper No. 91. FAO, Rome.

Howe, E. D. (1980). Distillation of sea water. In *Solar Energy Technology Handbook*, Part B: *Applications, Systems Design, and Economics*, ed. W. C. Dickinson & P. N. Cheremisinoff. Marcel Dekker, New York.

Howe, J. W. *et al.* (1977). *Energy for the Villages of Africa.* The Overseas Development Council, Washington, DC.

Howes, K. M. W. & Rummery, R. A. (1979). The impact of changes in energy costs on the rural sector of the Australian economy. Division of Land Management, CSIRO, Australia.

Hrabovszky, J. P. (1982). Energy, agriculture and rural development. Paper for presentation at the UN University Symposium on Energy Agriculture Futures: Policy Implications of Global Models, Paris, 22–25 February.

Hrabovszky, J. P. (1984). Energy, agriculture and rural development. In *Energy and Agriculture: Their Interacting Futures*, ed. M. Levy & J. L. Robinson. Harwood Academic Publishers for the United Nations University, New York.

Hubbard, H. M. (1984). Prospects for solar energy: an R&D manager's view. *Solar & Wind Technology*, **1**, 3–7.

Hubbert, M. K. (1956). *Nuclear Energy and the Fossil Fuels: Drilling and Production Practice.* American Petroleum Institute, New York.

Hughart, D. (1979). Prospects for traditional and non-conventional energy sources in developing countries. World Bank Staff Working Paper No. 346.

Hughes, D. E. *et al.* (Eds) (1982). *Anaerobic Digestion 1981.* Elsevier Biomedical Press, Amsterdam.

Hunt, D. (1983). *Farm Power and Machinery Management.* Iowa State University Press, Ames, IA.

Hunt, V. D. (1981). *Windpower: A Handbook on Wind Energy Conversion Systems.* Van Nostrand Reinhold, New York.

Hurley, J. (1981). Energy for rural development. In *Energy and the Developing Nations*, ed. P. Auer. Pergamon Press, Oxford, pp. 173–8.

IGT (annually since 1976). Energy from biomass and wastes. Proceedings of an annual conference held annually since 1976. Institute of Gas Technology, IIT Center, 3424 South State Street, Chicago, IL 60616.

IIASA (1981). Energy in a finite world: executive summary. International Institute for Applied Systems Analysis, A-2361 Laxenburg, Austria.

Inns, F. M. (1980). Animal power in agricultural production system. *World Animal Review*, No. 34. FAO, Rome, pp. 2–10.
INRA (1983). Les recherches de l'INRA dans le domaine de l'energie et de la biomasse. Institute National de la Recherche Agronomique.
ISES (1979). Biomass for energy. Proceedings of conference sponsored by the United Kingdom section of the International Solar Energy Society.
ITDG (1985). *Small Hydropower in China: A Survey.* Intermediate Technology Publications, London.
Jackson, C. I. (1978). *Human Settlements and Energy.* Pergamon Press, Oxford.
Jameel, M. (1982). Energy for agriculture in Pakistan. International Institute for Applied Systems Analysis, RR-82-20, Laxenburg, Austria.
Jewell, W. J. (Ed.) (1975). *Energy, Agriculture and Waste Management.* Ann Arbor Science Publishers, Ann Arbor, MI.
Jewell, W. J. *et al.* (1976). Bioconversion of Agricultural Wastes for Pollution Control and Energy Conservation, Final Report. Prepared under contract ERDA-NSF-74122A01 by Cornell University, Ithaca, New York. Available from National Technical Information Service, US Department of Commerce, Springfield, Virginia 22161.
Jewell, W. J., Koelsch, R. K., & Cummings, A. J. (1985). Cogeneration of electricity and heat from biogas. Cornell University, USDA Grant No. 59-2361-1-1-113-0/SERI Grant No. XB-0-90-3901.
Jin, Y., Lu, M., & Sun, C. (1983). Preliminary evaluation of energy consumption in China's agriculture—analysis of two specific cases. *Energy in Agriculture*, **1**, 289–301.
Jing, M. W. & Li, D. B. (1980). *An Introduction to the Development of Small Hydropower Generation in China.* UNIDO, New York.
Jones, T. F. (1988). *Wind Energy 1987: Wind Turbine Shipments and Applications.* STADIA, PO Box 442, Great Falls, VA 22066.
Julson, J. L. (1985). Performance and evaluation of variable-speed Darrieus wind turbine for agricultural purposes. In *Proceedings 5th Annual Solar and Biomass Energy Workshop*, USDA/DOE, 23–25 April. Atlanta, GA, pp. 261–3.
Justus, C. G., Hargraves, W. R., & Mikhail, A. (1976). Reference wind speed distribution and height profiles for wind turbine design and performance evaluation application. Technical Report, USERDA, Georgia Institute of Technology.
Kashkari, C. (1975). *Energy Resources, Demand and Conservation with Special Reference to India.* Tata McGraw-Hill, New Delhi, India.
Katzen, R. (1979). Grain motor fuel alcohol: technical and economic assessment study. Prepared for USDOE by Raphael Katzen Associates, Cincinnati, OH.
Kaufman, K. R. (1982). New developments in sunflower oil as an agricultural diesel fuel. AAAS Annual Meeting, Washington, DC.
Kaufman, K. R., Ziejewski, M., Marohl, M., & Jones, A. E. (1981). Sunflower oil and methyl ester as fuels for diesel engines. In *Proceedings of the Third International Conference on Energy Use Management*, Berlin, 26–30 October. Pergamon Press, Oxford, pp. 1635–43.
Kaupp, A. & Goss, J. (1983). Technical and economical problems in the gasification of rice hulls, Part 1. *Energy in Agriculture*, **1**, 201–35.
Kenney, V. P. & Erichsen, I. (1983). Conflict between fuel and food: the ethical dimension. *Energy in Agriculture*, **2**, 285–306.
Kepner, R. A., Basiner, R., & Barger, E. L. (1972). Principles of farm machinery. AVI, Westport, Connecticut.
Kiely, J. R. (1981). Long-term value of hydroelectric energy for developing nations. In *Energy and the Developing Nations*, ed. P. Auer. Pergamon Press, Oxford.
Kimura, K. I. (1976). Present technologies of solar heating, cooling and hot water supply in Japan. Tokyo, Department of Architecture, Waseda University. Reprinted from *Architectural Science Review*, **19**, No. 2. Surrey Hills, Research Publications Ptg Ltd.
King, A., Cleveland, H., & Streatfeild, G. (Eds) (1980). *Bioresources for Development.* Pergamon Press, Oxford.

Klass, D. L. & Emert, G. H. (Eds) (1981). *Fuel from Biomass and Wastes*. Ann Arbor Science Publishers, Ann Arbor, MI.

Klein, S. (1980). International applications of renewable energy resources. Statement before the Subcommittee on Energy Conservation and Supply, US Senate, 5 September.

Klein, S. (1985). Research networks for small-scale energy systems. *BOSTID Developments*, **5**, 6–10. National Research Council, Washington, DC.

Klein, S. (1985*a*). Small-scale energy systems for developing countries. *BOSTID Developments*, **5**, 6–9. National Research Council, 2101 Constitution Ave NW, Washington, DC.

Kline, C. K., Green, D. A. G., Donahue, R. L., & Stout, B. A. (1969). Agricultural mechanization in equatorial Africa. Institute of International Agriculture Research Report No. 6. Michigan State University, East Lansing, MI.

Klotz, L. H. (1980). *Water Power: Its Promises and Problems*. Center for Industrial and Institutional Development, University of New Hampshire, Durham, NH.

Kluter, H. H. & Soderholm, L. H. (Eds) (1979). Wind energy applications in agriculture. In *Proceedings USDA Wind Workshop*, Ames, Iowa.

Korshak, G. (1981). Small hydroelectric systems: a guide to development in California. California Energy Commission, Sacramento, CA 95825.

Kreider, J. F. & Kreith, F. (1975). *Solar Heating and Cooling*. Hemisphere Publishing Corporation, Washington, DC.

Kuether, D. O. & Duff, J. B. (1981). Energy requirements for alternative rice production systems in the tropics. IRRI Research Paper Series No. 59.

Kristoferson, L. A. & Bokalders, V. (1986). *Renewable Energy Technologies: Their Applications in Developing Countries*. Pergamon Press, Oxford.

Lacewell, R. D. & Person, J. B. Jr (1982). Economic analysis of an anaerobic digester based on the Del Valle hog farm system concepts. Report of Texas Energy and Natural Resources Advisory Council, Dept of Agricultural Economics, Texas A&M University, College Station, TX 77843.

Lansford, R. R., Supalla, R. J., Gilley, J. R., & Martin, D. L. (1980). Economics of wind energy for irrigation of pumping. Final Report DOE/SEA-7315-20741/81/2 prepared by Southwest Research and Development Company, Las Cruces, NM.

Larson, D. L. (1981). Operational evaluation of a 150-kW solar thermal power plant. In *Proceedings National Energy Symposium*, **I**, 151. American Society of Agricultural Engineers, St Joseph, MI.

Larson, D. L. & Sands, C. D. (1979). Solar powered deep well pumping experiment. American Society of Agricultural Engineers, Paper No. 79-4050, St Joseph, MI.

Leach, G. (1976). *Energy and Food Production*. IPC Science and Technology Press, Guildford, Surrey, UK.

Leach, G. (1979). Energy. Report prepared for the conference Agricultural Production: Research and Development Strategies for the 1980s, Bonn, Germany, 8–12 October.

Leach, G. & Slesser, M. (1973). *Energy Equivalents of Network Inputs to Food Producing Processes*. University of Strathclyde, Glasgow.

LePori, W. A. *et al.* (1981). Biomass energy conversion using fluidized bed technology. In *Proceedings of the Third International Conference on Energy Use Management*, Berlin, 26–30 October. Pergamon Press, Oxford, pp. A297–304.

LePori, W. A., Anthony, R. G., Lalk, T. R., & Craig, J. D. (1981*a*). Fluidized-bed combustion and gasification of biomass. In *Proceedings National Energy Symposium*, **II**, 330–4. American Society of Agricultural Engineers, St Joseph, MI.

Lettinga, G. (1981). Anaerobic digestion for energy saving and production. In *Proceedings International Conference on Biomass*, Brighton, UK, 4–7 November 1980. Applied Science Publishers, London, pp. 264–78.

Levy, M. & Robinson, J. L. (Eds) (1984). *Energy and Agriculture: Their Interacting Future*. Harwood Academic Publishers for the United Nations University, New York.

Lewis, D. A. (1982). The role of energy in UK agriculture. In *Energy Management and*

Agriculture, ed. D.W. Robinson & R.C. Mollan. *Proceedings 1st Summer School in Agriculture*, Dublin, Ireland. Elsevier Science Publishers, Amsterdam, pp. 43–65.

Liquid Fuels Trust Board (1982). Yields, costs, and availability of national oils/fats as diesel fuel substitutes. Prepared by Massey University, Agronomy Department, Greenock House, 102–112 Lambton Quay, 39 The Terrace, PO Box 17, Wellington, New Zealand.

Lockeretz, W. (1981). Macro-level implications of using soybean oil as a diesel fuel. In *Proceedings of the Third International Conference on Energy Use Management*, Berlin, 26–30 October. Pergamon Press, Oxford.

Lockeretz, W. (Ed.) (1977). *Agriculture and Energy*. Academic Press, New York.

Lockeretz, W. (Ed.) (1982). Agriculture as a producer and consumer of energy (AAAS Selected Symposium Series, No. 78). Westview Press, Boulder, CO.

Löf, G. (1961). Use of solar energy for heating purposes: solar cooking. *Proceedings of the United Nations Conference on New Sources of Energy*, **5**, 304–15.

Löf, G. (1978). Solar energy in the less developed countries. In *Renewable Energy Resources and Rural Applications in the Developing World*, ed. N. Brown. Westview Press, Boulder, CO, pp. 27–44.

Löf, G. *et al.* (1965). World distribution of solar radiation. Engineering Experiment Station, Report No. 21, University of Wisconsin, Madison, WI.

Löf, G. *et al.* (1966). *Solar Energy*, **10**, 27–37.

Loftness, R. L. (1984). *Energy Handbook*, 2nd edn. Von Nostrand Reinhold, New York.

Lyle, W. M. & Bordovsky, J. P. (1981). Low energy precision application (LEPA) irrigation system. *Trans. ASAE*, **24**, 1241–5.

Lyle, W. M. & Bordovsky, J. P. (1983). LEPA irrigation system evaluation. *Trans. ASAE*, **26**, 776–81.

Lysen, E. (unknown). *Introduction to Wind Energy*. CWD, PO Box 85, Amersfoort, The Netherlands.

McChesney, I. G. (1981). Liquid fuel demand in New Zealand agriculture. Occasional Paper No. 13. Joint Center for Environmental Sciences, Lincoln College, Canterbury, New Zealand.

McChesney, I. G., Sharp, B. M. H., & Hayward, J. A. (1981). Energy in New Zealand agriculture. Unpublished manuscript. Joint Center for Environmental Sciences, University of Canterbury and Lincoln College, Christchurch I, New Zealand.

McCutcheon, R. (1981). Vegetable oil as a diesel fuel—soybean oil. In *Proceedings of the Third International Conference on Energy Use Management*, Berlin, 26–30 October. Pergamon Press, Oxford, pp. 1679–86.

McFate, K. L. (Ed.) (in press). *Electrical Energy Use in Agriculture*. Energy in World Agriculture Series, Elsevier Science Publishers, Amsterdam.

McGowan, R. (1987). Current developments in photovoltaic irrigation in the developing world. In *Proceedings of the International Conference on Water and Water Policy in World Food Supplies*, ed. W. R. Jordan. Texas A&M University, 26–30 May 1985.

McIntosh, C. S., Smith, S. M., & Withers, R. V. (1984). Energy balance of an on-farm production and extraction of vegetable oil for fuel in the United States Inland Northwest. *Energy in Agriculture*, **3**, 155–66.

McKinney, J. D. *et al.* (1983). *Microhydropower Handbook*, Vols 1 and 2. DE83 006697 and 006698. Prepared for US Dept of Energy by EG&G, Inc., Idaho Falls, ID 83415. Available from NTIS. US Dept of Commerce, Springfield, VA 22161.

Makhijani, A. & Poole, A. (1975). *Energy and Agriculture in the Third World*. Ballinger, Cambridge, MA.

Manibog, F. R. (1982). Bangladesh: rural and renewable energy issues and prospects. World Bank Energy Department Paper No. 5.

Manibog, F. R. (1984). Improved cooking stoves in developing countries. *Ann. Rev. Energy*, **9**, 199–227.

Marier, D. (1985). Windfarm update. *Alternative Sources of Energy*, No. 72, March/April, pp. 12–19.

Marier, D. (Ed.) (1983). Hydropower. *Alternative Sources of Energy*, July/August.

Marsden, S. S. (1983). Methanol as a viable energy source in today's world. *Ann. Rev. Energy*, **8**, 333–54.

Masud, S. M., Lacewell, R. D., & Hiler, E. A. (1983). Economic implications of cotton gin trash and sorghum residues as alternative energy sources. *Energy in Agriculture*, **1**, 267–80.

Matthews, J. (1982). Energy and farm mechanisation. In *Energy Management and Agriculture*, ed. D. W. Robinson & R. C. Mollan. *Proceedings 1st Summer School in Agriculture*, Dublin, Ireland. Elsevier Science Publishers, Amsterdam, pp. 43–65.

Matthews, W. H. & Siddiqi, T. A. (1981). Energy and environmental issues in the developing countries. In *Energy and Environment in the Developing Countries*, ed. M. Chatterji. John Wiley and Sons, New York, pp. 53–67.

Meinel, A. B. & Meinel, M. P. (1976). *Applied Solar Energy*. Addison-Wesley, Reading, MA.

Mejer, G. J. (1981). Gas as an alternative fuel for agriculture. In *Proceedings of the Third International Conference on Energy Use Management*, Berlin, 26–30 October. Pergamon Press, Oxford, pp. 1687–95.

Merklein, H. A. & Hardy, W. C. (1977). *Energy Economics*. Gulf Publishing Co., Houston.

Merrikin, E. J. & Ward, J. A. (1981). On-farm production of fuel from vegetable oil. In *Proceedings of the Third International Conference on Energy Use Management*, Berlin, 26–30 October. Pergamon Press, Oxford, pp. 1697–705.

Merrill, R. & Gage, T. (Eds) (1974). *Energy Primer*. Portola Institute, Menlo Park, CA.

Metzler, R. (1982). The heat generator. Fachhochschule furtwangen, 7743 Furtwangen, Federal Republic of Germany.

MIC (1981). Assessment of Brazil's national alcohol program. Ministry of Industry and Commerce, Brasilia.

Midwest Plan Service (1980). *Low Temperature and Solar Grain Drying Handbook*. MWP-22. Iowa State University, Ames, IA 50011.

Miller, A. S., Mintzerand, I. M., & Hoagland, S. H. (1986). Growing power: bioenergy for development and industry. World Resources Institute Study 5, 1735 New York Avenue NW, Washington, DC 20006.

Miner, J. R. & Smith, R. J. (Eds) (1975). *Livestock Waste Management with Pollution Control*. North Central Reg. Pub. 222. Midwest Plan Service Handbook MWPS-19.

Mislin, H. & Bachofen, R. (Eds) (1982). *New Trends in Research and Utilization of Solar Energy through Biological Systems*. Birkhauser, Basel.

Mittal, J. P. & Dhawan, K. C. (1988). Research manual on energy requirements in the agricultural sector. ICAR/AICRP/ERAS-88-1. Punjab Agricultural University, Ludhiana, India 141004.

Mittelbach, M., Worgetter, M., Pernkopf, J., & Junek, H. (1983). Diesel fuel derived from vegetable oils: preparation and use of rape oil methyl ester. *Energy in Agriculture*, **2**, 369–84.

Molle, J. F. (1985). *Biomass News International*. A monthly newsletter. 254 rue de Vaugirard, 75640 Paris, Cedex 15, France.

Morrison, D. E. (1978). Equity impacts of some major energy alternatives. In *Energy Policy in the United States: Social and Behavioral Dimensions*, ed. S. Warkov. Praeger, New York, pp. 164–93.

Morrison, D. W. & Shove, G. C. (1981). Solar drying of large round hay bales. In *Proceedings National Energy Symposium*, **I**, 73–7. American Society of Agricultural Engineers, St Joseph, MI.

Moustafa, S. M. A. & Brusewitz, G. H. (1978). Direct use of solar energy for water desalination. *Solar Energy*, **22**, 141–8.

Mrema, G. (1983). Energy consumption in small holder rainfed agriculture in the semi-arid tropics. In *Proceedings Symposium on Energy in Agriculture, Inst. of Agr. Engr.* Harare, Zimbabwe, 11–12 May.

Mubayi, V., Palmedo, P., & Doernberg, A. (1981). A framework for energy policy and technology assessment in developing countries: a case study of Peru. In *Energy and Environment in the Developing Countries*, ed. M. Chatterji. John Wiley and Sons, New York, pp. 289–314.

Mudahar, M. & Hignett, T. (1982). Energy and fertilizer: policy implications and options for developing countries. International Fertilizer Development Center, PO Box 2040, Muscle Shoals, AL.

Munasinghe, M. (1986). *Rural Electrification for Development: Policy Analysis and Application.* Westview Press, Boulder, CO.

Munzinger, P. (1982). *Animal Traction in Africa.* GTZ Dag-Hammerskjold-Weg 1, D-6236 Eschborn, Germany.

Myers, C. (1983). Energy use for Durum wheat production in Tunisia—a case study of 23 farms. ASAE Paper No. 83-3021, American Society of Agricultural Engineering, St Joseph, MI.

Myers, C. & Mackson, C. J. (1981). Applications of solar energy in agriculture. FAO/ECE Working Party in Mechanization.

Nakarmi, A. M. & Backmann, A. (1982). *Multi-Purpose Power Unit with Horizontal Water Turbine.* Sahayogi Press, Kathmandu, Nepal.

NAS (1972). Solar energy in developing countries: perspectives and prospects. In *Proc. BOSTID.* National Academy of Sciences, Washington, DC.

NAS (1976). Energy for rural development: renewable resources and alternative technologies for developing countries. National Academy of Sciences, Board on Science and Technology for International Development.

NAS (1980). International Workshop on Energy Survey Methodologies for Developing Countries. In *Proc. BOSTID*, National Academy of Sciences. National Academy Press, Washington, DC.

NAS (1981). Supplement: *Energy for Rural Development.* National Academy of Sciences, Board on Science and Technology for International Development.

Nathan, R. A. (Ed.) (1978). Fuels from sugar crops. Prepared for USDOE, National Technical Information Service, US Department of Commerce, Springfield, VA 22161.

Nelson, V. (Ed.) (1982). *Proceedings National Conference American Wind Energy Association*, Amarillo, TX. American Wind Energy Association, Washington, DC 20036.

NewKirk, H. W. (1978). Solar technology applications: a survey of solar powered irrigation systems. UCID-17510, Rev. 1, Lawrence Livermore Laboratory, CA.

Newman, M. & Grey, J. (Eds) (1979). *Utilization of Alternative Fuels for Transportation.* American Institute of Aeronautics and Astronautics, 1290 Avenue of the Americas, New York.

Nilsson, C. (1981). The utilization of wood-products and straw for combustion. In *Proceedings of the Third International Conference on Energy Use Management*, Berlin, 26–30 October. Pergamon Press, Oxford, pp. 1731–40.

NOAA (1980). *Local Climatological Data, Annual Summaries, 1980.* NOAA National Climatic Center, Asheville, NC.

Nowakowska, J. & Wiebe, R. (1945). Bibliography on construction, design, economics, performance, and theory of portable and small stationary gas producers. AIC-103, USDA-Peoria, IL.

Noyes, R. (1980). *Small and Micro Hydroelectric Plants: Technology and Feasibility.* Noyes Data Corp., Park Ridge, NJ 07656.

NPC (1976). *Total Energy and Energy Substitution.* National Productivity Council, Productivity House, Lodi Road, New Delhi, India 110003.

NRAES (1982). *Solar Greenhouses for the Home.* Northeast Regional Agricultural Engineering Service, Riley Robb Hall, Cornell University, Ithaca, NY 14853.

NRC (1972). Atlas of nutritional data on United States and Canadian feeds. Publication ISBN 0-309-01919-2, Committee on Animal Nutrition, National Academy of Sciences, National Research Council, Washington, DC.

NRC (1976). *Energy for Rural Development.* Board on Science and Technology for International Development, National Academy of Sciences, Washington, DC.

NRC (1981). Workshop on energy and agriculture in developing countries. Board on Science and Technology for International Development, National Research Council, National Academy of Sciences, Washington, DC.

NRC (1981*a*). Supplement: *Energy for Rural Development*. Board on Science and Technology for International Development, National Academy of Sciences, Washington, DC.

NRC (1981*b*). Workshop on ethanol as an alternative source of fuels. Jakarta, Indonesia. Board on Science and Technology for International Development, National Research Council, National Academy of Sciences, Washington, DC.

NRC (1981*c*). *Food, Fuel and Fertilizer from Organic Wastes*. Board on Science and Technology for International Development, National Research Council, National Academy of Sciences, Washington, DC.

NRC (1982). *Diffusion of Biomass Energy Technologies in Developing Countries*. Board on Science and Technology for International Development, National Research Council, National Academy of Sciences, Washington, DC.

NRC (1983). *Alcohol Fuels: Options for Developing Countries*. Board on Science and Technology for International Development, National Research Council, National Academy of Sciences, Washington, DC.

NRC (1983*a*). *Producer Gas: Another Fuel for Motor Transport*. Board on Science and Technology for International Development, National Research Council, National Academy of Sciences, Washington, DC.

NRC (1984). *Diffusion of Biomass Energy Technologies in Developing Countries*, 2nd edn. Board on Science and Technology for International Development, National Research Council, National Academy of Sciences, Washington, DC.

Nuccitelli, G. & Ercoli, L. (1979). Rendimento di un impiato solare a collettori concentratori durante il periodo invernoprimavera. Instituto Sperimentale per la Meccanizzazione Agricola, Roma. *Estratto dagli Annali*, Vol. VII.

Odom, W. B. & Golightly, J. L. (1985). Energy integrated cotton farm system in Arizona. USDOE Contract No. DE-FC07-81CS40382.

Odum, H. T. (1984). Energy analysis of the environmental role in agriculture. In *Energy and Agriculture*, ed. G. Stanhill. Springer, Berlin, pp. 24–50.

Odum, H. T. & Odum, E. L. (1976). *Energy Basis for Man and Nature*. McGraw-Hill, New York.

OECD (1982). *The Energy Problem and the Agro-Food Sector*. Organization for Economic Cooperation and Development, 2, rue Andre-Pascal, 75775 Paris, Cedex 16, France.

Ofoli, R. Y. (1980). Energy balance analysis for fuel ethanol production. MS thesis, Michigan State University, East Lansing, MI.

Ofoli, R. Y. & Stout, B. A. (1980). Making ethanol for fuel on the farm. Extension Bulletin E-1438, Michigan State University, East Lansing, MI.

Ofoli, R. Y. & Stout, B. A. (1980*a*). Ethyl alcohol production for fuel: energy balance. In *Proceedings Energy from Biomass Conference*, Brighton, UK, 4–7 November.

Olver, E. F. (1983). Energy integrated farm systems program. Paper prepared by Energy from Agricultural Products and Residues Seminar, Austin, TX.

Openshaw, K. (1978). Woodfuel—a time for reassessment. *Natural Resources Forum* **3**(1).

Orth, H. W. (1977). Converting straw into energy. *Vortag Freisling*, **25**, 4.

Orth, H. W., Peters, H., & Kohler, U. (1976). Investigation into the technical combustion properties of straw. *Grundlagen der Landtechnik*, **26**, 230–4.

OSU (1982). Micro-hydropower: Oregon development guide. Extension Circular 1096, Oregon State University Extension Service, Corvallis, OR.

OTA (1979). *Gasohol: A Technical Memorandum*. Congress of the United States, Office of Technology Assessment, Washington, DC 20510.

OTA (1980). *Energy from Biological Processes*. US Government Printing Office, Washington, DC, Vols I and II.

Pachauri, R. K. (1982). Financing the energy needs of developing countries. *Ann. Rev. Energy*, **7**, 109–38.

Palz, W. & Chartier, P. (1980). *Energy from Biomass in Europe*. Applied Science Publishers, London.

Palz, W., Chartier, P. & Hall, D. O. (1981). Energy from biomass. In *Proceedings of International Biomass Conference*, Brighton, UK, 4–7 November 1980. Applied Science Publishers, London.

Papamarces, J. (1978). Developments in hydropower. *Power Engineering*, **82**, 12.

Parikh, J. K. (1980). *Energy Systems and Development*. Oxford University Press, Bombay.

Parikh, J. K. & Kramer, G. (1984). Modelling energy and agriculture interactions: an application to Bangladesh. IIASA Working Paper WP-84-10.

Park, G., Krauss, O., Weinberg, B., & Tedeshi, S. (1984). *Windmill Performance in Michigan*. Division of Engineering Research, Michigan State University, East Lansing, MI 48824.

Park, J. (1981). *The Wind Power Book*. Cheshire Books, 514 Bryant Street, Palo Alto, CA 94301.

Parker, S. P. (Ed.) (1981). *Encyclopedia of Energy*, 2nd edn. McGraw-Hill, New York.

Parnell, C. B. (1983). Systems analysis of cotton gin trash utilization alternatives, Vol. I. Final report submitted to National Science Foundation, Agricultural Engineering Dept, Texas A&M University, College Station, TX 77843-2117.

Parsons, R. A. (1984). On-farm biogas production. Northeast Regional Agricultural Engineering Service Report No. 20. Riley–Robb Hall, Cornell University, Ithaca, NY 14853.

Passmore, R. & Durnin, J. V. G. A. (1955). Human energy expenditure. *Physiological Reviews*, **34**, 801–40.

Patton, F. S. (1985). *Global Energy Perspectives—1985. International Conference on Water and Water Policy in World Food Supplies*. Texas A&M University, 26 May.

Paul, J. K. (1979). *Ethyl Alcohol Production and Use as a Motor Fuel*. Noyes Data Corp., Park Ridge, NJ 07656.

Paul, J. K. (Ed.) (1980). *Large and Small Scale Ethyl Alcohol Manufacturing Processes from Agricultural Raw Materials*. Noyes Data Corp., Park Ridge, NJ 07656.

Paul, S. (Ed.) (1982). *Bio-Energy Re-News. A Journal of Energy from Biomass and Recycling*. Published bi-monthly by India House Developments, 134 Palam Colony, New Delhi, India 110045.

Peart, R. M. (Ed.) (in press). *Analysis of Agricultural Energy Systems*. Energy in World Agriculture Series, Elsevier Science Publishers, Amsterdam.

Peill, J. (1980). Utilizing biomass (straw) for crop drying and heating. Report No. 80-030801801, Canadian Department of Agriculture—ERSI.

Pellizzi, G. (1977). Tecnologie per l'uso dell'energia solare in agricultura. *Genio Rurale*, No. 9: 17-241977.

Pellizzi, G. (1978). Solar energy in agriculture. Report presented to the Prima Mestra Convegno sull'Energia Solaire, Genova. University of Milan, Institute of Agricultural Engineering.

Pellizzi, G. (1980). New and renewable energy in agriculture. European Commission on Agriculture, main report.

Pellizzi, G. (1981). Isistemi energetici integrati al servizio dell'agricoltura. *Informatore Agrario*, **5**, 13873–81.

Peterson, W. D. (1983). Agricultural solar systems performance: Illinois solar demonstration program, In *Solar and Wind Systems Workshop Proceedings*, Great Plains Agr. Council, University of Nebraska, 18–19 October, pp. 17–28.

Pfeffer, J. T. (1977). Biological conversion of biomass to methane. Annual Progress Report, 1 June 1976–31 May 1977. Prepared under Contract No. EY-76-S-02-2917 by the University of Illinois, Urbana, Illinois. Available from the National Technical Information Service, US Department of Commerce, Springfield, VA 22161.

Pierotti, A., Keller, A. G., & Fritsch, A. J. (1977). *Energy and Food*. CSPI Energy Series X, Center for Science in the Public Interest, Washington, DC.

Pimentel, D. (1974). Energy use in world food production. Environmental Biology Report 74-1, Cornell University, Ithaca, NY.

Pimentel, D. (1980). *Handbook of Energy Utilization in Agriculture*. CRC Press, Boca Raton, FL.

Pimentel, D. (1983). Review article on biomass energy. *Solar Energy*, **30**, 1–31.

Pimentel, D. (1984). Research needs: energy and agriculture. Report to Energy Research Advisory Board. US Department of Energy, 1000 Independence Ave NW, Washington, DC 20585.

Pimentel, D. & Hall, C. (1984). *Food and Energy Resources*. Academic Press, New York.

Pimentel, D. & Pimentel, M. (1979). *Food, Energy and Society*. Edward Arnold Publishers, London.

Pimentel, D. *et al.* (1973). Food production and the energy crisis. *Science*, **182**, 443–9.

Pollard, W. G. (1976). The long-range prospects for solar energy. *American Scientist*, **64**, 424–9.

Polz, W. & Chartin, P. (1980). *Biomass Energy in Europe*. Applied Science Publishers, Barking, UK.

Posselius, J. H. & Stout, B. A. (1983). Crop residue availability for fuel. In *Liquid Fuel Systems*, ed. D. L. Wise. CRC Press, Boca Raton, FL, pp. 11–30.

Pratt, G., Lindley, J., Hirning, M., Giles, J., & Johnson, R. (1981). On-farm production of sunflower oil for fuel. In *Proceedings of the Third International Conference on Energy Use Management*, Berlin, 26–30 October. Pergamon Press, Oxford, pp. 1767–74.

Pratt, G. *et al.* (1986). Energy integrated dairy farm system in North Dakota. USDOE Contract No. DE-FC07-80CS0377.

Price, E. C. & Cheremisinoff, P. C. (1981). *Biogas Production and Utilization*. Ann Arbor Science Publishers, Ann Arbor, MI.

Prinsloo, M. & Hugo, F. J. C. (1981). On oil extraction for fuel purposes. In *Proceedings of the Third International Conference on Energy Use Management*, Berlin, 26–30 October. Pergamon Press, Oxford, pp. 1775–80.

Probstein, R. F. & Hicks, R. E. (1982). *Synthetic Fuels*. McGraw-Hill, New York.

Pryde, E. H. (1981). Vegetable oils vs. diesel fuel: chemistry and availability of vegetable oils. In *Alcohol and Vegetable Oil as Alternative Fuels, Proceedings of Regional Workshop*. USDA, Washington, DC.

Pyke, M. (1970). *Man and Food*. McGraw-Hill, New York.

Quenouille, M. H. *et al.* (1951). Statistical studies of recorded energy expenditure of man. Technical Communication No. 17. Commonwealth Bureau of Animal Nutrition, Rowett Institute, Scotland.

Quick, G. R. (1980). An in-depth look at farm fuel alternatives. *Power Farming Magazine*, February, pp. 10–17.

Quick, G. R. (1980*a*). Developments in use of vegetable oils as fuel for diesel engines. ASAE Paper No. 80-1525R, American Society of Agricultural Engineers, St Joseph, MI.

Rappaport, R. A. (1971). The flow of energy in an agricultural society. *Scientific American*, **224**, 116–22, 127–32.

Rask, N. (1979). *Using Agricultural Resources to Produce Food or Fuel—Policy Intervention or Market Choice?* Department of Agricultural Economics and Rural Sociology, Ohio State University.

Reddy, A. K. N. (1985). The energy and economic implications of agricultural technologies: an approach based on the technical options for the operations of crop production. ILO Working Paper. WEP 2-22/WP149.

Reed, T. B. (1979). *Problems and Opportunities for Solar Energy in Biomass, Pyrolysis, and Gasification*. Produced by Solar Energy Research Institute, Golden, CO, under a contract with the US Department of Energy.

Reed, T. B. (1980). Biomass gasification: yesterday, today, and tomorrow. In *Proc. BIO-ENERGY '80*. The Bioenergy Council, Washington, DC 20006.

Reed, T. B. (Ed.) (1981). *Biomass Gasification: Principles and Technology.* Noyes Data Corp., Park Ridge, NJ 07656.

Regan, R. W. (1975). *Methane Recovery from Farm Wastes: A Literature Review.* Penn State University, Dept of Agricultural Engineering.

Requillart, V. (ed.) (1989). *L'Énergie: un Nouveau Marché pour l'Agriculture?* INRA Publications, Versailles, France.

Revelle, R. (1976). Energy use in rural India. *Science*, **192**, 969–75.

Revelle, R. (1978). Requirements for energy in the rural areas of developing countries. In *Renewable Energy Resources and Rural Applications in the Developing World*, ed. N. Brown. Westview Press, Boulder, CO, pp. 11–26.

Revelle, R. (1980). Energy dilemmas in Asia: the needs for research and development. *Science*, **209**, 164–73.

Revelle, R. (1981). Energy and food in the poor countries. In *Energy and the Developing Nations*, ed. P. Auer. Pergamon Press, Oxford, pp. 110–22.

Riva, M. (1984). Applications of simplified solar energy systems in farming: results of the second year experiments. *Energy in Agriculture*, **3**, 109–20.

Robbins, F. V. & Spillman, C. K. (1981). Predicted performance of the KSU solar collector–storage system. In *Proceedings National Energy Symposium*, **I**, 160–5. American Society of Agricultural Engineers, St Joseph, MI.

Robinson, D. W. & Mollan, R. C. (Eds) (1982). Energy management and agriculture. In *Proceedings 1st Summer School in Agriculture*, Dublin, Ireland. Elsevier Science Publishers, Amsterdam.

Robinson, J. S. (1980). *Fuels from Biomass: Technology and Feasibility.* Noyes Data Corp., Park Ridge, NJ 07656.

Roller, W. L. (1975). Grown organic matter as a fuel raw material resource. Prepared by Ohio Agricultural Research and Development Center, Wooster, for the Lewis Research Center, NASA Report CR-2608.

Rose, W. T. (1981). Feasibility analysis for small wind energy conversion systems. MS thesis, Agricultural Engineering Department, Michigan State University, East Lansing, MI.

Rose, W. T., Asmussen, J. Jr, & Stout, B. A. (1980). Survey of commercially available wind–electric systems. AEIS 427, File No. 18.8. Agricultural Engineering Department, Michigan State University, East Lansing, MI 48824.

Rose, W. T., Asmussen, J. Jr, & Stout, B. A. (1981). A comparison of six methods for calculating the annual energy output of wind turbines. In *Proceedings of Wind Energy Technology*, 16–17 March 1981. Engineering Conferences, 1020 Engineering Building, University of Missouri–Columbia, Columbia, MO 65211.

Rotz, A., Cruz, M., Wilkinson, R., & Stout, B. A. (1980). Utilization of alcohol in spark ignition and diesel engines. Extension Bulletin E-1426, Michigan State University, East Lansing, MI.

Rui-Zhang, G. (1980). Chinese experience in small scale hydropower generation. In *United Nations Industrial Development Organization, Manual for Mini Hydro Generation Units.* UNIDO, Vienna, Austria.

SAE (1980). *Alcohols as Motor Fuels.* Progress in Technology Series No. 19, Society of Automotive Engineers Inc., 400 Commonwealth Drive, Warrendale, PA 15096.

Safley, L. M. & Westerman, P. W. (1981). Enhancing manure value through the addition of chemicals. ASAE Paper No. 81-2579, American Society of Agricultural Engineers, St Joseph, MI.

Sakr, I. A. (1973). An elliptical paraboloid solar cooker. International Congress on the Sun in the Service of Mankind, Paris.

Sarkanen, K. V. & Tillman, D. A. (1980). *Progress in Biomass Conversion.* Academic Press, New York.

Sasscer, D. S. & Morgan, T. O. (1986). Energy integrated dairy farm system in Puerto Rico. USDOE Contract No. DE-FC07-80CS40376.

Sassin, W. (1980). Energy. *Scientific American*, **243**, 119–22, 127–9.

Sathaye, J., Ghirardi, A., & Schipper, L. (1987). Energy demand in developing countries: a sectoral analysis of recent trends. *Ann. Rev. Energy*, **12**, 253–81.

Sathianathan, M. L. (1975). *Biogas: Achievements and Challenges*. Distributed by Lakshmi Bookstore, 72, Janpath, Ved Mansion, New Delhi, India 110001.

Schepens, G., Mahy, D., Buis, H., & Palz, W. (1986). *Solar Energy in Agriculture and Industry: Potential of Solar Heat in European Agriculture, an Assessment*. Kluwer Academic Publishers, Dordrecht.

Schroedder, F. (1981). Rapeseed oil as an alternative fuel for agriculture. In *Proceedings of the Third International Conference on Energy Use Management*, Berlin, 26–30 October. Pergamon Press, Oxford, pp. 1815–20.

Schultz, E. B. & Morgan, R. P. (1984). *Fuels and Chemicals from Oilseeds*. Westview Press, Boulder, CO.

Schulz, H. (1980). Regenerative energiequellen in der landwirtschaft. *Sonnenergie und Warmepump*, **5**, 16–23.

Schulz, H. (1980*a*). Warmwasser aus dem serpentinenkollektor. *Top. Agrar. H.*, **6**, 62–6.

Schulz, H. (1980*b*). Energiegewinnung aus sonne und wind. In *Energieeinsatz in der landwirtschaft. Bayerisches staatsministerium für ernahrung, Landwirtschaft und Forsten*, Munchen, pp. 151–67.

Schwartz, P. (1987). What happened to the energy crisis? The dilemma of an energy decision maker in a dynamic world. *Ann. Rev. Energy*, **12**, 397–414.

Scott, C. D. (Ed.) (1979). *Proceedings, Biotechnology in Energy Production and Conservation*. John Wiley, New York.

Sellers, W. D. (1965). *Physical Climatology*. Chicago University Press, Chicago, IL.

SERI (1979). *Generator Gas: The Swedish Experience from 1939–1945*. Translated from Swedish by the Solar Energy Research Institute, Golden, CO 80401.

SERI (1980). *Fuel from Farms: A Guide to Small Scale Ethanol Production*. Solar Energy Research Institute, SERI/SP-451-519, Golden, CO 80401.

SERI (1981). *Ethanol Fuel: Use, Production and Economics*. Solar Energy Research Institute, SERI/SP-751-1018, Golden, CO 80401.

SERI (1982). *A Selected Bibliography on Alcohol Fuels*. Solar Energy Research Institute, SERI/SP-290-1414, Golden, CO 80401.

SERI/UNESCO (1986). *International Directory of New and Renewable Energy Information Sources and Research Centers*. Solar Energy Research Institute, Document Distribution Service, Golden, CO 80401.

Sharma, M. P. & Singh, G. (1980). A low lift solar water pump. *Solar Energy*, **25**, 273–8.

Shelton, J. (1976). *The Woodburners Encyclopedia*. Vermont Crossroads Press, Waitsfield, VT 05673.

Shove, G. C. (Ed.) (1977). *Solar Grain Drying Conference Proceedings*, University of Illinois, Urbana, IL.

Silvestrini, V. (Ed.) (1978). *Solar Energy and Agriculture*. Casella Postale 65, 61029 Urbino, Italia.

Sims, R. E. H. & Meister, A. D. (1984). Introduction and economics of rapeseed oil and tallow esters as alternative fuels for compression ignition engines. *Energy in Agriculture*, **3**, 223–32.

Sims, R. E. H. *et al.* (1983). On-farm energy supply and conservation. Report No. 98, New Zealand Energy R&D Committee, University of Auckland.

Singh, G. & Chancellor, W. (1975). Energy inputs and agricultural production under various regimes of mechanization in northern India. *Trans. ASAE*, **18**, 252–9.

Singh, R. P. (Ed.) (1986). *Energy in Food Processing*. Energy in World Agriculture Series, Elsevier Science Publishers, Amsterdam.

Singhal, O. P. (1981). Design, development and testing of 2 tons/day solar energy operated paddy parboiling and drying unit for rural areas. In *Alternative Energy Sources II*, Vol. 2, ed. T. Nejat Veziroglu. Hemisphere Publishing Corp., Washington, DC. Distributed by McGraw-Hill International, pp. 919–28.

Sivard, R. L. (1981). *World Energy Survey*. World Priorities, Leesburg, VA 22075.

Siwatibau, S. (1981). Rural energy in Fiji: a survey of domestic rural energy use and potential. IDRC-157e. Ottawa.

Slesser, M. (1975). Accounting for energy. *Nature*, **25**, 171.

Slesser, M. (1984). Energy use in the food-producing sector of the European Economic Community. In *Energy and Agriculture*, ed. G. Stanhill. *Advanced Series in Agr. Sciences 14*. Springer, Berlin, pp. 132–53.

Slesser, M. & Lewis, C. (1979). *Biological Energy Resources*. E and F. N. Spon, London.

Smerdon, E. T. & Hiler, E. A. (1985). Energy in irrigation in developing countries. In *Proceedings of the International Conference on Water and Water Policy in World Food Supplies*, Texas A&M University, 26–30 May.

Smil, V. (1981). Of trees and straws. In *Energy and the Developing Nations*, ed. P. Auer. Pergamon Policy Studies, Pergamon Press, New York, pp. 126–43.

Smil, V. & Knowland, W. E. (Eds) (1980). *Energy in the Developing World*. Oxford University Press, Oxford.

Smil, V., Nachman, P., & Long, T. V. (1983). *Energy Analysis and Agriculture: An Application to US Corn Production*. Westview Press, Boulder, CO.

Smith, A. J. (1981). Draught animal research. World Animal Review No. 40, FAO, Rome, pp. 43–8.

Smith, C. B. (1981). *Energy Management Principles: Applications, Benefits, Savings*. Pergamon Press, New York.

Smith, N. (1981*b*). Wood: an ancient fuel with a new future. Paper No. 42, Worldwatch Institute, Washington, DC.

Smith, N. L. (1978). Opportunities for energy savings in crop production. Committee print for Subcommittee on Advanced Energy Technologies and Energy Conservation RD and D. US House of Representatives. Government Printing Office, Washington, DC.

Soderholm, L. H. (1983). Heating and cooling with wind energy. In *Proceedings of Solar and Wind Systems Workshop*, 18–19 October. Great Plains Agricultural Council Publications No. 108.

Soderholm, L. H. (1985). Maximizing wind-driven alternator performance. In *Proceedings of 5th Annual Solar and Biomass Energy Workshop*, USDA/DOE, 23–25 April, Atlanta, GA, pp. 249–52.

Sodha, M. S., Bansal, N. K., Kumar, A., Bansal, P. K., & Malik, M. A. (1987). *Solar Crop Drying*. CRC Press, Boca Raton, FL 33431.

Sofer, S. & Zaborsky, O. R. (1981). *Biomass Conversion Processes for Energy and Fuels*. Plenum Press, New York.

Soltes, E. J. (Ed.) (1983). *Wood and Agricultural Residues: Research on Use for Feed, Fuels and Chemicals*. Academic Press, New York.

Southwell, P. H. & Rothwell, T. M. (1977). Analysis of output/input energy ratios of food production in Ontario. Report on Contract No. OSW76-00048, School of Engineering, University of Guelph, Ontario.

Spear, P. J. (Ed.) (1983). *Energy in Agriculture, Symposium Proceedings*. The Institute of Agricultural Engineering, PO Box BW 330, Borrowdale, Harare, Zimbabwe.

Spedding, C. R. W. (1975). *The Biology of Natural Systems*. Academic Press, New York.

Spillman, C. K. (1978). *Principles of Collecting, Storing and Using Solar Energy*. Department of Agricultural Engineering, Kansas State University.

Spillman, C. K. (1980). *Low Temperature and Solar Grain Drying Handbook*. Midwest Plan Service, Iowa State University, Ames, IA 50011.

Spillman, C. K. *et al.* (1983). *Solar Livestock Housing Handbook*. Midwest Plan Service—23, Iowa State University, Ames, IA 50011.

Spillman, C. K., Robbins, F. V., & Hines, R. H. (1979). Solar energy for reducing fossil fuel usage in farrowing houses. Agricultural Experiment Station Paper No. 79-179A, Kansas State University, Manhattan, KS.

Splinter, W. E. & Schulte, D. D. (1987). Energy integrated swine farm system in Nebraska. USDOE Contract No. DE-FC07-80CS40378.

Stafford, D. A., Hawkes, D. L., & Horton, R. (1980). *Methane Production from Waste Organic Matter*. CRC Press, Boca Raton, FL 33431.

Stanhill, G. (1979). A comparative study of the Egyptian agro-ecosystem. *Agro-Ecosystems*, **5**, 213–20.

Stanhill, G. (Ed.) (1984). *Energy and Agriculture. Advanced Series in Agricultural Sciences 14.* Springer, Berlin.

Starkey, P. H. (1982). N'Dama cattle as draught animals in Sierra Leone. World Animal Review No. 42, FAO, Rome, pp. 19–26.

Starr, C. (1981). The energy strategy planning process for developing countries. In *Energy and the Developing Nations*, ed. P. Auer. Pergamon Press, Oxford, pp. 359–73.

Steinhart, C. & Steinhart, J. (1974). *Energy Sources, Use and Role in Human Affairs.* Duxbury Press, North Scituate, MA.

Stewart, G. A. *et al.* (1979). *The Potential for Liquid Fuels from Agriculture and Forestry in Australia.* CSIRO, Division of Chemical Technology, PO Box 1666, Canberra City, ACT 2601, Australia.

Stoeckel, A. & Patterson, P. (1980). The effects of changing oil prices on Australia's balance of payments. In *Proceedings Seminar on Energy: Crisis or Opportunity?* Marcus Oldham Farm Management College, Geelong, Australia, pp. 69–80.

Stone and Webster Engineering Corporation (1981). Report on turbogeneration equipment for low-head hydroelectric developments. Stone and Webster Engineering Corporation, PO Box 2325, Boston, MA 02107.

Stork, A. (1980). Solare Kalbertrank-anlage Wiesenfelden. *Sonnenenergie und Warmepump*, **5**, 58–60.

Stout, B. A. (1982). Energy for world agriculture. In *Energy Management and Agriculture*, ed. D. W. Robinson & R. C. Mollan. *Proceedings 1st Summer School in Agriculture*, Dublin, Ireland. Elsevier Science Publishers, Amsterdam, pp. 1–21.

Stout, B. A. (1982*a*). Agricultural biomass for fuel. *Experientia*, **38**, 145–51.

Stout, B. A. (1983). Energy balance of agricultural/forestry systems. In *Proceedings 1st International Conference on Energy and Agriculture*, Milan, Italy, 27–29 April, ed. B. Guerra. CESAT, Gruppo dell'Energio, Milan, pp. 1/2–1/22.

Stout, B. A. (1983*a*). Biomass energy profiles. Agricultural Services Bulletin No. 54, FAO, Rome.

Stout, B. A. (1984). *Energy Use and Management in Agriculture*. Breton Publishers, Boston, MA.

Stout, B. A. (1985). Agricultural energy—1985–2000. *Agr. Engr.*, **66**, 18–20.

Stout, B. A. & Downing, C. M. (1976). Agricultural mechanization policy. *Int. Labor Rev.*, **113**, 171–87.

Stout, B. A. *et al.* (1978). Energy—a vital resource for the US food system. Public Policy Issues Report, American Society of Agricultural Engineers, St Joseph, MI.

Stout, B. A., Myers, C. A., Hurand, A., & Faidley, L. W. (1979). *Energy for World Agriculture.* FAO Agriculture Series No. 7, FAO, Rome.

Stout, B. A., Clark, J. A., Maycock, P., & Asmussen, J. (1980). Overview of solar energy technologies. In *Proceedings of the Energy Seminar, National Society of Professional Engineers*, Detroit, MI, 23 July. National Society of Professional Engineers.

Stout, B. A., Butler, J. L., & Gavett, E. E. (1983). Energy use and management in US agriculture. In *Energy and Agriculture. Advanced Series in Agricultural Sciences 14*, ed. G. Stanhill. Springer, Berlin, pp. 167–89.

Stout, B. A. *et al.* (1984). Energy use and production in agriculture. CAST Report No. 99, Council for Agricultural Science and Technology, 250 Memorial Union, Iowa State University, Ames, IA 50011.

Stout, B. A. & Nehring, R. F. (1988). Agricultural energy: 1988 and the future. *National Forum*, Summer issue, 32–6.

Strehler, A. (1981). Heat from straw and wood. In *Proceedings of the Third International Conference on Energy Use Management*, Berlin, 26–30 October. Pergamon Press, Oxford, pp. 1855–62.

Suggs, L. W. & Splinter, W. E. (1961). Effect of environment on the allowable work load of man. *Trans. ASAE*, **4**, 48–51.

Surbrook, T. & Friedle, J. (1978). Wind power. Extension Bulletin E-1112, File No. 18-82. Cooperative Extension Service, Michigan State University, East Lansing, MI 48824.

Sweeten, J. M., Fulhage, C., & Humenik, F. J. (1981). Methane gas from swine manure. Leaflet No. L-1860, in *Pork Industry Handbook*. Agricultural Engineering Department, Texas A&M University, College Station, TX 77843-2117.

Swern, D. (1982). *Bailey's Industrial Oil and Fat Products*, 4th edn. Wiley–Interscience, New York.

Syed, S. (1982). Energy and the post-harvest food system in the developing countries—current use, saving and substitution potential. Fourth session of the Tech. Adv. Group of the ACC Task Force on Long-Term Dev. Objectives, United Nations, New York, 13–15 December.

Tabor, H. Z. (1966). A solar cooker for developing countries. *Solar Energy*, **19**, 153–7.

Tabor, H. Z. (1966*a*). Solar ponds. *Science Journal*, June, 56–61.

Tabor, H. Z. (1973). Solar energy for developing regions. Working Party on Solar Energy, Paris. UNESCO.

Taiganides, E. P. & Hazen, T. E. (1966). Properties of farm animal excreta. *Trans. ASAE*, **9**, 374–6.

Tatom, J. W., Welborn, H. W., & Hardy, R. D. (1981). Pyrolysis of waste biomass in developing countries. In *Proceedings of the Third International Conference on Energy Use Management*, Berlin, 26–30 October. Pergamon Press, Oxford, pp. 1871–87.

Taylor, R. P. (1982). Decentralized renewable energy development in China. The State of the Art. World Bank Staff Working Papers No. 535.

Tennyson, G. P. (1983). Potential of wind energy conversion systems in the Great Plains. In *Proceedings of Solar and Wind Systems Workshop*, 18–19 October. Great Plains Agricultural Council, Publication No. 108.

Thekaekara, M. P. (1976). Insolation data for solar energy conversion derived from satellite measurements of earth radiance. In *Sharing the Sun: Solar Technology in the Seventies*, ed. K. W. Boer. American Section, International Solar Energy Society, Florida.

Thomas, J. A. G. (Ed.) (1977). *Energy Analysis*. IPC Science and Technology Press Ltd, IPC House, 32 High Street, Guildford, Surrey GU1 3EW, UK.

Tillman, D., Sarkanen, K. V., & Anderson, L. L. (1977). *Fuels and Energy from Renewable Resources*. Academic Press, New York.

Trehan, R. K., Newman, L., & Park, W. R. (1980). Potential for energy farms in the Dominican Republic. Mitre Corp. Report MTR-80W165. Metrek Division, 1820 Dolby Madison Blvd, McClean, VA 22102.

Treharne, R. W. *et al.* (1980). *Fertilizer Production by Water Power*. NRECA, 1800 Massachusetts Avenue NW, Washington, DC.

Trotter, W. K., Heid, W. G. Jr, & McElroy, R. G. (1979). Solar energy for agriculture. Review of Research, USDA ESCS-67, Washington, DC 20250.

TRW (1980). Energy balances in the production and end-use of alcohols derived from biomass. Prepared by TRW Energy Systems Planning Division, McClean, VA.

UN (1984). *Energy Statistics Yearbook, 1982*. United Nations Statistical Office, New York.

UN Yearbook of World Energy Statistics, 1981. United Nations Statistical Office, New York.

US Army Corps of Engineers (1977). Estimate of national hydroelectric power potential at existing dams.

US Army Corps of Engineers (1978). Pacific Ocean Division. Summary report for hydroelectric power in Hawaii.

US National Alcohol Fuels Commission (1981). Fuel alcohol. Final report, Washington, DC.

UNEP (1980). The environmental aspects of production and use of energy. Part III, Renewable sources of energy. United Nations Environment Program ERS-7-80.

UNESCAP (1980). Renewable sources of energy, Vol. 1: Solar energy. United Nations

Economic and Social Commission for Asia and the Pacific. Technical cooperation among developing countries. St/ESCAP/134.

UNESCO (1986). *International Directory of New and Renewable Energy Information Sources and Research Centers*, 2nd edn. UNESCO, Paris. Published jointly with the Solar Energy Research Institute, USA.

Unger, S. G. (1975). Energy utilization in the leading energy-consuming food processing industries. *Food Technology*, **29**, 33–45.

UNIDO (1978). *Technology for Solar Energy Utilization*. Development and Transfer of Technology Series, United Nations Industrial Development Organization, Vienna, Austria.

USAID (1981). Energy assistance policy paper. US Agency for International Development, Washington, DC 20523.

USAID (1984). Defining the relations between energy and agriculture. Final Report. Development Sciences Inc., PO Box 444, Sagamore, MA 02561.

USDA (1974). The US food and fiber sector: energy use and outlook. Committee Print prepared for the Subcommittee on Agricultural Credit and Rural Electrification by the Economic Research Service, US Department of Agriculture, Washington, DC.

USDA (1977). *Solar Crop Drying Conference Proceedings*, North Carolina State University, Raleigh, NC, 30 June. Co-sponsored by the US Department of Agriculture.

USDA (1980). *Small-Scale Fuel Alcohol Production*. US Dept of Agriculture, Washington, DC.

USDA (1981). Alcohol and vegetable oil as alternative fuels. In *Proceedings of Regional Workshops held in Raleigh, NC; Sacramento, CA; and Peoria, IL*. SEA Extension, Washington, DC.

USDA (1981*a*). Vegetable oil as diesel fuel. Seminar II, Northern Agricultural Energy Center, US Dept of Agriculture, Peoria, IL.

USDA (1981*b*). *Energy Capabilities and Opportunities*. Science and Education Admin., US Dept of Agriculture.

USDA (1982). On-farm demonstration of solar heating of livestock shelters. Final report. US Department of Agriculture, Extension Service, Washington, DC.

USDA (1983). On-farm demonstration of solar drying of crops and grains. Final report. US Department of Agriculture, Extension Service, Washington, DC.

USDA (1983*a*). *3rd Annual Solar and Biomass Workshop Proceedings*, Atlanta, GA. Co-sponsored by US Department of Agriculture. Southern Agricultural Energy Center, Tifton, GA.

USDA (1983*b*). A biomass energy production and use plan for the United States, 1983–90. Agri. Econ. Report No. 505, US Department of Agriculture, Washington, DC.

USDA (1983*c*). Vegetable oil as diesel fuel. Seminar III, Northern Agricultural Energy Center, US Dept of Agriculture, Peoria, IL.

USDA (1984). Inputs—outlook and situation. Economic Research Service, US Department of Agriculture, IOS-5.

USDA (1984*a*). *4th Annual Solar and Biomass Workshop Proceedings*, Atlanta, GA. Co-sponsored by US Department of Agriculture. Southern Agricultural Energy Center, Tifton, GA.

USDA (1984*b*). *Quick Bibliography Series*. National Agricultural Library, US Department of Agriculture, Beltsville, MD.

USDA (1985). *5th Annual Solar and Biomass Workshop Proceedings*, Atlanta, GA. Co-sponsored by US Department of Agriculture. Southern Agricultural Energy Center, Tifton, GA.

USDA (1985*a*). Application of solar energy to agricultural production processes. Final report prepared for USDOE, Office of Solar Energy. Interagency Agreement No. DE-A101-79CS30130.

USDA (1986). Solar, biomass and wind energy workshop. US Department of Agriculture, Atlanta, GA, 25–27 February.

USDC (1980). *Solar Heating and Cooling of Buildings*. Vol. I, *Sizing, Installation and Operation of Systems*, and Vol. II, *Design of Systems*. US Department of Commerce, Washington, DC.

USDOE (various issues). Semi-annual reports: Rocky flats small wind systems. Test Center Activities, NTIS, Springfield, VA.

USDOE (1979). Report of the alcohol fuels policy review. US Dept of Energy, Washington, DC.

USDOE (1983). Energy projections to the year 2010. A technical report in support of the National Energy Plan. US Department of Energy, DOE/PE-0029/2.

USDOE (1984). *Alcohol Fuel Program Technical Review*. Available from Solar Energy Research Institute, 1617 Cole Blvd, Golden, CO 80401.

USDOE (1985). *Small Hydropower Development: The Process, Pitfalls and Experience*. Vol. 1, *Feasibility Studies Summary and Analysis*. Vol. 2, *Licensing Activities Summary and Analysis*. Prepared by Morrison-Knudsen Engineers, Inc., San Francisco, CA, for USDOE/EPRI.

Van Buren, A. (Ed.) (1976). *A Chinese Biogas Manual*. Translated by Intermediate Technology Publications Ltd, 9 King Street, Covent Garden, London WC2E 8HW, UK.

Van Ee, G. & Claar, P. (1981). Fundamentals of direct combustion and its potential in agricultural applications. Unpublished paper. Agr. Engr. Dept, Michigan State University, East Lansing, MI.

Van Vaerenbergh, R. (1985). Senior Officer. FAO Animal Production Service. Personal communications. Data on animals used for draft purposes in Africa and Asia.

Velebil, M. & Has, S. (undated). The use of solar energy in agriculture in Czechoslovakia. FAO/ECE/AGRI/WP.2/R.41. Research Institute of Agricultural Engineering, Prague.

Vergara, W. & Castello Branco, J. R. (1981). Energy use for ethanol production. In *Proceedings of the Third International Conference on Energy Use Management*, Berlin, 26–30 October. Pergamon Press, Oxford, pp. 1901–10.

Verma, S. R. (Ed.) (1987). Energy in production agriculture and food processing. In *Proceedings National Energy Conference*, Punjab Agricultural University, Ludhiana, India.

Veziroglu, T. N. (Ed.) (1981). *Alternative Energy Sources II*. Vol. 4. *Indirect Solar Energy*. Hemisphere Publishing Corp., Washington, DC, distributed by McGraw-Hill International, pp. 1485–892.

Vimal, O. P. & Bhatt, M. S. (1989). *Wood Energy Systems*. K.L. Publications, New Delhi, India.

VITA (1982). *Renewable Energy Dictionary*. Volunteers in Technical Assistance, 1815 North Lynn Street, Suite 220, Arlington, VA 22209.

Vogt, F. (Ed.) (1981). *Energy Conservation and Use of Renewable Energies in the Bio-Industries: Proceedings of the International Seminar on Energy Conservation and the Use of Solar and Other Renewable Energies in Agriculture, Horticulture and Fishculture*, Polytechnic of Central London, 15–19 September 1980, LC 80-49739. Pergamon Press, Oxford.

Vogt, F. (Ed.) (1982). *Energy Conservation and Use of Renewable Energies in the Bio-Industries: Proceedings of the International Seminar on Energy Conservation and Use of Renewable Energies in the Bio-Industries*, Trinity College, Oxford, UK, 6–10 September. Pergamon Press, Oxford.

Von Brakel, J. (1980). *Small-Scale Anaerobic Digesters (Biogas Plants): A Critical Review of the Pre-1970 Literature*. Delft University Press, Delft, The Netherlands.

Von Oppen, M. (1976). *The Sunbasket*. International Crops Research Institute for the Semi-Arid Tropics, Hyderabad, India.

Von Zabeltitz, Chr. (1978). Moglichkeiten des einsatzes von solar-energie zur gewachsausheizung. *Landtechnik*, **33**, 22–4.

Von Zabeltitz, Chr. (1979). Solarabsorber und warmepumpen im gartenbau. *Elektrizitat—KTBL, AEL, HEA—Symposium*, 12–13 September, pp. 131–41.

Vosper, F. C. (1985). Irrigation pumping with electrical wind systems. In *Proceedings of 5th*

Annual Solar and Biomass Energy Workshop, USDA/DOE, 23–25 April, Atlanta, GA, pp. 253–6.

Waddle, D. B. (1984). Development and utilization of renewable energy technologies in developing countries. ASAE Paper 84-5548, American Society of Agricultural Engineers, St Joseph, MI.

Walker, L. P. *et al.* (1985). Energy integrated dairy farm system in New York. USDOE Contract No. FC01-80CS40380.

Walsh, J. L. Jr, Ross, C. C., & Lamade, R. M. (1986). Energy integrated dairy farm system in Georgia. USDOE Contract No. DE-FC07-80CS40379.

Walters, F. C. (1984). *Fuel Alcohol Trade Directory 1983–84*. Fuel Alcohol USA, Box 9547, Kansas City, MO 64133.

Ward, G. M., Sutherland, T. M., & Sutherland, J. M. (1980). Animals as an energy source in the Third World. *Science*, **208**, 570–4.

Ward, R. F. (1981). Digesters for the Third World. In *Anaerobic Digestion 1981*, ed. D. E. Hughes *et al.* Elsevier Biomedical Press, Amsterdam, pp. 315–41.

Watson, P. R. (1981). *Animal Traction*. Peace Corps Information Collection and Exchange. Appropriate Technologies for Development, Manual No. M-12.

Wegley, H. L. *et al.* (1978). A siting handbook for small wind energy conversion systems. Battelle Pacific Northwest Laboratories, Richland, WA, PNL-2521 UC-60.

Weik, H. (1980). Warmegewinnung mit sonnenkollektoren. In *Energie sparen–Energie produzieren*. Landwirtschaftskammer, Schleswig-Holstein, Kiel, pp. 149–65.

Weisz, P. B. & Marshall, J. F. (1980). *Fuels from Biomass: A Critical Analysis of the Technology and Economics*. Marcel Dekker, New York.

Wenzlaff, R. (1981). Experiences on farm-scale digestion with piggery waste. In *Proceedings of the Third International Conference on Energy Use Management*, Berlin, 26–30 October. Pergamon Press, Oxford, pp. 1939–44.

Whittaker, R. H. & Likens, G. E. (1975). The biosphere and man. In *Primary Productivity of the Biosphere*, ed. N. Leith & R. H. Whittaker. Springer, Berlin.

Wienecke, F. (1979). Das solardach in der landwirtschaft. In *Elektrizitat—KTBL, AEL, HEA—Symposium*, 12–13 September, pp. 111–30.

Wigg, E. E. (1974). Methanol as a gasoline extender: a critique. *Science*, **186**, 29 November.

Wilbur, P. J. & Karaki, S. (1977). *Solar Cooling*. The Franklin Institute Press, Philadelphia, PA.

Williams, R. O. & Horsfield, B. (1977). Generation of low-Btu gas from agricultural residues—experiments with a laboratory scale gas producer. Report, Department of Agricultural Engineering, University of California, Davis, CA.

Wilson, C. L. (1977). *Energy Global Prospects 1985–2000. Report on Alternative Energy Strategies*. McGraw-Hill, New York.

Wise, D. L. (Ed.) (1981). *Fuel Gas Production from Biomass*, Vols I and II. CRC Press, Boca Raton, FL 33431.

Wise, D. L. (Ed.) (1983). *Bioenergy Systems* (5 vols). CRC Press, Boca Raton, FL 33431. Vol. I, *Fuel Gas Systems*; Vol. II, *Fuel Gas Developments*; Vol. III, *Liquid Fuel Systems*; Vol. IV, *Liquid Fuel Developments*; Vol. V, *Bioconversion Systems*.

Wise, D. L. (1987). *Global Bioconversions* (4 vols). CRC Press, Boca Raton, FL 33431.

Wise, D. L. (1987*a*). *Bioenvironmental System*. CRC Press, Boca Raton, FL 33431.

Wood, T. S. & Baldwin, S. (1985). Fuelwood and charcoal use in developing countries. *Ann. Rev. Energy*, **10**, 407–29.

Wolff, B. & Meyer, H. (1978). *Wind Energy*. The Franklin Institute Press, Philadelphia, PA.

Worgetter, M. (1981). Results of a long-term engine test based on rapeseed oil fuel. In *Proceedings of the Third International Conference on Energy Use Management*, Berlin, 26–30 October. Pergamon Press, Oxford, pp. 1955–62.

World Bank (1980). *Energy in the Developing Countries*. World Bank, Washington, DC.

World Bank (1980*a*). *Alcohol Production from Biomass in the Developing Countries*. World Bank, Washington, DC.

World Bank (1981). Energy pricing in developing countries: a review of the literature. World Bank Energy Department Paper No. 1, World Bank, Washington, DC.

World Bank (1983). *The Energy Transition in Developing Countries.* International Bank for Reconstruction and Development, World Bank, Washington, DC.

World Bank (1984). The UNDP/World Bank energy sector programmes. International Bank for Reconstruction and Development, World Bank, Washington, DC.

World Bank (1984*a*). A methodology for regional assessment of small-scale hydropower. World Bank Energy Department Paper No. 14, World Bank, Washington, DC.

World Bank (1984*b*). A survey of the future role of hydroelectric power in 100 developing countries. World Bank Energy Department Paper No. 17, World Bank, Washington, DC.

World Bank (1984*c*). *World Development Report 1984.* Published for the World Bank by Oxford University Press, Oxford.

World Bank (1984*d*). A survey of the future role of hydroelectric power in 100 developing countries. Energy Department Paper No. 17, World Bank, Washington, DC.

World Bank (1984*e*). A methodology for regional assessment of small-scale hydropower. Energy Department Paper No. 14, World Bank, Washington, DC.

World Bank (1985). *The World Bank Atlas, 1985.* International Bank for Reconstruction and Development, World Bank, Washington, DC.

WPD (1984). Wind energy directory. Wind Power Digest No. 26.

Yang, P. Y., Nagano, S. Y., Lin, J. K., & Wong, Y. T. (1981). Improved anaerobic digestion by application of constant concentration of digested sludge recycling. In *Proceedings of the Third International Conference on Energy Use Management*, Berlin, 26–30 October. Pergamon Press, Oxford, pp. 1963–9.

Index

Note that *italic numbers* refer to Figures and Tables; these will be found on the page number immediately following the Figure or Table number.